W0232298

The Immunoassay Kit Directory

Managing Editors
Dr Jane Kent
Dr Elizabeth Naysmith
Kluwer Academic Publishers
Edinburgh
Scotland

Consulting Editor
Professor Keith James
Department of Surgery
University of Edinburgh
Edinburgh
Scotland

Editorial Assistants
Linda Thomas
Jenny Kellett
Kluwer Academic Publishers, Lancaster

Series B: Infectious Diseases

Editor-in-Chief
Dr Hugh Young
Department of Medical Microbiology
University of Edinburgh
Edinburgh
Scotland

Editorial Advisory Board
A Linde, Stockholm, Sweden
D J Merry, Adelaide, Australia
A Rodriguez Torres, Valladolid, Spain
J Orfila, Amiens, France

Kluwer Academic Publishers

The Immunoassay Kit Directory

Series B: Infectious Diseases

Volume 1: Part 3
Enteric and Other Infections
December 1995

Guest Editors
Dr John Coia
Department of Clinical Microbiology
Western General Hospital NHS Trust
Crewe Road
Edinburgh
Scotland

and

Dr Heather Cubie
Regional Virus Laboratory
City Hospital
Greenbank Drive
Edinburgh
Scotland

Kluwer Academic Publishers
ISSN 1381-5067

The Immunoassay Kit Directory

Series B: Infectious Diseases

SUBSCRIPTION INFORMATION

The subscription price for Volume 1 (three parts) is:

US$ 486 / Dfl. 850 / £357 inclusive of shipping and handling

ORDERING INFORMATION

Subscription orders should be sent to:

Kluwer Academic Publishers
101 Philip Drive
Assinippi Park
Norwell, MA 02061, USA

or

Kluwer Academic Publishers
PO Box 322
3300 AH Dordrecht
The Netherlands

All requests for further information concerning The Immunoassay Kit Directory should be sent to:

Kluwer Academic Publishers
PO Box 55
Lancaster
LA1 1PE
UK

Telephone: +44 (0)1524 34996

Fax: +44 (0)1524 32144

COPYRIGHT

© 1995 by Kluwer Academic Publishers

All rights reserved. No part of this publication may be reproduced, stored in a retrieval system, or transmitted in any form or by any means, electronic, mechanical, photocopying, recording or otherwise, without prior permission from the publishers, Kluwer Academic Publishers, PO Box 55, Lancaster, LA1 1PE, UK

Published in the United Kingdom by Kluwer Academic Publishers bv, PO Box 55, Lancaster, LA1 1PE UK

ISSN: 1381-5067

ISBN-13:978-0-7923-8813-5   e-ISBN-13:978-94-009-0359-3
DOI:10.1007/978-94-009-0359-3

CONTENTS: VOLUME 1 PART 3

CONTENTS OF VOLUME 1 PART 1

CONTENTS OF VOLUME 1 PART 2

ABOUT THIS PUBLICATION

Introduction

This is a new series of the Immunoassay Kit Directory and deals with commercially available immunoassay kits in the area of infectious diseases. A companion publication, Series A: Clinical Chemistry, was started some years ago and is now in its third volume.

The aim of the publication is to provide a comprehensive, independent and easy-to-use reference source on the large and growing number of kits currently on the market for the diagnosis of clinically important infectious diseases.

Detailed information about each kit is provided; over 19 different major parameters are listed in a consistent manner to allow for easy comparison.

Each organism begins with a short introduction written by the guest editors.

Full details of the manufacturers and distributors are also included as well as several indexes.

Each volume of Series B will initially consist of three parts:

    Part 1:   Genitourinary infections
    Part 2:   Respiratory infections
    Part 3:   Enteric and other infections

Where appropriate, some organisms may be represented in more than one part.

About this part – Enteric and Other Infections

This, the third part of the Volume, covers kits for organisms that cause enteric infection as well as kits for organisms that cause other infections (often systemic) that are not primarily thought of as causing genitourinary infection (Part 1) or respiratory infection (Part 2). Biological diversity does not lend itself to rigid classification and inevitably there is a small degree of overlap in the organisms covered in each part of this first volume. For example, kits for detection of Adenovirus infection (excluding those test-ing only enteric samples) can be found in Part 2 whereas all Adenovirus kits are included in Part 3.

How is the publication organised?

The main body of the publication consists of the immunoassay kit entries. These are ranked as follows:

Alphabetically by organism, first by antigen detection and then by antibody detection. Antigen detection kits are then ranked alphabetically by assay type and then by manufacturer. Antibody detection kits are ranked first by assay type, second by antibody class and third by manufacturer.

Thus entries for Adenovirus will be sorted as shown below.

Adenovirus
1. Antigen detection
    EIA (non-competitive), by manufacturer
    Immunofluorescence assay (direct), by
        manufacturer
    Immunofluorescence assay (indirect), by
        manufacturer
    Partcile agglutination assay

2. Antibody detection
    EIA (non-competitive)
        IgG, by manufacturer
        IgM, by manufacturer

Two main indexes are provided at the back of the Directory to cross-reference many of these parameters. Firstly, an index by manufacturer, assay type, antibody/antigen detection and microorganism; secondly, by assay type, microorganism, antibody/antigen detection and manufacturer.

How frequently will the information be revised?

We will be revising the database of information that we are compiling on an on-going basis. Each major part of the Directory, such as this part on Enteric and Other Infections will be revised on a roughly annual basis.

© KLUWER ACADEMIC PUBLISHERS 1995, ISSN 1381-5067

How is the information collected?

Information about what kits are available has been gathered by a variety of means:

Direct contact with manufacturers

Research by the editorial board

Exhibitions and conferences

Whilst most manufacturers have been extremely helpful in giving us information there have been a very small number who, for one reason or another, have not supplied information. We hope that with the establishment of the Directory they will feel that they are now able to contribute.

No charge has been made to manufacturers for the inclusion of information.

What parameters have been used to select kits?

The basic parameter for the inclusion of a kit in the Directory is that it should be readily available to the average user on an international basis. Having said this, however, it may be that, because of regulatory or other restrictions, certain kits are available only in certain countries. Contact with the manufacturer will, we hope, clarify the availability and we would be grateful for any feedback that users can give us on this matter.

Information about the kits themselves has been compiled by contacting manufacturers for detailed information, often in the form of kit inserts. This has then been carefully edited and checked by the editorial staff to compile the basic entries that form the body of the directory. Manufacturers have then had the opportunity to see at least one set of proofs.

The parameters that have been used are shown in detail in the definitions section of the Directory.

Haemagglutination assays have not been included.

How can the user help?

This publication aims to be of practical use to the users of kits. We carried out extensive market research to define more closely the needs of the target audience and some of the results of this have been incorporated into the publication you now hold.

We very much would like to continue to incorporate the views and needs of users into the publication. Please feel free to contact us here with any comments at all, good or bad, about the publication. Letters to the Editor will be published. We will always welcome information about kits and their manufacturers that we may have overlooked.

What further developments will there be?

Now that Series A (Clinical Chemistry) is established and Series B (Infectious Diseases) is underway we already have plans for similar series to deal with other major areas of clinical practice and bio-medical research such as haematology, immunology and oncology.

A Final Note . . .

Many people have assisted in the preparation of the Directory and I would like to thank, in addition to those acknowledged on the first page, the manufacturers of kits who have kindly and generously provided us with information on their products, and all those involved in the provision of advice and information for the compilation of this issue.

Dr Peter L Clarke

Kluwer Academic Publishers, Lancaster

December 1995

© *KLUWER ACADEMIC PUBLISHERS 1995, ISSN 1381-5067*

ABBREVIATIONS AND DEFINITIONS

List of Abbreviations

Ab	antibody
ABTS	2,2-azinobis (3-ethylbenzothiazoline-6-sulphonic acid), diammonium salt
Ag	antigen
AGIg	anti-goat immunoglobulin
AGPC	anti-guinea pig complement
AGPIgG	anti-guinea pig immunoglobulin G
AHIg	anti-human immunoglobulin
AHIgA	anti-human immunoglobulin A
AHIgG	anti-human immunoglobulin G
AHIgM	anti-human immunoglobulin M
AMIg	anti-mouse immunoglobulin
AMIgG	anti-mouse immunoglobulin G
ANA	antinuclear antibody
AP	alkaline phosphatase
ARDS	adult respiratory distress syndrome
ARIg	anti-rabbit immunoglobulin
ARIgG	anti-rabbit immunoglobulin G
AU	arbitrary units
b	bovine
BCIP	5-bromo-4-chloro-3-indolyl phosphate
BHK	Baby hamster kidney cells
CAP	College of American Pathologists
CDC	USA Centers for Disease Control
CEV	Californian Encephalitis virus
CFT	complement fixation test
CIE	counter-current immunoelectrophoresis
CIN	cefsulodin irgasan novobiocin agar
CMV	Cytomegalovirus
4CN	4-chloro-naphthol
CNS	central nervous system
CoA	coagglutination
CPE	cytopathogenic effect
CSF	cerebrospinal fluid
DAB	diaminibenzidene
EBV-EA	Epstein-Barr virus early antigen
EBV-VCA	Epstein-Barr viral capsid antigen
EDTA	ethylenediamine tetra-acetic acid
EEEV	Eastern Equine Encephalitis virus
EIA	enzyme immunoassay
ELISA	enzyme-linked immunosorbent assay
FDA	US Food and Drugs Administration
FITC	fluorescein isothiocyanate
g	goat
gp	guinea pig
GPC	guinea pig complement
h	human
HHV-6	Human Herpes virus 6
hr	hour
HRP	horseradish peroxidase
HSV	Herpes simplex virus
IFA	immunfluorescent antibody
Ig	immunoglobulin
IgA	immunoglobulin A
IgG	immunoglobulin G
IgM	immunoglobulin M
3IP	3-Indoxyl phosphate
IU	International Units
m	mouse
MAb	monoclonal antibody
min	minute
MOMP	major outer membrane protein
4-MP	4-methylumbelliferyl phosphate
NA	not applicable
NAD	nicotinamide adenine dinucleotide
NADP	nicotinamide adenine dinucleotide phosphate
NBT	nitroblue tetrazolium
Neg	negative
OD	optical density
OPD	O-phenylenediamine
p	pig
PAb	polyclonal antibody
PBS	phosphate buffered saline
PCR	polymerase chain reaction
PMP	phenolphthalein monophosphate
PNP	p-nitrophenyl phosphate
POD	peroxidase
Pos	positive
r	rabbit
RB	reticulate body
RBC	red blood cells
rec	recombinant
RF	rheumatoid factor
RIA	radioimmunoassay
RT	room temperature
SAF	sodium acetate–acetic acid–formalin
sh	sheep
SLEV	St. Louis Encephalitis virus
staph	staphylococcus
strep	streptococcus
strept	streptavidin
syn	synthetic
temp	temperature
TM	transport medium
TMB	tetramethylbenzidine
WB	Western blot
WBC	white blood cells
WEEV	Western Equine Encephalitis virus
WHO	World Health Organisation

© KLUWER ACADEMIC PUBLISHERS 1995, ISSN 1381-5067

Outline of kit descriptions and definition of headings

Each entry is introduced by the full name of the organism(s) and is followed by a statement indicating whether the assay detects antigen and/or antibody. Assays which detect both antigen and antibody are entered and indexed under both antigen and antibody with appropriate cross-reference. For antibody detection the class(es) of immunoglobulin(s) to be detected are stated. For antigen detection the assay is for clinical specimens unless stated otherwise. The primary manufacturer and the catalogue number of the smallest kit available is given.

Definitions of headings

Summary: A schematic representation of the assay system showing the component being measured (**Ab** or **Ag**) in bold. This is presented in a linear form beginning with the solid phase but it does not necessarily represent the actual sequence of reactions – in a competitive assay the analyte and the competitor are placed above and below the line, e.g.

$$[\text{Well-Ag}]-\textbf{Ab}-[\text{AHIgG-AP}]-[\text{PMP}]-A_{550}$$

$$[\text{Well-Ag}]-\textbf{Ab}-[\text{AHIgG-FITC}]-\text{fluorescence}$$

$$\left.\begin{array}{c}\textbf{Ab}\\ [\text{Bead-Ab}]\end{array}\right\}[\text{Ag}]-[\text{Ab-}^{125}\text{I}]-\text{radioactivity}$$

Assay type: The assay type is written in full apart from enzyme immunoassay and radioimmunoassay which are abbreviated to EIA and RIA. These assays are also designated as competitive or non-competitive and whenever appropriate as amplified. The assay types and their corresponding detection systems are given in Table 1.

Detection: A description of the method used to measure the result of the assay, e.g. colorimetric, luminometric, radioisotopic, fluorometric, fluorescence microscopy, visual.

Format: The physical format in which the assay is presented, e.g. 'microtitre well Ab coated' or 'slide well Ag coated'.

Sample type: The samples recommended for use with the kit, e.g. plasma, serum. If heat inactivation of serum is not recommended this is cited as 'serum (do not heat inactivate)'.

Sample pre-treatment: Treatment performed on the sample before performing the assay, e.g. extracting antigen from swab, removal of IgG. Dilution of the sample is not considered as pre-treatment.

Sample volume: The volume of sample required to perform a single estimate. Diluent volume, where recommended is shown in parenthesis, e.g. 10 µl (+ 500 µl diluent) otherwise the recommended dilution is given, e.g. '20 µl of 1:10 dilution'.

Number of tests: The number of tests for the smallest kit followed by the number of tests for any larger kit(s), e.g. 96; 960.

Controls - standards run in assay: The number and type of controls and calibrators/standards run in assay, e.g.

Controls: Neg(1), Low Pos(2), High Pos(1)

Standards: Neg(1), Low Pos(1), High Pos(1)

Calibrators: Neg(1), Mid Pos(1), High Pos(1)

Incubation: A summary of the incubation steps giving time in hours and minutes and temperatures in °C or room temperature (RT), e.g.

2 hr (RT) + 18-22 hr (RT) + 15 min (45°C)

Washes: The total number of washing cycles. If there is a preliminary plate wash this is also mentioned, e.g. '3 (+ preliminary plate wash)'

Antibodies, antigens and labelled components: The antibodies, antigens and labelled components included in the kit. The type of antibody - polyclonal (PAb) or monoclonal (MAb), the source of the antibody or antigen and the phase to which it is bound are stated if known, e.g.

Anti-Chlamydia PAb(r) bound to particles

CMV (AD 169) bound to well

Anti-human IgG MAb (m) biotinylated

Streptavidin HRP conjugated

Substrate: The substrate of the conjugate, e.g. OPD

Controls - standards supplied: A description of the controls and calibrators/standards supplied with the kit, e.g.

Controls: Neg, Low Pos, High Pos (human serum)

Calibrators: 4 with assigned titre values (human serum)

Additional reagents required: Any reagents required for the assay but not provided with the kit, e.g. 'controls/calibrators not supplied but available separately', 'H_2SO_4' - common materials such as distilled water, saline and simple buffers such as Tris, PBS etc. are excluded.

Special equipment required: Dedicated equipment that is *essential* to the performance of the assay, e.g. 'FIAX fluorometer'. Equipment designed by the manufacturer for optional use in the assay may be listed as such, e.g. 'automated system and software (optional)'.

Scoring/comments and interpretation: General comment on whether equivocal/grey zones are recognised, if repeat or supplemental testing is recommended by the manufacturer, options available for quantitative testing and/or comparing acute and convalescent sera.

Number of references: The number of references given in the kit insert: these do not always include references giving performance data of the product itself.

Notes: Additional points of potential relevance to the kit used, e.g. kits designated 'for research use only'; certain types of swab unsuitable; recommended use of different incubation procedures; cross-reference to the appropriate section for kits detecting both antigen and antibody; etc.

Information not included

There are several parameters (outlined below) that, after careful consideration, we did not include. Although these are of interest it was not feasible to include them in a directory of this type.

Cost: The list price often bears little relationship to the discounted price paid by many laboratories. In addition list prices vary from country to country, supplier to supplier, and are subject to frequent change.

Sensitivity and specificity: These parameters are often quoted in product inserts but profound variation in the nature of the populations evaluated, the number of samples tested and the diversity of comparative methods make unqualified comparisons relatively meaningless and possibly misleading. In addition, as pointed out in the introductory article, the true utility of a test is given by the predictive value which is dependent on the prevalence of infection in the population being screened/tested.

Time to run tests: This can vary depending on pre-treatment, sample dilution, number of controls/standards and tests per batch, number and nature of wash cycles and the number and length of incubation stages. As the number and length of incubation stages is usually the most significant influence on assay running time a minimum comparative value for the time to run tests can be obtained from the details given in the **Incubation** field. If the number of wash cycles (**Washes** field) is also taken into account a more accurate estimate of the overall time required to run tests can be obtained.

Table 1   Assay types and corresponding detection systems

Assay type	Detection system(s)
EIA (competitive) EIA (non-competitive) EIA (non-competitive) amplified*	Colorimetric (occasionally visual, fluorometric or light microscopy)
Immunofluorescence assay (direct) Immunofluorescence assay (indirect)	Fluorometric or Fluorescence microscopy
Luminometric assay (competitive) Luminometric assay (non-competitive)	Luminometric
RIA (competitive) RIA (non-competitive)	Radioisotopic
Immunoblot assay Immunochromatographic assay Particle agglutination assay Particle agglutination assay (coagglutination)	Visual (occasionally colorimetric)

*This assay type was not used to detect any of the organisms in Part 3 of the Directory.

Haemagglutination assays have not been included.

STOP PRESS

The following information became available at too late a stage to include as full entries in the Directory:

Boehringer Mannheim Immunodiagnostics

Enzymun-Test® Toxo IgG	Cat. no. 1012 312
Enzymun-Test® Toxo IgM	Cat. no. 1554 140
Enzymun-Test® Rubella IgG	Cat. no. 1553 623
Enzymun-Test® Rubella IgM	Cat. no. 1554 123

Genzyme Virotech GmbH

Adenovirus antibody	Cat. no. EN 121.00
Aspergillus IgG/IgM	Cat. no. EN 141.00
Aspergillus IgA	Cat. no. EN 141.08
Borrelia burgdorferi antibody	Cat. no. EN 122.00
Borrelia garnii	Cat. no. EN 222.00
Brucella antibody	Cat. no. EN 101.00
Brucella IgA set	Cat. no. EN 101.08
Enterovirus IgG/IgM	Cat. no. EN 116.00
Enterovirus IgA set	Cat. no. EN 116.08
Epstein-Barr virus antibody	Cat. no. EN 102.00
Mumps virus antibody	Cat. no. EN 106.00
Tetanus antibody	Cat. no. EN 124.00

Laboratories Eurobio

Borrelia burgdorferi Immunoblot Kit	Cat. no. not given
Lyme IgG/IgM EIA ELIT® Kit	Cat. no. 901135
Rubella IgG EIA ELIT Kit®	Cat. no. 900277
Rubella IgM EIA ELIT® Kit	Cat. no. 900278
Clostridium difficile Toxin A EIA IDENT® Kit	Cat. no. 905266
Clostridium difficile Toxin B EIA INDENT® Kit	Cat. no. 905267
Salmonella Latex Test	Cat. no. 900536
E. coli enteropathogen Verotoxin Latex Test	Cat. no. 901128
Vibrio cholera Latex Test	Cat. no. 900629
Borrelia burgdorferi (Lyme) IFI IgG Test	Cat. no. 900427
Borrelia burgdorferi (Lyme) IFI IgM Test	Cat. no. 900450
Epstein-Barr Virus IFI (EBV CA) IgG Test	Cat. no. 904214
Epstein-Barr Virus IFI (EBV CA recombinant) IgM Test	Cat. no. 900564
Epstein-Barr Virus IFI (EEBV NA recombinant) Test	Cat. no. 904213
Epstein-Barr Virus IFI (EBV EA) Test	Cat. no. 904215
Mumps IFI IgG Test	Cat. no. 902096
Parvovirus B19 IFI Test	Cat. no. not given
Rubella Virus IFI Test	Cat. no. 902101

r-Biopharm GmbH

RIDA® Blot Borrelia	Cat. no. C0205
RIDASCREEN® Cryptosporidium	Cat. no. C1102
RIDASCREEN® Giardia	Cat. no. 1202
RIDA® Blot Yersinia	Cat. no. 1305

© KLUWER ACADEMIC PUBLISHERS 1995, ISSN 1381-5067

FULL CONTENTS LISTING

██

Antibody detection	EIA (non-competitive)	bioMerieux	651
Antibody detection	EIA (non-competitive)	GenBio	652
Antibody detection	EIA (non-competitive)	GenBio	652
Antibody detection	EIA (non-competitive)	GenBio	653
Antibody detection	EIA (non-competitive)	MRL Diagnostics	653
Antibody detection	EIA (non-competitive)	Seradyn	654
Antibody detection	EIA (non-competitive)	Sigma Diagnostics	654
Antibody detection (IgG)	EIA (non-competitive)	BAG-Biologische Analysensystem GmbH	655
Antibody detection (IgG)	EIA (non-competitive)	Dako A/S	655
Antibody detection (IgG)	EIA (non-competitive)	Immunobiological Laboratories	656
Antibody detection (IgG and/or IgM)	EIA (non-competitive)	Behringwerke AG	656
Antibody detection (IgG and/or IgM)	EIA (non-competitive)	BioWhittaker Inc	657
Antibody detection (IgG and/or IgM)	EIA (non-competitive)	BioWhittaker Inc	657
Antibody detection (IgG and/or IgM)	EIA (non-competitive)	Dako A/S	658
Antibody detection (IgG and/or IgM)	EIA (non-competitive)	Diagast Laboratories	658
Antibody detection (IgG and/or IgM))	EIA (non-competitive)	Immuno Pharmacology Research	659
Antibody detection (IgM)	EIA (non-competitive)	BAG-Biologische Analysensystem GmbH	659
Antibody detection (IgM)	EIA (non-competitive)	BioWhittaker Inc	660
Antibody detection (IgM)	EIA (non-competitive)	Dade International Inc (Bartels Division)	660
Antibody detection (IgM)	EIA (non-competitive)	Dako A/S	661
Antibody detection (IgM)	EIA (non-competitive)	Gull Laboratories Inc	661
Antibody detection (IgM)	EIA (non-competitive)	Immunobiological Laboratories	662
Antibody detection	Immunoblot assay	Seradyn	662
Antibody detection (IgG)	Immunoblot assay	Cambridge Diagnostics Ireland Ltd	663
Antibody detection (IgG)	Immunoblot assay	Immunetics	663
Antibody detection (IgG)	Immunoblot assay	Immunetics	664
Antibody detection (IgG and/or IgM)	Immunoblot assay	Cambridge Diagnostics Ireland Ltd	664
Antibody detection (IgG and/or IgM)	Immunoblot assay	Diagast Laboratories	665
Antibody detection (IgG and/or IgM)	Immunoblot assay	Scimedix Corporation	665
Antibody detection (IgM)	Immunoblot assay	Immunetics	666
Antibody detection (IgM)	Immunoblot assay	Immunetics	666
Antibody detection	Immunochromatographic assay	remel	667
Antibody detection	Immunofluorescence assay (indirect)	bioMerieux	667
Antibody detection	Immunofluorescence assay (indirect)	Diagast Laboratories	668
Antibody detection	Immunofluorescence assay (indirect)	Zeus Scientific Inc	668
Antibody detection (IgG)	Immunofluorescence assay (indirect)	LD, Labor Diagnostika GmbH	669
Antibody detection (IgG)	Immunofluorescence assay (indirect)	MRL Diagnostics	669
Antibody detection (IgM)	Immunofluorescence assay (indirect)	MRL Diagnostics	670

Brucella species

			page
Antibody detection (IgG)	EIA (non-competitive)	Alfa Biotech	672
Antibody detection (IgG)	EIA (non-competitive)	Clark Laboratories	672
Antibody detection (IgG)	EIA (non-competitive)	Immunobiological Laboratories	673
Antibody detection (IgG)	EIA (non-competitive)	PanBio	673
Antibody detection (IgG and/or IgM))	EIA (non-competitive)	Immuno Pharmacology Research	674

Antibody detection (IgG and/or IgM)	EIA (non-competitive)	Scimedix Corporation	674
Antibody detection (IgM)	EIA (non-competitive)	Alfa Biotech	675
Antibody detection (IgM)	EIA (non-competitive)	Clark Laboratories	675
Antibody detection (IgM)	EIA (non-competitive)	Immunobiological Laboratories	676
Antibody detection (IgM)	EIA (non-competitive)	PanBio	676

Campylobacter species

page

Antigen detection	Particle agglutination assay	Meridian Diagnostics Inc	678

Clostridium difficile

page

Antigen detection	Immunochromatographic assay	Meridian Diagnostics Inc	680
Antigen detection	Particle agglutination assay	Becton Dickinson	680
Antigen detection	Particle agglutination assay	Meridian Diagnostics Inc	681
Antigen detection	Particle agglutination assay	Microgen Bioproducts	681

Clostridium difficile (toxin A)

Antigen detection	EIA (non-competitive)	Alexon Inc	682
Antigen detection	EIA (non-competitive)	Becton Dickinson	682
Antigen detection	EIA (non-competitive)	bioMerieux	683
Antigen detection	EIA (non-competitive)	Dade International Inc (Bartels Division)	683
Antigen detection	EIA (non-competitive)	Meridian Diagnostics Inc	684
Antigen detection	EIA (non-competitive)	Shield Diagnostics	684

Clostridium difficile (toxin A and toxin B)

Antigen detection	EIA (non-competitive)	Cambridge Biotech Corporation	685
Antigen detection	EIA (non-competitive)	r-biopharm GmbH	685

Clostridium perfringens (enterotoxin, type A)

page

Antigen detection	Particle agglutination assay	Denka Seiken Co. Ltd	687

Clostridium tetani

page

Antibody detection (IgG)	EIA (non-competitive)	Immuno Pharmacology Research	689

Clostridium tetani (toxin)

Antibody detection (IgG)	EIA (non-competitive)	Gamma SA	689
Antibody detection (IgG)	EIA (non-competitive)	Immunobiological Laboratories	690

Cryptococcus neoformans

page

Antigen detection	EIA (non-competitive)	Meridian Diagnostics Inc	692
Antigen detection	Particle agglutination assay	bioMerieux	692
Antigen detection	Particle agglutination assay	Carter-Wallace Inc	693
Antigen detection	Particle agglutination assay	International Immunodiagnostics	693
Antigen detection	Particle agglutination assay	Meridian Diagnostics Inc	694
Antigen detection	Particle agglutination assay	Murex Diagnostics Limited	694

Cryptosporidium species

page

Antigen detection	EIA (non-competitive)	Alexon Inc	696

© KLUWER ACADEMIC PUBLISHERS 1995, ISSN 1381-5067

Antibody detection (IgG)	EIA (non-competitive)	Gull Laboratories Inc	720
Antibody detection (IgG)	EIA (non-competitive)	Immuno Concepts Inc	721
Antibody detection (IgM)	EIA (non-competitive)	Biotest Diagnostics	721
Antibody detection (IgG)	Immunofluorescence assay (indirect)	Gull Laboratories Inc	722
Antibody detection (IgG)	Immunofluorescence assay (indirect)	Immuno Concepts Inc	722
Antibody detection (IgG)	Immunofluorescence assay (indirect)	MRL Diagnostics	723

Epstein-Barr virus (EBNA)

Antibody detection	EIA (non-competitive)	Immuno Concepts Inc	723
Antibody detection (IgG)	EIA (non-competitive)	Biotest Diagnostics	724
Antibody detection (IgG)	EIA (non-competitive)	BioWhittaker Inc	724
Antibody detection (IgG)	EIA (non-competitive)	BioWhittaker Inc	725
Antibody detection (IgG)	EIA (non-competitive)	Diesse	725
Antibody detection (IgG)	EIA (non-competitive)	Gull Laboratories Inc	726
Antibody detection (IgG)	EIA (non-competitive)	IFCI Clone Systems	726
Antibody detection (IgG)	EIA (non-competitive)	Incstar	727
Antibody detection (IgG)	EIA (non-competitive)	Ortho Diagnostic Systems	727
Antibody detection (IgG)	EIA (non-competitive)	Sigma Diagnostics	728
Antibody detection (IgG and/or IgM)	EIA (non-competitive)	Meridian Diagnostics Inc	728
Antibody detection (IgM)	EIA (non-competitive)	Incstar	729
Antibody detection (IgM)	EIA (non-competitive)	Sigma Diagnostics	729
Antibody detection	Immunofluorescence assay (indirect)	Gull Laboratories Inc	730
Antibody detection	Immunofluorescence assay (indirect)	Immuno Concepts Inc	730

Epstein-Barr virus (VCA)

Antibody detection (IgA)	EIA (non-competitive)	Savyon Diagnostics Ltd	731
Antibody detection (IgG)	EIA (non-competitive)	Alfa Biotech	731
Antibody detection (IgG)	EIA (non-competitive)	Amico Laboratories Inc	732
Antibody detection (IgG)	EIA (non-competitive)	BAG-Biologische Analysensystem GmbH	732
Antibody detection (IgG)	EIA (non-competitive)	BioWhittaker Inc	733
Antibody detection (IgG)	EIA (non-competitive)	BioWhittaker Inc	733
Antibody detection (IgG)	EIA (non-competitive)	Clark Laboratories	734
Antibody detection (IgG)	EIA (non-competitive)	Dade International Inc (Bartels Division)	734
Antibody detection (IgG)	EIA (non-competitive)	Diesse	735
Antibody detection (IgG)	EIA (non-competitive)	Gull Laboratories Inc	735
Antibody detection (IgG)	EIA (non-competitive)	Human GmbH	736
Antibody detection (IgG)	EIA (non-competitive)	IFCI Clone Systems	736
Antibody detection (IgG)	EIA (non-competitive)	Immuno Concepts Inc	737
Antibody detection (IgG)	EIA (non-competitive)	Immunobiological Laboratories	737
Antibody detection (IgG)	EIA (non-competitive)	Incstar	738
Antibody detection (IgG)	EIA (non-competitive)	Melotec S.A.	738
Antibody detection (IgG)	EIA (non-competitive)	Ortho Diagnostic Systems	739
Antibody detection (IgG)	EIA (non-competitive)	PanBio	739
Antibody detection (IgG)	EIA (non-competitive)	Savyon Diagnostics Ltd	740
Antibody detection (IgG)	EIA (non-competitive)	Sigma Diagnostics	740
Antibody detection (IgM)	EIA (non-competitive)	Alfa Biotech	741
Antibody detection (IgM)	EIA (non-competitive)	Amico Laboratories Inc	741
Antibody detection (IgM)	EIA (non-competitive)	Amico Laboratories Inc	742
Antibody detection (IgM)	EIA (non-competitive)	BAG-Biologische Analysensystem GmbH	742
Antibody detection (IgM)	EIA (non-competitive)	BioWhittaker Inc	743
Antibody detection (IgM)	EIA (non-competitive)	Clark Laboratories	743
Antibody detection (IgM)	EIA (non-competitive)	Diesse	744
Antibody detection (IgM)	EIA (non-competitive)	Gull Laboratories Inc	744
Antibody detection (IgM)	EIA (non-competitive)	Human GmbH	745
Antibody detection (IgM)	EIA (non-competitive)	IFCI Clone Systems	745

© KLUWER ACADEMIC PUBLISHERS 1995, ISSN 1381-5067

Antibody detection (IgM)	EIA (non-competitive)	Immuno Concepts Inc	746
Antibody detection (IgM)	EIA (non-competitive)	Immunobiological Laboratories	746
Antibody detection (IgM)	EIA (non-competitive)	Incstar	747
Antibody detection (IgM)	EIA (non-competitive)	Ortho Diagnostic Systems	747
Antibody detection (IgM)	EIA (non-competitive)	PanBio	748
Antibody detection (IgM)	EIA (non-competitive)	Savyon Diagnostics Ltd	748
Antibody detection (IgM)	EIA (non-competitive)	Sigma Diagnostics	749
Antibody detection (IgG)	Immunofluorescence assay (indirect)	Bion Enterprises Ltd	749
Antibody detection (IgG)	Immunofluorescence assay (indirect)	Gull Laboratories Inc	750
Antibody detection (IgG)	Immunofluorescence assay (indirect)	Hemagen Diagnostics Inc	750
Antibody detection (IgG)	Immunofluorescence assay (indirect)	Immuno Concepts Inc	751
Antibody detection (IgG)	Immunofluorescence assay (indirect)	MRL Diagnostics	751
Antibody detection (IgG)	Immunofluorescence assay (indirect)	Stellar Bio Systems Inc	752
Antibody detection (IgG)	Immunofluorescence assay (indirect)	Zeus Scientific Inc	752
Antibody detection (IgM)	Immunofluorescence assay (indirect)	Bion Enterprises Ltd	753
Antibody detection (IgM)	Immunofluorescence assay (indirect)	Gull Laboratories Inc	753
Antibody detection (IgM)	Immunofluorescence assay (indirect)	Immuno Concepts Inc	754
Antibody detection (IgM)	Immunofluorescence assay (indirect)	MRL Diagnostics	754
Antibody detection (IgM)	Immunofluorescence assay (indirect)	Stellar Bio Systems Inc	755

Heterophile antigen

Antibody detection	Particle agglutination assay	BIOKIT SA	755
Antibody detection	Particle agglutination assay	Diesse	756
Antibody detection	Particle agglutination assay	Unipath Limited	756

Escherichia coli

page

Antigen detection	EIA (non-competitive)	LMD Laboratories Inc	758
Antigen detection	EIA (non-competitive)	Tecra Diagnostics	758
Antigen detection	Particle agglutination assay	Microgen Bioproducts	759
Antigen detection	Particle agglutination assay	PRO-LAB Diagnostics	759
Antigen detection	Particle agglutination assay	Unipath Limited	760

Escherichia coli (verotoxins)

Antigen detection	EIA (non-competitive)	Meridian Diagnostics Inc	760
Antigen detection	Particle agglutination assay	Denka Seiken Co. Ltd	761

Giardia lamblia

page

Antigen detection	EIA (non-competitive)	Alexon Inc	763
Antigen detection	EIA (non-competitive)	Alexon Inc	763
Antigen detection	EIA (non-competitive)	Cellabs Pty Ltd	764
Antigen detection	EIA (non-competitive)	LMD Laboratories Inc	764
Antigen detection	EIA (non-competitive)	Melotec S.A.	765
Antigen detection	EIA (non-competitive)	Seradyn	765
Antigen detection	Immunofluorescence assay (direct)	Cellabs Pty Ltd	766

Hantaviruses

page

Hantaan virus

Antibody detection (IgG)	EIA (non-competitive)	Progen	768
Antibody detection (IgM)	EIA (non-competitive)	Progen	768
Antibody detection	Immunofluorescence assay (indirect)	Progen	769

Puumala virus

Antibody detection (IgG)	EIA (non-competitive)	Progen	769

© KLUWER ACADEMIC PUBLISHERS 1995, ISSN 1381-5067

Antibody detection (IgG)	Immunofluorescence assay (indirect)	Stellar Bio Systems Inc	797

Human parvovirus B19 *page*

Antibody detection (IgG)	EIA (non-competitive)	Biotrin International Ltd	799
Antibody detection (IgG)	EIA (non-competitive)	Dako A/S	799
Antibody detection (IgG)	EIA (non-competitive)	Euro-Diagnostica B.V.	800
Antibody detection (IgG)	EIA (non-competitive)	Immunobiological Laboratories	800
Antibody detection (IgG)	EIA (non-competitive)	Laboratoire Eurobio	801
Antibody detection (IgG)	EIA (non-competitive)	MRL Diagnostics	801
Antibody detection (IgG)	EIA (non-competitive)	r-biopharm GmbH	802
Antibody detection (IgM)	EIA (non-competitive)	Biotrin International Ltd	802
Antibody detection (IgM)	EIA (non-competitive)	Dako A/S	803
Antibody detection (IgM)	EIA (non-competitive)	Euro-Diagnostica B.V.	803
Antibody detection (IgM)	EIA (non-competitive)	Immunobiological Laboratories	804
Antibody detection (IgM)	EIA (non-competitive)	Laboratoire Eurobio	804
Antibody detection (IgM)	EIA (non-competitive)	MRL Diagnostics	805
Antibody detection (IgM)	EIA (non-competitive)	r-biopharm GmbH	805
Antibody detection (IgG and/or IgM)	Immunoblot assay	Biotrin International Ltd	806
Antibody detection (IgG and/or IgM)	Immunoblot assay	r-biopharm GmbH	806
Antibody detection (IgM)	Immunoblot assay	Immunetics	807
Antibody detection (IgG)	Immunofluorescence assay (indirect)	Stellar Bio Systems Inc	807
Antibody detection (IgG and/or IgM)	Immunofluorescence assay (indirect)	Biotrin International Ltd	808
Antibody detection (IgM)	Immunofluorescence assay (indirect)	Stellar Bio Systems Inc	808

Leishmania species *page*

Antibody detection (IgG)	EIA (non-competitive)	Amico Laboratories Inc	810
Antibody detection (IgG)	EIA (non-competitive)	Immuno Pharmacology Research	810
Antibody detection (IgG)	EIA (non-competitive)	Melotec S.A.	811
Antibody detection (IgM)	EIA (non-competitive)	Amico Laboratories Inc	811
Antibody detection (IgM)	EIA (non-competitive)	Amico Laboratories Inc	812
Antibody detection	Immunofluorescence assay (indirect)	bioMerieux	812
Antibody detection	Immunofluorescence assay (indirect)	Immuno Pharmacology Research	813

Leptospira species *page*

Antibody detection (IgG and/or IgM))	EIA (non-competitive)	Immuno Pharmacology Research	815
Antibody detection (IgM)	EIA (non-competitive)	PanBio	815

Listeria species *page*

Antigen detection	EIA (non-competitive)	Tecra Diagnostics	817
Antigen detection	EIA (non-competitive)	Tecra Diagnostics	817
Antigen detection	Immunochromatographic assay	Unipath Limited	818
Antigen detection	Particle agglutination assay	Microgen Bioproducts	818

Mumps virus *page*

Antibody detection (IgG)	EIA (non-competitive)	Amico Laboratories Inc	820
Antibody detection (IgG)	EIA (non-competitive)	BAG-Biologische Analysensystem GmbH	820

Antibody detection (IgG)	EIA (non-competitive)	Behringwerke AG	821
Antibody detection (IgG)	EIA (non-competitive)	bioMerieux	821
Antibody detection (IgG)	EIA (non-competitive)	BioWhittaker Inc	822
Antibody detection (IgG)	EIA (non-competitive)	BioWhittaker Inc	822
Antibody detection (IgG)	EIA (non-competitive)	Chimica Diagnostica	823
Antibody detection (IgG)	EIA (non-competitive)	Clark Laboratories	823
Antibody detection (IgG)	EIA (non-competitive)	Denka Seiken Co. Ltd	824
Antibody detection (IgG)	EIA (non-competitive)	Gamma SA	824
Antibody detection (IgG)	EIA (non-competitive)	Human GmbH	825
Antibody detection (IgG)	EIA (non-competitive)	Immunobiological Laboratories	825
Antibody detection (IgG)	EIA (non-competitive)	Laboratoire Eurobio	826
Antibody detection (IgG)	EIA (non-competitive)	Medix Biotech Inc	826
Antibody detection (IgG)	EIA (non-competitive)	Melotec S.A.	827
Antibody detection (IgG)	EIA (non-competitive)	Radim	827
Antibody detection (IgM)	EIA (non-competitive)	Amico Laboratories Inc	828
Antibody detection (IgM)	EIA (non-competitive)	Amico Laboratories Inc	828
Antibody detection (IgM)	EIA (non-competitive)	BAG-Biologische Analysensystem GmbH	829
Antibody detection (IgM)	EIA (non-competitive)	Behringwerke AG	829
Antibody detection (IgM)	EIA (non-competitive)	Chimica Diagnostica	830
Antibody detection (IgM)	EIA (non-competitive)	Clark Laboratories	830
Antibody detection (IgM)	EIA (non-competitive)	Denka Seiken Co. Ltd	831
Antibody detection (IgM)	EIA (non-competitive)	Gamma SA	831
Antibody detection (IgM)	EIA (non-competitive)	Human GmbH	832
Antibody detection (IgM)	EIA (non-competitive)	Immunobiological Laboratories	832
Antibody detection (IgM)	EIA (non-competitive)	Laboratoire Eurobio	833
Antibody detection (IgM)	EIA (non-competitive)	Medix Biotech Inc	833
Antibody detection (IgM)	EIA (non-competitive)	Radim	834
Antibody detection (IgG)	Immunofluorescence assay (indirect)	Bion Enterprises Ltd	834
Antibody detection (IgG)	Immunofluorescence assay (indirect)	Hemagen Diagnostics Inc	835
Antibody detection (IgG)	Immunofluorescence assay (indirect)	Stellar Bio Systems Inc	835
Antibody detection (IgM)	Immunofluorescence assay (indirect)	Stellar Bio Systems Inc	836

Neisseria meningitidis

			page
Antigen detection	Particle agglutination assay	Becton Dickinson	838

Plasmodium falciparum

			page
Antigen detection	EIA (non-competitive)	Cellabs Pty Ltd	840
Antibody detection	EIA (non-competitive)	Cellabs Pty Ltd	840
Antibody detection	Immunofluorescence assay (indirect)	bioMerieux	841

Rickettsiae

			page

Rickettsia conorii

Antibody detection (IgG and/or IgM))	EIA (non-competitive)	Immuno Pharmacology Research	843
Antibody detection	Immunofluorescence assay (indirect)	bioMerieux	843
Antibody detection	Immunofluorescence assay (indirect)	Immuno Pharmacology Research	844

Spotted Fever bio-group and Typhus Fever bio-group

Antibody detection (IgG)	Immunofluorescence assay (indirect)	MRL Diagnostics	844
Antibody detection (IgM)	Immunofluorescence assay (indirect)	MRL Diagnostics	845

Rotavirus

			page
Antigen detection	EIA (non-competitive)	Abbott Laboratories	847
Antigen detection	EIA (non-competitive)	Abbott Laboratories	847
Antigen detection	EIA (non-competitive)	Alexon Inc	848
Antigen detection	EIA (non-competitive)	Cambridge Biotech Corporation	848
Antigen detection	EIA (non-competitive)	Dako A/S	849
Antigen detection	EIA (non-competitive)	LMD Laboratories Inc	849
Antigen detection	EIA (non-competitive)	Melotec S.A.	850
Antigen detection	EIA (non-competitive)	Microgen Bioproducts	850
Antigen detection	EIA (non-competitive)	r-biopharm GmbH	851
Antigen detection	EIA (non-competitive)	Sanofi Diagnostics Pasteur	851
Antigen detection	Immunochromatographic assay	Meridian Diagnostics Inc	852
Antigen detection	Particle agglutination assay	BIOKIT SA	852
Antigen detection	Particle agglutination assay	bioMerieux	853
Antigen detection	Particle agglutination assay	Meridian Diagnostics Inc	853
Antigen detection	Particle agglutination assay	Microgen Bioproducts	854
Antigen detection	Particle agglutination assay	Murex Diagnostics Limited	854
Antigen detection	Particle agglutination assay	Omega Diagnostics Ltd	855
Antigen detection	Particle agglutination assay	Orion Diagnostica	855
Antigen detection	Particle agglutination assay	Orion Diagnostica	856
Antibody detection (IgG)	EIA (non-competitive)	Amico Laboratories Inc	856
Antibody detection (IgG)	EIA (non-competitive)	Immuno Pharmacology Research	857
Antibody detection (IgM)	EIA (non-competitive)	Amico Laboratories Inc	857
Antibody detection (IgM)	EIA (non-competitive)	Amico Laboratories Inc	858

Rubella virus

			page
Antibody detection	EIA (non-competitive)	Centocor Inc	860
Antibody detection	EIA (non-competitive)	Centocor Inc	860
Antibody detection	EIA (non-competitive)	Centocor Inc	861
Antibody detection	EIA (non-competitive)	Diesse	861
Antibody detection	EIA (non-competitive)	GenBio	862
Antibody detection (IgG)	EIA (non-competitive)	Abbott Laboratories	862
Antibody detection (IgG)	EIA (non-competitive)	Alfa Biotech	863
Antibody detection (IgG)	EIA (non-competitive)	BAG-Biologische Analysensystem GmbH	863
Antibody detection (IgG)	EIA (non-competitive)	Behringwerke AG	864
Antibody detection (IgG)	EIA (non-competitive)	BIOKIT SA	864
Antibody detection (IgG)	EIA (non-competitive)	bioMerieux	865
Antibody detection (IgG)	EIA (non-competitive)	BioWhittaker Inc	865
Antibody detection (IgG)	EIA (non-competitive)	BioWhittaker Inc	866
Antibody detection (IgG)	EIA (non-competitive)	Bouty SpA	866
Antibody detection (IgG)	EIA (non-competitive)	Chimica Diagnostica	867
Antibody detection (IgG)	EIA (non-competitive)	Clark Laboratories	867
Antibody detection (IgG)	EIA (non-competitive)	Denka Seiken Co. Ltd	868
Antibody detection (IgG)	EIA (non-competitive)	Diamedix Corporation	868
Antibody detection (IgG)	EIA (non-competitive)	E. Merck	869
Antibody detection (IgG)	EIA (non-competitive)	Gull Laboratories Inc	869
Antibody detection (IgG)	EIA (non-competitive)	Human GmbH	870
Antibody detection (IgG)	EIA (non-competitive)	IFCI Clone Systems	870
Antibody detection (IgG)	EIA (non-competitive)	Immuno Pharmacology Research	871
Antibody detection (IgG)	EIA (non-competitive)	Incstar	871
Antibody detection (IgG)	EIA (non-competitive)	Kreatech Diagnostics	872
Antibody detection (IgG)	EIA (non-competitive)	Laboratoire Eurobio	872
Antibody detection (IgG)	EIA (non-competitive)	Labsystems Oy	873
Antibody detection (IgG)	EIA (non-competitive)	Labsystems Oy	873
Antibody detection (IgG)	EIA (non-competitive)	Medix Biotech Inc	874

Antibody detection (IgG)	EIA (non-competitive)	Melotec S.A.	874
Antibody detection (IgG)	EIA (non-competitive)	Menarini Diagnostics	875
Antibody detection (IgG)	EIA (non-competitive)	PanBio	875
Antibody detection (IgG)	EIA (non-competitive)	Radim	876
Antibody detection (IgG)	EIA (non-competitive)	Roche Diagnostic Systems	876
Antibody detection (IgG)	EIA (non-competitive)	Sanofi Diagnostics Pasteur	877
Antibody detection (IgG)	EIA (non-competitive)	Sigma Diagnostics	877
Antibody detection (IgG)	EIA (non-competitive)	Sorin Biomedica	878
Antibody detection (IgG)	EIA (non-competitive)	United Biotech Inc	878
Antibody detection (IgG)	EIA (non-competitive)	Zeus Scientific Inc	879
Antibody detection (IgM)	EIA (non-competitive)	Abbott Laboratories	879
Antibody detection (IgM)	EIA (non-competitive)	Alfa Biotech	880
Antibody detection (IgM)	EIA (non-competitive)	Amico Laboratories Inc	880
Antibody detection (IgM)	EIA (non-competitive)	Amico Laboratories Inc	881
Antibody detection (IgM)	EIA (non-competitive)	BAG-Biologische Analysensystem GmbH	881
Antibody detection (IgM)	EIA (non-competitive)	Behringwerke AG	882
Antibody detection (IgM)	EIA (non-competitive)	BIOKIT SA	882
Antibody detection (IgM)	EIA (non-competitive)	bioMerieux	883
Antibody detection (IgM)	EIA (non-competitive)	BioWhittaker Inc	883
Antibody detection (IgM)	EIA (non-competitive)	Bouty SpA	884
Antibody detection (IgM)	EIA (non-competitive)	Chimica Diagnostica	884
Antibody detection (IgM)	EIA (non-competitive)	Clark Laboratories	885
Antibody detection (IgM)	EIA (non-competitive)	Dade International Inc (Bartels Division)	885
Antibody detection (IgM)	EIA (non-competitive)	Denka Seiken Co. Ltd	886
Antibody detection (IgM)	EIA (non-competitive)	Diamedix Corporation	886
Antibody detection (IgM)	EIA (non-competitive)	Diesse	887
Antibody detection (IgM)	EIA (non-competitive)	Gull Laboratories Inc	887
Antibody detection (IgM)	EIA (non-competitive)	Human GmbH	888
Antibody detection (IgM)	EIA (non-competitive)	Human GmbH	888
Antibody detection (IgM)	EIA (non-competitive)	IFCI Clone Systems	889
Antibody detection (IgM)	EIA (non-competitive)	Immuno Pharmacology Research	889
Antibody detection (IgM)	EIA (non-competitive)	Incstar	890
Antibody detection (IgM)	EIA (non-competitive)	Kreatech Diagnostics	890
Antibody detection (IgM)	EIA (non-competitive)	Laboratoire Eurobio	891
Antibody detection (IgM)	EIA (non-competitive)	Labsystems Oy	891
Antibody detection (IgM)	EIA (non-competitive)	Medix Biotech Inc	892
Antibody detection (IgM)	EIA (non-competitive)	Melotec S.A.	892
Antibody detection (IgM)	EIA (non-competitive)	Menarini Diagnostics	893
Antibody detection (IgM)	EIA (non-competitive)	Radim	893
Antibody detection (IgM)	EIA (non-competitive)	Roche Diagnostic Systems	894
Antibody detection (IgM)	EIA (non-competitive)	Sigma Diagnostics	894
Antibody detection (IgM)	EIA (non-competitive)	Sigma Diagnostics	895
Antibody detection (IgM)	EIA (non-competitive)	Sorin Biomedica	895
Antibody detection (IgM)	EIA (non-competitive)	United Biotech Inc	896
Antibody detection (IgG)	Immunofluorescence assay (indirect)	Hemagen Diagnostics Inc	896
Antibody detection (IgG)	Immunofluorescence assay (indirect)	Stellar Bio Systems Inc	897
Antibody detection (IgM)	Immunofluorescence assay (indirect)	Stellar Bio Systems Inc	897
Antibody detection (IgG)	Luminometric immunoassay (non-competitive)	Diagnostic Products Corporation	898
Antibody detection (IgG)	Luminometric immunoassay (non-competitive)	Diagnostic Products Corporation	898
Antibody detection (IgG)	Luminometric immunoassay (non-competitive)	Johnson & Johnson Clinical Diagnostics Inc	899
Antibody detection (IgG)	Luminometric immunoassay (non-competitive)	Sanofi Diagnostics Pasteur	899
Antibody detection (IgM)	Luminometric immunoassay (non-competitive)	Johnson & Johnson Clinical Diagnostics Inc	900

Antibody detection (IgM)	Luminometric immunoassay (non-competitive)	Sanofi Diagnostics Pasteur	900
Antibody detection	Particle agglutination assay	Becton Dickinson	901
Antibody detection	Particle agglutination assay	BIOKIT SA	901
Antibody detection	Particle agglutination assay	Carter-Wallace Inc	902
Antibody detection	Particle agglutination assay	Orion Diagnostica	902
Antibody detection	Particle agglutination assay	S.A. Scientific	903
Antibody detection	Particle agglutination assay	Seradyn	903

Salmonella species

			page
Antigen detection	EIA (non-competitive)	Mast Diagnostics Ltd	905
Antigen detection	EIA (non-competitive)	Tecra Diagnostics	905
Antigen detection	EIA (non-competitive)	Tecra Diagnostics	906
Antigen detection	Immunochromatographic assay	PanBio	906
Antigen detection	Particle agglutination assay (coagglutination)	Boule Diagnostics AB	907
Antigen detection	Particle agglutination assay	Microgen Bioproducts	907
Antigen detection	Particle agglutination assay	Unipath Limited	908

Salmonella typhi

Antibody detection (IgG and/or IgM))	EIA (non-competitive)	Immuno Pharmacology Research	908

Schistosoma species

			page
Antibody detection (IgG)	EIA (non-competitive)	Amico Laboratories Inc	910
Antibody detection (IgM)	EIA (non-competitive)	Amico Laboratories Inc	910
Antibody detection (IgM)	EIA (non-competitive)	Amico Laboratories Inc	911

Taenia solium

			page
Antibody detection	EIA (non-competitive)	LMD Laboratories Inc	914
Antibody detection (IgG)	EIA (non-competitive)	Melotec S.A.	914
Antibody detection (IgG)	Immunoblot assay	Immunetics	915

Toxocara canis

			page
Antibody detection	EIA (non-competitive)	Cellabs Pty Ltd	917
Antibody detection	EIA (non-competitive)	LMD Laboratories Inc	917

Toxoplasma gondii

			page
Antigen detection	Immunofluorescence assay (direct)	Cellabs Pty Ltd	919
Antibody detection	EIA (competitive)	Diesse	919
Antibody detection	EIA (non-competitive)	LMD Laboratories Inc	920
Antibody detection (IgA)	EIA (non-competitive)	Abbott Laboratories	920
Antibody detection (IgA)	EIA (non-competitive)	Bouty SpA	921
Antibody detection (IgA)	EIA (non-competitive)	Chimica Diagnostica	921
Antibody detection (IgA)	EIA (non-competitive)	Diesse	922
Antibody detection (IgA)	EIA (non-competitive)	SFRI Laboratoire	922
Antibody detection (IgG)	EIA (non-competitive)	Abbott Laboratories	923
Antibody detection (IgG)	EIA (non-competitive)	Abbott Laboratories	923
Antibody detection (IgG)	EIA (non-competitive)	Alfa Biotech	924
Antibody detection (IgG)	EIA (non-competitive)	Amico Laboratories Inc	924
Antibody detection (IgG)	EIA (non-competitive)	Amico Laboratories Inc	925

Antibody detection (IgG)	EIA (non-competitive)	Behringwerke AG	925
Antibody detection (IgG)	EIA (non-competitive)	BIOKIT SA	926
Antibody detection (IgG)	EIA (non-competitive)	bioMerieux	926
Antibody detection (IgG)	EIA (non-competitive)	Biotecx Laboratories Inc	927
Antibody detection (IgG)	EIA (non-competitive)	BioWhittaker Inc	927
Antibody detection (IgG)	EIA (non-competitive)	BioWhittaker Inc	928
Antibody detection (IgG)	EIA (non-competitive)	Bouty SpA	928
Antibody detection (IgG)	EIA (non-competitive)	Centocor Inc	929
Antibody detection (IgG)	EIA (non-competitive)	Chimica Diagnostica	929
Antibody detection (IgG)	EIA (non-competitive)	Clark Laboratories	930
Antibody detection (IgG)	EIA (non-competitive)	Dade International Inc (Bartels Division)	930
Antibody detection (IgG)	EIA (non-competitive)	Denka Seiken Co. Ltd	931
Antibody detection (IgG)	EIA (non-competitive)	Diesse	931
Antibody detection (IgG)	EIA (non-competitive)	E. Merck	932
Antibody detection (IgG)	EIA (non-competitive)	Gull Laboratories Inc	932
Antibody detection (IgG)	EIA (non-competitive)	Human GmbH	933
Antibody detection (IgG)	EIA (non-competitive)	IFCI Clone Systems	933
Antibody detection (IgG)	EIA (non-competitive)	Immuno Pharmacology Research	934
Antibody detection (IgG)	EIA (non-competitive)	Immunobiological Laboratories	934
Antibody detection (IgG)	EIA (non-competitive)	Incstar	935
Antibody detection (IgG)	EIA (non-competitive)	Kreatech Diagnostics	935
Antibody detection (IgG)	EIA (non-competitive)	Labsystems Oy	936
Antibody detection (IgG)	EIA (non-competitive)	Labsystems Oy	936
Antibody detection (IgG)	EIA (non-competitive)	Medix Biotech Inc	937
Antibody detection (IgG)	EIA (non-competitive)	Melotec S.A.	937
Antibody detection (IgG)	EIA (non-competitive)	Menarini Diagnostics	938
Antibody detection (IgG)	EIA (non-competitive)	Organon Teknika NV	938
Antibody detection (IgG)	EIA (non-competitive)	Radim	939
Antibody detection (IgG)	EIA (non-competitive)	Roche Diagnostic Systems	939
Antibody detection (IgG)	EIA (non-competitive)	Sanofi Diagnostics Pasteur	940
Antibody detection (IgG)	EIA (non-competitive)	SFRI Laboratoire	940
Antibody detection (IgG)	EIA (non-competitive)	Sigma Diagnostics	941
Antibody detection (IgG)	EIA (non-competitive)	Sorin Biomedica	941
Antibody detection (IgG)	EIA (non-competitive)	United Biotech Inc	942
Antibody detection (IgM)	EIA (non-competitive)	Abbott Laboratories	942
Antibody detection (IgM)	EIA (non-competitive)	Abbott Laboratories	943
Antibody detection (IgM)	EIA (non-competitive)	Alfa Biotech	943
Antibody detection (IgM)	EIA (non-competitive)	Amico Laboratories Inc	944
Antibody detection (IgM)	EIA (non-competitive)	Amico Laboratories Inc	944
Antibody detection (IgM)	EIA (non-competitive)	Behringwerke AG	945
Antibody detection (IgM)	EIA (non-competitive)	BIOKIT SA	945
Antibody detection (IgM)	EIA (non-competitive)	bioMerieux	946
Antibody detection (IgM)	EIA (non-competitive)	BioWhittaker Inc	946
Antibody detection (IgM)	EIA (non-competitive)	Bouty SpA	947
Antibody detection (IgM)	EIA (non-competitive)	Centocor Inc	947
Antibody detection (IgM)	EIA (non-competitive)	Chimica Diagnostica	948
Antibody detection (IgM)	EIA (non-competitive)	Clark Laboratories	948
Antibody detection (IgM)	EIA (non-competitive)	Dade International Inc (Bartels Division)	949
Antibody detection (IgM)	EIA (non-competitive)	Denka Seiken Co. Ltd	949
Antibody detection (IgM)	EIA (non-competitive)	Diesse	950
Antibody detection (IgM)	EIA (non-competitive)	E. Merck	950
Antibody detection (IgM)	EIA (non-competitive)	Gull Laboratories Inc	951
Antibody detection (IgM)	EIA (non-competitive)	Human GmbH	951
Antibody detection (IgM)	EIA (non-competitive)	Human GmbH	952
Antibody detection (IgM)	EIA (non-competitive)	IFCI Clone Systems	952
Antibody detection (IgM)	EIA (non-competitive)	Immuno Pharmacology Research	953
Antibody detection (IgM)	EIA (non-competitive)	Incstar	953

Antibody detection (IgM)	EIA (non-competitive)	Kreatech Diagnostics	954
Antibody detection (IgM)	EIA (non-competitive)	Labsystems Oy	954
Antibody detection (IgM)	EIA (non-competitive)	Medix Biotech Inc	955
Antibody detection (IgM)	EIA (non-competitive)	Melotec S.A.	955
Antibody detection (IgM)	EIA (non-competitive)	Menarini Diagnostics	956
Antibody detection (IgM)	EIA (non-competitive)	Organon Teknika NV	956
Antibody detection (IgM)	EIA (non-competitive)	Radim	957
Antibody detection (IgM)	EIA (non-competitive)	Roche Diagnostic Systems	957
Antibody detection (IgM)	EIA (non-competitive)	Sanofi Diagnostics Pasteur	958
Antibody detection (IgM)	EIA (non-competitive)	SFRI Laboratoire	958
Antibody detection (IgM)	EIA (non-competitive)	Sigma Diagnostics	959
Antibody detection (IgM)	EIA (non-competitive)	Sigma Diagnostics	959
Antibody detection (IgM)	EIA (non-competitive)	Sorin Biomedica	960
Antibody detection (Total Ab and/ or IgG and/or IgM)	Immunofluorescence assay (indirect)	bioMerieux	960
Antibody detection	Immunofluorescence assay (indirect)	Zeus Scientific Inc	961
Antibody detection (IgG)	Immunofluorescence assay (indirect)	GenBio	961
Antibody detection (IgG)	Immunofluorescence assay (indirect)	Gull Laboratories Inc	962
Antibody detection (IgG)	Immunofluorescence assay (indirect)	Hemagen Diagnostics Inc	962
Antibody detection (IgG)	Immunofluorescence assay (indirect)	LD, Labor Diagnostika GmbH	963
Antibody detection (IgG)	Immunofluorescence assay (indirect)	Stellar Bio Systems Inc	963
Antibody detection (IgM)	Immunofluorescence assay (indirect)	GenBio	964
Antibody detection (IgM)	Immunofluorescence assay (indirect)	Gull Laboratories Inc	964
Antibody detection (IgM)	Immunofluorescence assay (indirect)	Stellar Bio Systems Inc	965
Antibody detection (IgG)	Luminometric immunoassay (non-competitive)	Diagnostic Products Corporation	965
Antibody detection (IgG)	Luminometric immunoassay (non-competitive)	Diagnostic Products Corporation	966
Antibody detection (IgG)	Luminometric immunoassay (non-competitive)	Johnson & Johnson Clinical Diagnostics Inc	966
Antibody detection (IgG)	Luminometric immunoassay (non-competitive)	Sanofi Diagnostics Pasteur	967
Antibody detection (IgM)	Luminometric immunoassay (non-competitive)	Johnson & Johnson Clinical Diagnostics Inc	967
Antibody detection (IgM)	Luminometric immunoassay (non-competitive)	Sanofi Diagnostics Pasteur	968
Antibody detection	Particle agglutination assay	BIOKIT SA	968
Antibody detection	Particle agglutination assay	Eiken Chemical Co. Ltd	969
Antibody detection	Particle agglutination assay	Laboratoire Fumouze	969
Antibody detection (IgA and/or IgM)	Particle agglutination assay	bioMerieux	970
Antibody detection (IgM)	Particle agglutination assay	bioMerieux	970

Trichinella spiralis

			page
Antibody detection	EIA (non-competitive)	LMD Laboratories Inc	972
Antibody detection	EIA (non-competitive)	Melotec S.A.	972

Trypanosoma cruzi

			page
Antibody detection (IgG)	EIA (non-competitive)	Cellabs Pty Ltd	974

Vibrio parahaemolyticus (haemolytic toxin)

			page
Antigen detection	Particle agglutination assay	Denka Seiken Co. Ltd	976

Adenoviruses

Natural history

The 47 serotypes of adenovirus show different degrees of homology within the hexon gene, allowing them to be grouped into 6 subgenera, with members of each subgenus showing >50% homology within the gene. Genotyping can be carried out by restriction fragment length poly-morphism (RFLP) and use of the enzyme *Sma I* allows differentiation into the subgenera A–F.

While it is common to find adenovirus in the stools of children they are not necessarily associated with diarrhoea or disease and indeed may be excreted for several months following infection. Pathogenic involvement should only be assumed if there is supporting clinical evidence. Adenovirus is asso-ciated with 4–15% of all hospitalised children with gastroenteritis. In some of these cases it may be a sign of systemic infection, as in babies where severe adenovirus infection can involve respiratory strains spreading through the body. Serotypes 1–7 (found within subgenera B, C and E) are recover-able from stools but are only occasionally asso-ciated with diarrhoea. In contrast, the fastidious subgenus F (serotypes 40 and 41) is an important cause of infantile diarrhoea in western countries and is found in 2/3 of the diarrhoeas associated with adenovirus, yet is not readily isolated in culture.

Adenovirus associated diarrhoea has an incubation period of 8–10 days, with viral shedding occurring over 3–13 days and is rarely associated with respiratory symptoms. It is thus less acute but of longer duration than infection associated with rotavirus. Infection with fastidious types can lead to dietary intolerance in young children. Outbreaks of infection can occur throughout the year, with large numbers of particles (typically 10^{11} particles/g faeces) being shed. These viruses however are not shed for long periods nor are they found in healthy controls. Following infection seroconversion occurs in 70% of patients and by aged 6–8 years, about half of children tested in western countries were found to have antibodies.

Diagnosis

Most adenoviruses will grow readily in standard cell cultures but the fastidious adenoviruses are difficult to grow, requiring specialist cell lines such as human embryo kidney (HEK) or Graham 293 cells. Care must be taken to ensure a high multiplicity of infection and trypsinisation of the cells prior to adsorption. Even so, high numbers of particles present in a stool sample may inhibit growth.

Both fluorescent antibody tests and ELISA are used to demonstrate the presence of adenovirus antigen directly in specimens with the former better suited to cellular specimens including respiratory secretions and ELISA for faecal material.

Comment

The ability to grow respiratory adenoviruses in culture readily and the doubtful significance of most adenoviruses other than types 40/41 in faecal samples creates practical problems for the labora-tory in deciding which tests to use. As a result different kit assays cover adenoviruses as a complete group and adenovirus types 40/41.

Reference

Sharp IR, Wadell G. Adenoviruses. In: Zuckerman AJ, Banatvala JE, Pattison JR. Principles and Practice of Clinical Virology, 3rd edn. Chichester: John Wiley & Sons. 1994:287–308.

Madeley CR. Viruses associated with acute diarrhoeal disease. In: Zuckerman AJ, Banatvala JE, Pattison JR. Principles and Practice of Clinical Virology, 3rd edn. Chichester: John Wiley & Sons. 1994:189–227.

See also Multipathogen Assays section under: Gastrointestinal pathogens

Adenovirus
ANTIGEN DETECTION

Manufacturer: Alfa Biotech
Cat. No./Trade name: 05773518/ADENOVIRUS ELISA System Antigen Detection Kit

SUMMARY

[Well-MAb]–**Ag**–[MAb-biotin]–[strept-POD]–[OPD]–A_{492}

Assay type: EIA (non-competitive)
Detection: Colorimetric A_{492}
Format: Microtitre well, Ab coated
Sample type: Nasopharyngeal aspirate, faecal sample
Sample pre-treatment:
   Dilute nasopharyngeal specimens with Mucolite solution provided. Dilute faecal samples with buffer, centrifuge and use supernatant
Sample volume: 100 µl of prediluted samples x 2
Number of tests: 96
Controls - standards run in assay:
   Controls: Neg (4), Pos (2)
Incubation:
   1 hr (37°C) + 30 min (37°C) + 10 min (37°C)
Washes: 2 (+ preliminary plate wash)

CONTENTS

Antibodies, antigens, labelled components:
   Anti-adenovirus MAb bound to well
   Anti-adenovirus MAb biotinylated
   Streptavidin POD conjugated
Substrate: OPD
Controls - standards supplied:
   Controls: Neg (use sample and conjugate diluent) and Pos (human faecal sample containing adenovirus)
Additional reagents:
   None
Special equipment required:
   None

INTERPRETATION

Comments on interpretation:
   Classification of sample is according to cut-off; no further testing
No. of references: 6

NOTES

120871.0

Adenovirus
ANTIGEN DETECTION

Manufacturer: Biotrin International Ltd
Cat. No./Trade name: V9ADE/Adenovirus Antigen EIA

SUMMARY

[Well-Ag]–**Ag**–[Ab-HRP]–[TMB]–A_{450}

Assay type: EIA (non-competitive)
Detection: Colorimetric A_{450}
Format: Microtitre well, Ab coated
Sample type: Faecal specimens, rectal swabs, cell cultures*
Sample pre-treatment:
   Direct swabs: add sample diluent and vortex
   Cell culture: from cultures showing CPE make cell suspension using diluent or use tissue culture fluid and vortex
Sample volume: 50 µl of prepared samples
Number of tests: 96
Controls - standards run in assay:
   Controls: Neg at least (1), Pos (1)
Incubation:
   10 min (RT) + 1 hr (RT) + 15 min (RT)
Washes: 1

CONTENTS

Antibodies, antigens, labelled components:
   Anti-adenovirus MAb (m) bound to well
   Anti-adenovirus Ab (g) HRP conjugated
Substrate: TMB
Controls - standards supplied:
   Controls: Neg and Pos
Additional reagents:
   None
Special equipment required:
   None

INTERPRETATION

Comments on interpretation:
   Equivocal: within cut-off ± 10%; retest to confirm or repeat test with fresh sample
No. of references: 5

NOTES

120805.0

*Ophthalmic, nasal, throat and rectal swabs, nasopharyngeal aspirate, nasal washings may be inoculated into cell cultures for testing

Adenovirus
ANTIGEN DETECTION

Manufacturer: Cambridge Biotech Corporation
Cat. No./Trade name: 6007/Adenoclone® EIA

SUMMARY

[Well-MAb]–**Ag**–[MAb-HRP]–[TMB]–A$_{450}$

Assay type: EIA (non-competitive)
Detection: Colorimetric A$_{450}$
Format: Microtitre well, Ab coated
Sample type: Faecal specimens, rectal swabs, tissue culture isolates*
Sample pre-treatment:
  Mix stool or swab with diluent or 2 drops of prepared sample
Sample volume: 100 µl of tissue culture fluid or few grams of faecal material
Number of tests: 48
Controls - standards run in assay:
  Controls: Neg (1), Pos (1)**
Incubation:
  1 hr (RT)
Washes: 1

CONTENTS

Antibodies, antigens, labelled components:
  Anti-adenovirus (group specific) MAb (m) bound to well
  Anti-adenovirus (group specific) MAb (m) HRP conjugated
Substrate: TMB
Controls - standards supplied:
  Controls: Neg (sample diluent) and Pos
Additional reagents:
  Adenovirus Pos culture
Special equipment required:
  None

INTERPRETATION

Comments on interpretation:
  Classification of sample is according to cut-off; no further testing
No. of references: 27

NOTES

120847.0

*Nasopharyngeal aspirates, secretions or washes, respiratory or opthalmic specimens, rectal swabs or stools can be inoculated into tissue culture
**A known adenovirus Pos culture and known Neg (uninfected cell supernatant) may be run from time to time to check that the tissue culture system is functioning correctly

Adenovirus
ANTIGEN DETECTION

Manufacturer: Cambridge Biotech Corporation
Cat. No./Trade name: 6006/Adenoclone® - Type 40/41 EIA

SUMMARY

[Well-MAb]–**Ag**–[MAb-HRP]–[TMB]–A$_{450}$

Assay type: EIA (non-competitive)
Detection: Colorimetric A$_{450}$
Format: Microtitre well, Ab coated
Sample type: Faecal specimens, rectal swab
Sample pre-treatment:
  Mix stool or swab with 1 ml diluent using sample transfer pipette provided
Sample volume: 100 µl of prepared sample
Number of tests: 48
Controls - standards run in assay:
  Controls: Neg (1), Pos (1)
Incubation:
  1 hr (RT) + 10 min (RT)
Washes: 1

CONTENTS

Antibodies, antigens, labelled components:
  Anti-adenovirus (group specific) MAb bound to well
  Anti-adenovirus (types 40 and 41) MAb HRP conjugated
Substrate: TMB
Controls - standards supplied:
  Controls: Neg (sample diluent) and Pos (adenovirus, AD41)
Additional reagents:
  None
Special equipment required:
  None

INTERPRETATION

Comments on interpretation:
  Classification of samples is according to cut-off; no further testing
No. of references: 17

NOTES

131178.0

Adenovirus
ANTIGEN DETECTION

Manufacturer: Dako A/S
Cat. No./Trade name: K6021/IDEIA® Adenovirus

SUMMARY

[Well-Ab]–**Ag**–[MAb-HRP]–[TMB]–A_{450}

Assay type: EIA (non-competitive)
Detection: Colorimetric A_{450}
Format: Microtitre well, Ab coated
Sample type: Faecal specimens, tissue culture fluid
Sample pre-treatment:
    Faecal specimens and swabs: prepare 10% suspension
    using sample diluent
Sample volume: 100 µl of prepared sample or culture fluid
Number of tests: 96
Controls - standards run in assay:
    Controls: Neg (1), Pos (1)
Incubation:
    1 hr (RT) + 10 min (RT)
Washes: 1

CONTENTS

Antibodies, antigens, labelled components:
    Anti-adenovirus MAb bound to well
    Anti-adenovirus MAb HRP conjugated
Substrate: TMB
Controls - standards supplied:
    Controls: Neg (sample diluent) and Pos
Additional reagents:
    None
Special equipment required:
    None

INTERPRETATION

Comments on interpretation:
    Equivocal: within cut-off ±10%; retest to confirm or
    repeat test with a fresh sample
No. of references: 30

NOTES

131227.0

Adenovirus
ANTIGEN DETECTION

Manufacturer: International
Immunodiagnostics
Cat. No./Trade name: /Adenovirus Test Kit - EIA

SUMMARY

[Well-MAb]–**Ag**–[MAb-HRP]–[TMB]–A_{450}

Assay type: EIA (non-competitive)
Detection: Colorimetric A_{450}
Format: Microtitre well, Ab coated
Sample type: Tissue culture isolates of respiratory/
    ophthalmic/enteric samples or direct testing of stool
    samples
Sample pre-treatment:
    Tissue cultures, use undiluted culture fluid.
    Stool sample or rectal swab: dilute with sample diluent
Sample volume: 100 µl cell culture fluid or diluted faecal
    sample
Number of tests: 96
Controls - standards run in assay:
    Controls: Neg (1), Pos (1)
Incubation:
    1 hr (RT) + 10 min (RT)
Washes: 2

CONTENTS

Antibodies, antigens, labelled components:
    Anti-adenovirus MAb (m) bound to well
    Anti-adenovirus MAb (m) HRP conjugated
Substrate: TMB
Controls - standards supplied:
    Controls: Neg (sample diluent) and Pos (adenovirus
    AD41)
Additional reagents:
    Adenovirus Pos control cultures (e.g. ATCC #VR-2,
    adenovirus type 2)
Special equipment required:
    None

INTERPRETATION

Comments on interpretation:
    Classification of samples is according to cut-off; no
    further testing
No. of references: 27

NOTES

120960.0

© *KLUWER ACADEMIC PUBLISHERS 1995, ISSN 1381-5067*

Adenovirus
ANTIGEN DETECTION

Manufacturer: r-biopharm GmbH
Cat. No./Trade name: RIDASCREEN® Adenovirus

SUMMARY

[Well-MAb]–**Ag**–[MAb-HRP]–[TMB]–A$_{450}$

Assay type: EIA (non-competitive)
Detection: Colorimetric A$_{450}$
Format: Microtitre well,
Sample type: Faecal specimens*
Sample pre-treatment:
　Mix specimen with Sample Diluent
Sample volume: 100 μl of prepared sample
Number of tests: 96
Controls - standards run in assay:
　Controls: Neg (1), Pos (1)
Incubation:
　1 hr (RT) + 10 min (RT)
Washes: 1

CONTENTS

Antibodies, antigens, labelled components:
　Anti-adenovirus (hexon) MAb bound to well
　Anti-adenovirus MAb HRP conjugated
Substrate: TMB
Controls - standards supplied:
　Controls: Neg (Sample Diluent) and Pos
Additional reagents:
　None
Special equipment required:
　None

INTERPRETATION

Comments on interpretation:
　Classification of sample is according to cut-off; no further
　testing
No. of references: 18

NOTES

131425.0

*This kit detects adenovirus in monkeys, dogs, pigs, mice
　and cattle in addition to humans and can be used for
　human or veterinary diagnostic purposes

Adenovirus
ANTIGEN DETECTION

Manufacturer: Argene-Biosoft
Cat. No./Trade name: 20-020/Anti-Adenovirus group
clone H60 Direct IFA kit

SUMMARY

[Slide]–**Ag**–[MAb-FITC]–fluorescence

Assay type: Immunofluorescence assay (direct)
Detection: Fluorescence microscopy
Format: Slide well, specimen coated
Sample type: Nasopharyngeal samples or infected cell
　culture
Sample pre-treatment:
　Nasopharyngeal samples: wash with PBS and centrifuge
　　to eliminate mucus
Sample volume: 30 μl
Number of tests: 80
Controls - standards run in assay:
　Controls: Neg (1), Pos (1)
Incubation:
　30 min (37°C) + 30 min (37°C)
Washes: 3

CONTENTS

Antibodies, antigens, labelled components:
　Anti-adenovirus MAb (m) FITC conjugated
Substrate:
Controls - standards supplied:
　Controls: Neg and Pos (slides)
Additional reagents:
　None
Special equipment required:
　None

INTERPRETATION

Comments on interpretation:
　Positive: ⩾1 cell showing bright green fluorescence
No. of references: 3

NOTES

121004.0

© KLUWER ACADEMIC PUBLISHERS 1995, ISSN 1381-5067

Adenovirus
ANTIGEN DETECTION

Manufacturer: Dako A/S
Cat. No./Trade name: K6100/IMAGEN® Adenovirus

SUMMARY

[Slide]–**Ag**–[MAb-FITC]–fluorescence

Assay type: Immunofluorescence assay (direct)
Detection: Fluorescence microscopy
Format: Slide well, specimen coated
Sample type: Nasopharyngeal aspirate or secretions, ophthalmic specimens, cell cultures
Sample pre-treatment:
Apply ocular material direct to slide respiratory and cell culture material; suspend in PBS, centrifuge, and apply to slide, air dry, fix with acetone
Sample volume: 25 μl of suspension or 1 swab
Number of tests: 50
Controls - standards run in assay:
Controls: Neg (as required), Pos (1)
Incubation:
15 min (37°C)
Washes: 1

CONTENTS

Antibodies, antigens, labelled components:
Anti-adenovirus MAb (m) FITC conjugated
Substrate:
Controls - standards supplied:
Controls: Pos (slide with human epithelial cells - HEp-2 infected with adenovirus)
Additional reagents:
Neg control slide prepared from uninfected cell culture (if required)
Special equipment required:
Teflon coated microscope slide (Cat. no. S6114)

INTERPRETATION

Comments on interpretation:
Negative: red counterstain and no fluorescence with at least 20 uninfected epithelial cells (clinical specimens) or 50 uninfected culture cells (culture confirmation) present; repeat if insufficient cells
Positive: red counterstain and ≥1 cell showing intracellular, nuclear and/or cytoplasmic apple green fluorescence
No. of references: 21

NOTES

120848.0

Adenovirus
ANTIGEN DETECTION

Manufacturer: Gull Laboratories Inc
Cat. No./Trade name: ADK001/Gull Adenovirus Kit

SUMMARY

[Slide]–**Ag**–[MAb-FITC]–fluorescence

Assay type: Immunofluorescence assay (direct)
Detection: Fluorescence microscopy
Format: Slide well, specimen coated
Sample type: Cell culture isolates*
Sample pre-treatment:
Apply cell suspension to slide well, dry, fix with acetone
Sample volume:
Number of tests: 20-40
Controls - standards run in assay:
Controls: Neg (1), Pos (1)**
Incubation:
30 min (37°C)
Washes: n1 (+5 min presoak)

CONTENTS

Antibodies, antigens, labelled components:
Anti-adenovirus MAb (m) FITC conjugated
Substrate:
Controls - standards supplied:
Controls: Neg slide (uninfected cells) and Pos slide (uninfected cells and cells infected with adenovirus)
Additional reagents:
Known Pos and Neg samples innoculated into cell cultures for use as controls
Special equipment required:
None

INTERPRETATION

Comments on interpretation:
Positive: red counterstained cells with bright apple green/ yellow fluorescence
No. of references: None

NOTES

120776.0

*Nasopharyngeal swabs, eye exudates, faecal samples, urine, other body fluids, tissue from biopsy or autopsy in TM are innoculated into conventional or shell vial cultures.
**Adenovirus control slides should be run periodically in parallel with test specimens and known Pos and Neg samples should be innoculated and processed with each batch of test specimens

Adenovirus
ANTIGEN DETECTION

Manufacturer: Argene-Biosoft
Cat. No./Trade name: 19-020/Anti-Adenovirus group clone H60 Indirect IFA kit

SUMMARY

[Slide]–**Ag**–[MAb]–[AMIg-FITC]–fluorescence

Assay type: Immunofluorescence assay (indirect)
Detection: Fluorescence microscopy
Format: Slide well, specimen coated
Sample type: Nasopharyngeal samples or infected cell culture
Sample pre-treatment:
    Nasopharyngeal samples: wash with PBS and centrifuge to eliminate mucus
Sample volume: 30 µl
Number of tests: 80, 650
Controls - standards run in assay:
    Controls: Neg (1), Pos (1)
Incubation:
    30 min (37°C) + 30 min (37°C)
Washes: 3

CONTENTS

Antibodies, antigens, labelled components:
    Anti-adenovirus MAb (m)
    Anti-mouse Ig MAb FITC conjugated
Substrate:
Controls - standards supplied:
    Controls: Neg and Pos (slides)
Additional reagents:
    None
Special equipment required:
    None

INTERPRETATION

Comments on interpretation:
    Positive: ≥1 cell showing bright green fluorescence
No. of references: 3

NOTES

120878.0

Adenovirus
ANTIGEN DETECTION

Manufacturer: BIOKIT SA
Cat. No./Trade name: adenogen

SUMMARY

[Latex-Ab]–**Ag**–agglutination

Assay type: Particle agglutination assay
Detection: Visual
Format: Latex particles, Ab coated (test done on slide)
Sample type: Faecal specimens, faecal swabs
Sample pre-treatment:
    Dilute with buffer, centrifuge and use supernatant
Sample volume: 50 µl of supernatant
Number of tests:
Controls - standards run in assay:
    Controls: Pos (1)
Incubation:
    2 min (RT) on rotary shaker
Washes:

CONTENTS

Antibodies, antigens, labelled components:
    Anti-adenovirus (types 40 & 41) Ab (r) bound to latex particles (Latex Reagent)
    Non-immune rabbit Ig Ab bound to latex particles (Latex Control)
Substrate:
Controls - standards supplied:
    Controls: Pos
Additional reagents:
    None
Special equipment required:
    None

INTERPRETATION

Comments on interpretation:
    Equivocal: equal agglutination with Latex Reagent and Latex Control; dilute sample 1:2 and retest
    Repeatably equivocal: consider as unsatisfactory and retest sample using another method
No. of references: 4

NOTES

131376.0

© KLUWER ACADEMIC PUBLISHERS 1995, ISSN 1381-5067

Adenovirus
ANTIGEN DETECTION

Manufacturer: bioMerieux
Cat. No./Trade name: 58843/Slidex Adeno-Kit

SUMMARY

[Latex-Ab]–**Ag**–agglutination

Assay type: Particle agglutination assay
Detection: Visual
Format: Latex particles, Ab coated (test done on card)
Sample type: Faecal specimens
Sample pre-treatment:
　Dilute sample, vortex and centrifuge
Sample volume: 50 μl of supernatant x 2
Number of tests: 30
Controls - standards run in assay:
　Controls: Pos (1)
Incubation:
　Read result within 2 min of mixing sample and reagent
Washes:

CONTENTS

Antibodies, antigens, labelled components:
　Anti-adenovirus MAb (m) bound to latex particles (Test Latex)
　Non-immune mouse Ig Ab bound to latex particles (Control Latex)
Substrate:
Controls - standards supplied:
　Controls: Pos
Additional reagents:
　None
Special equipment required:
　None

INTERPRETATION

Comments on interpretation:
　Positive: agglutination of sample with Test Latex and not with Control Latex
　Non-specific: agglutination with both Test Latex and Control Latex; repeat with a fresh sample
No. of references: 8

NOTES

131278.0

Adenovirus
ANTIGEN DETECTION

Manufacturer: Orion Diagnostica
Cat. No./Trade name: 67359/Adenolex®

SUMMARY

[Latex-Ab]–**Ag**–agglutination

Assay type: Particle agglutination assay
Detection: Visual
Format: Latex particles, Ab coated (test done on card)
Sample type: Faecal specimens
Sample pre-treatment:
　Filter, using Faecal Specimen Filtration Vial, or centrifuge diluted samples
Sample volume: 50 μl of prepared sample x 2
Number of tests: 20
Controls - standards run in assay:
　Controls: Pos (1)
Incubation:
　Read result within 4 min of mixing sample and reagents
Washes:

CONTENTS

Antibodies, antigens, labelled components:
　Anti-adenovirus Ab bound to latex particles (Test Latex)
　Non-immune Ig Ab bound to latex particles (Control Latex)
Substrate:
Controls - standards supplied:
　Controls: Pos
Additional reagents:
　None
Special equipment required:
　Faecal Specimen Filtration Vials Kit (Cat. no. 68312 – optional)

INTERPRETATION

Comments on interpretation:
　Positive: agglutination of sample with Test Latex and not with Control Latex
　Non-specific: agglutination of sample only with Control Latex; retest a fresh sample
No. of references: None

NOTES

131263.0

© *KLUWER ACADEMIC PUBLISHERS 1995, ISSN 1381-5067*

Adenovirus
ANTIGEN DETECTION

Manufacturer: Orion Diagnostica
Cat. No./Trade name: 67488/Diarlex® Adeno

SUMMARY

[Latex spot-Ab]–**Ag**–agglutination

Assay type: Particle agglutination assay
Detection: Visual
Format: Dry latex spot on card, Ab coated
Sample type: Serum
Sample pre-treatment:
　Filter, using Faecal Specimen Filtration Vial, or centrifuge
　diluted samples
Sample volume: 50 μl of prepared sample x 2
Number of tests: 20
Controls - standards run in assay:
　Controls: Pos (1)
Incubation:
　Read result within 2 min of mixing sample and reagents
Washes:

CONTENTS

Antibodies, antigens, labelled components:
　Anti-adenovirus Ab bound to dry latex spot on test card
　　(Adeno Latex)
　Non-immune Ig Ab bound to dry latex spot on test card
　　(Control Latex)
Substrate:
Controls - standards supplied:
　Controls: Pos
Additional reagents:
　None
Special equipment required:
　Faecal Specimen Filtration Vials Kit (Cat. no. 68312 –
　　optional)

INTERPRETATION

Comments on interpretation:
　Positive: agglutination of sample in dry Adeno Latex spot
　　and no agglutination in dry Control Latex spot on test
　　card
　Non-specific: agglutination of sample with Control Latex
　　only; retest a fresh sample
No. of references: None

NOTES

131266.0

Adenovirus
ANTIBODY DETECTION (IgG)

Manufacturer: Amico Laboratories Inc
Cat. No./Trade name: 1100G/AMIZYME®
ADENOVIRUS IgG ANTIBODY ASSAY

SUMMARY

[Well-Ag]–**Ab**–[AHIgG-HRP]–[ABTS]–A_{405}

Assay type: EIA (non-competitive)
Detection: Colorimetric A_{405}
Format: Microtitre well, Ag coated
Sample type: Serum
Sample pre-treatment:
　None
Sample volume: 5 μl (+ 245 μl diluent)
Number of tests: 96
Controls - standards run in assay:
　Controls: Neg (1)
　Calibrator: (1)
Incubation:
　25 min (RT) + 25 min (RT) + 25 min (RT)
Washes: 2

CONTENTS

Antibodies, antigens, labelled components:
　Adenovirus group Ag (Hexon) bound to well
　Anti-human IgG Ab (g) HRP conjugated
Substrate: ABTS
Controls - standards supplied:
　Controls: Neg
　Calibrator: IgG calibrator
Additional reagents:
　None
Special equipment required:
　None

INTERPRETATION

Comments on interpretation:
　Equivocal: net sample OD between 0.385 and 0.399;
　　retest to confirm
No. of references: 9

NOTES

120686.0

© KLUWER ACADEMIC PUBLISHERS 1995, ISSN 1381-5067

Adenovirus
ANTIBODY DETECTION (IgG)

Manufacturer: BAG-Biologische Analysensystem GmbH
Cat. No./Trade name: 5250 BAG - Adeno - EIA - G

SUMMARY

[Well-Ag]–**Ab**–[AHIgG-HRP]–[TMB]–A_{450}

Assay type: EIA (non-competitive)
Detection: Colorimetric A_{450}
Format: Microtitre well, Ag coated
Sample type: Serum
Sample pre-treatment:
   None
Sample volume: 100 µl of 1:101 dilution
Number of tests: 96
Controls - standards run in assay:
   Controls: Neg (1), cut-off (2), Pos (1)
Incubation:
   30 min (RT) + 30 min (RT) + 10 min (RT)
Washes: 2

CONTENTS

Antibodies, antigens, labelled components:
   Adenovirus (Adenoid 6) Ag bound to well
   Anti-human IgG Ab (sh) HRP conjugated
Substrate: TMB
Controls - standards supplied:
   Controls: Neg, cut-off and Pos (human serum)
Additional reagents:
   None
Special equipment required:
   None

INTERPRETATION

Comments on interpretation:
   Equivocal: within cut-off ±10%; data is not provided for
      further assessment of samples within this range
No. of references: None

NOTES

131604.0

Adenovirus
ANTIBODY DETECTION (IgG)

Manufacturer: Immunobiological Laboratories
Cat. No./Trade name: VE 57441/Adenovirus IgG ELISA

SUMMARY

[Well-Ag]–**Ab**–[AHIgG-HRP]–[TMB]–A_{450}

Assay type: EIA (non-competitive)
Detection: Colorimetric A_{450}
Format: Microtitre well, Ag coated
Sample type: Serum, plasma
Sample pre-treatment:
   None
Sample volume: 100 µl of 1:100 dilution x 2
Number of tests: 96
Controls - standards run in assay:
   Controls: Neg (2), low Pos (2), high Pos (2)
Incubation:
   1 hr (RT) + 30 min (RT) + 15 min (RT)
Washes: 2

CONTENTS

Antibodies, antigens, labelled components:
   Adenovirus Ag bound to well
   Anti-human IgG Ab HRP conjugated
Substrate: TMB
Controls - standards supplied:
   Controls: Neg, low Pos and high Pos (serum)
Additional reagents:
   None
Special equipment required:
   None

INTERPRETATION

Comments on interpretation:
   Negative: < cut-off −20%
   Positive: > cut-off +20%
   No data is provided for assessment of samples within
      cut-off ±20%
No. of references: 5

NOTES

120726.0

© KLUWER ACADEMIC PUBLISHERS 1995, ISSN 1381-5067

Adenovirus
ANTIBODY DETECTION (IgG)

Manufacturer: International Immunodiagnostics
Cat. No./Trade name: 108-160/Adenovirus total IgG

SUMMARY

[Well-Ag]–**Ab**–[AHIgG-HRP]–[TMB]–A$_{450}$

Assay type: EIA (non-competitive)
Detection: Colorimetric A$_{450}$
Format: Microtitre well, Ag coated
Sample type: Serum
Sample pre-treatment:
 None
Sample volume: 10 µl (+1 ml diluent)*
Number of tests: 96
Controls - standards run in assay:
 Controls: Neg (2), Pos (2)
Incubation:
 30 min (RT) + 30 min (RT) + 15 min (RT) (first two incubations on plate shaker)
Washes: 2

CONTENTS

Antibodies, antigens, labelled components:
 Adenovirus Ag bound to well
 Anti-human IgG Ab (g) HRP conjugated
Substrate: TMB
Controls - standards supplied:
 Controls: Neg and Pos (human serum)
Additional reagents:
 None
Special equipment required:
 None

INTERPRETATION

Comments on interpretation:
 Classification of samples is according to cut-off; no further testing
No. of references: 5

NOTES

120916.0

*Samples assayed in duplicate

Adenovirus
ANTIBODY DETECTION (IgG)

Manufacturer: Scimedix Corporation
Cat. No./Trade name: AD 96G/Adenovirus IgG EIA Test System

SUMMARY

[Well-Ag]–**Ab**–[AHIgG-POD]–[TMB]–A$_{450}$

Assay type: EIA (non-competitive)
Detection: Colorimetric A$_{450}$
Format: Microtitre well, Ag coated
Sample type: Serum
Sample pre-treatment:
 None
Sample volume: 100 µl of 1:100 dilution
Number of tests: 96
Controls - standards run in assay:
 Controls: Neg (1), Pos (1)
Incubation:
 1 hr (RT) + 30 min (RT) + 15 min (RT)
Washes: 2

CONTENTS

Antibodies, antigens, labelled components:
 Adenovirus Ag bound to well
 Anti-human IgG Ab (g) POD conjugated
Substrate: TMB
Controls - standards supplied:
 Controls: Neg and Pos (human)
Additional reagents:
 None
Special equipment required:
 None

INTERPRETATION

Comments on interpretation:
 Classification of samples is according to cut-off; no further testing
No. of references: 5

NOTES

131188.0

For research and investigational use only

© KLUWER ACADEMIC PUBLISHERS 1995, ISSN 1381-5067

Adenovirus
ANTIBODY DETECTION (IgM)

Manufacturer: Amico Laboratories Inc
Cat. No./Trade name: 1100M/AMIZYME® adenovirus
IgM antibody assay

SUMMARY

[Well-Ag]–**Ab**–[AHIgM-HRP]–[ABTS]–A_{405}

Assay type: EIA (non-competitive)
Detection: Colorimetric A_{405}
Format: Microtitre well, Ag coated
Sample type: Serum
Sample pre-treatment:
   None
Sample volume: 5 µl (+245 µl diluent)*
Number of tests: 96
Controls - standards run in assay:
   Controls: Neg (1)
   Calibrator: (1)
Incubation:
   25 min (RT) + 25 min (RT) + 25 min (RT)
Washes: 2

CONTENTS

Antibodies, antigens, labelled components:
   Adenovirus group Ag bound to well
   Anti-human IgM Ab (g) HRP conjugated
Substrate: ABTS
Controls - standards supplied:
   Controls: Neg
   Calibrator: IgG calibrator
Additional reagents:
   None
Special equipment required:
   None

INTERPRETATION

Comments on interpretation:
   Equivocal: sample OD between 0.285 and 0.299; retest
   to confirm
No. of references: 9

NOTES

120693.0

*Patient serum is diluted with IgM serum diluent to remove
interference from RF and IgG

Adenovirus
ANTIBODY DETECTION (IgM)

Manufacturer: BAG-Biologische
Analysensystem GmbH
Cat. No./Trade name: 5252 BAG - Adeno - EIA - M

SUMMARY

[Well-Ag]–**Ab**–[AHIgM-HRP]–[TMB]–A_{450}

Assay type: EIA (non-competitive)
Detection: Colorimetric A_{450}
Format: Microtitre well, Ag coated
Sample type: Serum (do not heat inactivate)
Sample pre-treatment:
   Treat diluted serum with BAG - IgM - Sep - System (Cat.
   no. 5540) to eliminate interference due to RF and IgG
Sample volume: 100 µl of treated sample
Number of tests: 96
Controls - standards run in assay:
   Controls: Neg (1), cut-off (2), Pos (1)
Incubation:
   30 min (RT) + 30 min (RT) + 10 min (RT)
Washes: 2

CONTENTS

Antibodies, antigens, labelled components:
   Adenovirus (Adenoid 6) Ag bound to well
   Anti-human IgM Ab (sh) HRP conjugated
Substrate: TMB
Controls - standards supplied:
   Controls: Neg, cut-off and Pos (human serum)
Additional reagents:
   None
Special equipment required:
   None

INTERPRETATION

Comments on interpretation:
   Equivocal: within cut-off ±10%; data is not provided for
   further assessment of samples within this range
No. of references: None

NOTES

131609.0

© KLUWER ACADEMIC PUBLISHERS 1995, ISSN 1381-5067

Adenovirus
ANTIBODY DETECTION (IgM)

Manufacturer: Immunobiological Laboratories
Cat. No./Trade name: VE 57451/Adenovirus IgM ELISA

SUMMARY

$$[\text{Well-Ag}]-\textbf{Ab}-[\text{AHIgM-HRP}]-[\text{TMB}]-A_{450}$$

Assay type: EIA (non-competitive)
Detection: Colorimetric A_{450}
Format: Microtitre well, Ag coated
Sample type: Serum, plasma
Sample pre-treatment:
　　It is advised to pre-incubate samples with an absorbent to remove interference due to RF, ANA and IgG (bought separately)
Sample volume: 100 μl of 1:100 dilution x 2
Number of tests: 96
Controls - standards run in assay:
　　Controls: Neg (2), low Pos (2), high Pos (2)
Incubation:
　　1 hr (RT) + 30 min (RT) + 15 min (RT)
Washes: 2

CONTENTS

Antibodies, antigens, labelled components:
　　Adenovirus Ag bound to well
　　Anti-human IgM Ab HRP conjugated
Substrate: TMB
Controls - standards supplied:
　　Controls: Neg, low Pos and high Pos (serum)
Additional reagents:
　　Absorbent for removal of RF and IgG in sample
Special equipment required:
　　None

INTERPRETATION

Comments on interpretation:
　　Negative: < cut-off -20%
　　Positive: > cut-off +20%
　　No data is provided on samples within cut-off ±20%
No. of references: 5

NOTES

120733.0

Adenovirus
ANTIBODY DETECTION (IgM)

Manufacturer: International Immunodiagnostics
Cat. No./Trade name: 108-161/Adenovirus total IgM

SUMMARY

$$[\text{Well-Ag}]-\textbf{Ab}-[\text{AHIgM-HRP}]-[\text{TMB}]-A_{450}$$

Assay type: EIA (non-competitive)
Detection: Colorimetric A_{450}
Format: Microtitre well, Ag coated
Sample type: Serum
Sample pre-treatment:
　　Preadsorb serum with Protein G (Cat. no. E201.1) to eliminate interference due to RF and IgG in sample
Sample volume: 10 μl (+1 ml diluent)*
Number of tests: 96
Controls - standards run in assay:
　　Controls: Neg (2), Pos (2)
Incubation:
　　30 min (RT) + 30 min (RT) + 15 min (RT) (first two incubations on plate shaker)
Washes: 2

CONTENTS

Antibodies, antigens, labelled components:
　　Adenovirus Ag bound to well
　　Anti-human IgM Ab (g) HRP conjugated
Substrate: TMB
Controls - standards supplied:
　　Controls: Neg and Pos (human serum)
Additional reagents:
　　None
Special equipment required:
　　None

INTERPRETATION

Comments on interpretation:
　　Classification of samples is according to cut-off; no further testing
No. of references: 5

NOTES

120928.0

*Samples assayed in duplicate

© KLUWER ACADEMIC PUBLISHERS 1995, ISSN 1381-5067

Adenovirus
ANTIBODY DETECTION (IgM)

Manufacturer: Scimedix Corporation
Cat. No./Trade name: AD 96M/Adenovirus IgM EIA Test System

SUMMARY

[Well-Ag]–**Ab**–[AHIgM-POD]–[TMB]–A_{450}

Assay type: EIA (non-competitive)
Detection: Colorimetric A_{450}
Format: Microtitre well, Ag coated
Sample type: Serum
Sample pre-treatment:
    None
Sample volume: 100 μl of 1:100 dilution
Number of tests: 96
Controls - standards run in assay:
    Controls: Neg (1), Pos (1)
Incubation:
    1 hr (RT) + 30 min (RT) + 15 min (RT)
Washes: 2

CONTENTS

Antibodies, antigens, labelled components:
    Adenovirus Ag bound to well
    Anti-human IgM Ab (g) POD conjugated
Substrate: TMB
Controls - standards supplied:
    Controls: Neg and Pos (human)
Additional reagents:
    None
Special equipment required:
    None

INTERPRETATION

Comments on interpretation:
    Classification of samples is according to cut-off; no
        further testing
No. of references: 5

NOTES

131189.0

For research and investigational use only

© KLUWER ACADEMIC PUBLISHERS 1995, ISSN 1381-5067

Arboviruses

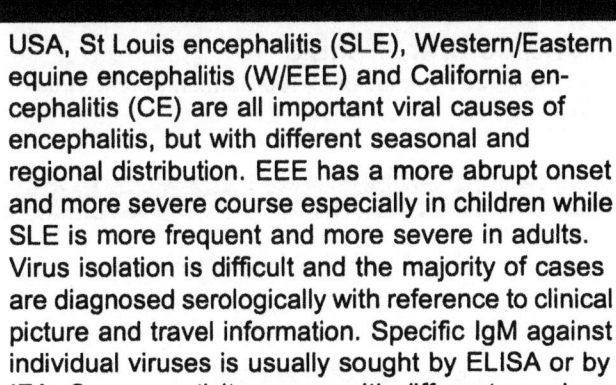

Natural history

Arbovirus infections involve alternate cycles of viral replication in blood-sucking arthropods and in vertebrate hosts. The vectors include mosquitos (more prevalent in tropical countries), ticks (more prevalent in temperate countries) and sandflies. Several virus families contain members which are arthropod-borne. These include Togaviridae (Genus Alphaviruses, also termed Group A arboviruses in the past); Flaviviridae (Genus Flavivirus, Group B arboviruses); Bunyaviridae and Reoviridae (see Table).

These viruses are responsible for an enormous burden of viral disease worldwide and show great diversity physically, biologically and in their disease manifestations. Classification is according to chemical and physical properties. Alphaviruses are 70 nm enveloped spherical viruses with positive sense, single stranded RNA genomes; Flaviviruses are similar but smaller (40 nm); Bunyaviridae are large enveloped negative sense RNA viruses (100 nm) while Orbiviruses are non-enveloped and have a unique double stranded RNA genome of 10 segments. Similarities between different members allow them to be grouped according to 'antigenic complexes'. There is considerable overlap in the disease patterns observed with different viruses. Three broadly distinct clinical syndromes are found: most commonly, fever with or without rash and usually mild; encephalitis with high fatality rate; and haemorrhagic fever, also frequently fatal. Dengue virus, for example, may be associated with more than one of these syndromes and is the most important arboviral cause of death and disease in man. Most arbovirus infections are however asymptomatic.

Diagnosis

(a) Encephalitis viruses: For differential diagnosis, all relevant arboviruses causing a similar disease may be examined together. Thus, in a patient in the USA, St Louis encephalitis (SLE), Western/Eastern equine encephalitis (W/EEE) and California encephalitis (CE) are all important viral causes of encephalitis, but with different seasonal and regional distribution. EEE has a more abrupt onset and more severe course especially in children while SLE is more frequent and more severe in adults. Virus isolation is difficult and the majority of cases are diagnosed serologically with reference to clinical picture and travel information. Specific IgM against individual viruses is usually sought by ELISA or by IFA. Cross-reactivity occurs with different members of the same antigenic complex. Classical techniques such as HI and CFT might still be used to show rising titres in paired samples.

(b) Flaviviruses: Again culture of virus is difficult and requires specialist reference facilities. Diagnosis is usually based on detection of specific IgM in serum or CSF, using ELISA or IFA. IgM antibody to dengue may last more than 3 months making interpretation more difficult in endemic areas. Cross-reaction with other flaviviruses can be extensive. In suspected cases of Japanese encephalitis, IgM can be found in CSF in the acute phase only.

Comments

The most effective means of control of arbovirus disease is by control of the arthropod vectors and by immunisation. Some vaccines are available, most notably, formalin inactivated vaccine for Japanese B encephalitis virus and live attenuated vaccine against yellow fever virus which was developed more than 50 years ago and control of dengue using engineered live attenuated virus is envisaged.

Reference

Jawetz E, Melnick JL, Adelberg EA. Arthropod-borne and rodent-borne viral diseases. In: Medical Microbiology, 19th edn. Connecticut: Appleton & Lange. 1991:488–505.

Family	Genus	Examples	Major symptoms
Togaviridae	*Alphavirus*	Equine encephalitis viruses e.g. WEE, EEE	Encephalitis
		Sindbis; Ross River virus;Chikungunya	Fever and arthropathy
Flaviviridae	*Flavivirus*	St Louis encephalitis; Japanese encephalitis; Tick-bourne encephalitis	Encephalitis
		Dengue fever; West Nile fever	Fever, rash and arthralgia
		Dengue; Yellow fever	Haemorrhagic fever
Bunyaviridae	*Bunyavirus*	Californian encephalitis virus	Encephalitis
	Phelobvirus	Sandfly fever; Rift valley fever virus	Fever, abdom. pain & leukopaenia
	Nairovirus	Crimean Congo haemorrhagic fever virus	Haemorrhagic fever
Reoviridae	*Orbivirus*	Colorado tick fever	Fever, photophobia & myalgia

Barmah Forest virus
ANTIBODY DETECTION (IgG)

Manufacturer: PanBio
Cat. No./Trade name: BFG-100/Barmah Forest Virus IgG Elisa Test

SUMMARY

[Well-Ag]–**Ab**–[AHIgG-HRP]–[TMB]–A_{450}

Assay type: EIA (non-competitive)
Detection: Colorimetric A_{450}
Format: Microtitre well, Ag coated
Sample type: Serum
Sample pre-treatment:
   None
Sample volume: 100 µl of 1:100 dilution
Number of tests: 96
Controls - standards run in assay:
   Controls: Neg (1), Pos (1)
   Calibrator: cut-off (2)
Incubation:
   20 min (RT) + 20 min (RT) + 10 min (RT)
Washes: 2

CONTENTS

Antibodies, antigens, labelled components:
   Barmah Forest virus Ag bound to well
   Anti-human IgG Ab (sh) HRP conjugated
Substrate: TMB
Controls - standards supplied:
   Controls: Neg and Pos (human serum)
   Calibrator: cut-off (human serum)
Additional reagents:
   None
Special equipment required:
   None

INTERPRETATION

Comments on interpretation:
   Equivocal: where the ratio of cut-off value:sample OD is between 0.9-1.1; retest to confirm
No. of references: 3

NOTES

131400.0

Barmah Forest virus
ANTIBODY DETECTION (IgM)

Manufacturer: PanBio
Cat. No./Trade name: BFM-200/Barmah Forest Virus IgM Elisa Test

SUMMARY

[Well-Ag]–**Ab**–[AHIgM-HRP]–[TMB]–A_{450}

Assay type: EIA (non-competitive)
Detection: Colorimetric A_{450}
Format: Microtitre well, Ag coated
Sample type: Serum
Sample pre-treatment:
   Mix diluted serum with absorbent solution, incubate 15 min (RT) to eliminate interference due to RF & IgG
Sample volume: 100 µl of prepared sample
Number of tests: 96
Controls - standards run in assay:
   Controls: Neg (1), Pos (1)
   Calibrator: cut-off (2)
Incubation:
   20 min (RT) + 20 min (RT) + 10 min (RT)
Washes: 2

CONTENTS

Antibodies, antigens, labelled components:
   Barmah Forest virus Ag bound to well
   Anti-human IgM Ab (sh) HRP conjugated
Substrate: TMB
Controls - standards supplied:
   Controls: Neg and Pos (human serum)
   Calibrator: cut-off (human serum)
Additional reagents:
   None
Special equipment required:
   None

INTERPRETATION

Comments on interpretation:
   Equivocal: where the ratio of cut-off value:sample OD is between 0.9-1.1; retest to confirm
No. of references: 3

NOTES

131406.0

© KLUWER ACADEMIC PUBLISHERS 1995, ISSN 1381-5067

Ross River virus
ANTIBODY DETECTION (IgG)

Manufacturer: PanBio
Cat. No./Trade name: RRG-100/Ross River Virus IgG Elisa Test

SUMMARY

[Well-Ag]–**Ab**–[AHIgG-HRP]–[TMB]–A_{450}

Assay type: EIA (non-competitive)
Detection: Colorimetric A_{450}
Format: Microtitre well, Ag coated
Sample type: Serum
Sample pre-treatment:
   None
Sample volume: 100 μl of 1:100 dilution
Number of tests: 96
Controls - standards run in assay:
   Controls: Neg (1), Pos (1)
   Calibrator: cut-off (2)
Incubation:
   Method A: 20 min (37°C) + 20 min (37°C) + 10 min (RT)
   Method B: 30 min (RT) + 30 min (RT) + 10 min (RT)*
Washes: 2

CONTENTS

Antibodies, antigens, labelled components:
   Ross River virus Ag bound to well
   Anti-human IgG Ab (sh) HRP conjugated
Substrate: TMB
Controls - standards supplied:
   Controls: Neg and Pos (human serum)
   Calibrator: cut-off (human serum)
Additional reagents:
   None
Special equipment required:
   None

INTERPRETATION

Comments on interpretation:
   Equivocal: where the ratio of cut-off value:sample OD is between 0.9-1.1; retest to confirm
No. of references: 4

NOTES

131399.0

*The preferred incubation method is Method A

Ross River virus
ANTIBODY DETECTION (IgM)

Manufacturer: PanBio
Cat. No./Trade name: RRM-200/Ross River Virus IgM Elisa Test

SUMMARY

[Well-Ag]–**Ab**–[AHIgM-HRP]–[TMB]–A_{450}

Assay type: EIA (non-competitive)
Detection: Colorimetric A_{450}
Format: Microtitre well, Ag coated
Sample type: Serum
Sample pre-treatment:
   Mix diluted serum with absorbent solution, incubate 15 min (RT) to eliminate interference due to RF & IgG
Sample volume: 100 μl of prepared sample
Number of tests: 96
Controls - standards run in assay:
   Controls: Neg (1), Pos (1)
   Calibrator: cut-off (2)
Incubation:
   Method A: 20 min (37°C) + 20 min (37°C) + 10 min (RT)
   Method B: 30 min (RT) + 30 min (RT) + 10 min (RT)*
Washes: 2

CONTENTS

Antibodies, antigens, labelled components:
   Ross River virus Ag bound to well
   Anti-human IgM Ab (sh) HRP conjugated
Substrate: TMB
Controls - standards supplied:
   Controls: Neg and Pos (human serum)
   Calibrator: cut-off (human serum)
Additional reagents:
   None
Special equipment required:
   None

INTERPRETATION

Comments on interpretation:
   Equivocal: where the ratio of cut-off value:sample OD is between 0.9-1.1; retest to confirm
No. of references: 4

NOTES

131405.0

*The preferred method of incubation is Method A

Encephalitis bio-group: EEEV, WEEV, SLEV and CEV
ANTIBODY DETECTION (IgG)

Manufacturer: MRL Diagnostics
Cat. No./Trade name: IF0300G/Arbovirus IFA (IgG)

SUMMARY

[Slide well-Ag]–**Ab**–[AHIgG-FITC]–fluorescence

Assay type: Immunofluorescence assay (indirect)
Detection: Fluorescence microscopy
Format: Slide well, Ag coated
Sample type: Serum
Sample pre-treatment:
  None
Sample volume: 25 μl of 1:16 dilution
Number of tests: 80
Controls - standards run in assay:
  Controls: Neg (1), SLEV Pos (1), Arbovirus polyvalent
    Pos (1)
Incubation:
  1 hr 30 min (37°C) + 30 min (RT)
Washes: 2

CONTENTS

Antibodies, antigens, labelled components:
  EEEV (New Jersey 60 strain), WEEV (Fleming strain),
  SLEV (TBH-28 strain) and CEV (La Cross strain)
  group Ags from infected and uninfected Vero cells
  bound to slide wells (4 spots per well: one spot for
  each group virus Ag)
Substrate:
Controls - standards supplied:
  Controls: Neg and SLEV Pos (human serum) and
  Arbovirus polyvalent Pos (mouse ascites)
Additional reagents:
  None
Special equipment required:
  None

INTERPRETATION

Comments on interpretation:
  Positive: ≥(1+) distinct cytoplasmic fluorescence with
  appropriate Ag spot in slide well identifies the relevant
  encephalitis group virus antibodies present
No. of references: 15

NOTES

131118.0

This kit can detect Abs to 4 major groups of encephalitis
pathogens and can differentiate between the groups

Encephalitis bio-group: EEEV, WEEV, SLEV and CEV
ANTIBODY DETECTION (IgM)

Manufacturer: MRL Diagnostics
Cat. No./Trade name: IF0300M/Arbovirus IFA (IgM)

SUMMARY

[Slide well-Ag]–**Ab**–[AHIgM-FITC]–fluorescence

Assay type: Immunofluorescence assay (indirect)
Detection: Fluorescence microscopy
Format: Slide well, Ag coated
Sample type: Serum
Sample pre-treatment:
  Treatment of serum to eliminate interference due to RF
  and IgG is advised
Sample volume: 25 μl of 1:16 dilution
Number of tests: 80
Controls - standards run in assay:
  Controls: Neg (1), SLEV Pos (1), Arbovirus polyvalent
    Pos (1)
Incubation:
  1 hr 30 min (37°C) + 30 min (RT)
Washes: 2

CONTENTS

Antibodies, antigens, labelled components:
  EEEV (New Jersey 60 strain), WEEV (Fleming strain),
  SLEV (TBH-28 strain) and CEV (La Cross strain)
  group Ags from infected and uninfected Vero cells
  bound to slide wells (4 spots per well: one spot for
  each group virus Ag)
Substrate:
Controls - standards supplied:
  Controls: Neg and SLEV Pos (human serum) and
  Arbovirus polyvalent Pos (mouse ascites)
Additional reagents:
  None
Special equipment required:
  None

INTERPRETATION

Comments on interpretation:
  Positive: ≥(1+) distinct cytoplasmic fluorescence with
  appropriate Ag spot in slide well identifies the relevant
  encephalitis group virus antibodies present
No. of references: 15

NOTES

131119.0

This kit can detect Abs to 4 major groups of encephalitis
pathogens and can differentiate between the groups

© KLUWER ACADEMIC PUBLISHERS 1995, ISSN 1381-5067

Dengue fever virus
ANTIBODY DETECTION (IgG)

Manufacturer: PanBio
Cat. No./Trade name: DEG-100

SUMMARY

[Well-Ag]–**Ab**–[AHIgG-HRP]–[TMB]–A_{450}

Assay type: EIA (non-competitive)
Detection: Colorimetric A_{450}
Format: Microtitre well, Ag coated
Sample type: Serum
Sample pre-treatment:
  None
Sample volume: 100 µl of 1:100 dilution
Number of tests: 96
Controls - standards run in assay:
  Controls: Neg (1), Pos (1)
  Calibrator: cut-off (2)
Incubation:
  Method A: 20 min (37°C) + 20 min (37°C) + 10 min (RT)
  Method B: 30 min (RT) + 30 min (RT) + 10 min (RT)*
Washes: 2

CONTENTS

Antibodies, antigens, labelled components:
  Dengue fever virus Ag bound to well
  Anti-human IgG Ab (sh) HRP conjugated
Substrate: TMB
Controls - standards supplied:
  Controls: Neg and Pos (human serum)
  Calibrator: cut-off (human serum)
Additional reagents:
  None
Special equipment required:
  None

INTERPRETATION

Comments on interpretation:
  Equivocal: where the ratio of cut-off value:sample OD is between 0.9-1.1; retest to confirm
No. of references: 5

NOTES

131398.0

*The preferred incubation method is Method A

Dengue fever virus
ANTIBODY DETECTION (IgM)

Manufacturer: PanBio
Cat. No./Trade name: DEM-200/Dengue IgM Capture Elisa

SUMMARY

[Well-Ag]–**Ab**–[AHIgM-HRP]–[TMB]–A_{450}

Assay type: EIA (non-competitive)
Detection: Colorimetric A_{450}
Format: Microtitre well, Ag coated
Sample type: Serum
Sample pre-treatment:
  Mix diluted serum with absorbent solution, incubate 15 min (RT) to eliminate interference due to RF & IgG
Sample volume: 100 µl of prepared sample
Number of tests: 96
Controls - standards run in assay:
  Controls: Neg (1), Pos (1)
  Calibrator: cut-off (2)
Incubation:
  1 hr (37°C) + 30 m in (37°C) + 10 min (RT)
Washes: 2

CONTENTS

Antibodies, antigens, labelled components:
  Dengue fever virus Ag bound to well
  Anti-human IgM Ab (sh) HRP conjugated
Substrate: TMB
Controls - standards supplied:
  Controls: Neg and Pos (human serum)
  Calibrator: cut-off (human serum)
Additional reagents:
  None
Special equipment required:
  None

INTERPRETATION

Comments on interpretation:
  Equivocal: where the ratio of cut-off value:sample OD is between 0.8-0.99; retest to confirm
No. of references: 5

NOTES

131404.0

Bacillus cereus

Bacillus cereus is an aerobic, Gram-positive, catalase-positive, bacillus, which may produce oval, central endospores. Vegetative cells occur singly or in short chains and the organism grows readily on nutrient agar and peptone media to yield granular or wrinkled colonies. The organism is motile producing a flocculent surface growth in broth media, and is haemolytic on blood agar. The organism grows well at 37°C and although vegetative cells are readily killed by moist heat at 55°C, the spores are relatively heat-resistant. *Bacillus* organisms are widely distributed in the environment and may be found as contaminants of laboratory media. *Bacillus cereus* has been implicated as the causal organism in outbreaks of food poisoning throughout the world. Two forms of enteric disease are recognised. The emetic form has an incubation period of 1–6 hours and is characterised by upper gastrointestinal symptoms and vomiting. It is commonly associated with reheated boiled rice in which the spores of the organism which survived the original boiling process germinate, and the rapidly-growing vegetative forms produce a potent emetic toxin. The diarrhoeal form has an incubation period of 10–12 hours and produces severe abdominal pain and profuse diarrhoea similar to the clinical presentation of *Clostridium perfringens* food poisoning. The disease is normally associated with meat and vegetable products and is due to the production of a distinct diarrhoeal toxin. *Bacillus cereus* has also been implicated as the aetiologic agent of bacteraemia, meningitis, pneumonia and empyema, usually in the compromised host. The organism has also been recently recognised as a primary ocular pathogen which may produce a post-traumatic endophthalmitis. Isolates from patients with more severe extraintestinal infections are often found to produce other toxins which may have purulent, pyrogenic, lethal and necrotic activities. Phospholipase activity may be important in the pathogenesis of ocular infections.

Diagnosis

Definitive diagnosis of *Bacillus cereus* infection may be made by isolation of the organism from wound swabs, tissue, blood and other body fluids as outlined above. Care is required in the interpretation of culture results due to the ubiquitous nature of the organism, and the possibility of contamination should be borne in mind. Conversely, due caution should be exercised before dismissing the significance of an isolate, particularly in an appropriate clinical setting.

The diagnosis of *Bacillus cereus* food poisoning is usually established by the isolation of the organism in numbers greater than 10^5 or more per gram from epidemiologically-incriminated foodstuffs. Isolation of the organism from stool samples alone is not usually regarded as sufficient for diagnosis in outbreaks, unless a suitable control-group is found to be culture-negative. *Bacillus cereus* enterotoxin has been partially purified and forms the basis of toxin detection assays which may increasingly play a role in detection and diagnosis.

References

Tuazon CU. Other *Bacillus* species. In: Mandell GL, Bennett JE, Dolin R, eds. Principles and Practice of Infectious Diseases, 4th edn. New York, Edinburgh, London: Churchill Livingstone. 1995:1890–4.

Turnbull PCB, Kramer JM. *Bacillus*. In: Balows A, Hausler WJ, Herrmann KL, Isenberg HD, Shadomy HJ, eds. Manual of Clinical Microbiology, 5th edn. Washington: AJM. 1991:296–303.

Bacillus cereus (diarrhoeal enterotoxin)
ANTIGEN DETECTION

Manufacturer: Tecra® Diagnostics
Cat. No./Trade name: BDEVIA48/TECRA® Bacillus Diarrhoeal Enterotoxin Visual Immunoassay

SUMMARY

[Well-Ab]–**Ag**–[Ab-enzyme]–[substrate]–green colour

Assay type: EIA (non-competitive)
Detection: Visual or colorimetric
Format: Microtitre well, Ab coated
Sample type: Food, food related samples
Sample pre-treatment:
  Sample preparation requires 30 min
Sample volume: 0.2 ml of prepared sample
Number of tests: 48
Controls - standards run in assay:
  Controls: Neg (1), Pos (1)
Incubation:
  2 hr (37°C) + 1 hr (20°C)
Washes: 2

CONTENTS

Antibodies, antigens, labelled components:
  Anti-*Bacillus cereus* diarrhoeal enterotoxin Ab bound to well
  Anti-*Bacillus cereus* diarrhoeal enterotoxin Ab enzyme conjugated
Substrate: Not specified
Controls - standards supplied:
  Controls: Neg (diluent) and Pos
Additional reagents:
  None
Special equipment required:
  None

INTERPRETATION

Comments on interpretation:
  Positive: green colouration in well confirmed by standard tests
No. of references: 2

NOTES

131724.0

Bacillus cereus (diarrhoeal enterotoxin)
ANTIGEN DETECTION

Manufacturer: Denka Seiken Co. Ltd
Cat. No./Trade name: BCET-RPLA Bacillus Cereus Enterotoxin (diarrhoeal type) Test Kit

SUMMARY

[Latex-Ab]–**Ag**–agglutination

Assay type: Particle agglutination assay
Detection: Visual
Format: Latex particles, Ab coated (test done in V-microtitre well)
Sample type: Faecal samples, culture isolates from food
Sample pre-treatment:
  Extract toxin from (a) food – mix with NaCl, centrifuge 30 min (4°C), filter; (b) culture isolates – innoculate isolated organism into Brain Heart Infusion, incubate 6–18 hr (32–37°C), centrifuge 20 min (4°C), filter
Sample volume: 25 µl of prepared sample (+ 25 µl diluent) x 2
Number of tests:
Controls - standards run in assay:
  Controls: Pos (1)
Incubation:
  20–24 hr (RT)
Washes:

CONTENTS

Antibodies, antigens, labelled components:
  Anti-*B. cereus* diarrhoeal enterotoxin Ab (r) bound to latex particles (Test Latex)
  Non-immune rabbit Ig Ab bound to latex particles (Control Latex)
Substrate:
Controls - standards supplied:
  Controls: Pos
Additional reagents:
  Brain heart medium
Special equipment required:
  None

INTERPRETATION

Comments on interpretation:
  Positive: greater agglutination of sample with Test Latex than with Control Latex
No. of references: 5

NOTES

131616.0

© KLUWER ACADEMIC PUBLISHERS 1995, ISSN 1381-5067

Borrelia burgdorferi

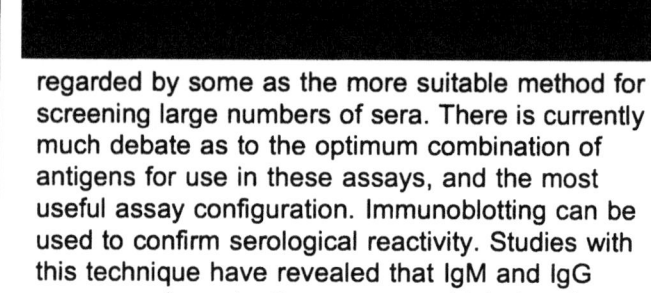

Borrelia burgdorferi is a spirochaete transmitted by the bite of Ixodid ticks, and is now recognised as the causal organism of Lyme disease and related disorders. Lyme disease is an endemic, inflammatory, multisystem disease which usually begins in summer with a distinctive skin lesion, erythema chronicum migrans (ECM), at the site of the tick bite. Weeks to months later, patients may develop migratory joint pains, myocarditis or meningoencephalitis, subsequent to the widespread migration of the organism from the site of the original lesion. Months to years later, often following a latent period, persistent chronic disease of the joints, skin and central nervous system may result. Lyme disease is a zoonosis, and it is now thought that a wide variety of wild and domestic animals, particularly deer, may act as reservoirs.

Borrelia burgdorferi, like other spirochaetes, has a central helical protoplasmic cylinder bounded by a cytoplasmic membrane. The organisms are highly motile and possess between 7 and 11 periplasmic flagellae which lie beneath a loosely attached outer membrane. The organism has a number of important membrane, structural and flagellar antigens including outer surface proteins OspA-F and the 41 kD flagellar antigen, which are of importance in pathogenic and diagnostic terms. The genetic regulation of the production of these and other structures involves a number of plasmids, some of which have an unusual linear structure. Regional variations in the clinical presentation of Lyme borreliosis are now thought to be attributable to genetic variation amongst isolates, and a small number of new species have been proposed on the basis of this genomic diversity.

Diagnosis

Unlike the tick-borne relapsing fever borreliae, the results of dark-field and conventional microscopy in the diagnosis of Lyme disease have been disappointing. The use of a variety of chemical and fluorescent antibody staining methods has yielded some improvements, but is still not entirely satisfactory. *Borrelia burgdorferi* can be cultivated *in vitro* on Barbour-Stoenner-Kelly (BSK) medium. Although this yields a definitive diagnosis, the technique is not widely available, and there has been significant variability in the results obtained.

As a result of these difficulties, serology is currently the principal method for the diagnosis of Lyme borreliosis. The two principal methods in routine use for serological detection are indirect immunofluorescence assay (IFA) and enzyme-linked immunosorbent assay (ELISA). The newer tests demonstrate similar specificity, but ELISA methods have generally shown increased sensitivity, and are regarded by some as the more suitable method for screening large numbers of sera. There is currently much debate as to the optimum combination of antigens for use in these assays, and the most useful assay configuration. Immunoblotting can be used to confirm serological reactivity. Studies with this technique have revealed that IgM and IgG responses in early disease are restricted to the genus-specific 41 kD flagellar antigen, with antibodies to other antigens developing only as the disease progresses.

Serology is less useful in early disease. A specific IgM response is frequently undetectable in the first two weeks of infection, reaching a peak at about 6 weeks post infection. Levels of specific serum IgG do not reach significant levels until this time. IgM levels can remain elevated, and may also reappear in late disease. Equally the IgG response may persist and may complicate the diagnosis in asymptomatic individuals, or in those with an unusual clinical presentation. Failure to mount a detectable antibody response may be due to a variety of causes. Antibiotic therapy may abrogate the immune response, or antibody production may occasionally be restricted to the site of active disease (as seen in some cases of CNS disease). It has also been suggested that the diversity of *Borrelia burgdorferi* and related organisms may account for the lack of sensitivity of antibody detection in some individuals, and this may again have implications for the development of diagnostic tests. Antigen detection methods and PCR-based tests remain largely within the realm of the research laboratory at this time. In all cases of suspected Lyme borreliosis care must be taken in the interpretation of laboratory results, and should be considered in the context of the clinical findings.

References

Steere AC. *Borrelia burgdorferi*. In: Mandell GL, Bennett JE, Dolin R, eds. Principles and Practice of Infectious Diseases, 4th edn. New York: Churchill Livingstone; 1995:2143–55.

Cutler SJ. Diagnostic tests for Lyme borreliosis. In: Bentley A, Davidson R, Heyderman R, Zuckerman M, eds. Infectious Diseases. London: Current Medical Literature; 1994:56–63.

© KLUWER ACADEMIC PUBLISHERS 1995, ISSN 1381-5067

Borrelia burgdorferi
ANTIBODY DETECTION

Manufacturer: Alexon Inc
Cat. No./Trade name: 700-96 ProSpecT® Lyme (Borrelia) Microplate Assay

SUMMARY

[Well-Ag]–**Ab**–[AHIgA,IgG&IgM-POD]–[ABTS]–A_{450}

Assay type: EIA (non-competitive)
Detection: Colorimetric A_{450}
Format: Microtitre well, Ag coated
Sample type:
Sample pre-treatment:
 Block serum after dilution by mixing with Blocking Solution (E. coli proteins), incubate 30–60 min (RT) prior to testing to improve specificity
Sample volume: 10 µl of treated sample
Number of tests: 96
Controls - standards run in assay:
 Controls: Neg (2), Pos (3)
Incubation:
 30 min (RT) + 30 min (RT) + 30 min (RT)
Washes: 2

CONTENTS

Antibodies, antigens, labelled components:
 B. burgdorferi (strain B31, P39) and *B. burgdorferi* (P39) Ags bound to well
 Anti-human IgA, IgG & IgM Abs (g) POD conjugated
Substrate: ABTS
Controls - standards supplied:
 Controls: Neg and Pos (human serum)
Additional reagents:
 None
Special equipment required:
 None

INTERPRETATION

Comments on interpretation:
 Equivocal: 0.12–0.16 unit values; retest to confirm
 Repeatably equivocal: report as such and advise retest a fresh sample in 2–4 weeks
No. of references: 12

NOTES

131299.0

Borrelia burgdorferi
ANTIBODY DETECTION

Manufacturer: bioMerieux
Cat. No./Trade name: 30 298/VIDAS Lyme IgG and IgM

SUMMARY

[Solid phase-Ag]–**Ab**–[AHIgG&IgM-AP]–[4MP]–fluorescence

Assay type: EIA (non-competitive)
Detection: Fluorometric
Format: Solid phase receptacle (SPR), Ag coated
Sample type: Serum (do not heat inactivate)
Sample pre-treatment:
 None
Sample volume: 100 µl
Number of tests: 60
Controls - standards run in assay:
 Controls: Neg (1), Pos (1)
 Standards: Pos (1)
Incubation:
 Automated - total time 35 min
Washes: Automated

CONTENTS

Antibodies, antigens, labelled components:
 B. burgdorferi (strain B31) Ag bound to solid phase receptacle
 Anti-human IgG & IgM MAbs (m) AP conjugated
Substrate:
Controls - standards supplied:
 Controls: Neg and Pos (human serum)
 Standards: Pos (human serum)
Additional reagents:
 None
Special equipment required:
 Vitek immunodiagnostic assay system (VIDAS)

INTERPRETATION

Comments on interpretation:
 Sample reactivity is calculated automatically
 Equivocal: within designated range; retest to confirm
 Repeatably equivocal: retest using a fresh sample in 2-3 weeks
No. of references: 9

NOTES

131312.0

© KLUWER ACADEMIC PUBLISHERS 1995, ISSN 1381-5067

Borrelia burgdorferi
ANTIBODY DETECTION

Manufacturer: GenBio
Cat. No./Trade name: 3110/ImmunoWell® Borrelia (Lyme) Test

SUMMARY

[Well-Ag]–**Ab**–[AHIgA,IgG&IgM-POD]–[ABTS]–A_{405}

Assay type: EIA (non-competitive)
Detection: Colorimetric A_{405}
Format: Microtitre well, Ag coated
Sample type: Serum
Sample pre-treatment:
   Add diluted serum to Borrelia Blocker (contains E. coli proteins). Incubate 30 min–1 hr (RT) to improve specificity
Sample volume: 100 µl of prepared sample
Number of tests: 96
Controls - standards run in assay:
   Controls: Neg (2), Pos (2) or more
Incubation:
   30 min (RT) + 30 min (RT) + 30 min (RT)
Washes: 2

CONTENTS

Antibodies, antigens, labelled components:
   B. burgdorferi (cell extract strain B31 and rec P39 protein) Ags bound to well
   Anti-human IgA, IgG & IgM Abs (g) POD conjugated
Substrate: ABTS
Controls - standards supplied:
   Controls: Neg and Pos (human serum)
Additional reagents:
   None
Special equipment required:
   None

INTERPRETATION

Comments on interpretation:
   Negative: normalized OD <0.12; if clinically suspicious of Lyme disease; retest fresh sample in 2–4 weeks time
   Borderline: normalized OD between 0.12 and 0.16; retest to confirm
   Repeatably equivocal: report as such and retest fresh sample in 2–4 weeks
   Positive: >0.16
No. of references: 12

NOTES

131417.0

Borrelia burgdorferi
ANTIBODY DETECTION

Manufacturer: GenBio
Cat. No./Trade name: 5025/ImmunoDot® Borrelia (Lyme) with Recombinant Protein

SUMMARY

[Dipstick-Ag]–**Ab**–[AHIgA,IgG&IgM–AP]–[BCIP]–dot

Assay type: EIA (non-competitive)
Detection: Visual
Format: Reaction vessel, Ag coated dipstick
Sample type: Serum
Sample pre-treatment:
   None
Sample volume: 10 µl (+2 ml diluent)*
Number of tests: 25, 50, 100
Controls - standards run in assay:
   Controls: Pos (1)
   Integral procedural: Neg (1), Pos (1)
Incubation: 5 min (44–48°C) + 5 min (44–48°C) + 15 min (44–48°C) + 5 min (44–48°C)
Washes: 4

CONTENTS

Antibodies, antigens, labelled components:
   B. burgdorferi (whole organism and flagellin, strain B31 and rec P39 protein) Ags bound to disptick as 4 discrete dots
   Anti-human IgA, IgG & IgM Abs AP conjugated
Substrate: BCIP
Controls - standards supplied:
   Integral Procedural Controls: Neg and Pos (bound to dipstick as separate dots)
Additional reagents:
   Positive Control Reagent (Cat. no. 2216)
Special equipment required:
   GenBio Workstation (Cat. no. 4001 or 4990)

INTERPRETATION

Comments on interpretation:
   Negative: all Borrelia Ag dots nonreactive; retest in 4–6 weeks if clinically suspicious of Lyme's disease
   Positive: when P39 and whole organism Ag dots (dots 3, 4, 5) are reactive; report as specific B. burgdorferi detected
   When other combinations of Ag dots are reactive or non-reactive; report as Borrelia Ab detected**
No. of references: 12

NOTES

131421.0

*Diluent contains E. coli extract to act as a blocking agent to improve specificity (incubate for 30 min–1 hr at 44–48°C)
**Cross-reactivity of whole organism and Borrelia flagellin Ag has been reported with infectious mononucleosis, other spirochetal infections, rheumatoid arthritis and autoimmune disorders

© KLUWER ACADEMIC PUBLISHERS 1995, ISSN 1381-5067

Borrelia burgdorferi
ANTIBODY DETECTION

Manufacturer: GenBio
Cat. No./Trade name: 4725/ImmunoDot® Lyme Test

SUMMARY

[Dipstick-Ag]–**Ab**–[AHIgA,IgG&IgM-AP]–[BCIP]–dot

Assay type: EIA (non-competitive)
Detection: Visual
Format: Reaction vessel, Ag coated dipstick
Sample type: Serum, heparinized whole blood
Sample pre-treatment:
  None
Sample volume: 10 μl serum or 20 μl whole blood (+2 ml diluent)
Number of tests: 25, 50, 100
Controls - standards run in assay:
  Controls: Pos (1)
  Integral procedural: Neg (1), Pos (1)
Incubation:
  5 min (44–48°C) + 5 min (44–48°C) + 15 min (44–48°C) + 5 min (44–48°C)
Washes: 4

CONTENTS

Antibodies, antigens, labelled components:
  B. burgdorferi (ATCC-35210, strain B31) Ags bound to dipstick as 4 discrete dots (2 flagella and 2 membrane concentration levels)
  Anti-human IgA, IgG & IgM Ab (g) AP conjugated
Substrate: BCIP
Controls - standards supplied:
  Integral Procedural Controls: Neg and Pos (bound to dipstick as separate dots)
Additional reagents:
  Positive Control Reagent (Cat. no. 2216)
Special equipment required:
  GenBio Workstation (Cat. no. 4001 or 4990)

INTERPRETATION

Comments on interpretation:
  Negative: all Borrelia Ag dots nonreactive; retest in 4–6 weeks if clinically suspicious of Lyme's disease
  Positive: if any 2 of the 4 Borrelia Ag dots are reactive
No. of references: 7

NOTES

131422.0

Cross-reactivity of Borrelia has been reported with infectious mononucleosis, other spirochetal infections, rheumatoid arthritis and auto-immune disorders

Borrelia burgdorferi
ANTIBODY DETECTION

Manufacturer: MRL Diagnostics
Cat. No./Trade name: EL0400P/Lyme Disease Elisa

SUMMARY

[Well-Ag]–**Ab**–[AHIgG&IgM-AP]–[OPD]–A$_{492}$

Assay type: EIA (non-competitive)
Detection: Colorimetric A$_{492}$
Format: Microtitre well, Ag coated
Sample type: Serum (do not heat inactivate)
Sample pre-treatment:
  None
Sample volume: 100 μl of 1:21 dilution
Number of tests: 96
Controls - standards run in assay:
  Controls: Neg (1), Pos (1)
  Calibrator: Cut-off (2)
Incubation:
  1 hr (RT) + 30 min (RT) + 10 min (RT)
Washes: 2 (+ preliminary plate wash)

CONTENTS

Antibodies, antigens, labelled components:
  B. burgdorfen (strain B31) Ag bound to well
  Anti-human IgG & IgM Abs (g) POD conjugated
Substrate: OPD
Controls - standards supplied:
  Controls: Neg and Pos (human serum)
  Calibrator: Cut-off (human serum)
Additional reagents:
  None
Special equipment required:
  None

INTERPRETATION

Comments on interpretation:
  Equivocal: when the ratio of sample OD:cut-off calibrator OD (Index value) is near 1; retest to confirm or repeat with a fresh sample
  Repeatably equivocal: report as negative and retest a fresh sample taken several weeks later to identify any rise in Ab titre. If still equivocal, report as negative
No. of references: 9

NOTES

131110.0

Borrelia burgdorferi
ANTIBODY DETECTION

Manufacturer: Seradyn
Cat. No./Trade name: 0370437/Color Vue® Borrelia (Lyme Test)

SUMMARY

[Well-Ag]–**Ab**–[AHIgA,IgG&IgM-POD]–[ABTS]–A_{450}

Assay type: EIA (non-competitive)
Detection: Colorimetric A_{450}
Format: Microtitre well, Ag coated
Sample type: Serum
Sample pre-treatment:
  Absorb diluted serum with Borrelia blocker provided and incubate 30–60 min (RT) to improve specificity
Sample volume: 10 µl (+200 µl diluent)
Number of tests: 96
Controls - standards run in assay:
  Controls: Neg (2), Pos (3)
Incubation:
  30 min (RT) + 30 min (RT) + 30 min (RT)
Washes: 2

CONTENTS

Antibodies, antigens, labelled components:
  rec *B. burgdorferi* (P39 protein) and *B. burgdorferi* (cell extract strain B31) Ags bound to well
  Anti-human IgA, IgM & IgG Ab (g) POD conjugated
Substrate: ABTS
Controls - standards supplied:
  Controls: Neg and Pos (human serum)
Additional reagents:
  None
Special equipment required:
  None

INTERPRETATION

Comments on interpretation:
  Equivocal: where cut-off is between 0.12 and 0.16; retest to confirm
  Repeatably equivocal: retest a fresh sample in 2–4 weeks
No. of references: 12

NOTES

131653.0

Borrelia burgdorferi
ANTIBODY DETECTION

Manufacturer: Sigma Diagnostics
Cat. No./Trade name: SIA 423/SIA® Lyme Disease

SUMMARY

[Well-Ag]–**Ab**–[AHIgG&IgM-AP]–[PMP]–A_{550}

Assay type: EIA (non-competitive)
Detection: Colorimetric A_{550}
Format: Microtitre well, Ag coated
Sample type: Serum (do not heat inactivate)
Sample pre-treatment:
  None
Sample volume: 10 µl (+200 µl diluent)
Number of tests: 96
Controls - standards run in assay:
  Controls: A (1), B (1)
  Calibrators: 1 (2), 2 (2), 3 (2)
Incubation:
  45 min (RT) + 45 min (RT) + 45 min (RT)
Washes: 2 (+ preliminary plate wash)

CONTENTS

Antibodies, antigens, labelled components:
  B. burgdorferi Ag bound to well
  Anti-human IgG & IgM Ab (g) AP conjugated
Substrate: PMP
Controls - standards supplied:
  Controls: A and B (human serum)
  Calibrators: 1, 2 and 3 (human serum)
Additional reagents:
  None
Special equipment required:
  Sigma EIA Multiwell Plate Reader (Cat. no. EQ104 – optional)

INTERPRETATION

Comments on interpretation:
  Equivocal: 0.18–0.19 SIA® Lyme Disease Value or 0.90–0.99 Index Value; retest to confirm
  Repeatably equivocal: report as such and retest using alternative method or fresh sample
No. of references: 8

NOTES

131439.0

© KLUWER ACADEMIC PUBLISHERS 1995, ISSN 1381-5067

Borrelia burgdorferi
ANTIBODY DETECTION (IgG)

Manufacturer: BAG-Biologische Analysensystem GmbH
Cat. No./Trade name: 5245 BAG - Borrelia - EIA - G

SUMMARY

[Well-Ag]–**Ab**–[AHIgG-HRP]–[TMB]–A_{450}

Assay type: EIA (non-competitive)
Detection: Colorimetric A_{450}
Format: Microtitre well, Ag coated
Sample type: Serum
Sample pre-treatment:
  None
Sample volume: 100 µl of 1:101 dilution
Number of tests: 96
Controls - standards run in assay:
  Controls: Neg (1), cut-off (2), Pos (1)
Incubation:
  30 min (RT) + 30 min (RT) + 10 min (RT)
Washes: 2

CONTENTS

Antibodies, antigens, labelled components:
  B. burgdorferi (Bavarian strain) Ag bound to well
  Anti-human IgG Ab (sh) HRP conjugated
Substrate: TMB
Controls - standards supplied:
  Controls: Neg, cut-off and Pos (human serum)
Additional reagents:
  None
Special equipment required:
  None

INTERPRETATION

Comments on interpretation:
  Equivocal: within cut-off ± 10%; data is not provided for further assessment of samples within this range
No. of references: None

NOTES

131603.0

Borrelia burgdorferi
ANTIBODY DETECTION (IgG)

Manufacturer: Dako A/S
Cat. No./Trade name: K6029/IDEIA® Borrelia burgdorferi IgG

SUMMARY

[Well-Ag]–**Ab**–[AHIgG-HRP]–[TMB]–A_{450}

Assay type: EIA (non-competitive)
Detection: Colorimetric A_{450}
Format: Microtitre well, Ag coated
Sample type: Serum
Sample pre-treatment:
  None
Sample volume: 10 µl (+2 ml diluent)*
Number of tests: 96
Controls - standards run in assay:
  Controls: Neg (1), cut-off (3), Pos (2)
Incubation:
  1 hr (RT) + 1 hr (RT) + 10 min (RT) (first two incubations with shaking)
Washes: 2

CONTENTS

Antibodies, antigens, labelled components:
  B. burgdorferi (flagellum) Ag bound to well
  Anti-human IgG Ab (r) HRP biotinylated
Substrate: TMB
Controls - standards supplied:
  Controls: Neg (sample diluent), cut-off and Pos (human serum)
Additional reagents:
  None
Special equipment required:
  None

INTERPRETATION

Comments on interpretation:
  Classification of samples is according to cut-off; no further testing
No. of references: 8

NOTES

131043.0

*Samples are assayed in duplicate

Borrelia burgdorferi
ANTIBODY DETECTION (IgG)

Manufacturer: Immunobiological Laboratories
Cat. No./Trade name: RE 57221/Borrelia 14kD-IgG Elisa Test

SUMMARY

[Well-Ag]–**Ab**–[AHIgG-HRP]–[TMB]–A_{450}

Assay type: EIA (non-competitive)
Detection: Colorimetric A_{450}
Format: Microtitre well, Ag coated
Sample type: Serum
Sample pre-treatment:
None
Sample volume: 100 µl of 1:100 dilution x 2
Number of tests: 96
Controls - standards run in assay:
Controls:
Qualititative: Neg (2), Pos 1:40 dilution (2)
Quantitative: Neg (2), Pos - dilutions: 1:5 (2), 1:10 (2), 1:20 (2), 1:40 (2), 1:80 (2)
Incubation:
1 hr (37°C) + 30 min (37°C) + 10 min (RT)
Washes: 2

CONTENTS

Antibodies, antigens, labelled components:
rec *B. burgdorferi* (14kD-flagellin protein) Ag bound to well
Anti-human IgG Ab HRP conjugated
Substrate: TMB
Controls - standards supplied:
Controls: Neg and Pos (serum)
Additional reagents:
None
Special equipment required:
None

INTERPRETATION

Comments on interpretation:
Equivocal:
Qualitative: between cut-off and cut-off —0.1; retest a fresh sample in 5–7 days
Quantitative: between 70–100 units; retest a fresh sample in 5–7 days
No. of references: 5

NOTES

131248.0

Borrelia burgdorferi
ANTIBODY DETECTION (IgG AND/OR IgM)

Manufacturer: Behringwerke AG
Cat. No./Trade name: Enzygnost® Borreliosis

SUMMARY

[Well-Ag]–**Ab**–[AHIgG/IgM-POD]–[TMB]–A_{450}

Assay type: EIA (non-competitive)
Detection: Colorimetric A_{450}
Format: Microtitre well, Ag coated
Sample type: Serum, plasma
Sample pre-treatment:
For IgM assays: add RF absorbent provided to diluted samples, incubate 15 min (RT) or overnight (2–8°C) to reduce interference due to RF and IgG
Sample volume: 20 µl of 1:21 dilution*
Number of tests: 96
Controls - standards run in assay:
Reference reagents: For IgG: N (2), P/N (2)
For IgM: N (2), P/P (2)
Incubation:
30 min (37°C) + 30 min (37°C) + 30 min (RT)
Washes: 2

CONTENTS

Antibodies, antigens, labelled components:
B. burgdorferi Ag bound to well
Anti-human IgG Ab (g) POD conjugated
Anti-human IgM Ab (g) POD conjugated
Substrate: TMB (bought separately)
Controls - standards supplied:
Reference reagents: P/N, P/P and N (anti-Borrelia Reference Reagents)
Additional reagents:
Supplementary Reagents for Enzygnost/TMB (Cat. no. OUVP)
Special equipment required:
Behring ELISA Processor II (optional)
Behring ELISA Processor III (optional)

INTERPRETATION

Comments on interpretation:
Equivocal: retest to confirm
Repeatably equivocal: report as equivocal and perform further tests
No. of references: 11

NOTES

131031.0

This kit can detect IgG and IgM antibodies and can differentiate between them
*If high Ab levels are expected a further serum dilution of 1:10 is required

© *KLUWER ACADEMIC PUBLISHERS 1995, ISSN 1381-5067*

Borrelia burgdorferi
ANTIBODY DETECTION (IgG AND/OR IgM)

Manufacturer: BioWhittaker Inc
Cat. No./Trade name: 30-649U/Lyme ELISA II

SUMMARY

[Well-Ag]–**Ab**–[AHIgG/IgM-AP]–[PMP]–A_{550}

Assay type: EIA (non-competitive)
Detection: Colorimetric A_{550}
Format: Microtitre well, Ag coated
Sample type: Serum (do not heat inactivate)
Sample pre-treatment:
    None
Sample volume: 10 µl (+200 µl diluent)
Number of tests: 192
Controls - standards run in assay:
    Controls: A (1), B (1)
    Calibrators: Neg (1), Pos 1 (1), Pos 2 (1)
Incubation:
    45 min (RT) + 45 min (RT) + 45 min (RT)
Washes: 2 (+preliminary plate wash)

CONTENTS

Antibodies, antigens, labelled components:
    B. burgdorferi Ag bound to well
    Anti-human IgG Ab (g) AP conjugated
    Anti-human IgM Ab (g) AP conjugated
Substrate: PMP
Controls - standards supplied:
    Controls: A and B
    Calibrators: Neg, Pos 1, Pos 2 (all human serum)
Additional reagents:
    None
Special equipment required:
    BioWhittaker automated system and software (optional)

INTERPRETATION

Comments on interpretation:
    Equivocal: Elisa value between 0.15–0.16 or Index value
        between 0.88–0.89; retest to confirm
    Repeatably equivocal: retest a fresh sample or use an
        alternative method
No. of references: 3

NOTES

131741.0

This kit can detect IgG and IgM antibodies and can
    differentiate between them

Borrelia burgdorferi
ANTIBODY DETECTION (IgG AND/OR IgM)

Manufacturer: BioWhittaker Inc
Cat. No./Trade name: 30-349U/Lyme Stat

SUMMARY

[Well-Ag]–**Ab**–[AHIgG/IgM-AP]–[PMP]–A_{550}

Assay type: EIA (non-competitive)
Detection: Colorimetric A_{550}
Format: Microtitre well, Ag coated
Sample type: Serum (do not heat inactivate)
Sample pre-treatment:
    None
Sample volume: 10 µl (+200 µl diluent)
Number of tests: 192
Controls - standards run in assay:
    Controls: Neg (1), Pos (1)
    Standards: Neg (1), low Pos (1), high Pos (1)
Incubation:
    15 min (RT) + 15 min (RT) + 15 min (RT)
Washes: 2 (+preliminary plate wash)

CONTENTS

Antibodies, antigens, labelled components:
    B. burgdorferi Ag bound to well
    Anti-human IgG Ab (g) AP conjugated
    Anti-human IgM Ab (g) AP conjugated
Substrate: PMP
Controls - standards supplied:
    Controls: Neg and Pos
    Standards: Neg, low Pos and high Pos (all human serum)
Additional reagents:
    None
Special equipment required:
    BioWhittaker automated system and software (optional)

INTERPRETATION

Comments on interpretation:
    Equivocal: Predicted Index value (PIV) between 0.80–
        0.90; retest to confirm
    Repeatably equivocal: retest a fresh sample or use
        alternative method
    Low Positive: PIV between 1.00–1.74
    Mid Positive: PIV between 1.75–3.89
    High Positive: PIV ⩾ 3.90
No. of references: 16

NOTES

131748.0

This kit can detect IgG and IgM antibodies and can
    differentiate between them

Borrelia burgdorferi
ANTIBODY DETECTION (IgG AND/OR IgM)

Manufacturer: Dako A/S
Cat. No./Trade name: K6028/IDEIA® Lyme Neuroborreliosis

SUMMARY

[Well-AHIgG/IgM]–**Ab**–[Ag-biotin]–[strept-POD]– [TMB]–A_{450}

Assay type: EIA (non-competitive)
Detection: Colorimetric A_{450}
Format: Microtitre well, Ab coated
Sample type: CSF and serum*
Sample pre-treatment:
  None
Sample volume: Serum 10 µl (+2 ml diluent); CSF 100 µl (+400 µl diluent)**
Number of tests: 96
Controls - standards run in assay:
  Controls: Neg (2), Pos IgG (2), Pos IgM (2)
Incubation:
  1 hr (RT) + 1 hr (RT) + 10 min (RT) (first two incubations with shaking)
Washes: 2

CONTENTS

Antibodies, antigens, labelled components:
  Anti-human IgG Ab (r) bound to well
  Anti-human IgM Ab (r) bound to well
  B. burgdorferi (flagellum) Ag biotinylated
  Streptavidin POD conjugated
Substrate: TMB
Controls - standards supplied:
  Controls: Neg (sample diluent) Pos IgG and Pos IgM (human serum)
Additional reagents:
  None
Special equipment required:
  None

INTERPRETATION

Comments on interpretation:
  Comparison of IgG and IgM Ab levels in CSF and serum must be made and considered in conjunction with clinical findings to aid diagnosis
No. of references: 13

NOTES

131044.0

This kit can detect IgG and IgM antibodies in CSF and serum and can differentiate between them
*CSF and serum samples must be collected at the same time and tested simultaneously to aid diagnosis of neuroborreliosis
**CSF and serum samples are assayed in duplicate

Borrelia burgdorferi
ANTIBODY DETECTION (IgG AND/OR IgM)

Manufacturer: Diagast Laboratories
Cat. No./Trade name: 69157/Elilyme - G/M

SUMMARY

[Well-Ag]–**Ab**–[AHIgG/IgM-POD]–[substrate]–A_{450}

Assay type: EIA (non-competitive)
Detection: Colorimetric A_{450}
Format: Microtitre well, Ag coated
Sample type: Serum, CSF
Sample pre-treatment:
  None
Sample volume: 10 µl (+1 ml diluent)
Number of tests: 24, 48*
Controls - standards run in assay:
  Controls: Neg (1), IgG threshold (1), IgG Pos (1), IgM threshold (1)
Incubation:
  30 min (RT) + 30 min (RT) + 15 min (RT)
Washes: 2

CONTENTS

Antibodies, antigens, labelled components:
  B. burgdorferi (P41 kDa, OspA, OspB and P100 kDa – strain B31) Ags bound to well
  Anti-human IgG Ab POD conjugated
  Anti-human IgM Ab POD conjugated
Substrate: Not specified
Controls - standards supplied:
  Controls: Neg, IgG Pos, IgG threshold and IgM threshold (with assigned DIAGAST titres)
Additional reagents:
  None
Special equipment required:
  None

INTERPRETATION

Comments on interpretation:
  Equivocal: within designated range; retest a second sample taken 15 days later**
No. of references: 5

NOTES

131055.0

This kit can detect IgG and IgM antibodies and can differentiate between them
*24 IgG and IgM combined tests or 48 of either IgG or IgM separate tests
**High levels of *Treponema pallidum* antibodies in serum can lead to false positive results

© KLUWER ACADEMIC PUBLISHERS 1995, ISSN 1381-5067

Borrelia burgdorferi
ANTIBODY DETECTION (IgG AND/OR IgM))

Manufacturer: Immuno Pharmacology Research
Cat. No./Trade name: LYME ELISA IgG/IgM

SUMMARY

[Well-Ag]–**Ab**–[AHIgG/IgM-AP]–[PNP]–A_{405}

Assay type: EIA (non-competitive)
Detection: Colorimetric A_{405}
Format: Microtitre well, Ag coated
Sample type: Serum
Sample pre-treatment:
 None
Sample volume: 100 µl of 1:100 dilution x 2
Number of tests: 48, 96
Controls - standards run in assay:
 Controls: Neg (1), Pos (1)
Incubation:
 1 hr (37°C) + 1 hr (37°C) + 30 min (37°C)
Washes: 2

CONTENTS

Antibodies, antigens, labelled components:
 B. burgdorferi (strain BITS 87) Ag bound to well
 Anti-human IgG Ab AP conjugated
 Anti-human IgM Ab AP conjugated
Substrate: PNP
Controls - standards supplied:
 Controls: Neg and Pos
Additional reagents:
 None
Special equipment required:
 None

INTERPRETATION

Comments on interpretation:
 Classification of samples is according to cut-off; no
 further testing
No. of references: 3

NOTES

131331.0

This kit can detect IgG and IgM antibodies and can
 differentiate between them

Borrelia burgdorferi
ANTIBODY DETECTION (IgM)

Manufacturer: BAG-Biologische Analysensystem GmbH
Cat. No./Trade name: 5246 BAG - Borrelia - EIA - M

SUMMARY

[Well-Ag]–**Ab**–[AHIgM-HRP]–[TMB]–A_{450}

Assay type: EIA (non-competitive)
Detection: Colorimetric A_{450}
Format: Microtitre well, Ag coated
Sample type: Serum (do not heat inactivate)
Sample pre-treatment:
 Treat diluted serum with BAG - IgM - Sep - System (Cat.
 no. 5540) to eliminate interference due to RF and IgG
Sample volume: 100 µl of treated sample
Number of tests: 96
Controls - standards run in assay:
 Controls: Neg (1), cut-off (2), Pos (1)
Incubation:
 30 min (RT) + 30 min (RT) + 10 min (RT)
Washes: 2

CONTENTS

Antibodies, antigens, labelled components:
 B. burgdorferi (Bavarian strain) Ag bound to well
 Anti-human IgM Ab (sh) HRP conjugated
Substrate: TMB
Controls - standards supplied:
 Controls: Neg, cut-off and Pos (human serum)
Additional reagents:
 None
Special equipment required:
 None

INTERPRETATION

Comments on interpretation:
 Equivocal: within cut-off ±10%; data is not provided for
 further assessment of samples within this range
No. of references: None

NOTES

131608.0

Borrelia burgdorferi
ANTIBODY DETECTION (IgM)

Manufacturer: BioWhittaker Inc
Cat. No./Trade name: 30-382U/Lyme Stat M

SUMMARY

[Well-Ag]–**Ab**–[AHIgM-AP]–[PMP]–A_{550}

Assay type: EIA (non-competitive)
Detection: Colorimetric A_{550}
Format: Microtitre well, Ag coated
Sample type: Serum (do not heat inactivate)
Sample pre-treatment:
Mix sample with Pretreatment Serum, incubate 30 min (RT) and centrifuge to eliminate interference due to IgG and RF
Sample volume: 25 µl of treated sample (+ 100 µl diluent)
Number of tests: 96
Controls - standards run in assay:
Calibrators: Neg (1), Pos 1 (1), Pos 2 (1)
Incubation:
45 min (RT) + 30 min (RT) + 30 min (RT) (all with shaking)
Washes: 2 (+ preliminary plate wash)

CONTENTS

Antibodies, antigens, labelled components:
B. burgdorferi Ag bound to well
Anti-human IgM Ab AP conjugated
Substrate: PMP
Controls - standards supplied:
Calibrators: Neg, Pos 1, Pos 2 (all human serum)
Rheumatoid Factor Control (human serum)
Additional reagents:
None
Special equipment required:
BioWhittaker automated system and software (optional)

INTERPRETATION

Comments on interpretation:
Equivocal: test value between 0.10–0.12; indicates probable positive, retest in duplicate to confirm
No. of references: 11

NOTES

131755.0

Borrelia burgdorferi
ANTIBODY DETECTION (IgM)

Manufacturer: Dade International Inc (Bartels Division)
Cat. No./Trade name: B1029-302/Bartels PRIMA System® Lyme Disease EIA

SUMMARY

[Well-Ag]–**Ab**–[AHIgM-enzyme]–[PNP]–A_{405}

Assay type: EIA (non-competitive)
Detection: Colorimetric A_{405}
Format: Microtitre well, Ag coated
Sample type: Serum
Sample pre-treatment:
None
Sample volume: 100 µl of 1:11 dilution*
Number of tests: 96
Controls - standards run in assay:
Controls: Neg (1), Pos (1)
Reference serum: (1)
Incubation:
30 min (37°C) + 30 min (37°C) + 30 min (37°C)
Washes: 2

CONTENTS

Antibodies, antigens, labelled components:
B. burgdorferi (JDI strain) Ag bound to well
Anti-human IgM Ab enzyme conjugated
Substrate: PNP
Controls - standards supplied:
Controls: Neg and Pos
Reference serum: 1
Additional reagents:
None
Special equipment required:
None

INTERPRETATION

Comments on interpretation:
Classification of samples is according to cut-off; no further testing
No. of references: 8

NOTES

131049.0

*Specimen diluent contains an absorbent to eliminate interference due to RF and IgG

© KLUWER ACADEMIC PUBLISHERS 1995, ISSN 1381-5067

Borrelia burgdorferi
ANTIBODY DETECTION (IgM)

Manufacturer: Dako A/S
Cat. No./Trade name: K6030/IDIEA® Borrelia burgdorferi IgM

SUMMARY

[Well-AHIgM]–**Ab**–[Ag-biotin]–[strept-POD]–[TMB]–A_{450}

Assay type: EIA (non-competitive)
Detection: Colorimetric A_{450}
Format: Microtitre well, Ab coated
Sample type: Serum
Sample pre-treatment:
    None
Sample volume: 10 µl (+2 ml diluent)*
Number of tests: 96
Controls - standards run in assay:
    Controls: Neg (1), cut-off (3), Pos (2)
Incubation:
    1 hr (RT) + 1 hr (RT) + 10 min (RT) (first two
        incubations with shaking)
Washes: 2

CONTENTS

Antibodies, antigens, labelled components:
    Anti-human IgM Ab bound to well
    B. burgdorferi (flagellum) Ag biotinylated
    Streptavidin POD conjugated
Substrate: TMB
Controls - standards supplied:
    Controls: Neg (sample diluent), cut-off and Pos (human
        serum)
Additional reagents:
    None
Special equipment required:
    None

INTERPRETATION

Comments on interpretation:
    Classification of samples is according to cut-off; no
        further testing
No. of references: 8

NOTES

131042.0

*Samples are assayed in duplicate

Borrelia burgdorferi
ANTIBODY DETECTION (IgM)

Manufacturer: Gull Laboratories Inc
Cat. No./Trade name: LDE150/Lyme IgM ELISA Test

SUMMARY

[Well-Ag]–**Ab**–[AHIgM-AP]–[PNP]–A_{405}

Assay type: EIA (non-competitive)
Detection: Colorimetric A_{405}
Format: Microtitre well, Ag coated
Sample type: Serum
Sample pre-treatment:
    None
Sample volume: 100 µl of 1:11 dilution*
Number of tests: 96
Controls - standards run in assay:
    Controls: Neg (1), Pos (1)
    Reference Serum: 1 (3)
Incubation:
    30 min (37°C) + 30 min (37°C) + 30 min (37°C)
Washes: 2

CONTENTS

Antibodies, antigens, labelled components:
    B. burgdorferi (JD1 strain) Ag bound to well
    Anti-human IgM Ab (caprine) AP conjugated
Substrate: PNP
Controls - standards supplied:
    Controls: Neg and Pos
    Reference serum: 1 (all human serum)
Additional reagents:
    None
Special equipment required:
    None

INTERPRETATION

Comments on interpretation:
    Equivocal: where sample OD is between mean OD of
        Reference Serum and mean OD of Reference Serum
        x0.9; retest to confirm
    Repeatably equivocal: retest using an alternative method
No. of references: 14

NOTES

131703.0

*Diluent contains an absorbent to eliminate interference due
    to RF and IgG in serum

© KLUWER ACADEMIC PUBLISHERS 1995, ISSN 1381-5067

Borrelia burgdorferi
ANTIBODY DETECTION (IgM)

Manufacturer: Immunobiological Laboratories
Cat. No./Trade name: RE 57211/Borrelia 14kD-IgM Elisa Test

SUMMARY

[Well-Ag]–**Ab**–[AHIgM-HRP]–[TMB]–A$_{450}$

Assay type: EIA (non-competitive)
Detection: Colorimetric A$_{450}$
Format: Microtitre well, Ag coated
Sample type: Serum, plasma
Sample pre-treatment:
 Mix diluted serum with RF-Absorbent to eliminate interference due to RF and IgG in sample
Sample volume: 100 μl of treated sample x 2
Number of tests: 96
Controls - standards run in assay:
 Controls:
 Qualitative: Neg (2), Pos 1:40 dilution (2)
 Quantitative: Neg (2), Pos - dilutions: 1:5 (2), 1:10 (2), 1:20 (2), 1:40 (2), 1:80 (2)
Incubation:
 1 hr (37°C) + 30 min (37°C) + 10 min (RT)
Washes:

CONTENTS

Antibodies, antigens, labelled components:
 rec *B. burgdorferi* (14kD-flagellin protein) Ag bound to well
 Anti-human IgM Ab HRP conjugated
Substrate: TMB
Controls - standards supplied:
 Controls: Neg and Pos (serum)
Additional reagents:
 None
Special equipment required:
 None

INTERPRETATION

Comments on interpretation:
 Qualitative: Equivocal: between cut-off and cut-off –0.1; retest a fresh sample in 5–7 days
 Quantitative: Equivocal: between 35–70 units; retest a fresh sample in 5–7 days
No. of references: 5

NOTES
131256.0

Borrelia burgdorferi
ANTIBODY DETECTION

Manufacturer: Seradyn
Cat. No./Trade name: 0370445/Color Spot® Borrelia (Lyme) with Recombinant Protein

SUMMARY

[Membrane-Ag]–**Ab**–[AHIgA,IgG&IgM-AP]–[BCIP]–dots

Assay type: Immunoblot assay
Detection: Visual
Format: Reaction vessel, membrane strip, Ag coated
Sample type: Serum
Sample pre-treatment:
 None
Sample volume: 10 μl (+2 ml diluent)*
Number of tests: 25, 100
Controls - standards run in assay:
 Controls: Neg (1), Pos (1) bound to membrane
Incubation:
 30 min–1 hr (46°C) + 5 min (46°C) + 5 min (46°C) + 15 min (46°C) + 5 min (46°C)
Washes: 5

CONTENTS

Antibodies, antigens, labelled components:
 rec *B. burgdorferi* (P39 protein), *B. burgdorferi* (cell extract strain B31) and *B. burgdorferi* (flagellin extract strain B31) bound to membrane as discrete dots
 Anti-human IgA, IgG & IgM Abs (g) AP conjugated
Substrate: BCIP
Controls - standards supplied:
 Controls: Neg and Pos (bound to membrane as discrete dots)
Additional reagents:
 Positive control (serum) available if required
Special equipment required:
 Seradyn Workstation (Cat. no. 0370460 or 0370510)

INTERPRETATION

Comments on interpretation:
 Positive: appearance of coloured dots on strip. For interpretation of dot pattern see manufacturer's instructions
No. of references: 12

NOTES
131654.0
*Diluent contains *E. coli* extract to absorb serum prior to testing to improve specificity

© KLUWER ACADEMIC PUBLISHERS 1995, ISSN 1381-5067

Borrelia burgdorferi
ANTIBODY DETECTION (IgG)

Manufacturer: Cambridge Diagnostics Ireland Ltd
Cat. No./Trade name: 8056/Human Lyme Western Blot IgG Kit

SUMMARY

[Strip-Ag]–**Ab**–[AHIgG-AP]–[BCIP]–bands

Assay type: Immunoblot assay
Detection: Visual
Format: Reaction tray, Ag coated nitrocellulose strip
Sample type: Serum, plasma
Sample pre-treatment:
    None
Sample volume: 20 µl
Number of tests: 30
Controls - standards run in assay:
    Controls: Neg (1), weak Pos (1), Pos (1)
Incubation:
    2 hr (22–26°C) + 30 min (RT) + 10–15 min (RT)* (all
        with rocking)
Washes: 2 (+ preliminary wash)

CONTENTS

Antibodies, antigens, labelled components:
    B. burgdorferi (strain B31 - fractionated proteins) Ag
        bound to nitrocellulose strip as discrete bands
    Anti-human IgG Ab (g) AP conjugated
Substrate: BCIP
Controls - standards supplied:
    Controls: Neg and Pos (diluted to provide weak Pos) – all
        human serum
Additional reagents:
    None
Special equipment required:
    None

INTERPRETATION

Comments on interpretation:
    Positive: appearance of coloured bands or strip. For
        interpretation of band pattern, see manufacturer's
        instructions
No. of references: 21

NOTES

131711.0

For research use only
*For first incubation: if temperature is 18–21°C, increase
    incubation time and if < 18°C use an incubator

Borrelia burgdorferi
ANTIBODY DETECTION (IgG)

Manufacturer: Immunetics
Cat. No./Trade name: LD0202G/Lyme Miniblot Kit for
IgG

SUMMARY

[Strip-Ag]–**Ab**–[AHIgG-AP]–[BCIP/NBT]–bands

Assay type: Immunoblot assay
Detection: Visual
Format: Cassette, Ag coated nitrocellulose strip*
Sample type: Serum, plasma, CSF, dried blood spot eluates
Sample pre-treatment:
    None
Sample volume: 50 µl of 1:100 dilution
Number of tests: 8, 24
Controls - standards run in assay:
    Controls: Neg (1), Pos (1)
Incubation:
    30 min (RT) + 15 min (RT)
Washes: 2

CONTENTS

Antibodies, antigens, labelled components:
    B. burgdorferi Ags bound to nitrocellulose strip as
        discrete bands
    Anti-human IgG Ab AP conjugated
Substrate: BCIP/NBT
Controls - standards supplied:
    Controls: Neg and Pos (human serum)
Additional reagents:
    None
Special equipment required:
    Starter Kit with Miniblot® C-Shell and Miniwash™
        Manifold (Cat. no. LD-101S)

INTERPRETATION

Comments on interpretation:
    Positive: coloured bands on membrane. For
        interpretation of band pattern see manufacturer's
        instructions
No. of references: 19

NOTES

131773.0

For research use only
*Membrane has 24 channels for parallel incubations of up to
    24 serum specimens. It is sealed within a disposable
    incubation cassette which is inserted in Miniblot C-Shell
    holder during assay

© KLUWER ACADEMIC PUBLISHERS 1995, ISSN 1381-5067

Borrelia burgdorferi
ANTIBODY DETECTION (IgG)

Manufacturer: Immunetics
Cat. No./Trade name: LD-0510G/Lyme Immunostrip Kit

SUMMARY

[Strip-Ag]–**Ab**–[AHIgG-AP]–[BCIP/NBT]–bands

Assay type: Immunoblot assay
Detection: Visual
Format: Reaction tray, Ag coated nitrocellulose strip
Sample type: Serum, plasma
Sample pre-treatment:
   None
Sample volume: 1 ml of 1:100 dilution
Number of tests: 24
Controls - standards run in assay:
   Integral Controls: Neg (1), Weak Pos (1), Pos (1)*
Incubation:
   15 min (RT) + 15 min (RT)
Washes: 2

CONTENTS

Antibodies, antigens, labelled components:
   B. burgdorferi (whole cell lysate) Ags bound to
      nitrocellulose strip as discrete bands
   Anti-human IgG Ab AP conjugated
Substrate: BCIP/NBT
Controls - standards supplied:
   Controls: Neg and Pos (human serum)
   Integral controls: (Pos, weak Pos and Neg) incorporated
      in strip*
Additional reagents:
   None
Special equipment required:
   None

INTERPRETATION

Comments on interpretation:
   Interpretation performed by comparison of test strip with
      lot-specific reference strip, according to CDC's most
      recent interpretive criteria
No. of references: 23

NOTES

131776.0

*Assay kit incorporates Guideline® Control Bands on each
   strip

Borrelia burgdorferi
ANTIBODY DETECTION (IgG AND/OR IgM)

Manufacturer: Cambridge Diagnostics
Ireland Ltd
Cat. No./Trade name: 8059/Human Lyme IgM Western
Blot Kit

SUMMARY

[Strip-Ag]–**Ab**–[AHIgG/IgM-AP]–[BCIP]–bands

Assay type: Immunoblot assay
Detection: Visual
Format: Reaction tray, Ag coated nitrocellulose strip
Sample type: Serum, plasma
Sample pre-treatment:
   None
Sample volume: 20 μl
Number of tests: 30
Controls - standards run in assay:
   Controls: Neg (1), weak IgM Pos (1), IgM Pos (1), IgG
      Pos (1)
Incubation:
   2 hr (22–26°C) + 30 min (RT) + 10–15 min (RT)* (all
      with rocking)
Washes: 2 (+ preliminary wash)

CONTENTS

Antibodies, antigens, labelled components:
   B. burgdorferi (strain B31 - fractionated proteins) Ag
      bound to nitrocellulose strip as discrete bands
   Anti-human IgG Ab (g) AP conjugated
   Anti-human IgM Ab (g) AP conjugated
Substrate: BCIP
Controls - standards supplied:
   Controls: Neg, IgM Pos (dilute to provide weak Pos), IgG
      Pos (human serum)
Additional reagents:
   None
Special equipment required:
   None

INTERPRETATION

Comments on interpretation:
   Positive: appearance of coloured bands or strip. For
      interpretation of band pattern, see manufacturer's
      instructions
No. of references: 21

NOTES

131712.0

For research use only
This kit can detect IgG and IgM antibodies and can
   differentiate between them
*For first incubation: if temperature is 18–21°C, increase
   incubation time and if < 18°C use an incubator

Borrelia burgdorferi
ANTIBODY DETECTION (IgG AND/OR IgM)

Manufacturer: Diagast Laboratories
Cat. No./Trade name: BLOTLYME

SUMMARY

[Strip-Ag]–**Ab**–[AHIgG/IgM-POD]–[4CN]–bands

Assay type: Immunoblot assay
Detection: Visual
Format: Reaction tray, Ag coated nitrocellulose strip
Sample type: Serum, CSF
Sample pre-treatment:
  None
Sample volume: Serum: 2 ml of 1:51 dilution
  CSF: 2 ml of 1:10 dilution
Number of tests: 10
Controls - standards run in assay:
  Controls: Neg (1), Pos (1) bound to membrane
Incubation:
  2 hr (RT) + 5 min (RT) + 1 hr (RT) + 5 min (RT) + 15 min (RT)
Washes: 5

CONTENTS

Antibodies, antigens, labelled components:
  B. burgdorferi (electrophoretic profile of cellular proteins) bound to nitrocellulose strip as discrete bands
  Anti-human IgG Ab POD conjugated
  Anti-human IgM Ab POD conjugated
Substrate: 4-chloro-napthol (4CN)
Controls - standards supplied:
  Controls: Neg and Pos
Additional reagents:
  None
Special equipment required:
  None

INTERPRETATION

Comments on interpretation:
  Positive: appearance of coloured bands on strip. For interpretation of band pattern, see manufacturer's instructions
No. of references: 7

NOTES

131670.0

This kit can detect IgG and IgM antibodies and can differentiate between them

Borrelia burgdorferi
ANTIBODY DETECTION (IgG AND/OR IgM)

Manufacturer: Scimedix Corporation
Cat. No./Trade name: W200/Lyme Disease Western Immunoblot

SUMMARY

[Strip-Ag]–**Ab**–[AHIgG/IgM-AP]–[BCIP]–visible bands

Assay type: Immunoblot assay
Detection: Visual
Format: Membrane strip, Ag coated
Sample type: Serum
Sample pre-treatment:
  Block strips prior to use with blocking buffer provided, incubate 20-45 min (RT)
Sample volume: 20 µl (+1000 µl diluent)
Number of tests: 8
Controls - standards run in assay:
  Controls: Pos (1) per run
Incubation:
  40 min (RT) + 30 min (RT) + 5 min (RT) - all with shaking
Washes: 2

CONTENTS

Antibodies, antigens, labelled components:
  B. burgdorferi Ag bound to membrane strip as discrete bands
  Anti-human IgG Ab (g) AP conjugated
  Anti-human IgM Ab (r) AP conjugated
Substrate: BCIP
Controls - standards supplied:
  Controls: Pos (IgG) and Pos (IgM)
Additional reagents:
  None
Special equipment required:
  Shaking platform (50-200 rpm)

INTERPRETATION

Comments on interpretation:
  Identify the Ab reactivity of the sample by comparing the bands on the Ag test strip with the bands on the Ag template strip provided
No. of references: 29

NOTES

131186.0

For research and investigational use only
This kit can detect IgG and IgM antibodies and can differentiate between them

Borrelia burgdorferi
ANTIBODY DETECTION (IgM)

Manufacturer: Immunetics
Cat. No./Trade name: LD-0202M/Lyme Miniblot Kit for IgM

SUMMARY

[Strip-Ag]–**Ab**–[AHIgM-AP]–[BCIP/NBT]–bands

Assay type: Immunoblot assay
Detection: Visual
Format: Cassette, Ag coated nitrocellulose strip*
Sample type: Serum, plasma, CSF, dried blood spot eluates
Sample pre-treatment:
  None
Sample volume: 50 µl of 1:50 dilution
Number of tests: 8, 24
Controls - standards run in assay:
  Controls: Neg (1), Pos (1)
Incubation:
  30 min (RT) + 15 min (RT)
Washes: 2

CONTENTS

Antibodies, antigens, labelled components:
  B. burgdorferi Ags bound to nitrocellulose strip as discrete bands
  Anti-human IgM Ab AP conjugated
Substrate: BCIP/NBT
Controls - standards supplied:
  Controls: Neg and Pos (human serum)
Additional reagents:
  None
Special equipment required:
  Starter Kit with Miniblot® C-Shell and Miniwash™ Manifold (Cat. no. LD-101S)

INTERPRETATION

Comments on interpretation:
  Positive: coloured bands on membrane. For interpretation of band pattern see manufacturer's instructions
No. of references: 19

NOTES

131774.0

For research use only
*Membrane has 24 channels for parallel incubations of up to 24 serum specimens. It is sealed within a disposable incubation cassette which is inserted in Miniblot C-Shell holder during assay

Borrelia burgdorferi
ANTIBODY DETECTION (IgM)

Manufacturer: Immunetics
Cat. No./Trade name: LD-0510M/Lyme Immunostrip Kit

SUMMARY

[Strip-Ag]–**Ab**–[AHIgM-AP]–[BCIP/NBT]–bands

Assay type: Immunoblot assay
Detection: Visual
Format: Reaction tray, Ag coated nitrocellulose strip
Sample type: Serum, plasma
Sample pre-treatment:
  None
Sample volume: 1 ml of 1:100 dilution
Number of tests: 24
Controls - standards run in assay:
  Controls: Neg (1), Weak Pos (1), Pos (1)*
Incubation:
  15 min (RT) + 15 min (RT)
Washes: 2

CONTENTS

Antibodies, antigens, labelled components:
  B. burgdorferi (whole cell lysate) Ags bound to nitrocellulose strip as discrete bands
  Anti-human IgM Ab AP conjugated
Substrate: BCIP/NBT
Controls - standards supplied:
  Controls: Neg and Pos (human serum)
  Integral controls: (Pos, Weak Pos and Neg) incorporated in strip*
Additional reagents:
  None
Special equipment required:
  None

INTERPRETATION

Comments on interpretation:
  Interpretation is performed by comparison of test strip with lot-specific reference strip, according to CDC's most recent interpretive criteria
No. of references: 23

NOTES

131822.0

*Assay kit incorporates Guideline® Control Bands on each strip

Borrelia burgdorferi
ANTIBODY DETECTION

Manufacturer: remel
Cat. No./Trade name: 24-240/Rapidot™ Lyme Disease Test

SUMMARY

[Membrane-Ag]–**Ab**–[AHIgG&IgM-HRP]–[substrate]–
purple colour

Assay type: Immunochromatographic assay
Detection: Visual
Format: Test device, Ag coated membrane
Sample type: Serum, plasma
Sample pre-treatment:
 None
Sample volume: 1 drop
Number of tests: 30
Controls - standards run in assay:
 None
Incubation:
 Read result 5 min after last reagent is added to Test
 device
Washes:

CONTENTS

Antibodies, antigens, labelled components:
 B. burgdorferi (strain B31) Ag bound to membrane (Test
 well)
 Anti-human IgG & IgM Ab HRP conjugated
Substrate: Not specified
Controls - standards supplied:
 Controls: Neg and Pos (human serum)
 Integral procedural controls incorporated in membrane
Additional reagents:
 None
Special equipment required:
 None

INTERPRETATION

Comments on interpretation:
 Positive: purple colour intensity of Test well is greater
 than colour in integral Neg control well
 Indeterminate: where result is not clear-cut; repeat test
 after filtering serum or plasma sample
No. of references: 17

NOTES

131713.0

Borrelia burgdorferi
ANTIBODY DETECTION

Manufacturer: bioMerieux
Cat. No./Trade name: 75941/Lyme-Spot IF

SUMMARY

[Slide-Ag]–**Ab**–[AHIg-FITC]–fluorescence

Assay type: Immunofluorescence assay (indirect)
Detection: Fluorescence microscopy
Format: Slide, Ag coated
Sample type: Serum
Sample pre-treatment:
 None
Sample volume: 10 µl of 1:80 and 1:160 dilutions
Number of tests: 100
Controls - standards run in assay:
 Controls: Neg (1), Pos (1) - for each dilution
Incubation:
 30 min (37°C) + 30 min (37°C)
Washes: 2

CONTENTS

Antibodies, antigens, labelled components:
 B. burgdorferi Ag (American strain B31 from culture)
 bound to slide
 Anti-human Ig Ab FITC conjugated
Substrate:
Controls - standards supplied:
 Controls: Neg (buffer) and Pos -serial dilution (Ab titre in
 U/ml, human serum)
Additional reagents:
 None
Special equipment required:
 None

INTERPRETATION

Comments on interpretation:
 Positive: (+1) to (+4) clear cut fluorescence of organism
No. of references: None

NOTES

131285.0

© KLUWER ACADEMIC PUBLISHERS 1995, ISSN 1381-5067

Borrelia burgdorferi
ANTIBODY DETECTION

Manufacturer: Diagast Laboratories
Cat. No./Trade name: 69150J/LYMIX

SUMMARY

[Slide-Ag]–**Ab**–[AHIgA,IgG&IgM-FITC]–fluorescence

Assay type: Immunofluorescence assay (indirect)
Detection: Fluorescence microscopy
Format: Slide well, Ag coated
Sample type: Serum, CSF
Sample pre-treatment:
Dilute serum with Adsorbent provided, containing non-pathogenic spirochetae extracts and incubate 15 min (RT) to improve specificity
Sample volume: Serum: 10 µl of 1:256 and 1:512 dilution
CSF: 10 µl of 1:5 dilution
Number of tests: 100
Controls - standards run in assay:
Controls: Neg (1), Pos (1)
Incubation:
15 min (37°C) + 30 min (37°C)
Washes: 4

CONTENTS

Antibodies, antigens, labelled components:
B. burgdorferi organisms bound to slide
Anti-human IgA, IgG & IgM Abs (r) FITC conjugated*
Substrate:
Controls - standards supplied:
Controls: Neg and Pos (serum)
Additional reagents:
None
Special equipment required:
None

INTERPRETATION

Comments on interpretation:
Positive: (2+) – (4+) fluorescence of organisms
No. of references: None

NOTES

131669.0

*An anti-IgM monospecific conjugate is also available (Cat. no. 69118N - Fluomum)

Borrelia burgdorferi
ANTIBODY DETECTION

Manufacturer: Zeus Scientific Inc
Cat. No./Trade name: 9350/Lyme Disease Antibody Test System

SUMMARY

[Slide-Ag]–**Ab**–[AHIg-FITC]–fluorescence

Assay type: Immunofluorescence assay (indirect)
Detection: Fluorescence microscopy
Format: Slide, Ag coated
Sample type: Serum
Sample pre-treatment:
None
Sample volume: 20 µl of 1:128 and 1:256 dilutions
Number of tests: 100
Controls - standards run in assay:
Controls: Neg (1), Pos (1)
Incubation:
30 min (RT) + 30 min (RT)
Washes: 2

CONTENTS

Antibodies, antigens, labelled components:
B. burgdorferi Ag (strain B31) bound to slide
Anti-human Ig Ab (r or g) FITC conjugated
Substrate:
Controls - standards supplied:
Controls: Neg and Pos (human serum)
Additional reagents:
None
Special equipment required:
None

INTERPRETATION

Comments on interpretation:
Positive: >(1+) fluorescence
No. of references: 16

NOTES

131297.0

Borrelia burgdorferi
ANTIBODY DETECTION (IgG)

Manufacturer: LD, Labor Diagnostika GmbH
Cat. No./Trade name: 61111/LD Lyme IFA Kit

SUMMARY

[Slide-Ag]–**Ab**–[AHIgG-FITC]–fluorescence

Assay type: Immunofluorescence assay (indirect)
Detection: Fluorescence microscopy
Format: Slide, Ag coated
Sample type: Serum
Sample pre-treatment:
   Sample may be treated with Lyme Absorbent (Cat. no. 1866) to remove nonspecific antibodies (optional)
Sample volume: 20 µl of 1:64 dilution
Number of tests: 122
Controls - standards run in assay:
   Controls: Neg (1), Pos (1)
Incubation:
   30 min (RT) + 30 min (RT)
Washes: 2

CONTENTS

Antibodies, antigens, labelled components:
   B. burgdorferi Ag bound to slide
   Anti-human IgG Ab (g) FITC conjugated
Substrate:
Controls - standards supplied:
   Controls: Neg and Pos
Additional reagents:
   Lyme Absorbent (Cat. no. 1866) (optional)
Special equipment required:
   None

INTERPRETATION

Comments on interpretation:
   Positive: green fluorescence at titre 1:64 for sera treated with Lyme Absorbent and at titre 1:256 for non-absorbed samples
No. of references: None

NOTES

131778.0

There may be some cross-reactivity in patient with Syphilis or Pinta

Borrelia burgdorferi
ANTIBODY DETECTION (IgG)

Manufacturer: MRL Diagnostics
Cat. No./Trade name: IF0400G/Lyme Disease IFA IgG

SUMMARY

[Slide-Ag]–**Ab**–[AHIgG-FITC]–fluorescence

Assay type: Immunofluorescence assay (indirect)
Detection: Fluorescence microscopy
Format: Slide well, Ag coated
Sample type: Serum
Sample pre-treatment:
   None
Sample volume: 15 µl of 1:16 dilution
Number of tests: 100
Controls - standards run in assay:
   Controls: Neg (1), Pos (1)
Incubation:
   30 min (37°C) + 30 min (37°C)
Washes: 2

CONTENTS

Antibodies, antigens, labelled components:
   B. burgdorferi (B31 strain – egg yolk suspension) bound to slide well
   Anti-human IgG Ab (g) FITC conjugated
Substrate:
Controls - standards supplied:
   Controls: Neg and Pos (human serum)
Additional reagents:
   None
Special equipment required:
   None

INTERPRETATION

Comments on interpretation:
   Positive: ⩾(1+) apple-green fluorescence
No. of references: 9

NOTES

131114.0

Borrelia burgdorferi
ANTIBODY DETECTION (IgM)

Manufacturer: MRL Diagnostics
Cat. No./Trade name: IF0400M

SUMMARY

[Well-Ag]–**Ab**–[AHIgM-FITC]–fluorescence

Assay type: Immunofluorescence assay (indirect)
Detection: Fluorescence microscopy
Format: Slide well, Ag coated
Sample type: Serum
Sample pre-treatment:
   None
Sample volume: 15 µl of 1:16 dilution*
Number of tests: 100
Controls - standards run in assay:
   Controls: Neg (1), Pos (1)
Incubation:
   1 hr 30 min (37°C) + 30 min (37°C)
Washes: 2

CONTENTS

Antibodies, antigens, labelled components:
   B. burgdorferi Ag (yolk sac preparation) bound to slide
      well
   Anti-human IgG Ab (g) FITC conjugated
Substrate:
Controls - standards supplied:
   Controls: Neg and Pos (human serum)
Additional reagents:
   None
Special equipment required:
   None

INTERPRETATION

Comments on interpretation:
   Positive: ≥(1+) fluorescence of organisms
No. of references: 9

NOTES

131122.0

*MRL IgM sample diluent contains a hyper-immune anti-
   human IgG precipitating Ig to eliminate interference due
   to RF and IgG in sample

© KLUWER ACADEMIC PUBLISHERS 1995, ISSN 1381-5067

Brucella species

Brucellae are small, non-sporing, non-motile, aerobic Gram-negative coccobacilli. These organisms are intracellular parasites which are responsible for abortion in various animals and may cause zoonotic infections of man which can result in a systemic, septicaemic febrile illness or localised infections of bones, joints, central nervous system, heart, lungs, skin, genitourinary system, and other tissues. The onset of the disease may be insidious and the illness may follow a chronic relapsing and remitting course. The four species known to be pathogenic in man, and their major animal hosts, are *B.abortus* (cattle), *B.suis* (pigs), *B.melitensis* (goats and sheep) and *B.canis* (dogs).

The organisms may be cultivated from infected animal products and from clinical samples such as blood, bone marrow, spleen, liver biopsies and other infected tissues. In view of the high risk of transmission of infection to laboratory workers, manipulation of cultures should be performed in a microbiological safety cabinet with appropriate safety precautions. Isolation may require prolonged incubation and is usually performed on blood or serum-enriched agar, in an atmosphere of enhanced CO_2. A variety of antibiotics and other agents may be incorporated in media to prevent overgrowth of other organisms. Non-haemolytic colonies, which demonstrate characteristic morphology on Gram stain, do not grow anaerobically, are oxidase-positive, and do not ferment glucose or lactose may be tested for agglutination with anti-smooth *Brucella* serum. Definitive identification and biotyping may be performed on the basis of tests for H_2S production, CO_2 requirement, agglutination with monospecific antisera, urease production, growth in the presence of basic fuchsin and thionin and phage typing.

Diagnosis

Definitive diagnosis may be made on the basis of isolation of the causative organism from clinical specimens as outlined above. However, in view of the insidious onset of the illness, and the protean manifestations with which the disease may present, the majority of laboratory confirmations of brucellosis are based upon serological tests. The results of serology must be interpreted with caution and only in the context of a detailed clinical history, particularly with reference to foreign travel and occupational exposure, and full physical examination. A variety of agglutination techniques have been employed in the diagnosis of brucellosis. The standard agglutination test (SAT) is well standardised, and has been widely employed. No single titre can be regarded as diagnostic, but most cases of active infection will have titres of 1:160 or greater.

A decline in IgG titres in response to therapy is regarded as a good prognostic sign. Equally a rise in titre may precede a relapse of the disease. Microagglutination techniques have also been used as a screening test for infection with the organism. Recently, a number of ELISA-based tests have become available, with the promise of increased sensitivity and the potential for automation. As yet these tests have not entirely replaced more traditional methods. Other tests for the serodiagnosis of *Brucella* infections include the Rose Bengal test, complement fixation and gel precipitation tests.

References

Moyer NP, Holcomb LA, Hausler WJ, Penner JL. In: Balows A, Hausler WJ, Herrmann KL, Isenberg HD, Shadomy HJ, eds. Manual of Clinical Microbiology, 5th edn. Washington: ASM; 1991:457–62.

Young EJ. *Brucella* species. In: Mandell GI, Bennett JE, Dolin R, eds. Principles and Practice of Infectious Diseases, 4th edn. New York: Churchill Livingstone; 1995:2053–9.

Brucella species
ANTIBODY DETECTION (IgG)

Manufacturer: Alfa Biotech
Cat. No./Trade name: 05772957/Brucella IgG Elisa System

SUMMARY

[Well-Ag]–**Ab**–[AHIgG-HRP]–[TMB]–A_{450}

Assay type: EIA (non-competitive)
Detection: Colorimetric A_{450}
Format: Microtitre well, Ag coated
Sample type: Serum
Sample pre-treatment:
    None
Sample volume: 100 µl of 1:100 dilution x 2
Number of tests: 96
Controls - standards run in assay:
    Controls: Neg (2), cut-off (4), Pos (2)
Incubation:
    20 min (37°C) + 20 min (37°C) + 20 min (37°C)
Washes: 2

CONTENTS

Antibodies, antigens, labelled components:
    Brucella species Ag bound to well
    Anti-human IgG Ab (g) HRP conjugated
Substrate: TMB
Controls - standards supplied:
    Controls: Neg, cut-off and Pos (human serum)
Additional reagents:
    None
Special equipment required:
    None

INTERPRETATION

Comments on interpretation:
    Equivocal: within cut-off ± 15%; retest to confirm
    Repeatably equivocal: retest a fresh sample in 2 weeks
No. of references: 4

NOTES

131098.0

Brucella species
ANTIBODY DETECTION (IgG)

Manufacturer: Clark Laboratories
Cat. No./Trade name: Brucella IgG Elisa Test

SUMMARY

[Well-Ag]–**Ab**–[AHIgG-HRP]–[OPD]–A_{490}

Assay type: EIA (non-competitive)
Detection: Colorimetric A_{490}
Format: Microtitre well, Ag coated
Sample type: Serum (do not heat inactivate)
Sample pre-treatment:
    None
Sample volume: 10 µl (+200 µl diluent)
Number of tests: 96
Controls - standards run in assay:
    Controls: Neg (1), Pos (1)
    Calibrators: 1 (2)
Incubation:
    20 min (RT) + 20 min (RT) + 10 min (RT)
Washes: 2

CONTENTS

Antibodies, antigens, labelled components:
    Brucella species Ag bound to well
    Anti-human IgG Ab (g) HRP conjugated
Substrate: TMB*
Controls - standards supplied:
    Controls: Neg and Pos (human serum or plasma)
    Calibrator: 1 (human serum or plasma)
Additional reagents:
    H_2SO_4
Special equipment required:
    None

INTERPRETATION

Comments on interpretation:
    Classification of samples is according to cut-off; no
        further testing
No. of references: None

NOTES

131657.0

*OPD can be supplied as an alternative

© KLUWER ACADEMIC PUBLISHERS 1995, ISSN 1381-5067

Brucella species
ANTIBODY DETECTION (IgG)

Manufacturer: Immunobiological Laboratories
Cat. No./Trade name: VE 57421/Brucellosis IgG EIA Kit

SUMMARY

[Well-Ag]–**Ab**–[AHIgG-HRP]–[TMB]–A_{450}

Assay type: EIA (non-competitive)
Detection: Colorimetric A_{450}
Format: Microtitre well, Ag coated
Sample type: Serum, plasma
Sample pre-treatment:
  It is advised to pre-incubate samples with an absorbent to remove interference due to RF, IgG and ANA (bought separately)
Sample volume: 100 μl of 1:100 dilution x 2
Number of tests: 96
Controls - standards run in assay:
  Controls: Neg (2), low Pos (2), high Pos (2)
Incubation:
  1 hr (RT) + 30 min (RT) + 15 min (RT)
Washes: 2

CONTENTS

Antibodies, antigens, labelled components:
  Brucella species Ag bound to well
  Anti-human IgG Ab HRP conjugated
Substrate: TMB
Controls - standards supplied:
  Controls: Neg, low Pos and high Pos (serum)
Additional reagents:
  Absorbent for removal of RF, IgG and ANA (bought separately)
Special equipment required:
  None

INTERPRETATION

Comments on interpretation:
  Equivocal: within cut-off ±20%; no data is provided for further assessment of samples within this range
No. of references: 9

NOTES

131247.0

Brucella species
ANTIBODY DETECTION (IgG)

Manufacturer: PanBio
Cat. No./Trade name: BAG-100/Brucella IgG Elisa Test

SUMMARY

[Well-Ag]–**Ab**–[AHIgG-HRP]–[TMB]–A_{450}

Assay type: EIA (non-competitive)
Detection: Colorimetric A_{450}
Format: Microtitre well, Ag coated
Sample type: Serum
Sample pre-treatment:
  None
Sample volume: 100 μl of 1:100 dilution
Number of tests: 96
Controls - standards run in assay:
  Controls: Neg (1), Pos (1)
  Calibrator: cut-off (2)
Incubation:
  20 min (RT) + 20 min (RT) + 10 min (RT)
Washes: 2

CONTENTS

Antibodies, antigens, labelled components:
  Brucella species Ag bound to well
  Anti-human IgG Ab (sh) HRP conjugated
Substrate: TMB
Controls - standards supplied:
  Controls: Neg and Pos (human serum)
  Calibrator: cut-off (human serum)
Additional reagents:
  None
Special equipment required:
  None

INTERPRETATION

Comments on interpretation:
  Equivocal: where the ratio of cut-off value:sample OD is between 0.9-1.1; retest to confirm
No. of references: 3

NOTES

131397.0

© KLUWER ACADEMIC PUBLISHERS 1995, ISSN 1381-5067

Brucella species
ANTIBODY DETECTION (IgG AND/OR IgM))

Manufacturer: Immuno Pharmacology Research
Cat. No./Trade name: BRUCELLA IgG/IgM ELISA

SUMMARY

[Well-Ag]–**Ab**–[AHIgG/IgM-AP]–[PNP]–A_{405}

Assay type: EIA (non-competitive)
Detection: Colorimetric A_{405}
Format: Microtitre well, Ag coated
Sample type: Serum
Sample pre-treatment:
   None
Sample volume: 100 μl of 1:100 dilution x 2
Number of tests: 48, 96
Controls - standards run in assay:
   Controls: Neg (1), Pos (1)
Incubation:
   1 hr (37°C) + 1 hr (37°C) + 30 min (37°C)
Washes: 2

CONTENTS

Antibodies, antigens, labelled components:
   Brucella species Ag bound to well
   Anti-human IgG Ab AP conjugated
   Anti-human IgM Ab AP conjugated
Substrate: PNP
Controls - standards supplied:
   Controls: Neg and Pos
Additional reagents:
   None
Special equipment required:
   None

INTERPRETATION

Comments on interpretation:
   Classification of samples is according to cut-off; no
      further testing
No. of references: 1

NOTES

131328.0

This kit can detect IgG and IgM antibodies and can
   differentiate between them

Brucella species
ANTIBODY DETECTION (IgG AND/OR IgM)

Manufacturer: Scimedix Corporation
Cat. No./Trade name: BR 96/Brucella IgG/IgM EIA Test System

SUMMARY

[Well-Ag]–**Ab**–[AHIgG/IgM-POD]–[TMB]–A_{450}

Assay type: EIA (non-competitive)
Detection: Colorimetric A_{450}
Format: Microtitre well, Ag coated
Sample type: Serum
Sample pre-treatment:
   None
Sample volume: 100 μl of 1:100 dilution
Number of tests: 96
Controls - standards run in assay:
   Controls: Neg (1), Pos (1)
Incubation:
   1 hr (RT) + 30 min (RT) + 15 min (RT)
Washes: 2

CONTENTS

Antibodies, antigens, labelled components:
   Brucella Ag bound to well
   Anti-human IgG Ab (g) POD conjugated
   Anti-human IgM Ab (g) POD conjugated
Substrate: TMB
Controls - standards supplied:
   Controls: Neg (IgG/IgM) and Pos (IgG/IgM) (human)
Additional reagents:
   None
Special equipment required:
   None

INTERPRETATION

Comments on interpretation:
   Classification of samples is according to cut-off; no
      further testing
No. of references: 5

NOTES

131190.0

For research and investigational use only
This kit can detect IgG and IgM antibodies and can
   differentiate between them

© KLUWER ACADEMIC PUBLISHERS 1995, ISSN 1381-5067

Brucella species
ANTIBODY DETECTION (IgM)

Manufacturer: Alfa Biotech
Cat. No./Trade name: 05772958/Brucella IgM Elisa System

SUMMARY

[Well-Ag]–**Ab**–[AHIgM-HRP]–[TMB]–A_{450}

Assay type: EIA (non-competitive)
Detection: Colorimetric A_{450}
Format: Microtitre well, Ag coated
Sample type: Serum
Sample pre-treatment:
  None
Sample volume: 100 µl of 1:100 dilution x 2
Number of tests: 96
Controls - standards run in assay:
  Controls: Neg (2), cut-off (4), Pos (2)
Incubation:
  20 min (37°C) + 20 min (37°C) + 20 min (37°C)
Washes: 2

CONTENTS

Antibodies, antigens, labelled components:
  Brucella Ag bound to well
  Anti-human IgM Ab (g) HRP conjugated
Substrate: TMB
Controls - standards supplied:
  Controls: Neg, cut-off and Pos (human serum)
Additional reagents:
  None
Special equipment required:
  None

INTERPRETATION

Comments on interpretation:
  Equivocal: within cut-off ± 15%; retest to confirm
  Repeatably equivocal: retest a fresh sample in 2 weeks
No. of references: 4

NOTES

131104.0

Brucella species
ANTIBODY DETECTION (IgM)

Manufacturer: Clark Laboratories
Cat. No./Trade name: Brucalla IgM Elisa Test

SUMMARY

[Well-Ag]–**Ab**–[AHIgM-HRP]–[OPD]–A_{490}

Assay type: EIA (non-competitive)
Detection: Colorimetric A_{490}
Format: Microtitre well, Ag coated
Sample type: Serum (do not heat inactivate)
Sample pre-treatment:
  Add Absorbent Solution provided to diluted serum and incubate 20 min (RT) to eliminate interference due to RF and IgG
Sample volume: 100 µl of treated sample
Number of tests: 96
Controls - standards run in assay:
  Controls: Neg (1), Pos (1)
  Calibrators: 1 (2)
Incubation:
  20 min (RT) + 20 min (RT) + 10 min (RT)
Washes: 2

CONTENTS

Antibodies, antigens, labelled components:
  Brucella species Ag bound to well
  Anti-human IgM Ab (g) HRP conjugated
Substrate: TMB*
Controls - standards supplied:
  Controls: Neg and Pos (human serum or plasma)
  Calibrator: 1 (human serum or plasma)
Additional reagents:
  H_2SO_4
Special equipment required:
  None

INTERPRETATION

Comments on interpretation:
  Classification of samples is according to cut-off; no further testing
No. of references: None

NOTES

131663.0

*OPD can be supplied as an alternative

© *KLUWER ACADEMIC PUBLISHERS 1995, ISSN 1381-5067*

Brucella species
ANTIBODY DETECTION (IgM)

Manufacturer: Immunobiological Laboratories
Cat. No./Trade name: VE 57431/Brucellosis IgM EIA Kit

SUMMARY

[Well-Ag]–**Ab**–[AHIgM-HRP]–[TMB]–A_{450}

Assay type: EIA (non-competitive)
Detection: Colorimetric A_{450}
Format: Microtitre well, Ag coated
Sample type: Serum, plasma
Sample pre-treatment:
    None
Sample volume: 100 µl of 1:100 dilution x 2
Number of tests: 96
Controls - standards run in assay:
    Controls: Neg (2), low Pos (2), high Pos (2)
Incubation:
    1 hr (RT) + 30 min (RT) + 15 min (RT)
Washes: 2

CONTENTS

Antibodies, antigens, labelled components:
    Brucella species Ag bound to well
    Anti-human IgM Ab HRP conjugated
Substrate: TMB
Controls - standards supplied:
    Controls: Neg, low Pos and high Pos (serum)
Additional reagents:
    None
Special equipment required:
    None

INTERPRETATION

Comments on interpretation:
    Equivocal: within cut-off ±20%; no data is provided for
        further assessment of samples within this range
No. of references: 9

NOTES

131255.0

Brucella species
ANTIBODY DETECTION (IgM)

Manufacturer: PanBio
Cat. No./Trade name: BAM-200/Brucella IgM Elisa Test

SUMMARY

[Well-Ag]–**Ab**–[AHIgM-HRP]–[TMB]–A_{450}

Assay type: EIA (non-competitive)
Detection: Colorimetric A_{450}
Format: Microtitre well, Ag coated
Sample type: Serum
Sample pre-treatment:
    Mix diluted serum with absorbent solution, incubate 15
        min (RT) to eliminate interference due to RF & IgG
Sample volume: 100 µl of prepared sample
Number of tests: 96
Controls - standards run in assay:
    Controls: Neg (1), Pos (1)
    Calibrator: cut-off (2)
Incubation:
    20 min (RT) + 20 min (RT) + 10 min (RT)
Washes: 2

CONTENTS

Antibodies, antigens, labelled components:
    Brucella species Ag bound to well
    Anti-human IgM Ab (sh) HRP conjugated
Substrate: TMB
Controls - standards supplied:
    Controls: Neg and Pos (human serum)
    Calibrator: cut-off (human serum)
Additional reagents:
    None
Special equipment required:
    None

INTERPRETATION

Comments on interpretation:
    Equivocal: where the ratio of cut-off value:sample OD is
        between 0.9-1.1; retest to confirm
No. of references: 3

NOTES

131403.0

Campylobacter species

Campylobacters are comma shaped, non-spore forming Gram-negative rods, motile by means of a single polar flagellum at one or both ends of the cell. Infections with these organisms are amongst the most common bacterial infections of humans worldwide. Infections may manifest themselves as both diarrhoeal and systemic illness. Gastrointestinal infections are associated with diarrhoea, which is often bloody, abdominal pain, fever, and vomiting and are most commonly the result of *C.jejuni* or *C.coli* infection, although *C.lari*, *C.hyointestinalis* and *C.upsaliensis* may also be enteropathogenic. Although transient bacteraemia is occasionally seen in enteric infections, serious extraintestinal illness is less common and usually occurs in the compromised host. This may result in a septicaemic illness which may be complicated by meningitis, vascular infections and abscesses, and is most frequently observed in *C.fetus* infections although *C.jejuni* and other campylobacters may be implicated. *Campylobacter* infections are essentially zoonotic in nature, and are most commonly acquired through consumption of contaminated water or food, particularly undercooked meat and unpasteurised milk, although case to case spread can occur amongst family members and within institutions.

Campylobacters and related organisms are strictly microaerophilic and require an atmosphere of 5–10% CO_2 for growth. Cultures are often incubated at 42°C which favours the isolation of the thermophilic *C.jejuni*, *C.coli* and *C.lari*, although this may result in the failure to isolate less thermotolerant campylobacters. A variety of selective media for the isolation of these organisms have been described which contain blood or charcoal to remove toxic oxygen metabolites, and a variety of antibiotics to prevent overgrowth with the more rapidly-growing organisms of the faecal flora. The organisms form small, flat, non-haemolytic grey or colourless colonies after 48–72 h incubation. Presumptive identification may be made on the basis of a positive oxidase test, Gram stain morphology and characteristic darting motility in a wet preparation. Further differentiation and biotyping may be performed on the basis of biochemical tests and serotyping schemes for epidemiological studies have been described.

Diagnosis

Definitive diagnosis of *Campylobacter* infection is by means of isolation of the organism as described above. The organisms are most commonly cultured from blood or faeces and occasionally from other extraintestinal sites. The organisms may also be recovered from contaminated food and water. A variety of serodiagnostic tests have been described, including CFT and ELISA-based systems, but are not in widespread use. There is currently considerable interest in the food industry in the development and application of novel methods for the rapid detection of Campylobacter in meat and other food products.

References

Skirrow MB. Epidemiology of *Campylobacter* enteritis. Int J Food Microbiol. 1991;12:9–16.

Penner JL. *Campylobacter, Helicobacter* and related spiral bacteria. In: Balows A, Hausler WJ, Herrmann KL, Isenberg HD, Shadomy HJ, eds. Manual of Clinical Microbiology, 5th edn. Washington: ASM; 1991:402–9.

Campylobacter species
ANTIGEN DETECTION

Manufacturer: Meridian Diagnostics Inc
Cat. No./Trade name: 203050/Meritec® Campy (jcl)

SUMMARY

[Latex-Ab]–**Ag**–agglutination

Assay type: Particle agglutination assay
Detection: Visual
Format: Latex particles, Ab coated (test done on card)
Sample type: Culture isolates (from faecal specimens or rectal swabs)
Sample pre-treatment:
  None
Sample volume: 1 colony (+ 1 drop of Extraction Reagent)
Number of tests: 50
Controls - standards run in assay:
  Controls: Pos (1)
Incubation:
  5 min (RT) with rotation
Washes:

CONTENTS

Antibodies, antigens, labelled components:
  Anti-Campylobacter (*C. jejuni*, *C. coli* and *C. laridis*) Ab bound to latex particles (Latex Reagent)
Substrate:
Controls - standards supplied:
  Controls: Pos
Additional reagents:
  None
Special equipment required:
  None

INTERPRETATION

Comments on interpretation:
  Positive: agglutination of sample and Latex Reagent
No. of references: 20

NOTES

131388.0

© KLUWER ACADEMIC PUBLISHERS 1995, ISSN 1381-5067

Clostridium difficile

Clostridium difficile is the causal organism of a predominantly endogenous gastrointestinal infection which is characterised by its association with the prior administration of antibiotics. It is now well recognised that the spectrum of disease produced may be very wide, ranging from mild diarrhoeal illness to severe life-threatening pseudo-membranous colitis (PMC). Systemic infection with this organism is rare and, in common with other clostridial diseases, pathogenesis is thought to be the result of the action of two major protein exotoxins A and B which respectively result in loss of fluid from the gut mucosa and a direct local cytopathic effect. Between 2–3% of healthy adults carry the organism in their gastrointestinal tract, but this figure rises to 36% in the hospitalised population. Neonates and elderly patients may have even higher rates of carriage, although in the former group this does not seem to cause clinical disease. It is thought that antibiotic therapy results in an alteration of the gut flora which permits the overgrowth of *Clostridium difficile* which is already present in the patient's gut, or which may be acquired nosocomially from the hospital environment.

Clostridium difficile is an anaerobic, Gram-positive bacillus which forms oval sub-terminal spores. The ability of these to resist heat and desiccation, and hence their ability to persist in the hospital environment, are thought to account for the epidemiology of the infection. Isolation of the organism is normally performed on cycloserine-cefoxitin fructose agar (CCFA), and yields yellowish, irregular flat colonies after 48 hours anaerobic incubation. Isolation rates may be increased by the use of alcohol or heat shock prior to culture. Presumptive identification is based upon characteristic morphology, odour and yellow-green fluorescence under long wavelength ultraviolet light. Definitive identification may be based upon characterisation by biochemical testing and GLC profiles of short-chain fatty acid metabolic products. A variety of typing methods have been described for the epidemiological investigation of outbreaks of infection.

Diagnosis

Although *Clostridium difficile* may be cultured from faeces as outlined above, the isolation of the organism does not establish the diagnosis of *Clostridium difficile* mediated gastrointestinal disease. Asymptomatic carriage can occur as outlined above, and strains may be non-toxigenic. In view of the central role of *Clostridium difficile* toxins in pathogenesis, it is now recommended that toxin detection should play a major role in the investigation of suspected infection in the symptomatic patient. Toxin detection alone is adequate in the investigation of sporadic cases, but in outbreak and epidemiological investigations culture for the organism is also required.

Toxin detection has formerly relied upon the use of cytotoxicity assays, with the demonstration of inhibition of the cytopathic effect by *Clostridium sordellii* antitoxin. Whilst cell culture-based systems are still in routine use, a number of commercial toxin detection systems are based predominantly upon ELISA technology are now available. These techniques are particularly useful in those laboratories where cell culture facilities are unavailable.

References

Allen SD, Baron SJ. *Clostridium*. In: Balows A, Hausler WJ, Herrmann KL, Isenberg HD, Shadomy HJ, eds. Manual of Clinical Microbiology, 5th edn. Washington: ASM;1991:505–21.

DH/PHLS Joint Working Group. *Clostridium difficile* Infection:Prevention and Management. London: PHLS;1994:1–49.

Clostridium difficile
ANTIGEN DETECTION

Manufacturer: Meridian Diagnostics Inc
Cat. No./Trade name: 706050/ImmunoCard® C. difficile

SUMMARY

[Membrane-Ab]–**Ag**–[Ab-AP]–[BCIP]–blue dot

Assay type: Immunochromatographic assay
Detection: Visual
Format: Reaction unit, Ab coated membrane
Sample type: Faecal specimens
Sample pre-treatment:
None
Sample volume: 2 mm portion (solid sample), 25 µl (liquid sample)
Number of tests: 50
Controls - standards run in assay:
Controls: Neg (1), Pos (1)
Incubation:
3 min (RT) + 3 min (RT)
Washes: 1

CONTENTS

Antibodies, antigens, labelled components:
Anti-*Cl. difficile* Ab bound to membrane (Test port)
Anti-*Cl. difficile* Ab AP conjugated
Substrate: BCIP
Controls - standards supplied:
Controls: Neg (buffer) and Pos
Integral procedural controls incorporated in membrane (Control port)
Additional reagents:
None
Special equipment required:
None

INTERPRETATION

Comments on interpretation:
Negative: blue colour in Control port but no colour in Test port
Positive: blue colour in Control and Test ports
Invalid: no colour in Control port; repeat test
No. of references: 28

NOTES

131382.0

Clostridium difficile
ANTIGEN DETECTION

Manufacturer: Becton Dickinson
Cat. No./Trade name: 4364001/Culturette Brand CDT

SUMMARY

[Latex-Ab]–**Ag**–agglutination

Assay type: Particle agglutination assay
Detection: Visual
Format: Latex particles, Ab coated (test done on slide)
Sample type: Faecal specimens
Sample pre-treatment:
Mix with sample buffer and centrifuge
Sample volume: 1 drop of prepared sample x 2
Number of tests: 25
Controls - standards run in assay:
Controls: Pos (1)
Incubation:
Read result 3 min after mixing sample and reagents
Washes:

CONTENTS

Antibodies, antigens, labelled components:
Anti-*Cl. difficile* Ab (r) bound to latex particles (Detection Latex)
Non-immune rabbit Ig Ab bound to latex particles (Control Latex)
Substrate:
Controls - standards supplied:
Controls: Pos
Additional reagents:
None
Special equipment required:
None

INTERPRETATION

Comments on interpretation:
Positive: agglutination with the Detection Latex and no agglutination with the Control Latex
No. of references: 9

NOTES

131206.0

Clostridium difficile
ANTIGEN DETECTION

Manufacturer: Meridian Diagnostics Inc
Cat. No./Trade name: 204030/Meritec® - C. difficile

SUMMARY

[Latex-Ab]–**Ag**–agglutination

Assay type: Particle agglutination assay
Detection: Visual
Format: Latex particles, Ab coated (test done on card)
Sample type: Faecal specimens
Sample pre-treatment:
   Mix with diluent, vortex, centrifuge
Sample volume: 50 µl of prepared sample x 2
Number of tests: 30
Controls - standards run in assay:
   Controls: Pos (1)
Incubation:
   3 min (RT) with rotation
Washes:

CONTENTS

Antibodies, antigens, labelled components:
   Anti-*Cl. difficile* Ab (r) bound to latex particles (Detection Latex)
   Non-immune rabbit Ig Ab bound to latex particles (Control Latex)
Substrate:
Controls - standards supplied:
   Controls: Pos
Additional reagents:
   None
Special equipment required:
   None

INTERPRETATION

Comments on interpretation:
   Negative: no agglutination of sample with Detection Latex or Control Latex
   Positive: agglutination of sample with Detection Latex and no agglutination with Control Latex
No. of references: 19

NOTES

131389.0

Clostridium difficile
ANTIGEN DETECTION

Manufacturer: Microgen Bioproducts
Cat. No./Trade name: M41/MicroScreen® C. Difficile Latex Slide Agglutination Test

SUMMARY

[Latex-Ab]–**Ag**–agglutination

Assay type: Particle agglutination assay
Detection: Visual
Format: Latex particles, Ab coated (test done on slide)
Sample type: Broth culture or colonies from solid media culture (isolated from faecal specimens)
Sample pre-treatment:
   None
Sample volume: 1 drop of broth or 2-3 colonies (+ 1 drop saline)
Number of tests: 50
Controls - standards run in assay:
   Controls: Pos (1)
Incubation:
   Read result within 2 min of mixing sample and reagents
Washes:

CONTENTS

Antibodies, antigens, labelled components:
   Anti-*Cl. difficile* (cell wall) IgG Ab (r) bound to latex (Test Latex)
Substrate:
Controls - standards supplied:
   Controls: Pos
Additional reagents:
   None
Special equipment required:
   None

INTERPRETATION

Comments on interpretation:
   Positive: agglutination of sample with Test Latex
No. of references: 8

NOTES

131194.0

Clostridium difficile (toxin A)
ANTIGEN DETECTION

Manufacturer: Alexon Inc
Cat. No./Trade name: 720-48/ProSpecT® Clostridium difficile Microplate Assay

SUMMARY

[Well-Ab]–**Ag**–[Ab]–[ARIg-POD]–[TMB]–A_{450}

Assay type: EIA (non-competitive)
Detection: Colorimetric A_{450}
Format: Microtitre well, Ab coated
Sample type: Faecal specimens
Sample pre-treatment:
 Mix specimens with diluent, centrifuge
Sample volume: 100 µl of diluted sample
Number of tests: 48, 96
Controls - standards run in assay:
 Controls: Neg (3), Pos (1)
Incubation:
 1 hr 30 min (37°C) + 30 min (37°C) + 15 min (RT)
Washes: 2

CONTENTS

Antibodies, antigens, labelled components:
 Anti-*Cl. difficile* toxin A IgG Ab (m) bound to wells
 Anti-*Cl. difficile* toxin A Ig Ab (r)
 Anti-rabbit Ig Ab (g) POD conjugated
Substrate: TMB
Controls - standards supplied:
 Controls: Neg and Pos
Additional reagents:
 None
Special equipment required:
 None

INTERPRETATION

Comments on interpretation:
 Equivocal: within designated range; retest to confirm
No. of references: 16

NOTES

131301.0

Clostridium difficile (toxin A)
ANTIGEN DETECTION

Manufacturer: Becton Dickinson
Cat. No./Trade name: Culturette® Brand Toxin CD

SUMMARY

[Well-Ab]–**Ag**–[Ab-HRP]–[TMB]–A_{450}

Assay type: EIA (non-competitive)
Detection: Colorimetric A_{450}
Format: Microtitre well, Ab coated
Sample type: Faecal specimens
Sample pre-treatment:
 Mix sample with sample buffer and vortex
Sample volume: 10 µl of prepared sample
Number of tests: 96
Controls - standards run in assay:
 Controls: Neg (1), low Pos (1), high Pos (1)
Incubation:
 1 hr (37°C) + 10 min (RT)
Washes: 1

CONTENTS

Antibodies, antigens, labelled components:
 Anti-*Cl. difficile* toxin A Ab (r) bound to well
 Anti-*Cl. difficile* toxin A Ab (r) HRP conjugated
Substrate: TMB
Controls - standards supplied:
 Controls: Neg, low Pos and high Pos
Additional reagents:
 None
Special equipment required:
 None

INTERPRETATION

Comments on interpretation:
 Equivocal: where OD is > 0.07 but < 0.100; retest to confirm
 Repeatably equivocal: repeat test using a fresh specimen
No. of references: 11

NOTES

131205.0

© *KLUWER ACADEMIC PUBLISHERS 1995, ISSN 1381-5067*

Clostridium difficile (toxin A)
ANTIGEN DETECTION

Manufacturer: bioMerieux
Cat. No./Trade name: 30 103/VIDAS C. difficile TOXIN A

SUMMARY

[Solid phase-Ab]–**Ag**–[Ab]-[AMIg-AP]–[4MP]–
fluorescence

Assay type: EIA (non-competitive)
Detection: Fluorometric
Format: Solid phase receptacle (SPR), Ab coated
Sample type: Faecal specimens
Sample pre-treatment:
Mix with Sample Treatment Reagent provided, centrifuge
and use supernatant for assay
Sample volume: 300 μl of prepared sample
Number of tests: 30
Controls - standards run in assay:
Controls: Neg (1), Pos (1)
Incubation:
Automated: total time 2 hr 30 min
Washes: Automated

CONTENTS

Antibodies, antigens, labelled components:
Anti-*Cl. difficile* toxin A Ab (r) bound to solid phase
receptacle
Anti-*Cl. difficile* toxin A Ab (m)
Anti-mouse Ig Ab (r) AP conjugated
Substrate:
Controls - standards supplied:
Controls: Neg and Pos
Additional reagents:
None
Special equipment required:
Vitek immunodiagnostic assay system (VIDAS)

INTERPRETATION

Comments on interpretation:
This is an automated assay
Results designated equivocal; retest original or fresh
sample to confirm
No. of references: None

NOTES

131313.0

Clostridium difficile (toxin A)
ANTIGEN DETECTION

Manufacturer: Dade International Inc (Bartels Division)
Cat. No./Trade name: B1029-69/Bartels Clostridium
difficile Toxin A EIA

SUMMARY

[Well-Ab]–**Ag**–[Ab]–[ARIg-POD]–[TMB]–A$_{450}$

Assay type: EIA (non-competitive)
Detection: Colorimetric A$_{450}$
Format: Microtitre well, Ab coated
Sample type: Faecal samples
Sample pre-treatment:
Mix specimen with buffer, vortex, remove supernatant
and centrifuge
Sample volume: 100 μl of prepared sample
Number of tests: 96
Controls - standards run in assay:
Controls: Neg (1), Pos (1)
Incubation:
1 hr 30 min (34–37°C) + 30 min (34–37°C) + 15 min
(RT)
Washes: 2

CONTENTS

Antibodies, antigens, labelled components:
Anti-*Cl. difficile* toxin A IgG Ab (m) bound to well
Anti-*Cl. difficile* toxin A Ig Ab (r)
Anti-rabbit Ig Ab (g) POD conjugated
Substrate: TMB
Controls - standards supplied:
Controls: Neg and Pos
Additional reagents:
None
Special equipment required:
None

INTERPRETATION

Comments on interpretation:
Classification of samples is according to cut-off; no
further testing
No. of references: 4

NOTES

131050.0

© KLUWER ACADEMIC PUBLISHERS 1995, ISSN 1381-5067

Clostridium difficile (toxin A)
ANTIGEN DETECTION

Manufacturer: Meridian Diagnostics Inc
Cat. No./Trade name: 601096/Premier® C. difficile Toxin A

SUMMARY

[Well-Ab]–**Ag**–[MAb-AP]–[TMB]–A_{450}

Assay type: EIA (non-competitive)
Detection: Colorimetric A_{450}
Format: Microtitre well, Ab coated
Sample type: Faecal specimens
Sample pre-treatment:
  Mix specimen with sample diluent and vortex
Sample volume: 50 µl of prepared sample
Number of tests: 96
Controls - standards run in assay:
  Controls: Neg (1), Pos (1)
Incubation:
  (a) 2 hr (35–39°C) + 10 min (RT)
  (b) 1 hr (37°C) on plate shaker + 10 min (RT)
Washes: 1

CONTENTS

Antibodies, antigens, labelled components:
  Anti-*Cl. difficile* toxin A Ab bound to well
  Anti-*Cl. difficile* toxin A MAb HRP conjugated
Substrate: TMB
Controls - standards supplied:
  Controls: Neg and Pos
Additional reagents:
  None
Special equipment required:
  Stat Fax® - 2220 Incubator/shaker (optional)

INTERPRETATION

Comments on interpretation:
  Equivocal: where sample OD is ⩾0.100 but <0.150; retest to confirm
  Repeatably equivocal: repeat test with a fresh specimen
No. of references: 20

NOTES

131386.0

Clostridium difficile (toxin A)
ANTIGEN DETECTION

Manufacturer: Shield Diagnostics
Cat. No./Trade name: FCDT 100/CD-TOX®

SUMMARY

[Well-Ab]–**Ag**–[Ab-POD]–[TMB]–A_{450}

Assay type: EIA (non-competitive)
Detection: Colorimetric A_{450}
Format: Microtitre well, Ab coated
Sample type: Faecal specimen
Sample pre-treatment:
  Mix specimen with diluent, vortex
Sample volume: 50 µl of treated sample
Number of tests: 96
Controls - standards run in assay:
  Controls: Neg (1), low Pos (2), high Pos (1)
Incubation:
  1 hr (37°C) + 10 min (RT)*
Washes: 1

CONTENTS

Antibodies, antigens, labelled components:
  Anti-*Cl. difficile* toxin A Ab (r) bound to well
  Anti-*Cl. difficile* toxin A Ab (r) POD conjugated
Substrate: TMB
Controls - standards supplied:
  Controls: Neg, low Pos and high Pos
Additional reagents:
  None
Special equipment required:
  None

INTERPRETATION

Comments on interpretation:
  Classification of samples is according to cut-off; no further testing
No. of references: 12

NOTES

131714.0

*Shaking is recommended at 37°C but not essential

Clostridium difficile (toxin A and toxin B)
ANTIGEN DETECTION

Manufacturer: Cambridge Biotech Corporation
Cat. No./Trade name: 96087/Cytoclone® A+B EIA

SUMMARY

[Well-MAb]–**Ag**–[Ab-biotin]–[strept-HRP]–[TMB]–A$_{450}$

Assay type: EIA (non-competitive)
Detection: Colorimetric A$_{450}$
Format: Microtitre well, Ab coated
Sample type: Faecal sample
Sample pre-treatment:
  Add sample diluent to specimen
Sample volume: 100 µl of prepared sample
Number of tests: 96
Controls - standards run in assay:
  Controls: Neg (1), Pos (1)
Incubation:
  1 hr (37°C) + 15 min (RT) + 15 min (RT)
Washes: 2

CONTENTS

Antibodies, antigens, labelled components:
  Anti-*Cl. difficile* toxin A and toxin B MAbs (m) bound to well
  Anti-*Cl. difficile* toxin A and toxin B Abs (g) biotinylated Streptavidin HRP conjugated
Substrate: TMB
Controls - standards supplied:
  Controls: Neg (sample diluent) and Pos (denatured toxins)
Additional reagents:
  None
Special equipment required:
  None

INTERPRETATION

Comments on interpretation:
  Equivocal: cut-off between 0.2 and 0.25; retest to confirm
No. of references: 15

NOTES

131183.0

This kit can detect *Cl. difficile* toxin A and toxin B but does not differentiate between them

Clostridium difficile (toxin A and toxin B)
ANTIGEN DETECTION

Manufacturer: r-biopharm GmbH
Cat. No./Trade name: RIDASCREEN® Clostridium difficile Toxin A/B

SUMMARY

[Well-MAb]–**Ag**–[Ab-biotin]–[strept-POD]–[TMB]–A$_{450}$

Assay type: EIA (non-competitive)
Detection: Colorimetric A$_{450}$
Format: Microtitre well, Ab coated
Sample type: Faecal specimens
Sample pre-treatment:
  Add sample diluent to specimen and mix
Sample volume: 50 µl of diluted sample
Number of tests: 48
Controls - standards run in assay:
  Controls: Neg (1), Pos (1)
Incubation:
  1 hr 30 min (RT) + 30 min (RT) + 30 min (RT)
Washes: 2

CONTENTS

Antibodies, antigens, labelled components:
  Anti-*Cl. difficile* toxin A and toxin B Abs bound to well
  Anti-*Cl. difficile* toxin A and toxin B Abs biotinylated Streptavidin POD conjugated
Substrate: TMB
Controls - standards supplied:
  Controls: Neg (sample diluent) and Pos (denatured toxin A and B)
Additional reagents:
  None
Special equipment required:
  None

INTERPRETATION

Comments on interpretation:
  Classification of sample is according to cut-off; no further testing
No. of references: None

NOTES

131153.0

This kit can detect *Cl. difficile* toxin A and toxin B but does not differentiate between them

© KLUWER ACADEMIC PUBLISHERS 1995, ISSN 1381-5067

Clostridium perfringens

Clostridium perfringens, formerly known as *C.Welchii*, is one of the numerous species of clostridia which have now been described. This organism is almost universally distributed in soil and colonises the intestinal tract of man and all other animals. *Clostridium perfringens* is the most frequent clinical isolate of the genus. The organism is non-motile and the oval, central spores are only rarely observed in routine laboratory cultures. All types produce a potent lecithinase which is detected by means of the Nagler reaction. Although an anaerobe, the organism is relatively aerotolerant and is readily isolated on routine anaerobic media producing a characteristic double zone of haemolysis on blood agar and a typical 'stormy clot' in milk. At least twelve toxins may be elaborated by *C.perfringens* and five types (A–E) may be recognised on the basis of serotyping of the four major lethal toxins. The organism may also elaborate a number of enzymes which are also involved in pathogenesis. Infection manifests itself in the form of three distinct clinical syndromes which are the result of the production of a range of potent protein exotoxins: clostridial myonecrosis (gas gangrene), food poisoning and enteritis necroticans. Clostridial myonecrosis is a fulminant infection with gangrene, marked systemic toxaemia and profuse gas production within the tissues. The condition arises as a result of soil or faecal contamination at the sites of trauma, post abdominal surgery and soft tissue lesions resulting from vascular insufficiency or as the result of malignancy. The main pathogenic effects are thought to be mediated by phospholipase C and a potent haemolysin. In addition hyaluronidase, collagenase, DNase and neuraminidase enzymes may all contribute to the striking clinical presentation. The condition is rapidly progressive and fatal without prompt, widespread surgical resection, with high dose parenteral antibiotics playing a secondary role. *C.perfringens* is frequently implicated in less severe soft tissue sepsis and bacteraemia, particularly following abdominal surgery.

Enterotoxin-producing strains of *C.perfringens* type A may cause food poisoning which arises following the consumption of large numbers of the organism. The most frequent vehicles are cooked meats, stews and gravies which become grossly contaminated following prolonged periods of cooling and storage at raised ambient temperatures. Spores which have survived the cooking process germinate and the organisms may then multiply rapidly. The incubation period is 7–15 hours and the disease is characterised by diarrhoea and abdominal cramps and, less commonly, nausea and vomiting. The disease is generally of short duration. Enteritis

necroticans is a necrotising infection of the small intestine caused by the β-toxin of *C.perfringens* type C. Although important in veterinary medicine, it is uncommon in man with a characteristic geographical and sociocultural distribution in Papua New Guinea and other parts of the Pacific and south-east Asia, where particular dietary and cultural practices predispose to development of the condition.

Diagnosis

Definitive diagnosis of *C.perfringens* infection is by culture of the organism from wound swabs, tissue samples and aspirates, and blood cultures as outlined above. In the context of post-surgical sepsis, the organism may often be isolated as part of a polymicrobial infection. The diagnosis of gas gangrene is essentially clinical, and isolation of the causal organism serves mainly to support the clinical findings.

C.perfringens food-poisoning is usually diagnosed by a combination of criteria which may include isolation of the organism in large numbers from the incriminated food and the detection of large numbers of spores in faecal samples from cases. However large numbers of spores are not uncommon in asymptomatic individuals. More recently, the availability of reverse passive latex agglutination and ELISA-based systems for the direct detection of the enterotoxin in stools have provided a useful adjunct to diagnosis. In a number of outbreaks phage-typing, bacteriocin-typing, plasmid profiles and chromosomal DNA analysis have been employed to demonstrate the identity of patient and food isolates.

Reference

Lorber B. Gas gangrene and other *Clostridium*-associated diseases. In: Mandell GL, Bennett JE, Dolin R, eds. Principles and Practice of Infectious Diseases, 4th edn. New York, Edinburgh, London: Churchill Livingstone. 1995:2182–95.

Clostridium perfringens (enterotoxin, type A)
ANTIGEN DETECTION

Manufacturer: Denka Seiken Co. Ltd
Cat. No./Trade name: PET-RPLA Clostridium perfringens Enterotoxin Test Kit

SUMMARY

[Latex-Ab]–**Ag**–agglutination

Assay type: Particle agglutination assay
Detection: Visual
Format: Latex particles, Ab coated (test done in V-microtitre well)
Sample type: Faecal specimens, culture isolates from clinical samples
Sample pre-treatment:
Extract toxin from (a) faecal specimens – mix with PBS, centrifuge 20 min (4°C), filter; (b) culture isolates – innoculate isolated organism into Cooked Meat Medium, then into medium to promote enterotoxin
Sample volume: 25 μl of prepared sample (+25 μl diluent) x 2
Number of tests:
Controls - standards run in assay:
Controls: Pos (1)
Incubation:
20–24 hr (RT)
Washes:

CONTENTS

Antibodies, antigens, labelled components:
Anti-*Cl. perfringens* enterotoxin type A Ab bound to latex particles (Test Latex)
Non-immune rabbit Ig Ab bound to latex particles (Control Latex)
Substrate:
Controls - standards supplied:
Controls: Pos
Additional reagents:
Cooked meat medium
Special equipment required:
None

INTERPRETATION

Comments on interpretation:
Positive: greater agglutination of sample with Test Latex than with Control Latex
No. of references: 4

NOTES

131615.0

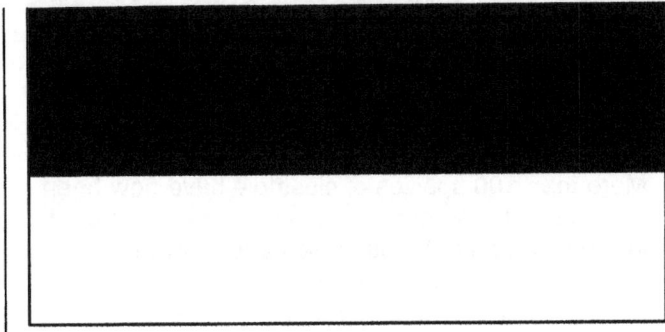

Clostridium tetani

More than 100 species of clostridia have now been described. These organisms are widely distributed in the environment, both in soil and in water, and may also colonise the intestinal tract of man and other animals. *Clostridium tetani* is the causal organism of tetanus, which is an exogenous infection of humans arising as the result of inoculation of spores of the organism at the site of major or minor trauma. The clinical manifestations are entirely the consequence of the action of the potent, plasmid-encoded neurotoxin, tetanospasmin, which is elaborated by the organism and released after autolysis of the cell. This heat-labile protein toxin interferes with the secretion of inhibitory neurotransmitters in motor neurones, resulting in the characteristic prolonged spastic paralysis which is the key feature of the disease. Non-toxigenic, and hence non-pathogenic, isolates lacking the toxin-encoding plasmid can occur. Although the disease is still not uncommon in developing countries, it is entirely preventable by immunisation with inactivated tetanus toxoid.

Clostridium tetani is a Gram-positive strictly anaerobic bacillus. The organism is motile by means of peritrichous flagellae, and forms characteristic round, terminal spores which result in a typical 'drumstick' appearance. Spore formation may be enhanced by alcohol or heat 'shock'. Many strains may produce swarming growth on culture on anaerobic blood agar at 37°C, and subculture from the swarming edge is often required which yields slightly raised, translucent, glistening grey colonies with an irregular edge and a narrow zone of haemolysis. The organism may be further characterised by biochemical testing. Demonstration of the toxin is not usually performed in the routine laboratory, and has most commonly been performed by animal inoculation.

Diagnosis

The diagnosis of tetanus is essentially clinical. In established disease the dramatic clinical presentation leaves little doubt of the underlying aetiology. Diagnostic efforts are then directed at the exclusion of the limited number of intoxicants which may mimic tetanus. Initial presentation may be some time after the initial injury which may itself have been overlooked due to its trivial nature. In the light of the foregoing, it is unusual for the clinical laboratory to be asked to isolate the organism from wound specimens. Equally the demonstration of the organism by microscopy or culture as a coincidental finding in a clinical specimen submitted to the laboratory does not of itself establish a diagnosis of tetanus.

Reference

Hatheway CL. Toxigenic clostridia. Clin Microbiol Rev. 1990;3:66–98.

Collee JG, van Heyningen S. Systemic toxigenic diseases. In: Duerden BI, Drasar BS, eds. Anaerobes in Human Disease. London: Arnold; 1991:372–84.

© KLUWER ACADEMIC PUBLISHERS 1995, ISSN 1381-5067

Clostridium tetani
ANTIBODY DETECTION (IgG)

Manufacturer: Immuno Pharmacology Research
Cat. No./Trade name: ELISA TETANUS IgG

SUMMARY

[Well-Ag]–**Ab**–[AHIgG-AP]–[PNP]–A_{405}

Assay type: EIA (non-competitive)
Detection: Colorimetric A_{405}
Format: Microtitre well, Ag coated
Sample type: Serum
Sample pre-treatment:
   None
Sample volume: 100 μl of 1:50 dilution x 2
Number of tests: 48, 96
Controls - standards run in assay:
   Controls: Neg (1), Pos (1)
Incubation:
   1 hr (37°C) + 1 hr (37°C) + 30 min (37°C)
Washes: 2

CONTENTS

Antibodies, antigens, labelled components:
   Cl. tetani Ag bound to well
   Anti-human IgG Ab AP conjugated
Substrate: PNP
Controls - standards supplied:
   Controls: Neg and Pos
Additional reagents:
   Set of Standards for quantitative assay (optional)
Special equipment required:
   None

INTERPRETATION

Comments on interpretation:
   Classification of samples is according to cut-off; no
      further testing
   A set of standards is available separately to perform a
      quantitative assay if required
No. of references: 8

NOTES

131315.0

Clostridium tetani (toxin)
ANTIBODY DETECTION (IgG)

Manufacturer: Gamma SA
Cat. No./Trade name: Tetanus Toxoid IgG Antibody Test

SUMMARY

[Well-Ag]–**Ab**–[AHIgG-AP]–[PNP]–A_{405}

Assay type: EIA (non-competitive)
Detection: Colorimetric A_{405}
Format: Microtitre well, Ag coated
Sample type: Serum
Sample pre-treatment:
   None
Sample volume: 100 μl of diluted sample
Number of tests: 192
Controls - standards run in assay:
   Controls: R3 (1), R4 (1), R5 (1)
Incubation:
   1 hr (37°C) + 1 hr (37°C) + 30 min (37°C)
Washes: 2 (+ preliminary plate wash)

CONTENTS

Antibodies, antigens, labelled components:
   Cl. tetani toxin Ag bound to well
   Anti-human IgG Ab (r) AP conjugated
Substrate: PNP
Controls - standards supplied:
   Controls: R3, R4 and R5 (5, 10 and 22 IU)
Additional reagents:
   None
Special equipment required:
   None

INTERPRETATION

Comments on interpretation:
   This is a quantitative assay. Antibody levels are
      measured with reference to Standard curve
No. of references: 0

NOTES

131196.0

© *KLUWER ACADEMIC PUBLISHERS 1995, ISSN 1381-5067*

Clostridium tetani (toxin)
ANTIBODY DETECTION (IgG)

Manufacturer: Immunobiological Laboratories
Cat. No./Trade name: /Tetanus Toxoid IgG Elisa Test (HRP)

SUMMARY

[Well-Ag]–**Ab**–[AHIgG-HRP]–[TMB]–A_{450}

Assay type: EIA (non-competitive)
Detection: Colorimetric A_{450}
Format: Microtitre well, Ag coated
Sample type: Serum
Sample pre-treatment:
    None
Sample volume: 100 µl of 1:100 dilution x 2
Number of tests: 96
Controls - standards run in assay:
    Standards: 1 (2), 2 (2), 3 (2), 4 (2), 5 (2), 6 (2), 7 (2), 8 (2)
Incubation:
    1 hr (RT) + 30 min (RT) + 15 min (RT)
Washes: 2

CONTENTS

Antibodies, antigens, labelled components:
    Cl. tetani toxin Ag bound to well
    Anti-human IgG Ab HRP conjugated
Substrate: TMB
Controls - standards supplied:
    Standards: 8 (0, 0.1, 0.5, 1, 5, 10, 25 and 50 IU/ml)
Additional reagents:
    None
Special equipment required:
    None

INTERPRETATION

Comments on interpretation:
    Results are expressed quantitatively with reference to
        the Standards provided
No. of references: 5

NOTES

131254.0

Cryptococcus neoformans

Cryptococcus neoformans is a budding yeast-like fungus which is typically capsulate. It is ubiquitous in the environment throughout the world, but there appears to be high natural resistance and clinical infection with the organism is rare and is usually associated with a degree of immunocompromise,, such as corticosteroid therapy, diabetes, organ transplantation and, recently AIDS. Based on serological, biochemical and genetic differences, two varieties are distinguished: *C.neoformans* var. *neoformans* and *C.neoformans* var. *gatii*. Clinical isolates are usually capsulate, though strains with scanty or virtually absent capsules are occasionally implicated in disease, especially among patients with AIDS.

The route of entry in many cases appears to be the respiratory tract, and in a minority of cases infection appears to be confined mainly to the lungs. The other main site of infection is the central nervous system, resulting in insidiously developing meningoencephalitis. Disseminated infection occurs mainly in the severely immunocompromised, and is associated with poor prognosis. Lesions in the skin and elsewhere are usually associated with disseminated infections. The biological response of the host to infection is very variable, relatively heavy infections being sometimes associated with remarkably little local or systemic inflammatory or immune response. No known exotoxins are elaborated by the organism. Though the polysaccharide capsule is produced in the tissues, often in large amounts, and has been shown *in vitro* to have a number of biological activities such as complement activation and inhibition of phagocytosis, its role *in vivo* in pathogenesis is not clear.

Clinical infections often have an insidious onset and a prolonged and variable course requiring long-term therapy and monitoring.

Diagnosis

The load of organisms at the site of infection is often unrelated to the clinical or radiological signs. Repeated cultures may be needed to confirm the diagnosis. Direct examination of centrifuged CSF deposits using India ink to demonstrate the capsule is positive in about half the cases of meningoencephalitis, and infection may sometimes be indicated by examination of sputum and tissue samples after conventional staining techniques. However, culture is essential to avoid false positive results, to improve sensitivity and to secure isolates for confirmation of identity and sensitivity testing. CSF, blood, sputum and tissue samples should be cultured repeatedly and in sufficient quantities to improve the diagnostic yield. Detection of capsular polysaccharides in CSF or serum is of some value in diagnosis and in the monitoring of disease progression during therapy, but there are discrepancies in the results from the different tests available. False positive reactions may occasionally occur and confirmation must always be sought by culture. Detection of serum antibodies and skin tests for delayed hypersensitivity reactions are of little value in the diagnosis of infection in individual patients.

References

Kovacs JA, Kovacs AA, Polis M et al. Cryptococcosis in the acquired immunodeficiency syndrome. Ann Intern Med. 1985;103:533–8.

McDonnell JM, Hutchinson GM. Pulmonary cryptococcosis. Hum Pathol. 1985;16:121–8.

Cryptococcus neoformans
ANTIGEN DETECTION

Manufacturer: Meridian Diagnostics Inc
Cat. No./Trade name: 602096/Premier® Cryptococcal Antigen

SUMMARY

[Well-Ab]–**Ag**–[MAb-HRP]–[TMB]–A_{450}

Assay type: EIA (non-competitive)
Detection: Colorimetric A_{450}
Format: Microtitre well, Ab coated
Sample type: Serum, CSF
Sample pre-treatment:
    CSF: centrifuge, use supernatant
Sample volume: 50 µl
Number of tests: 192
Controls - standards run in assay:
    Controls: Neg (1), Pos (1)
Incubation:
    10 min (RT) + 10 min (RT) + 10 min (RT)
Washes: 2

CONTENTS

Antibodies, antigens, labelled components:
    Anti-*C. neoformans* PAb bound to well
    Anti-*C. neoformans* MAb HRP conjugated
Substrate: TMB
Controls - standards supplied:
    Controls: Neg (sample diluent) and Pos
Additional reagents:
    None
Special equipment required:
    None

INTERPRETATION

Comments on interpretation:
    Equivocal: sample OD between ≥0.100 and <0.150; retest to confirm
    Repeatably equivocal: retest fresh sample
No. of references: 10

NOTES

120712.0

Cryptococcus neoformans
ANTIGEN DETECTION

Manufacturer: bioMerieux
Cat. No./Trade name: 58861/Slidex Crypto-Kit

SUMMARY

[Latex-Ab]–**Ag**–agglutination

Assay type: Particle agglutination assay
Detection: Visual
Format: Latex particles, Ab coated (test done on card)
Sample type: CSF, Serum
Sample pre-treatment:
    CSF: heat 5 min (80–100°C), centrifuge
    Serum: heat 30 min (56°C) or 5 min (80°C)
Sample volume: 30 µl of treated sample x 2
Number of tests: 80
Controls - standards run in assay:
    Controls: Pos (1)
Incubation:
    Read result within 5 min of mixing sample and reagent
Washes:

CONTENTS

Antibodies, antigens, labelled components:
    Anti-*C. neoformans* Ab (r) bound to latex particles (Test Latex)
    Non-immune rabbit Ig Ab bound to latex particles (Control Latex)
Substrate:
Controls - standards supplied:
    Controls: Pos
Additional reagents:
    None
Special equipment required:
    None

INTERPRETATION

Comments on interpretation:
    Positive: agglutination of sample with Test Latex and not with Control Latex and confirmed by other tests
    Non-specific: agglutination with both Test Latex and Control Latex; repeat with a fresh specimen
No. of references: 7

NOTES

131277.0

Cryptococcus neoformans
ANTIGEN DETECTION

Manufacturer: Carter-Wallace Inc
Cat. No./Trade name: Crypto-LA Test Fumouze

SUMMARY

[Latex-Ab]–**Ag**–agglutination

Assay type: Particle agglutination assay
Detection: Visual
Format: Latex particles, Ab coated
Sample type: CSF, serum
Sample pre-treatment:
CSF: centrifuge or filter
All samples: inactivate for 30 min (56°C)
Sample volume: 50 µl of prepared sample x 2
Number of tests:
Controls - standards run in assay:
Controls: Pos (1)
Incubation:
10 min (RT) on rotator
Washes:

CONTENTS

Antibodies, antigens, labelled components:
Anti-*C. neoformans* Ab bound to latex particle (Latex reagent)
Non-immune rabbit Ig Ab bound to latex particles (Latex control)
Substrate:
Controls - standards supplied:
Controls: Pos
Additional reagents:
None
Special equipment required:
None

INTERPRETATION

Comments on interpretation:
Equivocal: agglutination of sample with Latex Reagent and Latex Control; retest using quantitative procedure. If difference of titre between Latex Reagent and Latex Control is higher than 2 dilutions, result is considered Pos; retest using fresh sample in 2 weeks
Positive: agglutination of sample with Latex Reagent and no agglutination with Latex Control
No. of references: 5

NOTES

131236.0

Cryptococcus neoformans
ANTIGEN DETECTION

Manufacturer: International Immunodiagnostics
Cat. No./Trade name: CR1004/LATEX-CRYPTO antigen detection system

SUMMARY

[Latex-Ab]–**Ag**–agglutination

Assay type: Particle agglutination assay
Detection: Visual
Format: Latex particles, Ab coated (test done on slide)
Sample type: Serum, CSF
Sample pre-treatment:
Mix serum or CSF with Detacher Enzyme, heat in water bath (56°C) for 30 min. Add Enzyme Inhibitor
Sample volume: 25 µl of treated sample
Number of tests: 45-100
Controls - standards run in assay:
Controls: Neg (1), low Pos (1)
Incubation:
10 min (RT) on rotator
Washes:

CONTENTS

Antibodies, antigens, labelled components:
Anti-*C. neoformans* (capsular) Ab (r) bound to latex particles
Substrate:
Controls - standards supplied:
Controls: Neg (human serum), low Pos (capsular Ag) and anti-globulin control (anti-rabbit Ig Ab (g)) used to validate Detacher Enzyme once a month
Additional reagents:
None
Special equipment required:
None

INTERPRETATION

Comments on interpretation:
Positive: (1+)-(4+) agglutination of sample and latex reagent
Sera from patients with pulmonary or osteolytic lesions should also be evaluated using an antibody detection test, e.g. LA-CRYPTO antibody test
No. of references: 21

NOTES

120943.0

© KLUWER ACADEMIC PUBLISHERS 1995, ISSN 1381-5067

Cryptococcus neoformans
ANTIGEN DETECTION

Manufacturer: Meridian Diagnostics Inc
Cat. No./Trade name: 140100/CALAS®

SUMMARY

[Latex-Ab]–**Ag**–agglutination

Assay type: Particle agglutination assay
Detection: Visual
Format: Latex particles, Ab coated (test done on slide)
Sample type: CSF, serum
Sample pre-treatment:
  CSF: inactivate by heating 5 min (100°C) to limit
    nonspecific interference. Do not routinely pretreat
    with Pronase
  Serum: pretreat with Pronase
Sample volume: 25 µl of pretreated sample x 2
Number of tests: 100
Controls - standards run in assay:
  Controls: Neg (2), Pos (2), anti-globulin control (2)
Incubation:
  10 min (RT) with rotation
Washes:

CONTENTS

Antibodies, antigens, labelled components:
  Anti-*C. neoformans* Ab (r) bound to latex (Test Latex)
  Rabbit Ig bound to latex (Contol latex)
Substrate:
Controls - standards supplied:
  Controls: Neg (human serum), Pos (*C. neoformans* Ag)
    and anti-globulin control (anti-rabbit serum (g))
Additional reagents:
  None
Special equipment required:
  None

INTERPRETATION

Comments on interpretation:
  Positive: ≥(2+) agglutination of sample with Test Latex
    and not with Control Latex
No. of references: 14

NOTES

120747.0

Cryptococcus neoformans
ANTIGEN DETECTION

Manufacturer: Murex Diagnostics Limited
Cat. No./Trade name: /MUREX CRYPTOCOCCUS TEST

SUMMARY

[Latex-MAb]–**Ag**–agglutination

Assay type: Particle agglutination assay
Detection: Visual
Format: Latex particles, Ab coated (test done on reaction card)
Sample type: Serum, CSF
Sample pre-treatment:
  CSF: heat 5 min (100°C)
  Serum: treat with Protease, heat 10 min (100°C)
Sample volume: 50 µl of treated sample
Number of tests:
Controls - standards run in assay:
  Controls: Neg (1), low Pos (1), high Pos (1) as required
Incubation:
  Reagent and sample are mixed and results read
    immediately after 5 min rotation
Washes:

CONTENTS

Antibodies, antigens, labelled components:
  Anti-*C. neoformans* IgM MAb bound to latex particles
Substrate:
Controls - standards supplied:
  Controls: Neg, low Pos, high Pos (human serum)
Additional reagents:
  None
Special equipment required:
  None

INTERPRETATION

Comments on interpretation:
  Negative: no agglutination after 5 min rotation. If clinical
    symptoms suggest infection then retest at 1:10
    dilution to eliminate possible prozone effect
  Positive: agglutination of sample with Test Latex within 5
    min
No. of references: 15

NOTES

120770.0

© *KLUWER ACADEMIC PUBLISHERS 1995, ISSN 1381-5067*

Cryptosporidium species

Coccidian organisms of the genus *Cryptosporidium* are intracellular protozoan parasites which are associated with gastrointestinal disease of man and other animals. Although their role in veterinary disease has been known for over 50 years, the first cases of human diarrhoeal disease were not described until 1976. Since that time sporadic and epidemic infections have been reported from throughout the world and the organism is recognised as one of the commonest enteric pathogens. *Cryptosporidium* infection may occur after person-to-person, zoonotic, and environmental, particularly waterborne, transmission. Infection is more common in less developed countries which is thought to be a consequence of overcrowding, poor sanitation and increased animal contact. However the epidemiology of human cryptosporidiosis was transformed by the advent of the global AIDS pandemic. Unlike infection in the immunocompetent host, which produces a profuse self-limiting watery diarrhoea with cramping abdominal pain of around two weeks duration, infection in the compromised host results in protracted severe diarrhoea which may be life threatening. *Cryptosporidium* infection is now acknowledged as one of the commonest causes of diarrhoeal illness in the AIDS patient. A variety of other clinical manifestations including cholecystitis, hepatitis, pancreatitis, reactive arthritis and respiratory symptoms have been associated with the organism, particularly in the compromised host.

Human cryptosporidiosis is caused by *Cryptosporidium parvum* which infects the brush border of columnar epithelial cells in the gastrointestinal tract. Infection follows ingestion of oocysts which excyst in the gut with the release of four motile sporozoites which attach to the epithelial cell wall. Subsequent development occurs in an extracytoplasmic intracellular vacuole. Ensuing asexual and sexual phases result in the intraluminal release of merozoites which may establish further cycles of autoinfection, and the production of further infectious oocysts. The precise mechanism of pathogenesis of diarrhoeal disease, with characteristic profuse watery secretions, is unknown.

Diagnosis

The organism cannot be cultured *in vitro* and initially infection was demonstrated by the presence of the various developmental stages of the organism in biopsy material. A number of microscopic methods for the detection of oocysts in faecal material, duodenal aspirates, bile or respiratory secretions have subsequently been developed, initially by means of Giemsa staining, and subsequently by a variety of modified acid-fast procedures. More recently fluorescent staining techniques such as auramine-phenol have been employed in many centres. These procedures are used either for direct examination of specimens, or following concentration procedures.

The availability of polyclonal and monoclonal antibodies reactive with a variety of *Cryptosporidium* antigens has facilitated the development of immunofluorescence and ELISA-based tests for the detection of the organism in clinical samples. Serologic tests using antigens derived from organisms or infected clinical material have also been developed based on indirect fluorescent antibody and ELISA technology. These latter tests have been predominantly utilised in the epidemiological setting, as persistence of serum antibodies limits their diagnostic usefulness. The detection of oocysts in environmental samples, particularly waters, is a specialised procedure requiring sophisticated concentration techniques, and is generally not undertaken in the routine diagnostic laboratory.

References

Current WL, Garcia LS. Cryptosporidiosis. Clinical Microbiology Reviews. 1991;4:325–58.

See also Multipathogen Assays section under: Gastrointestinal pathogens

© KLUWER ACADEMIC PUBLISHERS 1995, ISSN 1381-5067

Cryptosporidium species
ANTIGEN DETECTION

Manufacturer: Alexon Inc
Cat. No./Trade name: 540-24/ProSpecT®
Cryptosporidium Microplate Assay

SUMMARY

[Well-Ab]–**Ag**–[Ab-HRP]–[TMB]–A$_{450}$

Assay type: EIA (non-competitive)
Detection: Colorimetric A$_{450}$
Format: Microtitre well, Ab coated
Sample type: Faecal specimens
Sample pre-treatment:
    Mix specimens with dilution buffer
Sample volume: 200 μl of diluted sample
Number of tests: 24, 96
Controls - standards run in assay:
    Controls: Neg (1), Pos (1)
Incubation:
    1 hr (RT) + 30 min (RT) + 10 min (RT)
Washes: 2

CONTENTS

Antibodies, antigens, labelled components:
    Anti-Cryptosporidium species (CSA) Ab (r) bound to well
    Anti-Cryptosporidium species (CSA) Ab (r) HRP
      conjugated
Substrate: TMB
Controls - standards supplied:
    Controls: Neg (human) and Pos (bovine and human)
Additional reagents:
    None
Special equipment required:
    None

INTERPRETATION

Comments on interpretation:
    Classification of samples is according to cut-off; no
      further testing
No. of references: 13

NOTES

131303.0

Cryptosporidium species
ANTIGEN DETECTION

Manufacturer: Alexon Inc
Cat. No./Trade name: 340-20/ProSpecT®
Cryptosporidium Rapid Assay

SUMMARY

[Membrane-Ab]–**Ag**–[Ab-biotin]–[strept–HRP]–[TMB]–
blue spot

Assay type: EIA (non-competitive)
Detection: Visual
Format: Test device, Ab coated membrane
Sample type: Faecal specimens
Sample pre-treatment:
    Mix specimens with dilution buffer
Sample volume: 4 drops of diluted sample
Number of tests: 20
Controls - standards run in assay:
    Controls: Pos (1) on membrane
Incubation:
    2 min (RT) + 2 min (RT) + 2 min (RT) + 3 min (RT)
Washes: 3

CONTENTS

Antibodies, antigens, labelled components:
    Anti-Cryptosporidium species (CSA) Ab bound to
      membrane (test area)
    Anti-Cryptosporidium species (CSA) Ab (r) biotinylated
    Streptavidin HRP conjugated
Substrate: TMB
Controls - standards supplied:
    Controls: Integral Pos controls incorporated on
      membrane
Additional reagents:
    None
Special equipment required:
    None

INTERPRETATION

Comments on interpretation:
    Negative: circular blue spot in Pos control area and no
      blue spot in specimen test area
    Positive: circular blue spot in Pos control area and
      specimen test area
No. of references: 13

NOTES

131305.0

Cryptosporidium species
ANTIGEN DETECTION

Manufacturer: LMD Laboratories Inc
Cat. No./Trade name: CP-1/Cryptosporidium Antigen
Detection ELISA Kit

SUMMARY

[Well-Ab]–**Ag**–[Ab]–[AMIg-biotin]–[strept-HRP]–
[TMB]–A_{450}

Assay type: EIA (non-competitive)
Detection: Colorimetric A_{450}
Format: Microtitre well, Ab coated
Sample type: Faecal specimen, rectal swab
Sample pre-treatment:
 Mix sample with wash buffer and vortex
Sample volume: 100 µl of prepared sample
Number of tests: 48, 96
Controls - standards run in assay:
 Controls: Neg (1), Pos (1)
Incubation:
 45 min (37°C) + 20 min (37°C) + 10 min (37°C) + 10
 min (37°C) + 10 min (RT)
Washes: 4

CONTENTS

Antibodies, antigens, labelled components:
 Anti-Cryptosporidium Ab bound to well
 Anti-Cryptosporidium Ab (m)
 Anti-mouse Ig Ab biotinylated
 Streptavidin HRP conjugated
Substrate: TMB
Controls - standards supplied:
 Controls: Neg and Pos
Additional reagents:
 None
Special equipment required:
 None

INTERPRETATION

Comments on interpretation:
 Classification of samples is according to cut-off; no
 further testing
No. of references: 17

NOTES

131180.0

Cryptosporidium species
ANTIGEN DETECTION

Manufacturer: Melotec S.A.
Cat. No./Trade name: Melotest Cryptosporidium Ag

SUMMARY

[Well-Ab]–**Ag**–[Ab]–[AMIg-biotin]–[strept-HRP]–
[TMB]–A_{450}

Assay type: EIA (non-competitive)
Detection: Colorimetric A_{450}
Format: Microtitre well, Ab coated
Sample type: Faecal sample
Sample pre-treatment:
 Dilute sample with diluent (1:15) and further dilute
  supernatant (1:3) before use
Sample volume: 100 µl of prepared sample x 2
Number of tests: 96
Controls - standards run in assay:
 Controls: Neg (2), Pos (1)
Incubation:
 45 min (37°C) + 20 min (37°C) + 10 min (37°C) + 10
 min (37°C) + 10 min (RT)
Washes: 4

CONTENTS

Antibodies, antigens, labelled components:
 Anti-Cryptosporidium Ab bound to well
 Anti-Cryptosporidium Ab (m)
 Anti-mouse Ig Ab biotinylated
 Streptavidin HRP conjugated
Substrate: TMB
Controls - standards supplied:
 Controls: Neg and Pos
Additional reagents:
 None
Special equipment required:
 None

INTERPRETATION

Comments on interpretation:
 Equivocal: ratio of sample OD:cut-off value between 0.9
  and 1.1; retest a fresh sample
 Repeatably equivocal: retest a fresh sample taken in 2-4
  weeks
No. of references: 8

NOTES

131138.0

Cryptosporidium species
ANTIGEN DETECTION

Manufacturer: Seradyn
Cat. No./Trade name: 0369553/Color Vue® Cryptosporidium

SUMMARY

[Well-Ab]–**Ag**–[Ab]–[Anti-Ab-biotin]–[strept-HRP]–[TMB]–A_{450}

Assay type: EIA (non-competitive)
Detection: Colorimetric A_{450}
Format: Microtitre well, Ab coated
Sample type: Faecal specimen
Sample pre-treatment:
   Mix sample with diluted wash buffer
Sample volume: 50 μl of prepared sample
Number of tests: 48, 96
Controls - standards run in assay:
   Controls: Neg (1), Pos (1)
Incubation:
   45 min (37°C) + 20 min (37°C) + 10 min (37°C) + 10 min (37°C) + 10 min (RT)
Washes: 4

CONTENTS

Antibodies, antigens, labelled components:
   Anti-Cryptosporidium Ab bound to well
   Anti-Cryptosporidium MAb
   Anti-2nd antibody biotinylated
   Streptavidin HRP conjugated
Substrate: TMB
Controls - standards supplied:
   Controls: Neg and Pos
Additional reagents:
   None
Special equipment required:
   None

INTERPRETATION

Comments on interpretation:
   Classification of samples is according to cut-off; no further testing
No. of references: 17

NOTES

131651.0

Cryptosporidium species
ANTIGEN DETECTION

Manufacturer: Cellabs Pty Ltd
Cat. No./Trade name: KR1/CRYPTO-CEL® IF Test

SUMMARY

[Slide]–**Ag**–[MAb-FITC]–fluorescence

Assay type: Immunofluorescence assay (direct)
Detection: Fluorescence microscopy
Format: Slide, specimen coated
Sample type: Faecal specimens
Sample pre-treatment:
   Specimens may be concentrated prior to use by filtration or with Faecal Parasite Concentration Kit*. Otherwise dilute samples with 10% formalin or PBS, stand before use, apply to slide, dry, fix
Sample volume: 20 μl of prepared specimen
Number of tests: 50
Controls - standards run in assay:
   Controls: Pos (1)
Incubation:
   30 min (37°C) + 10 min (37°C)
Washes: 1

CONTENTS

Antibodies, antigens, labelled components:
   Anti-Cryptosporidium (oocysts) MAb (m) FITC conjugated
Substrate:
Controls - standards supplied:
   Controls: Pos (slide)
Additional reagents:
   None
Special equipment required:
   None

INTERPRETATION

Comments on interpretation:
   Positive: green fluorescence of ≥ one cyst
No. of references: 14

NOTES

131411.0

This kit detects oocysts of *C. parvum*, *C. baileyi* and *C. muris* but does not differentiate between them
*Kit purchased from Evergreen Scientific, CA 90058, USA (Cat. no. 240-3074-030)

Cryptosporidium species
ANTIGEN DETECTION

Manufacturer: Shield Diagnostics
Cat. No./Trade name: FICR100/DETECT-IF

SUMMARY

[Slide]–**Ag**–[MAb-FITC]–fluorescence

Assay type: Immunofluorescence assay (direct)
Detection: Fluorescence microscopy
Format: Slide, specimen coated
Sample type: Faecal specimen, environmental water
   samples
Sample pre-treatment:
   Faecal specimen: dilute with stool diluent
   Water samples: filter, centrifuge and suspend on cold
      sucrose gradient
   Slurry, sewage: see HMSO publication (ISBN:
      0117522821)
   Apply all to slide, fix
Sample volume: 20 µl of prepared samples
Number of tests:
Controls - standards run in assay:
   Controls: Neg (1), Pos (1)
Incubation:
   30 min (37°C)
Washes: 1

CONTENTS

Antibodies, antigens, labelled components:
   Anti-Cryptosporidium (oocysts) MAb (m) FITC
      conjugated
Substrate:
Controls - standards supplied:
   Controls: Neg and Pos (slides)
Additional reagents:
   None
Special equipment required:
   None

INTERPRETATION

Comments on interpretation:
   Equivocal: <5 apple-green fluorescent oocysts; retest
      using further sample dilutions and/or fresh specimen
   Positive: >5 apple-green fluorescent round/oval oocysts
No. of references: 12

NOTES

131716.0

© KLUWER ACADEMIC PUBLISHERS 1995, ISSN 1381-5067

Echinococcus species

Echinococcus spp. are the aetiological agents of hydatid disease, a zoonotic infection of worldwide distribution, in man and other intermediate hosts. Many questions regarding the factors which determine susceptibility/resistance to hydatid disease, and the factors which influence the viability and fertility of hydatid cysts, remain to be answered. Two species *E.granulosus* and *E.multilocularis* are principally responsible for human disease, although other species have been implicated. The definitive host for the former is the dog, with sheep or cattle acting as intermediate hosts following ingestion of contaminated dog faeces. Man is also infected coincidentally by contact with contaminated dog faeces. The eggs are extremely persistent in the environment. Following ingestion oncospheres penetrate the mesenteric vessels and are carried throughout the body, forming hydatid cysts mainly in the liver and lungs. These fluid filled cysts enlarge and contain numerous protoscolices, developed from an inner germinal layer within brood capsules, which may cause multiple daughter cysts within the peritoneum and adjacent organs if the cyst is ruptured. Most of the symptoms of echinococcosis arise from the compression of adjacent host structures by the gradually enlarging hydatid cyst. The preferred form of treatment is surgical removal of the intact cyst.

E.multilocularis is restricted in distribution to the Northern hemisphere. The definitive hosts for this organism are foxes and dogs, with small rodents acting as intermediate hosts. Infection is mainly acquired by ingestion of contaminated fruit and vegetables. Unlike the well defined cysts of *E.granulosus*, the germinal membrane is not confined within a single cyst, and scolices develop in an uncontrolled manner, often with extensive local invasion and infiltration, particularly in the liver. Unlike classical hydatid disease, this renders little possibility of effective surgical resection.

Diagnosis

As the clinical manifestations of hydatid disease in man are variable, the diagnosis of the condition presents complex problems for clinicians. Since the parasitological diagnosis of the disease is difficult, the specific diagnosis of the condition relies heavily on immunodiagnostic procedures. Tests using a variety of hydatid antigens and employing complement fixation, CIE, haemagglutination, latex agglutination and ELISA technologies have been developed. These tests are generally more sensitive and specific for liver cyst disease than for lung or other organ involvement. Recent approaches to diagnosis have often employed a combination of two or three serological tests to diagnose the condition, as a single test may fail to detect all cases. ELISA tests for the detection of circulating hydatid antigen in the serum have also been utilised. Other immunochemical and recombinant DNA techniques are being applied to improve diagnosis of hydatid disease in man and *Echinococcus* infection in dogs, and also in the development of vaccines against infection with taeniid cestode larvae.

References

Lightowlers MW. Immunology and molecular biology of Echinococcus infections. International Journal for Parasitology. 1990;20:471–8.

Gottstein B. Molecular and Immunological Diagnosis of Echinococcosis. Clinical Microbiology Reviews. 1992;5:248–61.

Echinococcus species
ANTIBODY DETECTION

Manufacturer: LMD Laboratories Inc
Cat. No./Trade name: EG-8/Echinococcus Serology Microwell ELISA Kit

SUMMARY

[Well-Ag]–**Ab**–[Protein A-HRP]–[TMB]–A_{450}

Assay type: EIA (non-competitive)
Detection: Colorimetric A_{450}
Format: Microtitre well, Ag coated
Sample type: Serum (do not heat inactivate)
Sample pre-treatment:
　None
Sample volume: 2 µl (+254 µl diluent)
Number of tests: 48, 96
Controls - standards run in assay:
　Controls: Neg (1), weak Pos (1), strong Pos (1)
Incubation:
　10 min (RT) + 5 min (RT) + 5 min (RT)
Washes: 2

CONTENTS

Antibodies, antigens, labelled components:
　Echinococcus species (cysts) Ag bound to well
　Protein A HRP conjugated
Substrate: TMB
Controls - standards supplied:
　Controls: Neg (human serum), weak Pos and strong Pos
　　(rabbit serum)
Additional reagents:
　None
Special equipment required:
　None

INTERPRETATION

Comments on interpretation:
　Classification of samples is according to cut-off; no
　　further testing
No. of references: 7

NOTES

131645.0

Cross-reactivity with *Taenia solium* may occur

Echinococcus species
ANTIBODY DETECTION (IgG)

Manufacturer: Immuno Pharmacology Research
Cat. No./Trade name: ELISA HYDATIDE

SUMMARY

[Well-Ag]–**Ab**–[AHIgG-AP]–[PNP]–A_{405}

Assay type: EIA (non-competitive)
Detection: Colorimetric A_{405}
Format: Microtitre well, Ag coated
Sample type: Serum
Sample pre-treatment:
　None
Sample volume: 100 µl of 1:100 dilution x 2
Number of tests: 48, 96
Controls - standards run in assay:
　Controls: Neg (1), Pos (1)
Incubation:
　1 hr (37°C) + 1 hr (37°C) + 30 min (37°C)
Washes: 2

CONTENTS

Antibodies, antigens, labelled components:
　Echinococcus species Ag bound to well
　Anti-human IgG Ab AP conjugated
Substrate: PNP
Controls - standards supplied:
　Controls: Neg and Pos
Additional reagents:
　None
Special equipment required:
　None

INTERPRETATION

Comments on interpretation:
　Classification of samples is according to cut-off; no
　　further testing
No. of references: 7

NOTES

131314.0

Echinococcus species
ANTIBODY DETECTION (IgG)

Manufacturer: Melotec S.A.
Cat. No./Trade name: Melotest Echinococcosis

SUMMARY

[Well-Ag]–**Ab**–[AHIgG-HRP]–[TMB]–A_{450}

Assay type: EIA (non-competitive)
Detection: Colorimetric A_{450}
Format: Microtitre well, Ag coated
Sample type: Serum, plasma
Sample pre-treatment:
   None
Sample volume: 100 μl of 1:64 dilution x 2
Number of tests: 96
Controls - standards run in assay:
   Controls: Neg (1), low Pos (2), high Pos (1)
Incubation:
   10 min (RT) + 5 min (RT) + 10 min (RT)
Washes: 2

CONTENTS

Antibodies, antigens, labelled components:
   Echinococcus species (cysts) Ag bound to well
   Anti-human IgG Ab HRP conjugated
Substrate: TMB
Controls - standards supplied:
   Controls: Neg, low Pos and high Pos (sera)
Additional reagents:
   None
Special equipment required:
   None

INTERPRETATION

Comments on interpretation:
   Equivocal: ratio of sample OD:cut-off value between 0.9
      and 1.1; retest a fresh sample
   Repeatably equivocal: retest a fresh sample taken in 2-4
      weeks
No. of references: 7

NOTES

131134.0

Significant cross-reactivity has been reported between
   Taenia solium and Echinococcus

Echinococcus granulosus
ANTIBODY DETECTION (IgG)

Manufacturer: Immunetics
Cat. No./Trade name: Human Hydatid Disease Western Blot Kit

SUMMARY

[Strip-Ag]–**Ab**–[AHIgG-AP]–[BCIP/NBT]–bands

Assay type: Immunoblot assay
Detection: Visual
Format: Reaction tray, Ag coated nitrocellulose strip
Sample type: Serum or plasma
Sample pre-treatment:
   None
Sample volume: 1 ml of 1:100 dilution
Number of tests: 24
Controls - standards run in assay:
   Controls: Neg (1), Pos (1)
Incubation:
   30 min (RT) + 15 min (RT)
Washes: 2

CONTENTS

Antibodies, antigens, labelled components:
   Echinococcus granulosus Ags bound to nitrocellulose
      strip as discrete bands
   Anti-human IgG Ab AP conjugated
Substrate: BCIP/NBT
Controls - standards supplied:
   Controls: Neg and Pos (human serum)
Additional reagents:
   None
Special equipment required:
   None

INTERPRETATION

Comments on interpretation:
   Positive: coloured bands on membrane. For
      interpretation of band pattern see manufacturer's
      instructions
No. of references: 3

NOTES

131820.0

For research use only

Entamoeba histolytica

Entamoeba histolytica is a protozoan parasite of the human colon. Motile trophozoite forms invade the bowel wall and also inhabit the lumen. The trophozoites are capable of undergoing encystment within the bowel lumen and may then be shed in the faeces. It is these cysts which are implicated in further transmission as a result of the consumption of contaminated water or vegetables. After ingestion, cysts pass through the stomach to reach the small intestine, where they can transform into the trophozoite form. The organism adheres to the intestinal mucosa by means of specific receptors and produces cytotoxic factors which result in the characteristic inflammation and ulceration of the gut wall. The range of illness which may result is very wide, ranging from asymptomatic carriage, through diarrhoeal illness which may be chronic and mild in nature, more severe dysenteric illness with fever and severe abdominal pain, to severe massive infection with perforation of the gut and amoebic peritonitis or the development of toxic megacolon. Extraintestinal infection can occur, and most commonly involves the liver, although the lungs and rarely the brain and other sites may be involved.

Infection is extremely common, with in excess of 10 million cases occurring annually worldwide and an estimated 10% of the world's population thought to be infected at any one time. Rates of infection are highest in developing countries, where poverty, poor sanitation and overcrowding result in an increased risk of transmission. Morbidity and mortality due to *E.histolytica* vary from area to area and person to person. The high rate of asymptomatic carriage is now thought to reflect the existence of pathogenic and non-pathogenic zymodemes, which are strains of *E.histolytica* that can be differentiated by isoenzyme analysis, monoclonal antibodies, and DNA probes. Whether pathogenicity is a genotypic trait or can be changed by environmental influences has yet to be resolved.

Diagnosis

In intestinal infection, the diagnosis is made by identifying either cysts or trophozoites of *E.histolytica* in the stool, or trophozoites in fresh material from lesions observed at sigmoidoscopy. Fresh stool specimens may be examined directly, and following concentration. Some technical expertise is required in the differentiation of *E.histolytica* from other non-pathogenic species. Although a variety of media have now been described which permit the *in vitro* culture of amoebae, few diagnostic laboratories routinely attempt to culture *E.histolytica* from clinical specimens. In extraintestinal infection there may be no evidence of active amoebic colitis, and the stools may be negative for the organism.

Serological tests are currently of most value in the diagnosis of invasive intestinal amoebiasis and extraintestinal amoebiasis. The indirect haemagglutination test will be positive in approximately 70-80% of cases with biopsy-proven invasive amoebic colitis, while the test is usually negative in asymptomatic cyst passers. In extra-intestinal disease this figure rises to more than 90%. Various other serological tests including immunodiffusion and ELISA-based systems have been described. There has also been considerable interest in the development of antigen detection systems, mainly employing ELISA technology, which may be particularly useful in endemic areas where the persistence of amoebic antibodies may complicate serodiagnostic procedures. The advent of novel nucleic acid technology may facilitate the development of rapid tests for the differentiation of pathogenic and non-pathogenic zymodemes.

Reference

Bruckner DA. Amoebiasis. Clinical Microbiology Reviews. 1992;5:356–69.

Entamoeba histolytica
ANTIGEN DETECTION

Manufacturer: Alexon Inc
Cat. No./Trade name: 560-48/ProSpecT® Entamoeba histolytica Microplate Assay

SUMMARY

[Well-Ab]–**Ag**–[Ab-HRP]–[TMB]–A_{450}

Assay type: EIA (non-competitive)
Detection: Colorimetric A_{450}
Format: Microtitre well, Ab coated
Sample type: Faecal specimens
Sample pre-treatment:
　Mix specimens with dilution buffer
Sample volume: 200 µl of diluted sample
Number of tests: 48, 96
Controls - standards run in assay:
　Controls: Neg (1), Pos (1)
Incubation:
　1 hr (RT) + 30 min (RT) + 10 min (RT)
Washes: 2

CONTENTS

Antibodies, antigens, labelled components:
　Anti-*Entamoeba histolytica* (EHSA) Ab bound to well
　Anti-*Entamoeba histolytica* (EHSA) Ab HRP conjugated
Substrate: TMB
Controls - standards supplied:
　Controls: Neg and Pos (human)
Additional reagents:
　None
Special equipment required:
　None

INTERPRETATION

Comments on interpretation:
　Classification of samples is according to cut-off; no further testing
No. of references: 7

NOTES

131302.0

Entamoeba histolytica/dispar
ANTIGEN DETECTION

Manufacturer: Cellabs Pty Ltd
Cat. No./Trade name: KE2/ENTAMOEBA CELISA® - Screen

SUMMARY

[Well-Ab]–**Ag**–[MAb-HRP]–[TMB]–A_{450}

Assay type: EIA (non-competitive)
Detection: Colorimetric A_{450}
Format: Microtitre well, Ab coated
Sample type: Faecal specimens
Sample pre-treatment:
　Mix with diluent and vortex
Sample volume: 100 µl of prepared sample
Number of tests: 48
Controls - standards run in assay:
　Controls: Neg (1), Pos (1)
Incubation:
　2 hr (RT) + 10 min (RT)
Washes: 1

CONTENTS

Antibodies, antigens, labelled components:
　Anti-*Entamoeba histolytica/dispar* (adhesin) Ab bound to well
　Anti-*Entamoeba histolytica/dispar* (adhesin) MAb (m) HRP conjugated
Substrate: TMB
Controls - standards supplied:
　Controls: Neg (diluent) and Pos
Additional reagents:
　None
Special equipment required:
　None

INTERPRETATION

Comments on interpretation:
　Classification of samples is according to cut-off; no further testing
No. of references: 10

NOTES

131412.0

This kit detects but does not differentiate between *Entamoeba histolytica* and *Entamoeba dispar*

Entamoeba histolytica
ANTIGEN DETECTION

Manufacturer: Cellabs Pty Ltd
Cat. No./Trade name: KE1/ENTAMOEBA CELISA® - Path

SUMMARY

$$[\text{Well-Ag}]-\textbf{Ag}-[\text{MAb-HRP}]-[\text{TMB}]-A_{450}$$

Assay type: EIA (non-competitive)
Detection: Colorimetric A_{450}
Format: Microtitre well, Ab coated
Sample type: Faecal specimens
Sample pre-treatment:
  Mix with diluent and vortex
Sample volume: 100 µl of prepared sample
Number of tests: 48
Controls - standards run in assay:
  Controls: Neg (1), Pos (1)
Incubation:
  2 hr (RT) + 10 min (RT)
Washes: 1

CONTENTS

Antibodies, antigens, labelled components:
  Anti-*Entamoeba histolytica* (adhesin) Ab bound to well
  Anti-*Entamoeba histolytica* (adhesin) MAb (m) HRP
    conjugated
Substrate: TMB
Controls - standards supplied:
  Controls: Neg (diluent) and Pos
Additional reagents:
  None
Special equipment required:
  None

INTERPRETATION

Comments on interpretation:
  Classification of samples is according to cut-off; no
    further testing
No. of references: 7

NOTES

131780.0

Entamoeba histolytica
ANTIBODY DETECTION

Manufacturer: Diamedix Corporation
Cat. No./Trade name: 783-120/Amebiasis Microassay

SUMMARY

$$[\text{Well-Ag}]-\textbf{Ab}-[\text{AHIg-AP}]-[\text{PNP}]-A_{405}$$

Assay type: EIA (non-competitive)
Detection: Colorimetric A_{405}
Format: Microtitre well, Ag coated
Sample type: Serum (do not heat inactivate)
Sample pre-treatment:
  None
Sample volume: 10 µl (+240 µl diluent)
Number of tests: 96
Controls - standards run in assay:
  Controls: Neg (1), Pos (1)
  Calibrator: 1 (1)
Incubation:
  15 min (RT) + 15 min (RT) + 15 min (RT)
Washes: 2

CONTENTS

Antibodies, antigens, labelled components:
  Entamoeba histolytica (strain HK9 and NIH 200) Ag
    bound to well
  Anti-human Ig Ab AP conjugated
Substrate: PNP
Controls - standards supplied:
  Controls: Neg and Pos
  Calibrator: 1 (assigned Ab titre)
Additional reagents:
  None
Special equipment required:
  None

INTERPRETATION

Comments on interpretation:
  Equivocal: when ratio sample OD:calibrator OD x100 is
    between 40 and 50 AU/ml; retest with this assay or
    another serological test or retest a fresh specimen
  Repeatably equivocal: report as negative
No. of references: 5

NOTES

131054.0

© *KLUWER ACADEMIC PUBLISHERS 1995, ISSN 1381-5067*

Entamoeba histolytica
ANTIBODY DETECTION

Manufacturer: LMD Laboratories Inc
Cat. No./Trade name: EH-012/E. histolytica Serology Microwell ELISA Kit

SUMMARY

[Well-Ag]–**Ab**–[Protein A-HRP]–[TMB]–A_{450}

Assay type: EIA (non-competitive)
Detection: Colorimetric A_{450}
Format: Microtitre well, Ag coated
Sample type: Serum (do not heat inactivate)
Sample pre-treatment:
  None
Sample volume: 5 μl (+315 μl diluent)
Number of tests: 48, 96
Controls - standards run in assay:
  Controls: Neg (1), weak Pos (1), strong Pos (1)
Incubation:
  10 min (RT) + 5 min (RT) + 5 min (RT)
Washes: 2

CONTENTS

Antibodies, antigens, labelled components:
  Entamoeba histolytica (HK-9) Ag bound to well
  Protein A HRP conjugated
Substrate: TMB
Controls - standards supplied:
  Controls: Neg (human serum), weak Pos and strong Pos
  (rabbit serum)
Additional reagents:
  None
Special equipment required:
  None

INTERPRETATION

Comments on interpretation:
  Classification of samples is according to cut-off; no
  further testing
No. of references: 5

NOTES

131646.0

Entamoeba histolytica
ANTIBODY DETECTION (IgG)

Manufacturer: Amico Laboratories Inc
Cat. No./Trade name: 8000G/AMIZYME® Amebiasis IgG Assay

SUMMARY

[Well-Ag]–**Ab**–[AHIgG-HRP]–[ABTS]–A_{405}

Assay type: EIA (non-competitive)
Detection: Colorimetric A_{405}
Format: Microtitre well, Ag coated
Sample type: Serum
Sample pre-treatment:
  None
Sample volume: 5 μl (+245 μl diluent)
Number of tests: 96
Controls - standards run in assay:
  Controls: Neg (1)
  Calibrators: I (1)
Incubation:
  25 min (RT) + 25 min (RT) + 25 min (RT)
Washes: 2

CONTENTS

Antibodies, antigens, labelled components:
  Entamoeba histolytica (strain NIH 200) Ag bound to well
  Anti-human IgG Ab (g) HRP conjugated
Substrate: ABTS
Controls - standards supplied:
  Controls: Neg
  Calibrator: 1
Additional reagents:
  None
Special equipment required:
  None

INTERPRETATION

Comments on interpretation:
  Equivocal: where sample OD is between 0.439 and
  0.449 (95–99 U/ml); retest to confirm
No. of references: 3

NOTES

131783.0

Entamoeba histolytica
ANTIBODY DETECTION (IgG)

Manufacturer: Clark Laboratories
Cat. No./Trade name: Entamoeba histolytica IgG Elisa Test

SUMMARY

[Well-Ag]–**Ab**–[AHIgG-HRP]–[OPD]–A_{490}

Assay type: EIA (non-competitive)
Detection: Colorimetric A_{490}
Format: Microtitre well, Ag coated
Sample type: Serum (do not heat inactivate)
Sample pre-treatment:
  None
Sample volume: 10 µl (+200 µl diluent)
Number of tests: 96
Controls - standards run in assay:
  Controls: Neg (1), Pos (1)
  Calibrators: 1 (2)
Incubation:
  20 min (RT) + 20 min (RT) + 10 min (RT)
Washes: 2

CONTENTS

Antibodies, antigens, labelled components:
  Entamoeba histolytica Ag bound to well
  Anti-human IgG Ab (g) HRP conjugated
Substrate: TMB*
Controls - standards supplied:
  Controls: Neg and Pos (human serum or plasma)
  Calibrator: 1 (human serum or plasma)
Additional reagents:
  H_2SO_4
Special equipment required:
  None

INTERPRETATION

Comments on interpretation:
  Classification of samples is according to cut-off; no further testing
No. of references: None

NOTES

131658.0

*OPD can be supplied as an alternative

Entamoeba histolytica
ANTIBODY DETECTION (IgG)

Manufacturer: Melotec S.A.
Cat. No./Trade name: Melotest Amebiasis

SUMMARY

[Well-Ag]–**Ab**–[AHIgG-HRP]–[TMB]–A_{450}

Assay type: EIA (non-competitive)
Detection: Colorimetric A_{450}
Format: Microtitre well, Ag coated
Sample type: Serum, plasma
Sample pre-treatment:
  None
Sample volume: 100 µl of 1:64 dilution x 2
Number of tests: 96
Controls - standards run in assay:
  Controls: Neg (1), low Pos (2), high Pos (1)
Incubation:
  10 min (RT) + 5 min (RT) + 10 min (RT)
Washes: 2

CONTENTS

Antibodies, antigens, labelled components:
  Entamoeba histolytica Ag bound to well
  Anti-human IgG Ab HRP conjugated
Substrate: TMB
Controls - standards supplied:
  Controls: Neg, low Pos and high Pos (sera)
Additional reagents:
  None
Special equipment required:
  None

INTERPRETATION

Comments on interpretation:
  Equivocal: ratio of sample OD:cut-off value between 0.9 and 1.1; retest a fresh sample
  Repeatably equivocal: retest a fresh sample taken in 2-4 weeks
No. of references: 5

NOTES

131135.0

© KLUWER ACADEMIC PUBLISHERS 1995, ISSN 1381-5067

Entamoeba histolytica
ANTIBODY DETECTION (IgM)

Manufacturer: Amico Laboratories Inc
Cat. No./Trade name: 8000M/AMIZYME® Amebiasis
IgM Assay

SUMMARY

[Well-Ag]–**Ab**–[AHIgM-HRP]–[ABTS]–A_{405}

Assay type: EIA (non-competitive)
Detection: Colorimetric A_{405}
Format: Microtitre well, Ag coated
Sample type: Serum
Sample pre-treatment:
   None
Sample volume: 5 µl (+245 µl diluent)*
Number of tests: 96
Controls - standards run in assay:
   Controls: Neg (1)
   Calibrators: I (1)
Incubation:
   25 min (RT) + 25 min (RT) + 25 min (RT)
Washes: 2

CONTENTS

Antibodies, antigens, labelled components:
Entamoeba histolytica (strain NIH 200) Ag bound to well
Anti-human IgM Ab (g) HRP conjugated
Substrate: ABTS
Controls - standards supplied:
   Controls: Neg
   Calibrator: 1
Additional reagents:
   None
Special equipment required:
   None

INTERPRETATION

Comments on interpretation:
   Equivocal: where sample OD is between 0.275–0.299
     (92–99 U/ml); retest to confirm
No. of references: 3

NOTES

131791.0

*Sample diluted in IgM serum diluent provided to eliminate
interference due to RF and IgG

Entamoeba histolytica
ANTIBODY DETECTION (IgM)

Manufacturer: Amico Laboratories Inc
Cat. No./Trade name: 8000MC/AMIZYME® Amebiasis
IgM Capture Assay

SUMMARY

[Well-Ag]–**Ab**–[AHIgM]–[AMIg-HRP]–[ABTS]–A_{405}

Assay type: EIA (non-competitive)
Detection: Colorimetric A_{405}
Format: Microtitre well, Ag coated
Sample type: Serum
Sample pre-treatment:
   None
Sample volume: 5 µl (+245 µl diluent)
Number of tests: 96
Controls - standards run in assay:
   Controls: Neg (1)
   Calibrators: I (1)
Incubation:
   25 min (RT) + 25 min (RT) + 25 min (RT) + 25 min
   (RT)
Washes: 3

CONTENTS

Antibodies, antigens, labelled components:
Entamoeba histolytica (strain NIH 200) Ag bound to well
Anti-human IgM MAb (m)
Anti-mouse Ig Ab (g) HRP conjugated
Substrate: ABTS
Controls - standards supplied:
   Controls: Neg
   Calibrator: 1
Additional reagents:
   None
Special equipment required:
   None

INTERPRETATION

Comments on interpretation:
   Equivocal: where sample OD is between 0.285–0.299;
     retest to confirm
No. of references: 2

NOTES

131801.0

Entamoeba histolytica
ANTIBODY DETECTION (IgM)

Manufacturer: Clark Laboratories
Cat. No./Trade name: Entamoeba histolytica IgM Elisa Test

SUMMARY

[Well-Ag]–**Ab**–[AHIgM-HRP]–[OPD]–A$_{490}$

Assay type: EIA (non-competitive)
Detection: Colorimetric A$_{490}$
Format: Microtitre well, Ag coated
Sample type: Serum (do not heat inactivate)
Sample pre-treatment:
   Add Absorbent Solution provided to diluted serum and incubate 20 min (RT) to eliminate interference due to RF and IgG
Sample volume: 100 µl of treated sample
Number of tests: 96
Controls - standards run in assay:
   Controls: Neg (1), Pos (1)
   Calibrators: 1 (2)
Incubation:
   20 min (RT) + 20 min (RT) + 10 min (RT)
Washes: 2

CONTENTS

Antibodies, antigens, labelled components:
   Entamoeba histolytica Ag bound to well
   Anti-human IgM Ab (g) HRP conjugated
Substrate: TMB*
Controls - standards supplied:
   Controls: Neg and Pos (human serum or plasma)
   Calibrator: 1 (human serum or plasma)
Additional reagents:
   H$_2$SO$_4$
Special equipment required:
   None

INTERPRETATION

Comments on interpretation:
   Classification of samples is according to cut-off; no further testing
No. of references: None

NOTES

131664.0

*OPD can be supplied as an alternative

Entamoeba histolytica
ANTIBODY DETECTION

Manufacturer: Laboratoire Fumouze
Cat. No./Trade name: Bichro-Latex Amibe® (Amoeba) Fumouze

SUMMARY

[Latex-Ag]–**Ab**–red agglutination

Assay type: Particle agglutination assay
Detection: Visual
Format: Latex particles, Ag coated (test done on slide)
Sample type: Serum
Sample pre-treatment:
   None
Sample volume: 20 µl (+2 drops diluent)
Number of tests: 25
Controls - standards run in assay:
   Controls: Neg (1), Pos (1)
Incubation:
   Read result within 5 min of mixing sample and reagent
Washes:

CONTENTS

Antibodies, antigens, labelled components:
   Entamoeba histolytica Ag bound to latex particle (Latex reagent)
Substrate:
Controls - standards supplied:
   Controls: Neg and Pos (animal serum)
Additional reagents:
   None
Special equipment required:
   None

INTERPRETATION

Comments on interpretation:
   Positive: red agglutination which forms a red edge round a green central area
No. of references: None

NOTES

131234.0

Enterococcus species (Streptococcus beta-haemolytic, group D)

The enterococci are facultatively anaerobic, catalase-negative, Gram-positive cocci which appear as pairs and short chains on microscopy. Until recently they were classified as streptococci of Lancefield group D along with true streptococci such as *S.bovis* and *S.equinus*. More than a dozen enterococcal species have now been either proposed or accepted, but the most important species from a clinical standpoint are *E.faecalis* and *E.faecium*. The natural habitat of these organisms is the gastrointestinal tract of man and other animals, where they constitute an important component of the normal gut flora, although their ability to survive in harsher environments results in their common isolation from soil, food and water. Enterococci are associated with a variety of clinical conditions including urinary tract infection, bacteraemia associated with intra-abdominal or hepato-biliary surgery, surgical wound sepsis and infective endocarditis. Bacteraemia with these organisms is much more common in the elderly and the compromised patient, and they may be important in infections related to percutaneous intravascular devices. Nosocomial enterococcal infection is increasingly important in intensive care and high-dependency settings, and these organisms are now the third most common cause of nosocomial infection in the United States. There has been much recent concern at the increase in infections due to multiple antibiotic-resistant strains, particularly vancomycin-resistant enterococci (VRE), which are refractory to antibiotic therapy.

Enterococci are somewhat less fastidious than streptococci in their cultural requirements and grow well on a wide variety of routine laboratory media after 18 hours aerobic or anaerobic incubation. On blood agar the organisms yield larger colonies than do streptococci and are white or grayish in colour. Some isolates, particularly those of *E.faecalis*, may produce beta-haemolysis. Enterococci can grow in 6.5% NaCl broth and in the presence of bile salts, which may be useful in enrichment and differentiation from other organisms. The organisms are reactive in agglutination tests for the Lancefield group D antigen, and hydrolyse aesculin and L-pyrrolidinoyl-β-naphthylamide (PYR). The identity of presumptive isolates may be confirmed by further biochemical testing.

Diagnosis

Enterococci may be cultured from a wide variety of clinical specimens including blood, urine, wound swabs and aspirates, and CSF, as outlined above. The organism is commonly isolated as part of the normal faecal flora on enteric media. The interpretation of the clinical significance of enterococcal isolates from other non-sterile body sites may be difficult due to their ubiquitous nature and the frequent involvement of these organisms as a component of 'synergistic' infections. Direct Lancefield group D antigen detection may occasionally be performed on blood and other clinical samples, but cross reactions may occur and sensitivity is low. The presence of enterococci in food and water samples is of some use as a surrogate marker of faecal contamination.

Reference

Moellering RC. *Enterococcus* specie, *Streptococcus bovis* and *Leuconostoc* species. In: Mandell GL, Bennett JE, Dolin R, eds. Principles and Practice of Infectious Diseases, 4th edn. New York, Edinburgh, London: Churchill Livingstone; 1995:1826–35.

Enterococcus species (Streptococcus beta-haemolytic, group D)

ANTIGEN DETECTION

Manufacturer: bioMerieux
Cat. No./Trade name: 58817/Slidex Strepto D

SUMMARY

[Latex-Ab]–**Ag**–agglutination

Assay type: Particle agglutination assay
Detection: Visual
Format: Latex particles, Ab coated (test done on card)
Sample type: Broth culture or colony from solid media culture
Sample pre-treatment:
  Incubate 10-15 min (37°C)
Sample volume: 50 µl of prepared sample x 2
Number of tests: 50
Controls - standards run in assay:
  Controls: Pos (1)
Incubation:
  Read result within 2 min of mixing sample and reagent
Washes:

CONTENTS

Antibodies, antigens, labelled components:
  Anti-group D Strep Ab bound to latex particles (Test Latex)
Substrate:
Controls - standards supplied:
  Controls: Pos
Additional reagents:
  None
Special equipment required:
  None

INTERPRETATION

Comments on interpretation:
  Positive: agglutination of sample with Test Latex
No. of references: 3

NOTES

131280.0

Enterococcus species (Streptococcus beta-haemolytic, group D)

ANTIGEN DETECTION

Manufacturer: Boule Diagnostics AB
Cat. No./Trade name: /Phadebact® Strep D Test

SUMMARY

[Staph-Ab]–**Ag**–coagglutination

Assay type: Particle agglutination assay
  (coagglutination)
Detection: Visual
Format: Staphylococci, Ab coated (test done on slide)
Sample type: Colonies from solid culture media or broth culture*
Sample pre-treatment:
  None
Sample volume: 1-2 colonies or 1 drop broth culture x 2
Number of tests: 50
Controls - standards run in assay:
  None
Incubation:
  Read result within 1 min of mixing sample and reagent
Washes:

CONTENTS

Antibodies, antigens, labelled components:
  Anti-group D Strep Ab bound to staphylococci (Test Reagent)
  Non-immune rabbit Ig Ab bound to staphylococci (Control Reagent)
Substrate:
Controls - standards supplied:
  Controls: Neg
Additional reagents:
  Phadebact® Strep D Pos Control Reagent
Special equipment required:
  None

INTERPRETATION

Comments on interpretation:
  Positive: co-agglutination of sample with Test Reagent and not with Control Reagent
No. of references: None

NOTES

131392.0

*Sample for culture can be taken from any part of the body where viable organisms are present

© KLUWER ACADEMIC PUBLISHERS 1995, ISSN 1381-5067

Enterococcus species (Streptococcus beta-haemolytic, group D)
ANTIGEN DETECTION

Manufacturer: Diagnostic Products Corporation
Cat. No./Trade name: GK5GD/PathoDx® Strep D

SUMMARY

[Latex-Ab]–**Ag**–agglutination

Assay type: Particle agglutination assay
Detection: Visual
Format: Latex particles, Ab coated (test done on card)
Sample type: Colonies from plate culture or enrichment broth culture
Sample pre-treatment:
    None
Sample volume: 5 colonies or 50 µl broth
Number of tests: 50
Controls - standards run in assay:
    Controls: Neg (1), Pos (1) once per day
Incubation:
    1–3 min (RT) with rocking
Washes:

CONTENTS

Antibodies, antigens, labelled components:
    Anti-group D Strep IgG Ab (r) bound to latex particles (Test Latex)
    Non-immune rabbit IgG Ab bound to latex particles (Control Latex)
Substrate:
Controls - standards supplied:
    Controls: Neg (groups A, B & C Strep) and Pos (group D Strep)
Additional reagents:
    None
Special equipment required:
    None

INTERPRETATION

Comments on interpretation:
    Positive: agglutination of sample with Test Latex and no agglutination with Control Latex
No. of references: 11

NOTES

131737.0

© KLUWER ACADEMIC PUBLISHERS 1995, ISSN 1381-5067

Epstein-Barr virus

Natural history

Epstein-Barr Virus (EBV) was discovered in 1964 within the lesion now called Burkitt's lymphoma. This is a common tumour of African children living within the malarial belt of Africa. Thereafter, EBV was shown to be the aetiological agent of infectious mononucleosis (IM), an acute lymphoproliferative disease resulting from primary infection. More recently it has been implicated in other diseases including anaplastic nasopharyngeal carcinoma, a tumour geographically restricted to Eastern countries especially China, X-linked lymphoproliferative syndrome, and in immunocompromised patients, chronic IM, non-Hodgkin's lymphoma and oral hairy leukoplakia.

EBV is the prototype for the Lymphocryptovirus genus of the gamma-herpesvirus subfamily. It has a typical herpesvirus structure of inner protein core wrapped with double-stranded DNA, surrounded by an icosahedral coat of 162 capsomeres and an irregular outer envelope to give a particle of 150–200 nm in size. The genome consists of alternating unique (U) segments and internal tandem repeats (IR) with terminal repeat (TR) segments at each end. Two EBV types (types 1 and 2) circulate in most populations and are identical except for their small nuclear proteins (EBNAs). The virus infects only cells expressing the CD21 receptor, i.e. some squamous epithelial cells and mature B-lymphocytes. In the epithelial cells of the salivary glands there is full replication and release of new virus particles. In B-lymphocytes, latent infection is established with restricted gene expression and immortalisation of the cell.

Primary infection in childhood is generally sub-clinical and leads to a life-long carrier state with periodic virus-shedding from the salivary glands. This acts as a source of infectious virus for spread by close contact (hence the popular name of 'kissing disease'). When infection is delayed until adolescence or young adulthood, symptoms occur in about 50% of cases and can be quite severe and debilitating. IM begins with swelling of the cervical glands and sore throat, accompanied by fever, malaise, anorexia and abdominal pain. Generalised lymphadenopathy is common and can last several weeks. The sore throat can last up to two weeks and there is often a distinct exudate which can make swallowing very difficult and lead to obstruction. Jaundice occurs in 5–10% of cases and a skin rash appears in patients who are treated with ampicillin. Return to full health after IM can take a long time particularly in those over 25 years of age and the frequency of sore throats, fever and lymphadenopathy is increased in many patients in the year following infection. Lymphoproliferative disease can occur post transplant and in chronically immunosuppressed patients including those with HIV and is potentially fatal.

Diagnosis

IM is named after the atypical large mononuclear cells found in the peripheral blood during infection. These cells are CD8 T-lymphocytes involved in the immune response to the virus. Diagnosis is usually by a combination of the blood picture with serological tests. These include the detection of EBV specific IgM antibodies to viral capsid antigen (VCA) expressed in lymphoblastoid cells, by either IFA or ELISA and the detection of non-specific heterophile antibody in the Paul Bunnel reaction. Heterophile antibodies are frequently absent in children under 10 years of age and positive reactions can occasionally be obtained with other infectious agents such as CMV or Toxoplasma. Thus specific tests for antibodies to EBV are essential. IgG antibodies to VCA can be used as a marker of past infection and IgA to VCA is hugely raised in cases of nasopharyngeal carcinoma. Antibodies to other EBV proteins can give a better indication of the timing of infection. Thus, antibodies to the nuclear antigen, EBNA, are slow to develop and a patient who has anti-VCA IgG but no anti-EBNA can be considered to have had EBV in the relatively recent past. Similarly antibodies to the early antigen (EA) complex can also give an idea of the timing of infection.

References

Hotchin NA, Crawford DH. The diagnosis of Epstein-Barr Virus-associated disease. In: Morgan-Capner P, ed., Current Topics in Clinical Virology. PHLS. 1991:115–40.

Liebowitz D, Kieff E. Epstein Barr Virus. In: Roizman B, Whitley RJ, Lopez C, eds., The Human Herpesviruses. 1993:107–72.

See also Multipathogen Assays section under: Infectious Mononucleosis Syndrome Test TECH Screening Panel

Epstein-Barr virus
ANTIBODY DETECTION (IgG)

Manufacturer: Behringwerke AG
Cat. No./Trade name: /Enzygnost® Anti-EBV/IgG

SUMMARY

[Well-Ag]–**Ab**–[AHIgG-POD]–[TMB]–A$_{450}$

Assay type: EIA (non-competitive)
Detection: Colorimetric A$_{450}$
Format: Microtitre well, Ag coated
Sample type: Serum, plasma
Sample pre-treatment:
   None
Sample volume: 20 µl of 1:21 dilution x 2*
Number of tests: 48
Controls - standards run in assay:
   Reference reagent: P/N (2)*
Incubation:
   1 hr (37°C) + 1 hr (37°C) + 30 min (RT)
Washes: 2

CONTENTS

Antibodies, antigens, labelled components:
   EBV Ag from lymphoblastoid cells bound to well
   Control Ag from uninfected lymphoblastoid cells bound to well
   Anti-human IgG Ab (r) POD conjugated
Substrate: TMB (bought separately)
Controls - standards supplied:
   Reference reagent: P/N (Anti EBV Reference Reagent)
Additional reagents:
   Supplementary Reagents for Enzygnost®/TMB (Code No. OUVP)
Special equipment required:
   Behring ELISA Processor II (optional)
   Behring ELISA Processor III (optional)

INTERPRETATION

Comments on interpretation:
   Equivocal: where sample OD is between 0.10 and 0.20; retest to confirm
   Repeatably equivocal: report as equivocal
   For quantitative procedure see manufacturer's instructions
No. of references: 22

NOTES

131022.0

*Samples and controls assayed simultaneously in Ag coated and Control Ag coated wells

Epstein-Barr virus
ANTIBODY DETECTION (IgG)

Manufacturer: Bouty SpA
Cat. No./Trade name: 20914/BEIA EBV IgG

SUMMARY

[Well-Ag]–**Ab**–[AHIgG-HRP]–[TMB]–A$_{450}$

Assay type: EIA (non-competitive)
Detection: Colorimetric A$_{450}$
Format: Microtitre well, Ag coated
Sample type: Serum, plasma (do not heat inactivate)
Sample pre-treatment:
   None
Sample volume: 10 µl (+200 µl diluent)*
Number of tests: 96
Controls - standards run in assay:
   Controls: Neg (1), cut-off (2), Pos (1)
Incubation:
   30 min (RT) + 30 min (RT) + 15 min (RT)
Washes: 2

CONTENTS

Antibodies, antigens, labelled components:
   EBV Ag bound to well
   Anti-human IgG Ab HRP conjugated
Substrate: TMB
Controls - standards supplied:
   Controls: Neg, cut-off and Pos (human serum)
Additional reagents:
   None
Special equipment required:
   None

INTERPRETATION

Comments on interpretation:
   Equivocal: if ratio of OD sample: cut-off value is between 0.95 and 1.05; retest to confirm
No. of references: 4

NOTES

131244.0

*Samples can be assayed singly or in duplicate

Epstein-Barr virus
ANTIBODY DETECTION (IgG)

Manufacturer: Chimica Diagnostica
Cat. No./Trade name: 41800/Epstein-Barr Virus IgG

SUMMARY

[Well-Ag]–**Ab**–[AHIgG-POD]–[TMB]–A_{450}

Assay type: EIA (non-competitive)
Detection: Colorimetric A_{450}
Format: Microtitre well, Ag coated
Sample type: Serum
Sample pre-treatment:
 None
Sample volume: 100 μl of 1:101 dilution x 2
Number of tests: 96
Controls - standards run in assay:
 Controls: Neg (2), cut-off (2), Pos (2)
Incubation:
 30 min (RT) + 30 min (RT) + 15 min (RT)
Washes: 2

CONTENTS

Antibodies, antigens, labelled components:
 EBV Ag bound to well
 Anti-human IgG Ab (r) POD conjugated
Substrate: TMB
Controls - standards supplied:
 Controls: Neg, cut-off and Pos
Additional reagents:
 None
Special equipment required:
 None

INTERPRETATION

Comments on interpretation:
 Equivocal: where Index (ratio of OD sample:OD cut-off
  control) is between > 0.9 and < 1.1; report as
  equivocal
No. of references: 2

NOTES

131358.0

Epstein-Barr virus
ANTIBODY DETECTION (IgG)

Manufacturer: Immuno Pharmacology Research
Cat. No./Trade name: EPSTEIN-BARR IgG ELISA

SUMMARY

[Well-Ag]–**Ab**–[AHIgG-AP]–[PNP]–A_{405}

Assay type: EIA (non-competitive)
Detection: Colorimetric A_{405}
Format: Microtitre well, Ag coated
Sample type: Serum
Sample pre-treatment:
 None
Sample volume: 100 μl of 1:100 dilution x 2
Number of tests: 48, 96
Controls - standards run in assay:
 Controls: Neg (1), Pos (1)
Incubation:
 1 hr (37°C) + 1 hr (37°C) + 30 min (37°C)
Washes: 2

CONTENTS

Antibodies, antigens, labelled components:
 EBV Ag bound to well
 Anti-human IgG Ab AP conjugated
Substrate: PNP
Controls - standards supplied:
 Controls: Neg and Pos
Additional reagents:
 None
Special equipment required:
 None

INTERPRETATION

Comments on interpretation:
 Equivocal: where sample OD is between 0.150 and
  0.300; retest to confirm
 Repeatably equivocal: retest a fresh specimen in 15 days
No. of references: None

NOTES

131318.0

© KLUWER ACADEMIC PUBLISHERS 1995, ISSN 1381-5067

Epstein-Barr virus
ANTIBODY DETECTION (IgG)

Manufacturer: Savyon Diagnostics Ltd
Cat. No./Trade name: 121-01/SeroELISA® EBV IgG

SUMMARY

[Well-Ag]–**Ab**–[AHIgG-HRP]–[TMB]–A_{450}

Assay type: EIA (non-competitive)
Detection: Colorimetric A_{450}
Format: Microtitre well, Ag coated
Sample type: Serum (do not heat inactivate)
Sample pre-treatment:
 None
Sample volume: 50 µl of 1:256 dilution
Number of tests: 96
Controls - standards run in assay:
 Controls: Neg (2), low Pos (2), high Pos (2)
Incubation:
 30 min (37°C) + 30 min (37°C) + 15 min (RT)
Washes: 2

CONTENTS

Antibodies, antigens, labelled components:
 EBV (VCA, EA and EBNA) Ags bound to well
 Anti-human IgG Ab HRP conjugated
Substrate: TMB
Controls - standards supplied:
 Controls: Neg, low Pos and Pos (human serum)
Additional reagents:
 None
Special equipment required:
 None

INTERPRETATION

Comments on interpretation:
 Equivocal: between cut-off x0.9 and cut-off x1.1; retest
 initial sample with a fresh sample taken 14 days later
 Repeatably equivocal: report as negative
No. of references: 31

NOTES

131141.0

Epstein-Barr virus
ANTIBODY DETECTION (IgM)

Manufacturer: Behringwerke AG
Cat. No./Trade name: Enzygnost® Anti-EBV/IgM

SUMMARY

[Well-Ag]–**Ab**–[AHIgM-POD]–[TMB]–A_{450}

Assay type: EIA (non-competitive)
Detection: Colorimetric A_{450}
Format: Microtitre well, Ag coated
Sample type: Serum, plasma
Sample pre-treatment:
 Add RF absorbent provided to diluted sample, incubate
  15 min (RT) or overnight (2–8°C) to eliminate
  interference due to RF and IgG
Sample volume: 150 µl of treated sample*
Number of tests: 48
Controls - standards run in assay:
 Reference reagents: P/N (2), PP (2)*
Incubation:
 1 hr (37°C) + 1 hr (37°C) + 30 min (RT)
Washes: 2

CONTENTS

Antibodies, antigens, labelled components:
 EBV Ag from lymphoblastoid cells bound to well
 Control Ag from uninfected lymphoblastoid cells bound to
  well
 Anti-human IgM Ab (g) POD conjugated
Substrate: TMB (bought separately)
Controls - standards supplied:
 Reference reagent: P/P and P/N (Anti-EBV Reference
  Reagents)
Additional reagents:
 Supplementary reagents for Enzygnost/TMB (Cat. no.
  OUVP)
Special equipment required:
 Behring ELISA Processor II (optional)
 Behring ELISA Processor III (optional)

INTERPRETATION

Comments on interpretation:
 Equivocal: where sample OD is between 0.10 and 0.20;
  retest to confirm
 Repeatably equivocal: report as equivocal and retest a
  fresh specimen at a later date
No. of references: 34

NOTES

131026.0

*Samples and controls are assayed simultaneously in Ag
 coated and control Ag coated wells

© KLUWER ACADEMIC PUBLISHERS 1995, ISSN 1381-5067

Epstein-Barr virus
ANTIBODY DETECTION (IgM)

Manufacturer: Bouty SpA
Cat. No./Trade name: 20916/BEIA EBV IgM Capture

SUMMARY

[Well-AHIgA]–**Ab**–[Ag-biotin]–[strept-HRP]–[TMB]–A_{450}

Assay type: EIA (non-competitive)
Detection: Colorimetric A_{450}
Format: Microtitre well, Ab coated
Sample type: Serum, plasma
Sample pre-treatment:
   None
Sample volume: 10 µl (+800 µl diluent)*
Number of tests: 96
Controls - standards run in assay:
   Controls: Neg (1), cut-off (2), Pos (1)
Incubation:
   1 hr (RT) + 1 hr (RT) + 30 min (RT)
Washes: 2

CONTENTS

Antibodies, antigens, labelled components:
   Anti-human IgM Ab bound to well
   EBV Ag biotinylated
   Streptavidin HRP conjugated
Substrate: TMB
Controls - standards supplied:
   Controls: Neg, cut-off and Pos (human serum)
Additional reagents:
   None
Special equipment required:
   None

INTERPRETATION

Comments on interpretation:
   Equivocal: if ratio of OD sample:cut-off value is between
      0.95 and 1.05; retest to confirm
No. of references: 5

NOTES

131245.0

*Samples can be assayed singly or in duplicate

Epstein-Barr virus
ANTIBODY DETECTION (IgM)

Manufacturer: Chimica Diagnostica
Cat. No./Trade name: 41850/Epstein-Barr Virus IgM

SUMMARY

[Well-Ag]–**Ab**–[AHIgM-POD]–[TMB]–A_{450}

Assay type: EIA (non-competitive)
Detection: Colorimetric A_{450}
Format: Microtitre well, Ag coated
Sample type: Serum
Sample pre-treatment:
   None
Sample volume: 100 µl of 1:101 dilution x 2*
Number of tests: 96
Controls - standards run in assay:
   Controls: Neg (2), cut-off (2), Pos (2)
Incubation:
   30 min (RT) + 30 min (RT) + 15 min (RT)
Washes: 2

CONTENTS

Antibodies, antigens, labelled components:
   EBV Ag bound to well
   Anti-human IgM Ab (r) POD conjugated
Substrate: TMB
Controls - standards supplied:
   Controls: Neg, cut-off and Pos
Additional reagents:
   None
Special equipment required:
   None

INTERPRETATION

Comments on interpretation:
   Equivocal: where Index (ratio of OD sample:OD cut-off
      control) is between >0.9 and <1.1; report as
      equivocal
No. of references: 2

NOTES

131363.0

*Diluent contains AHIgG to eliminate interference due to RF
   and IgG in sample

© KLUWER ACADEMIC PUBLISHERS 1995, ISSN 1381-5067

Epstein-Barr virus
ANTIBODY DETECTION (IgM)

Manufacturer: Immuno Pharmacology Research
Cat. No./Trade name: EPSTEIN-BARR IgM ELISA

SUMMARY

[Well-Ag]–**Ab**–[AHIgM-AP]–[PNP]–A_{405}

Assay type: EIA (non-competitive)
Detection: Colorimetric A_{450}
Format: Microtitre well, Ag coated
Sample type: Serum
Sample pre-treatment:
    None
Sample volume: 100 µl of 1:100 dilution x 2
Number of tests: 48
Controls - standards run in assay:
    Controls: Neg (1), Pos (1)
Incubation:
    1 hr (37°C) + 1 hr (37°C) + 30 min (37°C)
Washes: 2

CONTENTS

Antibodies, antigens, labelled components:
    EBV Ag bound to well
    Anti-human IgM Ab AP conjugated
Substrate: PNP
Controls - standards supplied:
    Controls: Neg and Pos
Additional reagents:
    None
Special equipment required:
    None

INTERPRETATION

Comments on interpretation:
    Equivocal: where OD is between 0.150 and 0.300; retest
        to confirm
    Repeatably equivocal: retest a fresh specimen in 15 days
No. of references: None

NOTES

131323.0

Epstein-Barr virus
ANTIBODY DETECTION (IgM)

Manufacturer: Savyon Diagnostics Ltd
Cat. No./Trade name: 122-01/SeroELISA® EBV TRUE-IgM

SUMMARY

[Well-Ag]–**Ab**–[AHIgM-HRP]–[TMB]–A_{450}

Assay type: EIA (non-competitive)
Detection: Colorimetric A_{450}
Format: Microtitre well, Ag coated
Sample type: Serum (do not heat inactivate)
Sample pre-treatment:
    None
Sample volume: 50 µl of 1:512 dilution*
Number of tests: 96
Controls - standards run in assay:
    Controls: Neg (2), Pos (2)
Incubation:
    30 min (37°C) + 30 min (37°C) + 15 min (RT)
Washes: 2

CONTENTS

Antibodies, antigens, labelled components:
    EBV (VCA, EB and EBNA) Ags bound to well
    Anti-human IgM Ab HRP conjugated
Substrate: TMB
Controls - standards supplied:
    Controls: Neg and Pos (human serum)
Additional reagents:
    None
Special equipment required:
    None

INTERPRETATION

Comments on interpretation:
    Equivocal: between cut-off x0.9 and cut-off x1.1; retest
        initial sample with a fresh sample taken 14 days later
    Repeatably equivocal: report as negative
No. of references: 28

NOTES

131142.0

*Samples and controls are diluted with diluent containing an
    IgG neutralizing reagent to eliminate interference due to
    RF and IgG

Epstein-Barr virus
ANTIBODY DETECTION (IgG AND/OR IgM)

Manufacturer: Scimedix Corporation
Cat. No./Trade name: W500/Epstein Barr Virus Western Immunoblot

SUMMARY

[Strip-Ag]–**Ab**–[AHIgG/IgM-AP]–[BCIP]–visible bands

Assay type: Immunoblot assay
Detection: Visual
Format: PVDF membrane strips, Ag coated
Sample type: Serum
Sample pre-treatment:
Block strips prior to use with blocking buffer provided, incubate 20-45 min (RT)
Sample volume: 20 µl (+1000 µl diluent)
Number of tests: 8
Controls - standards run in assay:
Controls: Pos (1) per run
Incubation:
40 min (RT) + 30 min (RT) + 5 min (RT) - all with shaking
Washes: 2

CONTENTS

Antibodies, antigens, labelled components:
EBV Ag bound to PVDF membrane strip
Anti-human IgG Ab (r) AP conjugated
Anti-human IgM Ab (r) AP conjugated
Substrate: BCIP
Controls - standards supplied:
Controls: Pos (IgG) and Pos (IgM)
Additional reagents:
Positive control serum
Special equipment required:
Shaking platform (50-200 rpm)

INTERPRETATION

Comments on interpretation:
Identify the Ab reactivity of the sample by comparing the bands on the Ag test strip with the bands on the Ag template strip provided
No. of references: None

NOTES

131185.0

For research and investigational use only
This kit can detect IgG and IgM antibodies and can differentiate between them

---

Epstein-Barr virus
ANTIBODY DETECTION (IgM)

Manufacturer: Quidel Corporation
Cat. No./Trade name: 0901/Cards O.S. Mono 20 Test

SUMMARY

[Membrane-Ag]–**Ab**–[AHIgM-latex]–fluorescence

Assay type: Immunochromatographic assay
Detection: Visual
Format: Reaction unit, Ag coated membrane
Sample type: Serum, plasma, whole blood
Sample pre-treatment:
None
Sample volume: 50 µl
Number of tests: 20
Controls - standards run in assay:
Controls: Neg (1), Pos (1)
Integral controls: Neg (1), Pos (1) on membrane
Incubation:
Read result as soon as blue line appears in "Test Complete" window
Washes: 1

CONTENTS

Antibodies, antigens, labelled components:
EBV Ag (from bovine serum) bound to membrane
Anti-human IgM Ab bound to latex beads
Substrate:
Controls - standards supplied:
Controls: Neg and Pos
Integral controls incorporated in membrane
Additional reagents:
None
Special equipment required:
None

INTERPRETATION

Comments on interpretation:
Negative: horizontal blue line in "Read Result" window
Positive: horizontal and verticle blue line in "Read Result" window
No. of references: 11

NOTES

131080.0

Epstein-Barr virus (EA)
ANTIBODY DETECTION (IgG)

Manufacturer: Biotest Diagnostics
Cat. No./Trade name: Biotest Anti-EBV recombinant EA IgG

SUMMARY

[Well-Ag]–**Ab**–[AHIgG-POD]–[OPD]–A_{492}

Assay type: EIA (non-competitive)
Detection: Colorimetric A_{492}
Format: Microtitre well, Ag coated
Sample type: Serum, plasma (do not heat inactivate)
Sample pre-treatment:
  None
Sample volume: 10 µl of 1:21 dilution
Number of tests: 96
Controls - standards run in assay:
  Controls: Neg (2), Pos (2)
Incubation:
  30 min (40°C) + 30 min (40°C) + 15 min (RT)
Washes: 2

CONTENTS

Antibodies, antigens, labelled components:
  rec EBV-EA (p54/p138) Ag bound to well
  Anti-human IgG MAb POD conjugated
Substrate: OPD
Controls - standards supplied:
  Controls: Neg and Pos (human plasma)
Additional reagents:
  H_2SO_4
Special equipment required:
  None

INTERPRETATION

Comments on interpretation:
  Classification of samples is according to cut-off; no
  further testing
No. of references: 0

NOTES

131035.0

Epstein-Barr virus (EA)
ANTIBODY DETECTION (IgG)

Manufacturer: Gull Laboratories Inc
Cat. No./Trade name: EAE101/EBV-EA (D) ELISA Test

SUMMARY

[Well-Ag]–**Ab**–[AHIgG-AP]–[PNP]–A_{405}

Assay type: EIA (non-competitive)
Detection: Colorimetric A_{405}
Format: Microtitre well, Ag coated
Sample type: Serum
Sample pre-treatment:
  None
Sample volume: 100 µl of 1:21 dilution
Number of tests: 96
Controls - standards run in assay:
  Controls: Neg (1), Pos (1)
  Calibrators: 1 (1), 2 (1), 3 (1)*
  Reference serum: 1 (3)
Incubation:
  30 min (37°C) + 30 min (37°C) + 30 min (37°C)
Washes: 2

CONTENTS

Antibodies, antigens, labelled components:
  EBV-EA (D component) bound to well
  Anti-human IgG Ab (caprine) AP conjugated
Substrate: PNP
Controls - standards supplied:
  Controls: Neg and Pos
  Calibrators: 3
  Reference serum: 1 (all human serum)
Additional reagents:
  None
Special equipment required:
  None

INTERPRETATION

Comments on interpretation:
  Equivocal: where sample OD is between mean OD of
    Reference Serum x 0.91 and mean OD of Reference
    Serum x 0.99; retest to confirm
  Repeatably equivocal: retest a fresh specimen or use an
    alternative method
No. of references: 10

NOTES

131696.0

*Omit calibrators for qualitative assay

© KLUWER ACADEMIC PUBLISHERS 1995, ISSN 1381-5067

Epstein-Barr virus (EA)
ANTIBODY DETECTION (IgG)

Manufacturer: Immuno Concepts Inc
Cat. No./Trade name: EA Antibody Test System

SUMMARY

[Slide-Ag]–**Ab**–[AHIgG-HRP]–[4CN]–purple stain

Assay type: EIA (non-competitive)
Detection: Light microscopy
Format: Slide well, Ag coated
Sample type: Serum
Sample pre-treatment:
  None
Sample volume: 100 µl of 1:10 dilution
Number of tests: 13*
Controls - standards run in assay:
  Controls: Neg (1), Pos (1)
Incubation:
  30 min (RT) + 30 min (RT) + 30 min (RT)
Washes: 2

CONTENTS

Antibodies, antigens, labelled components:
  EBV-EA infected and uninfected lymphoblastoid cells
    from Burkitt's lymphoma bound to slide well
  Anti-human IgG Ab (g) HRP conjugated
Substrate: 4-chloro-1-naphthol (4CN)
Controls - standards supplied:
  Controls: Neg, Pos and titrable control (human serum)
Additional reagents:
  None
Special equipment required:
  None

INTERPRETATION

Comments on interpretation:
  Positive: light/dark blue purple staining of nucleus and
    cytoplasm in 5–10% of cells per field at ≥ 1:10
    dilution
No. of references: 17

NOTES

131811.0

*13 wells per slide

Epstein-Barr virus (EA)
ANTIBODY DETECTION (IgM)

Manufacturer: Biotest Diagnostics
Cat. No./Trade name: Biotest Anti-EBV recombinant IgM Elisa

SUMMARY

[Well-Ag]–**Ab**–[AHIgM-POD]–[OPD]–A_{492}

Assay type: EIA (non-competitive)
Detection: Colorimetric A_{492}
Format: Microtitre well, Ag coated
Sample type: Serum, plasma (do not heat inactivate)
Sample pre-treatment:
  None
Sample volume: 10 µl of 1:21 dilution
Number of tests: 96
Controls - standards run in assay:
  Controls: Neg (2), Pos (2)
Incubation:
  1 hr (40°C) + 30 min (40°C) + 15 min (RT)
Washes: 2

CONTENTS

Antibodies, antigens, labelled components:
  EBV-EA (p54/p138) Ag bound to well
  Anti-human IgM MAb POD conjugated
Substrate: OPD
Controls - standards supplied:
  Controls: Neg and Pos (human plasma)
Additional reagents:
  H_2SO_4
Special equipment required:
  None

INTERPRETATION

Comments on interpretation:
  Classification of samples is according to cut-off; no
    further testing
No. of references: None

NOTES

131380.0

Epstein-Barr virus (EA)
ANTIBODY DETECTION (IgG)

Manufacturer: Gull Laboratories Inc
Cat. No./Trade name: EA101/EBV-EA Test

SUMMARY

[Slide-Ag]–**Ab**–[AHIgG-FITC]–fluorescence

Assay type: Immunofluorescence assay (indirect)
Detection: Fluorescence microscopy
Format: Slide well, Ag coated
Sample type: Serum
Sample pre-treatment:
  None
Sample volume: 15 µl of 1:10 dilution
Number of tests: 100
Controls - standards run in assay:
  Controls: Neg (1), Pos (1)
Incubation:
  30 min (RT) + 30 min (RT)
Washes: 2

CONTENTS

Antibodies, antigens, labelled components:
  EBV-EA infected and uninfected Raji cells from Burkitt's
    lymphoma bound to slide well
  Anti-human IgG Ab (caprine) FITC conjugated
Substrate:
Controls - standards supplied:
  Controls: Neg and Pos (human serum)
Additional reagents:
  None
Special equipment required:
  None

INTERPRETATION

Comments on interpretation:
  Positive: yellow-green cytoplasmic or whole cell
    fluorescence in 15–25% of cells per field
No. of references: 16

NOTES

131705.0

Epstein-Barr virus (EA)
ANTIBODY DETECTION (IgG)

Manufacturer: Immuno Concepts Inc
Cat. No./Trade name: EA Antibody Test System

SUMMARY

[Slide-Ag]–**Ab**–[AHIgG-FITC]–fluorescence

Assay type: Immunofluorescence assay (indirect)
Detection: Fluorescence microscopy
Format: Slide well, Ag coated
Sample type: Serum
Sample pre-treatment:
  None
Sample volume: 100 µl of 1:10 dilution
Number of tests: 13*
Controls - standards run in assay:
  Controls: Neg (1), Pos (1)
Incubation:
  30 min (RT) + 30 min (RT)
Washes: 2

CONTENTS

Antibodies, antigens, labelled components:
  EBV-EA infected and uninfected lymphoblastoid cells
    from Burkitt's lymphoma bound to slide well
  Anti-human IgG Ab (g) FITC conjugated
Substrate:
Controls - standards supplied:
  Controls: Neg, Pos and titratable control (human serum)
Additional reagents:
  None
Special equipment required:
  None

INTERPRETATION

Comments on interpretation:
  Positive: apple-green fluorescence of nucleus and
    cytoplasm of 5–10% of cells per field at ⩾1:10
    dilution
No. of references: 17

NOTES

131684.0

*13 wells per slide

© *KLUWER ACADEMIC PUBLISHERS 1995, ISSN 1381-5067*

Epstein-Barr virus (EA)
ANTIBODY DETECTION (IgG)

Manufacturer: MRL Diagnostics
Cat. No./Trade name: IF0700/EBV Early Antigen IFA

SUMMARY

[Slide well-Ag]–**Ab**–[AHIgG-FITC]–fluorescence

Assay type: Immunofluorescence assay (indirect)
Detection: Fluorescence microscopy
Format: Slide well, Ag coated
Sample type: Serum
Sample pre-treatment:
  None
Sample volume: 15 µl of 1:10 dilution
Number of tests: 120
Controls - standards run in assay:
  Controls: Neg (1), Pos (1)
Incubation:
  30 min (37°C) + 30 min (37°C)
Washes: 2

CONTENTS

Antibodies, antigens, labelled components:
  EBV-EA infected lymphocytes and uninfected cells
    bound to slide well
  Anti-human IgG Ab (g) FITC conjugated
Substrate:
Controls - standards supplied:
  Controls: Neg and Pos (human serum)
Additional reagents:
  None
Special equipment required:
  None

INTERPRETATION

Comments on interpretation:
  Positive: ≥(1+) apple-green fluorescence
No. of references: 9

NOTES

131117.0

Epstein-Barr virus (EBNA)
ANTIBODY DETECTION

Manufacturer: Immuno Concepts Inc
Cat. No./Trade name: EBNA Antibody Test System

SUMMARY

[Slide-Ag]–**Ab**–[GPC]–[AGPC-HRP]–[4CN]–purple stain

Assay type: EIA (non-competitive)
Detection: Light microscopy
Format: Slide well, Ag coated
Sample type: Serum
Sample pre-treatment:
  None
Sample volume: 100 µl of 1:10 dilution
Number of tests: 13*
Controls - standards run in assay:
  Controls: Neg (1), Pos (1)
Incubation:
  30 min (RT) + 30 min (RT) + 30 min (RT)
Washes: 2

CONTENTS

Antibodies, antigens, labelled components:
  EBV-EBNA infected Raji cells and uninfected cells bound
    to slide well
  Guinea pig complement (GPC)
  Anti-guinea pig complement (AGPC) HRP conjugated
Substrate: 4-chloro-1-naphthol (4CN)
Controls - standards supplied:
  Controls: Neg, Pos and titrable control (human serum)
Additional reagents:
  None
Special equipment required:
  None

INTERPRETATION

Comments on interpretation:
  Positive: light/dark blue purple staining of 50% of cells
    per field at ≥1:10 dilution
No. of references: 9

NOTES

131812.0

*13 wells per slide

Epstein-Barr virus (EBNA)
ANTIBODY DETECTION (IgG)

Manufacturer: Biotest Diagnostics
Cat. No./Trade name: Biotest Anti-EBV recombinant EBNA IgG

SUMMARY

[Well-Ag]–**Ab**–[AHIgG-POD]–[OPD]–A_{492}

Assay type: EIA (non-competitive)
Detection: Colorimetric A_{492}
Format: Microtitre well, Ag coated
Sample type: Serum, plasma (do not heat inactivate)
Sample pre-treatment:
  None
Sample volume: 10 μl of 1:21 dilution
Number of tests: 96
Controls - standards run in assay:
  Controls: Neg (2), Pos (2)
Incubation:
  Method 1: 30 min (40°C) + 30 min (40°C) + 15 min (RT)*
  Method 2: 1 hr (40°C) + 30 min (40°C) + 15 min (RT)
Washes:

CONTENTS

Antibodies, antigens, labelled components:
  rec EBV-EBNA-1 (p72) Ag bound to well
  Anti-human IgG MAb POD conjugated
Substrate: OPD
Controls - standards supplied:
  Controls: Neg and Pos (human plasma)
Additional reagents:
  H_2SO_4
Special equipment required:
  None

INTERPRETATION

Comments on interpretation:
  Classification of samples is according to cut-off. Results negative by Method 1 should be confirmed by Method 2
No. of references: 0

NOTES

131036.0

*Method 1 is for diagnosis of acute EBV infection. Method 2 is for determining EBNA immune status

---

Epstein-Barr virus (EBNA)
ANTIBODY DETECTION (IgG)

Manufacturer: BioWhittaker Inc
Cat. No./Trade name: 30-681U-EBNA Elisa

SUMMARY

[Well-Ag]–**Ab**–[AHIgG-AP]–[PMP]–A_{550}

Assay type: EIA (non-competitive)
Detection: Colorimetric A_{550}
Format: Microtitre well, Ag coated
Sample type: Serum (do not heat inactivate)
Sample pre-treatment:
  None
Sample volume: 10 μl (+200 μl diluent)
Number of tests: 192
Controls - standards run in assay:
  Controls: A (1), B (1)
  Calibrators: Neg (1), Pos 1 (1), Pos 2 (1)
Incubation:
  45 min (RT) + 45 min (RT) + 45 min (RT)
Washes: 2 (+preliminary plate wash)

CONTENTS

Antibodies, antigens, labelled components:
  EBV-EBNA Ag bound to well
  Anti-human IgG Ab (g) AP conjugated
Substrate: PMP
Controls - standards supplied:
  Controls: A and B
  Calibrators: Neg, Pos 1, Pos 2 (all human serum)
Additional reagents:
  None
Special equipment required:
  BioWhittaker automated system and software (optional)

INTERPRETATION

Comments on interpretation:
  Equivocal: Elisa value between 0.15–0.16 or Index value between 0.88–0.89; retest to confirm
  Repeatably equivocal: retest a fresh sample or use an alternative method
No. of references: 3

NOTES

131745.0

Epstein Barr virus (EBNA)
ANTIBODY DETECTION (IgG)

Manufacturer: BioWhittaker Inc
Cat. No./Trade name: 30-381U/EBNA Stat

SUMMARY

[Well-Ag]–**Ab**–[AHIgG-AP]–[PMP]–A_{550}

Assay type: EIA (non-competitive)
Detection: Colorimetric A_{550}
Format: Microtitre well, Ag coated
Sample type: Serum (do not heat inactivate)
Sample pre-treatment:
　None
Sample volume: 10 μl (+200 μl diluent)
Number of tests: 192
Controls - standards run in assay:
　Controls: Neg (1), Pos (1)
　Standards: Neg (1), low Pos (1), high Pos (1)
Incubation:
　15 min (RT) + 15 min (RT) + 15 min (RT)
Washes: 2 (+preliminary plate wash)

CONTENTS

Antibodies, antigens, labelled components:
　EBV-EBNA Ag bound to well
　Anti-human IgG Ab (g) AP conjugated
Substrate: PMP
Controls - standards supplied:
　Controls: Neg and Pos
　Standards: Neg, low Pos and high Pos (all human serum)
Additional reagents:
　None
Special equipment required:
　BioWhittaker automated system and software (optional)

INTERPRETATION

Comments on interpretation:
　Equivocal: Predicted Index value (PIV) between 0.80–
　　0.90; retest to confirm
　Repeatably equivocal: retest a fresh sample or use
　　alternative method
　Low Positive: PIV between 1.00–1.74
　Mid Positive: PIV between 1.75–3.89
　High Positive: PIV ⩾3.90
No. of references: 16

NOTES

131752.0

Epstein-Barr virus (EBNA)
ANTIBODY DETECTION (IgG)

Manufacturer: Diesse
Cat. No./Trade name: 91057/ENZYWELL Epstein Barr EBNA IgG

SUMMARY

[Well-Ag]–**Ab**–[AHIgG-POD]–[TMB]–A_{450}

Assay type: EIA (non-competitive)
Detection: Colorimetric A_{450}
Format: Microtitre well, Ag coated
Sample type: Serum (do not heat inactivate)
Sample pre-treatment:
　None
Sample volume: 50 μl of 1:26 dilution x 2*
Number of tests: 96
Controls - standards run in assay:
　Controls: Neg (2), cut-off (3), Pos (2)
Incubation:
　45 min (37°C) + 45 min (37°C) + 15 min (RT)
Washes: 2

CONTENTS

Antibodies, antigens, labelled components:
　rec EBV-EBNA Ag bound to well
　Anti-human IgG MAb POD conjugated
Substrate: TMB
Controls - standards supplied:
　Controls: Neg, cut-off and Pos (human serum)
Additional reagents:
　None
Special equipment required:
　None

INTERPRETATION

Comments on interpretation:
　Equivocal: within cut-off ±10%; retest to confirm
　Repeatably equivocal: repeat with a fresh sample
No. of references: None

NOTES

131443.0

*Sorbent G solution (provided) is also added to sample wells
　to neutralize non-specific reactions

© KLUWER ACADEMIC PUBLISHERS 1995, ISSN 1381-5067

Epstein-Barr virus (EBNA)
ANTIBODY DETECTION (IgG)

Manufacturer: Gull Laboratories Inc
Cat. No./Trade name: ENE101/EBNA IgG ELISA Test

SUMMARY

[Well-Ag]–**Ab**–[AHIgG-AP]–[PNP]–A_{405}

Assay type: EIA (non-competitive)
Detection: Colorimetric A_{405}
Format: Microtitre well, Ag coated
Sample type: Serum
Sample pre-treatment:
 None
Sample volume: 100 µl of 1:21 dilution
Number of tests: 96
Controls - standards run in assay:
 Controls: Neg (1), Pos (1)
 Calibrators: 1 (1), 2 (1), 3 (1)*
 Reference serum: 1 (3)
Incubation:
 30 min (37°C) + 30 min (37°C) + 30 min (37°C)
Washes: 2

CONTENTS

Antibodies, antigens, labelled components:
 EBV-EBNA-1 Ag bound to well
 Anti-human IgG Ab (caprine) AP conjugated
Substrate: PNP
Controls - standards supplied:
 Controls: Neg and Pos
 Calibrators: 3
 Reference serum: 1 (all human serum)
Additional reagents:
 None
Special equipment required:
 None

INTERPRETATION

Comments on interpretation:
 Equivocal: where sample OD is between mean OD of
 Reference Serum x0.91 and mean OD of Reference
 Serum x0.99; retest to confirm
 Repeatably equivocal: retest a fresh specimen or use an
 alternative method
No. of references: 16

NOTES

131697.0

*Omit calibrators for qualitative assay

Epstein-Barr virus (EBNA)
ANTIBODY DETECTION (IgG)

Manufacturer: IFCI Clone Systems
Cat. No./Trade name: EIAGEN Epstein Barr EBNA IgG

SUMMARY

[Well-Ag]–**Ab**–[AHIgG-POD]–[TMB]–A_{450}

Assay type: EIA (non-competitive)
Detection: Colorimetric A_{450}
Format: Microtitre well, Ag coated
Sample type: Serum
Sample pre-treatment:
 None
Sample volume: 40 µl (+1 ml diluent)*
Number of tests: 96
Controls - standards run in assay:
 Controls: Neg (2), cut-off (3), Pos (2)
Incubation:
 45 min (37°C) + 45 min (37°C) + 15 min (RT)
Washes: 2

CONTENTS

Antibodies, antigens, labelled components:
 EBV-EBNA Ag bound to well
 Anti-human IgG MAb POD conjugated
Substrate: TMB
Controls - standards supplied:
 Controls: Neg, cut-off and Pos (human serum)
Additional reagents:
 None
Special equipment required:
 None

INTERPRETATION

Comments on interpretation:
 Equivocal: within cut-off ±10%; retest to confirm
 Repeatably equivocal: retest a fresh sample
No. of references: None

NOTES

131672.0

*Samples are assayed in duplicate and Sorbitol G Reagent
 (provided) is added to diluted serum to neutralize non-
 specific reactions

Epstein-Barr virus (EBNA)
ANTIBODY DETECTION (IgG)

Manufacturer: Incstar
Cat. No./Trade name: 7580/Epstein Barr EBNA IgG

SUMMARY

[Well-Ag]–**Ab**–[AHIgG-HRP]–[TMB]–A_{450}

Assay type: EIA (non-competitive)
Detection: Colorimetric A_{450}
Format: Microtitre well, Ag coated
Sample type: Serum
Sample pre-treatment:
  None
Sample volume: 100 µl of 1:101 dilution
Number of tests: 96
Controls - standards run in assay:
  Controls: Neg (1), low Pos (1), high Pos (1)
  Calibrators: I-Neg (1), II-cut-off (1), III-low Pos (1), IV-high Pos (1)
Incubation:
  1 hr (37°C) + 1 hr (37°C) + 30 min (RT)
Washes: 2

CONTENTS

Antibodies, antigens, labelled components:
  EBV-EBNA (1-peptide) Ag bound to well
  Anti-human IgG Ab (g) HRP conjugated
Substrate: TMB
Controls - standards supplied:
  Controls: Neg, low Pos and high Pos (human serum)
  Calibrators: I-Neg, II-cut-off, III-low Pos, IV-high Pos (human serum)
Additional reagents:
  None
Special equipment required:
  None

INTERPRETATION

Comments on interpretation:
  Classification of samples is according to cut-off; no further testing
No. of references: 13

NOTES

131209.0

Epstein-Barr virus (EBNA)
ANTIBODY DETECTION (IgG)

Manufacturer: Ortho Diagnostic Systems
Cat. No./Trade name: 520020/ORTHO® Epstein-Barr Virus EBNA-IgG Antibody ELISA Test

SUMMARY

[Well-Ag]–**Ab**–[AHIgG-POD]–[OPD]–A_{490}

Assay type: EIA (non-competitive)
Detection: Colorimetric A_{490}
Format: Microtitre well, Ag coated
Sample type: Serum
Sample pre-treatment:
  None
Sample volume: 10 µl (+200 µl diluent)
Number of tests: 192
Controls - standards run in assay:
  Controls: Neg (2), cut-off (3), Pos (2)
Incubation:
  30 min (37°C) + 1 hr (37°C) + 30 min (RT)
Washes: 2

CONTENTS

Antibodies, antigens, labelled components:
  EBV-EBNA (protein 70 kD) Ag bound to well
  Anti-human IgG Ab (g) POD conjugated
Substrate: OPD
Controls - standards supplied:
  Controls: Neg, cut-off and Pos (human serum)
Additional reagents:
  H_2SO_4
Special equipment required:
  None

INTERPRETATION

Comments on interpretation:
  Classification of samples is according to cut-off; no further testing
No. of references: 14

NOTES

131220.0

© KLUWER ACADEMIC PUBLISHERS 1995, ISSN 1381-5067

Epstein-Barr virus (EBNA)
ANTIBODY DETECTION (IgG)

Manufacturer: Sigma Diagnostics
Cat. No./Trade name: SIA 132/SIA® Epstein-Barr EBNA IgG

SUMMARY

[Well-Ag]–**Ab**–[AHIgG-AP]–[PNP]–A_{405}

Assay type: EIA (non-competitive)
Detection: Colorimetric A_{405}
Format:
Sample type: Serum (do not heat inactivate)
Sample pre-treatment:
  None
Sample volume: 10 µl (+500 µl diluent)
Number of tests: 96
Controls - standards run in assay:
  Controls: low Pos (1), high Pos 1
  Calibrators: 1 (2), 2 (2), 3 (2)
Incubation:
  30 min (RT) + 30 min (RT) + 45 min (RT)
Washes: 2

CONTENTS

Antibodies, antigens, labelled components:
  rec EBV-EBNA-1 (polypeptide 28 kDa) Ag bound to well
  Anti-human IgG Ab (sh or g) AP conjugated
Substrate: PNP
Controls - standards supplied:
  Controls: low Pos and high Pos (human serum)
  Calibrators: 1, 2 and 3 (IgG Ab titre in AU/ml – human serum)
Additional reagents:
  Neg control (Cat. no. N3519 - optional)
Special equipment required:
  EIA Microwell Plate reader (Cat. no. EQ104 – optional) or EIA Microwell Strip reader (Cat. no. EQ103 – optional)

INTERPRETATION

Comments on interpretation:
  Negative: <56 AU/ml
  Positive: ≥56 AU/ml
No. of references: 11

NOTES

131432.0

Epstein-Barr virus (EBNA)
ANTIBODY DETECTION (IgG AND/OR IgM)

Manufacturer: Meridian Diagnostics Inc
Cat. No./Trade name: 776500/MONOLERT® Rapid Elisa Test

SUMMARY

[Well-Ag]–**Ab**–[AHIgG/IgM-HRP]–[ABTS]–green colour

Assay type: EIA (non-competitive)
Detection: Visual
Format: Well, Ag coated (test done in paddle device)*
Sample type: Serum, plasma
Sample pre-treatment:
  None
Sample volume: 50 µl x 2
Number of tests: 100
Controls - standards run in assay:
  Controls: Neg (2), Pos (2)
Incubation:
  2 min (RT) + 2 min (RT) + 2 min (RT)
Washes: 2

CONTENTS

Antibodies, antigens, labelled components:
  EBV-EBNA-1 (peptide) Ag bound to well
  Anti-human IgG MAb (m) HRP conjugated
  Anti-human IgM MAb (m) HRP conjugated
Substrate: ABTS
Controls - standards supplied:
  None
Additional reagents:
  Meridian Reagent Control Sera (Cat. no. 776520)
Special equipment required:
  None

INTERPRETATION

Comments on interpretation:
  Immune status:
  No previous exposure: no colour in IgG or IgM well
  Previous exposure: green colour in IgG well > green colour in IgM well
  Acute/current infection: green colour in IgM well ≥ green colour in IgG well
No. of references: 15

NOTES

131383.0

This kit can detect IgG and IgM antibodies and can differentiate between them
*Each paddle-shaped device has two Ag coated wells – one paddle per patient sample or Pos and Neg controls (in duplicate)

© KLUWER ACADEMIC PUBLISHERS 1995, ISSN 1381-5067

Epstein-Barr virus (EBNA)
ANTIBODY DETECTION (IgM)

Manufacturer: Incstar
Cat. No./Trade name: 5580/Epstein Barr EBNA IgM

SUMMARY

[Well-Ag]–**Ab**–[AHIgM-AP]–[PNP]–A_{405}

Assay type: EIA (non-competitive)
Detection: Colorimetric A_{405}
Format: Microtitre well, Ag coated
Sample type: Serum
Sample pre-treatment:
  Dilute serum with Serum and Control Pretreatment Reagent and incubate 30 min (RT) to eliminate interference due to RF and IgG
Sample volume: 200 μl of treated sample
Number of tests: 96
Controls - standards run in assay:
  Controls: low Pos (1), high Pos (1)
  Calibrators: I-Neg (2), II-low Pos (2), III-high Pos (2)
Incubation:
  30 min (RT) + 30 min (RT) + 45 min (RT)
Washes: 2

CONTENTS

Antibodies, antigens, labelled components:
  EBV-EBNA (1 peptide) Ag bound to well
  Anti-human IgM Ab (sh) AP conjugated
Substrate: PNP
Controls - standards supplied:
  Controls: low Pos and high Pos (human serum)
  Calibrators: I-Neg, II-low Pos, III-high Pos (human serum
Additional reagents:
  NaOH
Special equipment required:
  None

INTERPRETATION

Comments on interpretation:
  The assay can be run in parallel with Epstein Barr EBNA IgG Kit and interpretation of results depends on relative levels of IgG Ab and IgM Ab in sample considered in conjunction with clinical data - see manufacturer's notes
No. of references: 21

NOTES

131211.0

Epstein-Barr virus (EBNA)
ANTIBODY DETECTION (IgM)

Manufacturer: Sigma Diagnostics
Cat. No./Trade name: SIA 133/SIA® Epstein Barr EBNA IgM

SUMMARY

[Well-Ag]–**Ab**–[AHIgM-AP]–[PNP]–A_{405}

Assay type: EIA (non-competitive)
Detection: Colorimetric A_{405}
Format: Microtitre well, Ag coated
Sample type: Serum (do not heat inactivate)
Sample pre-treatment:
  Add Pretreatment Solution to sample, incubate 30 min (RT) to eliminate interference due to RF and IgG
Sample volume: 200 μl of treated sample
Number of tests: 96
Controls - standards run in assay:
  Controls: Neg (1), Pos (1)
  Calibrators: 1 (2), 2 (2), 3 (2)
Incubation:
  30 min (RT) + 30 min (RT) + 45 min (RT)
Washes: 2

CONTENTS

Antibodies, antigens, labelled components:
  rec EBV-EBNA-1 (polypeptide 28 kDa) Ag bound to well
  Anti-human IgM Ab (sh or g) AP conjugated
Substrate: PNP
Controls - standards supplied:
  Controls: Neg and Pos (human serum)
  Calibrators: 1, 2 and 3 (IgM Ab titre in AU/ml – human serum)
Additional reagents:
  None
Special equipment required:
  Sigma EIA Microwell Plate Reader (Cat. no. EQ104 – optional) or EIA Microwell Strip Reader (Cat. no. EQ103 – optional)

INTERPRETATION

Comments on interpretation:
  Negative: < 56 AU/ml
  Positive: ≥ 56 AU/ml
No. of references: 11

NOTES

131431.0

© KLUWER ACADEMIC PUBLISHERS 1995, ISSN 1381-5067

Epstein-Barr virus (EBNA)
ANTIBODY DETECTION

Manufacturer: Gull Laboratories Inc
Cat. No./Trade name: EN100/EBV-NA Test

SUMMARY

[Slide-Ag]–**Ab**–[GPC]–[AGPC-FITC]–fluorescence

Assay type: Immunofluorescence assay (indirect)
Detection: Fluorescence microscopy
Format: Slide well
Sample type: Serum
Sample pre-treatment:
   Heat serum for 30 min (56°C) to inactivate endogenous
   complement
Sample volume: 15 µl of 1:10 dilution
Number of tests: 100
Controls - standards run in assay:
   Controls: Neg (1), Pos (1)
Incubation:
   30 min (37°C) + 30 min (37°C)
Washes: 2

CONTENTS

Antibodies, antigens, labelled components:
   EBV-EBNA infected and uninfected Raji cells from
      Burkitt's lymphoma bound to slide well
   Guinea pig complement (GPC)
   Anti-Guinea pig complement (AGPC) FITC conjugated
Substrate:
Controls - standards supplied:
   Controls: Neg and Pos (human serum)
Additional reagents:
   None
Special equipment required:
   None

INTERPRETATION

Comments on interpretation:
   Positive: apple-green fluorescence of nucleus of 20–40%
      of cells per field at ≥1:10 dilution
No. of references: 9

NOTES

131709.0

Epstein-Barr virus (EBNA)
ANTIBODY DETECTION

Manufacturer: Immuno Concepts Inc
Cat. No./Trade name: EBNA Antibody Test System

SUMMARY

[Slide-Ag]–**Ab**–[GPC]–[AGPC-FITC]–fluorescence

Assay type: Immunofluorescence assay (indirect)
Detection: Fluorescence microscopy
Format: Slide well, Ag coated
Sample type: Serum
Sample pre-treatment:
   Heat serum for 30 min (56°C) to inactivate endogenous
   complement
Sample volume: 100 µl of 1:10 dilution
Number of tests: 13*
Controls - standards run in assay:
   Controls: Neg (1), Pos (1)
Incubation:
   30 min (RT) + 30 min (RT) + 30 min (RT)
Washes: 3

CONTENTS

Antibodies, antigens, labelled components:
   EBV-EBNA infected Raji cells and uninfected cells bound
      to slide well
   Guinea pig complement (GPC)
   Anti-Guinea pig complement (AGPC) FITC conjugated
Substrate:
Controls - standards supplied:
   Controls: Neg, Pos and titratable control (human serum)
Additional reagents:
   None
Special equipment required:
   None

INTERPRETATION

Comments on interpretation:
   Positive: apple-green fluorescence of nucleus of 50% of
      cells per field at ≥1:10 dilution
No. of references: 9

NOTES

131710.0

*13 wells per slide

© KLUWER ACADEMIC PUBLISHERS 1995, ISSN 1381-5067

Epstein-Barr virus (VCA)
ANTIBODY DETECTION (IgA)

Manufacturer: Savyon Diagnostics Ltd
Cat. No./Trade name: 015-01-096/IPAZyme® EBV/VCA IgA

SUMMARY

[Well-Ag]–**Ab**–[AHIgA-HRP]–[TMB]–blue precipitate

Assay type: EIA (non-competitive)
Detection: Light microscopy
Format: Slide well, Ag coated
Sample type: Serum
Sample pre-treatment:
    None
Sample volume: 10 μl of 1:64, 1:512 and 1:1024
Number of tests: 144
Controls - standards run in assay:
    Controls: Neg (1), Pos (1)
Incubation:
    1 hr (37°C) + 45 min (37°C) + 10 min (RT)
Washes: 2

CONTENTS

Antibodies, antigens, labelled components:
    EBV-VCA infected cells (PB HR-1 Burkitt's lymphoma
        cell line) bound to slide well
    Anti-human IgA Ab HRP conjugated
Substrate: TMB
Controls - standards supplied:
    Controls: Neg and Pos (human serum)
Additional reagents:
    None
Special equipment required:
    None

INTERPRETATION

Comments on interpretation:
    Positive: blue precipitate in cells
No. of references: 25

NOTES

131817.0

Epstein-Barr virus (VCA)
ANTIBODY DETECTION (IgG)

Manufacturer: Alfa Biotech
Cat. No./Trade name: 05772966/Epstein Barr Virus VCA IgG Elisa System

SUMMARY

[Well-Ag]–**Ab**–[AHIgG-HRP]–[TMB]–A_{450}

Assay type: EIA (non-competitive)
Detection: Colorimetric A_{450}
Format: Microtitre well, Ag coated
Sample type: Serum
Sample pre-treatment:
    None
Sample volume: 100 μl of 1:100 dilution x 2
Number of tests: 96
Controls - standards run in assay:
    Controls: Neg (2), cut-off (4), Pos (2)
Incubation:
    20 min (37°C) + 20 min (37°C) + 20 min (37°C)
Washes: 2

CONTENTS

Antibodies, antigens, labelled components:
    EBV-VCA Ag bound to well
    Anti-human IgG Ab (g) HRP conjugated
Substrate: TMB
Controls - standards supplied:
    Controls: Neg, cut-off and Pos (human serum)
Additional reagents:
    None
Special equipment required:
    None

INTERPRETATION

Comments on interpretation:
    Equivocal: within cut-off $\pm$15%; retest to confirm
    Repeatably equivocal: retest a fresh sample in 2 weeks
No. of references: 4

NOTES

131100.0

Epstein-Barr virus (VCA)
ANTIBODY DETECTION (IgG)

Manufacturer: Amico Laboratories Inc
Cat. No./Trade name: 5200G/AMIZYME® Epstein Barr Virus IgG Assay

SUMMARY

[Well-Ag]–**Ab**–[AHIgG-HRP]–[ABTS]–A_{405}

Assay type: EIA (non-competitive)
Detection: Colorimetric A_{405}
Format: Microtitre well, Ag coated
Sample type: Serum
Sample pre-treatment:
　None
Sample volume: 5 μl (+245 μl diluent)
Number of tests: 96
Controls - standards run in assay:
　Controls: Neg (1)
　Calibrators: I (1)
Incubation:
　25 min (RT) + 25 min (RT) + 25 min (RT)
Washes: 2

CONTENTS

Antibodies, antigens, labelled components:
　EBV-VCA (HR1) Ag bound to well
　Anti-human IgG Ab (g) HRP conjugated
Substrate: ABTS
Controls - standards supplied:
　Controls: Neg
　Calibrator: 1
Additional reagents:
　None
Special equipment required:
　None

INTERPRETATION

Comments on interpretation:
　Equivocal: where sample OD is between 0.375–0.399
　(95–99 U/ml); retest to confirm
No. of references: 14

NOTES

131784.0

Epstein-Barr virus (VCA)
ANTIBODY DETECTION (IgG)

Manufacturer: BAG-Biologische Analysensystem GmbH
Cat. No./Trade name: 5200 BAG - EBV (VCA) - EIA - G

SUMMARY

[Well-Ag]–**Ab**–[AHIgG-HRP]–[TMB]–A_{450}

Assay type: EIA (non-competitive)
Detection: Colorimetric A_{450}
Format: Microtitre well, Ag coated
Sample type: Serum
Sample pre-treatment:
　None
Sample volume: 100 μl of 1:101 dilution
Number of tests: 96
Controls - standards run in assay:
　Controls: Neg (1), cut-off (2), Pos (1)
Incubation:
　30 min (RT) + 30 min (RT) + 10 min (RT)
Washes: 2

CONTENTS

Antibodies, antigens, labelled components:
　EBV-VCA (strain P3-HR-1) Ag bound to well
　Anti-human IgG Ab (sh) HRP conjugated
Substrate: TMB
Controls - standards supplied:
　Controls: Neg, cut-off and Pos (human serum)
Additional reagents:
　None
Special equipment required:
　None

INTERPRETATION

Comments on interpretation:
　Equivocal: within cut-off ±10%; data is not provided for
　further assessment of samples within this range
No. of references: None

NOTES

131602.0

© KLUWER ACADEMIC PUBLISHERS 1995, ISSN 1381-5067

Epstein-Barr virus (VCA)
ANTIBODY DETECTION (IgG)

Manufacturer: BioWhittaker Inc
Cat. No./Trade name: 30-690U/EB VCA Elisa II

SUMMARY

$[Well-Ag]-Ab-[AHIgG-AP]-[PMP]-A_{550}$

Assay type: EIA (non-competitive)
Detection: Colorimetric A_{550}
Format: Microtitre well, Ag coated
Sample type: Serum (do not heat inactivate)
Sample pre-treatment:
  None
Sample volume: 10 µl (+200 µl diluent)
Number of tests: 192
Controls - standards run in assay:
  Controls: A (1), B (1)
  Calibrators: Neg (1), Pos 1 (1), Pos 2 (1)
Incubation:
  45 min (RT) + 45 min (RT) + 45 min (RT)
Washes: 2 (+preliminary plate wash)

CONTENTS

Antibodies, antigens, labelled components:
  EBV-VCA Ag bound to well
  Anti-human IgG Ab (g) AP conjugated
Substrate: PMP
Controls - standards supplied:
  Controls: A and B
  Calibrators: Neg, Pos 1, Pos 2 (all human serum)
Additional reagents:
  None
Special equipment required:
  BioWhittaker automated system and software (optional)

INTERPRETATION

Comments on interpretation:
  Equivocal: Elisa value between 0.15–0.16 or Index value
    between 0.88–0.89; retest to confirm
  Repeatably equivocal: retest a fresh sample or use an
    alternative method
No. of references: 3

NOTES

131746.0

Epstein-Barr virus (VCA)
ANTIBODY DETECTION (IgG)

Manufacturer: BioWhittaker Inc
Cat. No./Trade name: 30-361U/EB-VCA Stat

SUMMARY

$[Well-Ag]-Ab-[AHIgG-AP]-[PMP]-A_{550}$

Assay type: EIA (non-competitive)
Detection: Colorimetric A_{550}
Format: Microtitre well, Ag coated
Sample type: Serum (do not heat inactivate)
Sample pre-treatment:
  None
Sample volume: 10 µl (+200 µl diluent)
Number of tests: 192
Controls - standards run in assay:
  Controls: Neg (1), Pos (1)
  Standards: Neg (1), low Pos (1), high Pos (1)
Incubation:
  15 min (RT) + 15 min (RT) + 15 min (RT)
Washes: 2 (+preliminary plate wash)

CONTENTS

Antibodies, antigens, labelled components:
  EBV-VCA Ag bound to well
  Anti-human IgG Ab (g) AP conjugated
Substrate: PMP
Controls - standards supplied:
  Controls: Neg and Pos
  Standards: Neg, low Pos and high Pos (all human serum)
Additional reagents:
  None
Special equipment required:
  BioWhittaker automated system and software (optional)

INTERPRETATION

Comments on interpretation:
  Equivocal: Predicted Index value (PIV) between 0.80–
    0.90; retest to confirm
  Repeatably equivocal: retest a fresh sample or use
    alternative method
  Low Positive: PIV between 1.00–1.74
  Mid Positive: PIV between 1.75–3.89
  High Positive: PIV $\geqslant$ 3.90
No. of references: 16

NOTES

131753.0

© *KLUWER ACADEMIC PUBLISHERS 1995, ISSN 1381-5067*

Epstein-Barr virus (VCA)
ANTIBODY DETECTION (IgG)

Manufacturer: Clark Laboratories
Cat. No./Trade name: Epstein-Barr Virus IgG Elisa Test

SUMMARY

[Well-Ag]–**Ab**–[AHIgG-HRP]–[OPD]–A_{490}

Assay type: EIA (non-competitive)
Detection: Colorimetric A_{490}
Format: Microtitre well, Ag coated
Sample type: Serum (do not heat inactivate)
Sample pre-treatment:
   None
Sample volume: 10 µl (+200 µl diluent)
Number of tests: 96
Controls - standards run in assay:
   Controls: Neg (1), Pos (1)
   Calibrators: 1 (2)
Incubation:
   20 min (RT) + 20 min (RT) + 10 min (RT)
Washes: 2

CONTENTS

Antibodies, antigens, labelled components:
   EBV-VCA Ag bound to well
   Anti-human IgG Ab (g) HRP conjugated
Substrate: TMB*
Controls - standards supplied:
   Controls: Neg and Pos (human serum or plasma)
   Calibrator: 1 (human serum or plasma)
Additional reagents:
   H_2SO_4
Special equipment required:
   None

INTERPRETATION

Comments on interpretation:
   Classification of samples is according to cut-off; no
   further testing
No. of references: None

NOTES

131659.0

*OPD can be supplied as an alternative

Epstein-Barr virus (VCA)
ANTIBODY DETECTION (IgG)

Manufacturer: Dade International Inc
(Bartels Division)
Cat. No./Trade name: B1029-360/Bartels PRIMA
System® Epstein Barr Virus IgG EIA

SUMMARY

[Well-Ag]–**Ab**–[AHIgG-enzyme]–[PNP]–A_{405}

Assay type: EIA (non-competitive)
Detection: Colorimetric A_{405}
Format: Microtitre well, Ag coated
Sample type: Serum
Sample pre-treatment:
   None
Sample volume: 100 µl of 1:21 dilution
Number of tests: 96
Controls - standards run in assay:
   Controls: Neg (1), Pos (1)
   Reference serum: (1)
Incubation:
   30 min (37°C) + 30 min (37°C) + 30 min (37°C)
Washes: 2

CONTENTS

Antibodies, antigens, labelled components:
   EBV VCA (gp 125) from infected cells from P3HR-1 cell
      bound to well
   Anti-human IgG Ab enzyme conjugated
Substrate: PNP
Controls - standards supplied:
   Controls: Neg and Pos
   Reference serum: 1
Additional reagents:
   None
Special equipment required:
   None

INTERPRETATION

Comments on interpretation:
   Classification of samples is according to cut-off; no
   further testing
No. of references: 7

NOTES

131047.0

© KLUWER ACADEMIC PUBLISHERS 1995, ISSN 1381-5067

Epstein-Barr virus (VCA)
ANTIBODY DETECTION (IgG)

Manufacturer: Diesse
Cat. No./Trade name: 91055/ENZYWELL Epstein Barr VCA IgG

SUMMARY

[Well-Ag]–**Ab**–[AHIgG-POD]–[TMB]–A$_{450}$

Assay type: EIA (non-competitive)
Detection: Colorimetric A$_{450}$
Format: Microtitre well, Ag coated
Sample type: Serum (do not heat inactivate)
Sample pre-treatment:
    None
Sample volume: 50 μl of 1:26 dilution x 2*
Number of tests: 96
Controls - standards run in assay:
    Controls: Neg (2), cut-off (3), Pos (2)
Incubation:
    45 min (37°C) + 45 min (37°C) + 15 min (RT)
Washes: 2

CONTENTS

Antibodies, antigens, labelled components:
    EBV-VCA Ag bound to well
    Anti-human IgG MAb POD conjugated
Substrate: TMB
Controls - standards supplied:
    Controls: Neg, cut-off and Pos (human serum)
Additional reagents:
    None
Special equipment required:
    None

INTERPRETATION

Comments on interpretation:
    Equivocal: within cut-off ± 10%; retest to confirm
    Repeatably equivocal: repeat with a fresh sample
No. of references: None

NOTES

131444.0

*Sorbent G solution (provided) is also added to sample wells to neutralize non-specific reactions

Epstein-Barr virus (VCA)
ANTIBODY DETECTION (IgG)

Manufacturer: Gull Laboratories Inc
Cat. No./Trade name: EBQ100/EBV IgG ELISA Test

SUMMARY

[Well-Ag]–**Ab**–[AHIgG-AP]–[PNP]–A$_{405}$

Assay type: EIA (non-competitive)
Detection: Colorimetric A$_{405}$
Format: Microtitre well, Ag coated
Sample type: Serum
Sample pre-treatment:
    None
Sample volume: 100 μl of 1:21 dilution
Number of tests: 96
Controls - standards run in assay:
    Controls: Neg (1), Pos (1)
    Reference Serum: 1 (3)
Incubation:
    30 min (37°C) + 30 min (37°C) + 30 min (37°C)
Washes: 2

CONTENTS

Antibodies, antigens, labelled components:
    EBV-VCA (gp 125 from P3HR-1 cells) Ag bound to well
    Anti-human IgG Ab (caprine) AP conjugated
Substrate: PNP
Controls - standards supplied:
    Controls: Neg and Pos
    Reference serum: 1 (all human serum)
Additional reagents:
    None
Special equipment required:
    None

INTERPRETATION

Comments on interpretation:
    Equivocal: where sample OD is between mean OD of Reference Serum and mean OD of Reference Serum x 0.9; retest to confirm
    Repeatably equivocal: retest using an alternative method
No. of references: 10

NOTES

131699.0

© KLUWER ACADEMIC PUBLISHERS 1995, ISSN 1381-5067

Epstein-Barr virus (VCA)
ANTIBODY DETECTION (IgG)

Manufacturer: Human GmbH
Cat. No./Trade name: 51204/Epstein-Barr-Virus - IgG - Elisa

SUMMARY

[well-Ag]–**Ab**–[AHIgG-HRP]–[TMB]–A_{450}

Assay type: EIA (non-competitive)
Detection: Colorimetric A_{450}
Format: Microtitre well, Ag coated
Sample type: Serum
Sample pre-treatment:
  None
Sample volume: 100 µl of 1:20 dilution
Number of tests: 96
Controls - standards run in assay:
  Controls: Neg (2), Pos (2)
Incubation:
  1 hr (RT) + 30 min (RT) + 15 min (RT)
Washes: 2

CONTENTS

Antibodies, antigens, labelled components:
  EBV-VCA Ag bound to well
  Anti-human IgG Ab (r) HRP conjugated
Substrate: TMB
Controls - standards supplied:
  Controls: Neg and Pos (human)
Additional reagents:
  None
Special equipment required:
  None

INTERPRETATION

Comments on interpretation:
  Equivocal: within cut-off $\pm 15\%$; retest in parallel with a
    fresh sample taken 7–14 days later
No. of references: 5

NOTES

131344.0

Epstein-Barr virus (VCA)
ANTIBODY DETECTION (IgG)

Manufacturer: IFCI Clone Systems
Cat. No./Trade name: EIAGEN Epstein Barr VCA IgG

SUMMARY

[Well-Ag]–**Ab**–[AHIgG-POD]–[TMB]–A_{450}

Assay type: EIA (non-competitive)
Detection: Colorimetric A_{450}
Format: Microtitre well, Ag coated
Sample type: Serum
Sample pre-treatment:
  None
Sample volume: 40 µl (+ 1 ml diluent)*
Number of tests: 96
Controls - standards run in assay:
  Controls: Neg (2), cut-off (3), Pos (2)
Incubation:
  45 min (37°C) + 45 min (37°C) + 15 min (RT)
Washes: 2

CONTENTS

Antibodies, antigens, labelled components:
  EBV-VCA Ag bound to well
  Anti-human IgG MAb POD conjugated
Substrate: TMB
Controls - standards supplied:
  Controls: Neg, cut-off and Pos (human serum)
Additional reagents:
  None
Special equipment required:
  None

INTERPRETATION

Comments on interpretation:
  Equivocal: within cut-off $\pm 10\%$; retest to confirm
  Repeatably equivocal: retest a fresh sample
No. of references: None

NOTES

131671.0

*Samples are assayed in duplicate and Sorbitol G Reagent
  (provided) is added to diluted serum to neutralize non-
  specific reactions

Epstein-Barr virus (VCA)
ANTIBODY DETECTION (IgG)

Manufacturer: Immuno Concepts Inc
Cat. No./Trade name: /EBV-VCA IgG Antibody Test System

SUMMARY

[Slide-Ag]–**Ab**–[AHIgG-HRP]–[4CN]–purple stain

Assay type: EIA (non-competitive)
Detection: Light microscopy
Format: Slide well, Ag coated
Sample type: Serum
Sample pre-treatment:
   None
Sample volume: 100 µl of 1:10 dilution
Number of tests: 7, 13*
Controls - standards run in assay:
   Controls: Neg (1), Pos (1)
Incubation:
   30 min (RT) + 30 min (RT) + 30 min (RT)
Washes: 2

CONTENTS

Antibodies, antigens, labelled components:
   EBV-VCA infected and uninfected lymphoblastoid cells from Burkitt's lymphoma bound to slide well
   Anti-human IgG Ab (g) HRP conjugated
Substrate: 4-chloro-1-naphol (4CN)
Controls - standards supplied:
   Controls: Neg, Pos and titratable control (human serum)
Additional reagents:
   None
Special equipment required:
   None

INTERPRETATION

Comments on interpretation:
   Positive: light/dark blue purple staining of nucleus and cytoplasm of 10–20% of cells per field at ≥1:10 dilution
No. of references: 14

NOTES

131682.0

*7 or 13 wells per slide

Epstein-Barr virus (VCA)
ANTIBODY DETECTION (IgG)

Manufacturer: Immunobiological Laboratories
Cat. No./Trade name: VE 57091/Epstein Barr VCA IgG Elisa Test (HRP)

SUMMARY

[Well-Ag]–**Ab**–[AHIgG-HRP]–[TMB]–A_{450}

Assay type: EIA (non-competitive)
Detection: Colorimetric A_{450}
Format: Microtitre well, Ag coated
Sample type: Serum, plasma
Sample pre-treatment:
   None
Sample volume: 100 µl of 1:20 dilution x 2
Number of tests: 96
Controls - standards run in assay:
   Controls: Neg (2), low Pos (2), high Pos (2)
Incubation:
   20 min (RT) + 20 min (RT) + 10 min (RT)
Washes: 2

CONTENTS

Antibodies, antigens, labelled components:
   EBV-VCA Ag bound to well
   Anti-human IgG Ab HRP conjugated
Substrate: TMB
Controls - standards supplied:
   Controls: Neg, low Pos and high Pos (serum)
Additional reagents:
   None
Special equipment required:
   None

INTERPRETATION

Comments on interpretation:
   Classification of samples is according to cut-off; no further testing
No. of references: 9

NOTES

131249.0

Epstein-Barr virus (VCA)
ANTIBODY DETECTION (IgG)

Manufacturer: Incstar
Cat. No./Trade name: 7590 Epstein Barr VCA IgG

SUMMARY

[Well-Ag]–**Ab**–[AHIgG-HRP]–[TMB]–A_{450}

Assay type: EIA (non-competitive)
Detection: Colorimetric A_{450}
Format: Microtitre well, Ag coated
Sample type: Serum, plasma
Sample pre-treatment:
  None
Sample volume: 100 µl of 1:505 dilution x 2
Number of tests:
Controls - standards run in assay:
  Controls: Neg (2)
  Calibrators: 1 (2), 2 (2), 3 (2), 4 (2)
Incubation:
  1 hr (37°C) + 1 hr (37°C) + 30 min (RT)
Washes: 2

CONTENTS

Antibodies, antigens, labelled components:
  EBV-VCA (p18 peptide) Ag bound to well
  Anti-human IgG IgG Ab (g) HRP conjugated
Substrate: TMB
Controls - standards supplied:
  Controls: Neg (human serum)
  Calibrators: 4 (10, 25, 50 and 100 IU/ml)
Additional reagents:
  None
Special equipment required:
  ETI-System Reader and ETI-System Washer (optional)

INTERPRETATION

Comments on interpretation:
  Equivocal: within cut-off ±10%; retest to confirm
No. of references: 14

NOTES

131214.0

Epstein-Barr virus (VCA)
ANTIBODY DETECTION (IgG)

Manufacturer: Melotec S.A.
Cat. No./Trade name: /Melotest EBV-VCA IgG

SUMMARY

[Well-Ag]–**Ab**–[AHIgG-HRP]–[TMB]–A_{450}

Assay type: EIA (non-competitive)
Detection: Colorimetric A_{450}
Format: Microtitre well, Ag coated
Sample type: Serum, plasma
Sample pre-treatment:
  None
Sample volume: 100 µl of 1:20 dilution x 2
Number of tests: 96
Controls - standards run in assay:
  Controls: Neg (1), low Pos (2), high Pos (1)
Incubation:
  20 min (RT) + 20 min (RT) + 10 min (RT)
Washes: 2

CONTENTS

Antibodies, antigens, labelled components:
  EBV-VCA Ag bound to well
  Anti-human IgG Ab HRP conjugated
Substrate: TMB
Controls - standards supplied:
  Controls: Neg, low Pos and high Pos (sera)
Additional reagents:
  None
Special equipment required:
  None

INTERPRETATION

Comments on interpretation:
  Equivocal: ratio of sample OD:cut-off value between 0.9
    and 1.1; retest a fresh sample
  Repeatably equivocal: retest a fresh sample taken in 2-4
    weeks
No. of references: 8

NOTES

131127.0

© KLUWER ACADEMIC PUBLISHERS 1995, ISSN 1381-5067

Epstein-Barr virus (VCA)
ANTIBODY DETECTION (IgG)

Manufacturer: Ortho Diagnostic Systems
Cat. No./Trade name: 520010/ORTHO® Epstein-Barr Virus VCA-IgG Antibody ELISA Test

SUMMARY

[Well-Ag]–**Ab**–[AHIgG-POD]–[OPD]–A_{490}

Assay type: EIA (non-competitive)
Detection: Colorimetric A_{490}
Format: Microtitre well, Ag coated
Sample type: Serum
Sample pre-treatment:
  None
Sample volume: 10 µl (+200 µl diluent)
Number of tests: 192
Controls - standards run in assay:
  Controls: Neg (2), cut-off (3), Pos (2)
Incubation:
  30 min (37°C) + 1 hr (37°C) + 30 min (RT)
Washes: 2

CONTENTS

Antibodies, antigens, labelled components:
  EBV-VCA (protein 125 kD) Ag bound to well
  Anti-human IgG Ab (g) POD conjugated
Substrate: OPD
Controls - standards supplied:
  Controls: Neg, cut-off and Pos (human serum)
Additional reagents:
  H_2SO_4
Special equipment required:
  None

INTERPRETATION

Comments on interpretation:
  Classification of samples is according to cut-off; no
    further testing
No. of references: 16

NOTES

131218.0

Epstein-Barr virus (VCA)
ANTIBODY DETECTION (IgG)

Manufacturer: PanBio
Cat. No./Trade name: EBG-100/EBV IgG Elisa Test

SUMMARY

[Well-Ag]–**Ab**–[AHIgG-HRP]–[TMB]–A_{450}

Assay type: EIA (non-competitive)
Detection: Colorimetric A_{450}
Format: Microtitre well, Ag coated
Sample type: Serum
Sample pre-treatment:
  None
Sample volume: 100 µl of 1:100 dilution
Number of tests: 96
Controls - standards run in assay:
  Controls: Neg (1), Pos (1)
  Calibrator: cut-off (2)
Incubation:
  20 min (37°C) + 20 min (37°C) + 10 min (RT)
Washes: 2

CONTENTS

Antibodies, antigens, labelled components:
  EBV-VCA Ag bound to well
  Anti-human IgG Ab (sh) HRP conjugated
Substrate: TMB
Controls - standards supplied:
  Controls: Neg and Pos (human serum)
  Calibrator: cut-off (human serum)
Additional reagents:
  None
Special equipment required:
  None

INTERPRETATION

Comments on interpretation:
  Equivocal: where the ratio of cut-off value:sample OD is
    between 0.9-1.1; retest to confirm
No. of references: None

NOTES

131396.0

© KLUWER ACADEMIC PUBLISHERS 1995, ISSN 1381-5067

Epstein-Barr virus (VCA)
ANTIBODY DETECTION (IgG)

Manufacturer: Savyon Diagnostics Ltd
Cat. No./Trade name: 013-01-144/IPAzyme® EBV/VCA IgG

SUMMARY

[Slide-Ag]–**Ab**–[AHIgG-HRP]–[TMB]–blue precipitate

Assay type: EIA (non-competitive)
Detection: Light microscopy
Format: Slide well, Ag coated
Sample type: Serum
Sample pre-treatment:
   None
Sample volume: 10 µl of 1:64, 1:512 and 1:1024 dilutions
Number of tests: 144
Controls - standards run in assay:
   Controls: Neg (1), Pos (1)
Incubation:
   45 min (37°C) + 45 min (37°C) + 10 min (RT)
Washes: 2

CONTENTS

Antibodies, antigens, labelled components:
   EBV-VCA infected cells (P3HR-1 Burkitt's lymphoma cell line) bound to slide well
   Anti-human IgG Ab HRP conjugated
Substrate: TMB
Controls - standards supplied:
   Controls: Neg and Pos (human serum)
Additional reagents:
   None
Special equipment required:
   None

INTERPRETATION

Comments on interpretation:
   Positive: blue precipitate in cells
No. of references: 19

NOTES

131143.0

Epstein-Barr virus (VCA)
ANTIBODY DETECTION (IgG)

Manufacturer: Sigma Diagnostics
Cat. No./Trade name: SIA 130/SIA® Epstein-Barr VCA IgG

SUMMARY

[Well-Ag]–**Ab**–[AHIgG-AP]–[PNP]–A_{405}

Assay type: EIA (non-competitive)
Detection: Colorimetric A_{405}
Format: Microtitre well, Ag coated
Sample type: Serum (do not heat inactivate)
Sample pre-treatment:
   None
Sample volume: 10 µl (+500 µl diluent)
Number of tests: 96
Controls - standards run in assay:
   Controls: low Pos (1), high Pos (1)
   Calibrators: 1 (2), 2 (2), 3 (2)
Incubation:
   30 min (RT) + 30 min (RT) + 45 min (RT)
Washes: 2

CONTENTS

Antibodies, antigens, labelled components:
   EBV-VCA Ag bound to well
   Anti-human IgG Ab (sh or g) AP conjugated
Substrate: PNP
Controls - standards supplied:
   Controls: low Pos and high Pos (human serum)
   Calibrators: 1, 2 and 3 (IgG Ab titre in AU/ml – human serum)
Additional reagents:
   Neg control (Cat. no. N9373 - optional)
Special equipment required:
   EIA Microwell Plate reader (Cat. no. EQ104 – optional) or EIA Microwell Strip reader (Cat. no. EQ103 – optional)

INTERPRETATION

Comments on interpretation:
   Equivocal: <110 but >100 AU/ml; retest in parallel with a fresh sample taken 7–14 days later
   Repeatably equivocal: report as such and test by other methods
No. of references: 11

NOTES

131429.0

© KLUWER ACADEMIC PUBLISHERS 1995, ISSN 1381-5067

Epstein-Barr virus (VCA)
ANTIBODY DETECTION (IgM)

Manufacturer: Alfa Biotech
Cat. No./Trade name: 05772978/Epstein Barr VCA IgM Elisa System

SUMMARY

[Well-AHIgM]–**Ab**–[Ag-biotin]–[strept-HRP]–[TMB]–A_{450}

Assay type: EIA (non-competitive)
Detection: Colorimetric A_{450}
Format: Microtitre well, Ab coated
Sample type: Serum
Sample pre-treatment:
  None
Sample volume: 100 µl of 1:100 dilution x2
Number of tests: 96
Controls - standards run in assay:
  Controls: Neg (2), cut-off (4), Pos (2)
Incubation:
  20 min (37°C) + 20 min (37°C) + 20 min (37°C) + 20 min (37°C)
Washes: 2

CONTENTS

Antibodies, antigens, labelled components:
  Anti-human IgM Ab bound to well
  EBV-VCA Ag biotinylated
  Streptavidin HRP conjugated
Substrate: TMB
Controls - standards supplied:
  Controls: Neg, cut-off and Pos (human serum)
Additional reagents:
  None
Special equipment required:
  None

INTERPRETATION

Comments on interpretation:
  Equivocal: within cut-off ± 15%; retest to confirm
  Repeatably equivocal: retest a fresh sample in 2 weeks
No. of references: 3

NOTES

131105.0

Epstein-Barr virus (VCA)
ANTIBODY DETECTION (IgM)

Manufacturer: Amico Laboratories Inc
Cat. No./Trade name: 5200M/AMIZYME® Epstein Barr Virus IgM Assay

SUMMARY

[Well-Ag]–**Ab**–[AHIgM-HRP]–[ABTS]–A_{405}

Assay type: EIA (non-competitive)
Detection: Colorimetric A_{405}
Format: Microtitre well, Ag coated
Sample type: Serum
Sample pre-treatment:
  None
Sample volume: 5 µl (+245 µl diluent)*
Number of tests: 96
Controls - standards run in assay:
  Controls: Neg (1)
  Calibrators: I (1)
Incubation:
  25 min (RT) + 25 min (RT) + 25 min (RT)
Washes: 2

CONTENTS

Antibodies, antigens, labelled components:
  EBV-VCA (HR1) Ag bound to well
  Anti-human IgM Ab (g) HRP conjugated
Substrate: ABTS
Controls - standards supplied:
  Controls: Neg
  Calibrator: 1
Additional reagents:
  None
Special equipment required:
  None

INTERPRETATION

Comments on interpretation:
  Equivocal: where sample OD is between 0.275–0.299 (92–99 U/ml); retest to confirm
No. of references: 15

NOTES

131792.0

*Sample diluted in IgM serum diluent provided to eliminate interference due to RF and IgG

© *KLUWER ACADEMIC PUBLISHERS 1995, ISSN 1381-5067*

Epstein-Barr virus (VCA)
ANTIBODY DETECTION (IgM)

Manufacturer: Amico Laboratories Inc
Cat. No./Trade name: 5200MC/AMIZYME® Epstein Barr Virus IgM Capture Assay

SUMMARY

[Well-Ag]–**Ab**–[AHIgM]–[AMIg-HRP]–[ABTS]–A_{405}

Assay type: EIA (non-competitive)
Detection: Colorimetric A_{405}
Format: Microtitre well, Ag coated
Sample type: Serum
Sample pre-treatment:
    None
Sample volume: 5 μl (+245 μl diluent)
Number of tests: 96
Controls - standards run in assay:
    Controls: Neg (1)
    Calibrators: I (1)
Incubation:
    25 min (RT) + 25 min (RT) + 25 min (RT) + 25 min (RT)
Washes: 3

CONTENTS

Antibodies, antigens, labelled components:
    EBV-VCA (HR1) Ag bound to well
    Anti-human IgM MAb (m)
    Anti-mouse Ig Ab (g) HRP conjugated
Substrate: ABTS
Controls - standards supplied:
    Controls: Neg
    Calibrator: 1
Additional reagents:
    None
Special equipment required:
    None

INTERPRETATION

Comments on interpretation:
    Equivocal: where sample OD is between 0.285–0.299; retest to confirm
No. of references: 2

NOTES

131802.0

Epstein-Barr virus (VCA)
ANTIBODY DETECTION (IgM)

Manufacturer: BAG-Biologische Analysensystem GmbH
Cat. No./Trade name: 5201 BAG - EBV (VCA) - EIA - M

SUMMARY

[Well-Ag]–**Ab**–[AHIgM-HRP]–[TMB]–A_{450}

Assay type: EIA (non-competitive)
Detection: Colorimetric A_{450}
Format: Microtitre well, Ag coated
Sample type: Serum (do not heat inactivate)
Sample pre-treatment:
    Treat diluted serum with BAG - IgM - Sep - System (Cat. no. 5540) to eliminate interference due to RF and IgG
Sample volume: 100 μl of treated sample
Number of tests: 96
Controls - standards run in assay:
    Controls: Neg (1), cut-off (2), Pos (1)
Incubation:
    30 min (RT) + 30 min (RT) + 10 min (RT)
Washes: 2

CONTENTS

Antibodies, antigens, labelled components:
    EBV-VCA (P3-HR-1) Ag bound to well
    Anti-human IgM Ab (sh) HRP conjugated
Substrate: TMB
Controls - standards supplied:
    Controls: Neg, cut-off and Pos (human serum)
Additional reagents:
    None
Special equipment required:
    None

INTERPRETATION

Comments on interpretation:
    Equivocal: within cut-off ±10%; data is not provided for further assessment of samples within this range
No. of references: None

NOTES

131607.0

© *KLUWER ACADEMIC PUBLISHERS 1995, ISSN 1381-5067*

Epstein-Barr virus (VCA)
ANTIBODY DETECTION (IgM)

Manufacturer: BioWhittaker Inc
Cat. No./Trade name: 30-362U/EB VCA Stat M

SUMMARY

[Well-Ag]–**Ab**–[AHIgM-AP]–[PMP]–A_{550}

Assay type: EIA (non-competitive)
Detection: Colorimetric A_{550}
Format: Microtitre well, Ag coated
Sample type: Serum (do not heat inactivate)
Sample pre-treatment:
Mix sample with Pretreatment Serum, incubate 30 min (RT) and centrifuge to eliminate interference due to IgG and RF
Sample volume: 25 µl of treated sample (+ 100 µl diluent)
Number of tests: 96
Controls - standards run in assay:
Calibrators: Neg (1), Pos 1 (1), Pos 2 (1)
Incubation:
45 min (RT) + 30 min (RT) + 30 min (RT) (all with shaking)
Washes: 2 (+ preliminary plate wash)

CONTENTS

Antibodies, antigens, labelled components:
EBV-VCA Ag bound to well
Anti-human IgM Ab (g) AP conjugated
Substrate: PMP
Controls - standards supplied:
Calibrators: Neg, Pos 1, Pos 2 (all human serum)
Rheumatoid Factor Control (human serum)
Additional reagents:
None
Special equipment required:
BioWhittaker automated system and software (optional)

INTERPRETATION

Comments on interpretation:
Equivocal: test value between 0.10–0.12; indicates probable positive, retest in duplicate to confirm
No. of references: 11

NOTES

131757.0

Epstein-Barr virus (VCA)
ANTIBODY DETECTION (IgM)

Manufacturer: Clark Laboratories
Cat. No./Trade name: Epstein Barr Virus IgM Elisa Test

SUMMARY

[Well-Ag]–**Ab**–[AHIgM-HRP]–[OPD]–A_{490}

Assay type: EIA (non-competitive)
Detection: Colorimetric A_{490}
Format: Microtitre well, Ag coated
Sample type: Serum (do not heat inactivate)
Sample pre-treatment:
Add Absorbent Solution provided to diluted serum and incubate 20 min (RT) to eliminate interference due to RF and IgG
Sample volume: 100 µl of treated sample
Number of tests: 96
Controls - standards run in assay:
Controls: Neg (1), Pos (1)
Calibrators: 1 (2)
Incubation:
20 min (RT) + 20 min (RT) + 10 min (RT)
Washes: 2

CONTENTS

Antibodies, antigens, labelled components:
EBV-VCA Ag bound to well
Anti-human IgM Ab (g) HRP conjugated
Substrate: TMB*
Controls - standards supplied:
Controls: Neg and Pos (human serum or plasma)
Calibrator: 1 (human serum or plasma)
Additional reagents:
H_2SO_4
Special equipment required:
None

INTERPRETATION

Comments on interpretation:
Classification of samples is according to cut-off; no further testing
No. of references: None

NOTES

131665.0

*OPD can be supplied as an alternative

© KLUWER ACADEMIC PUBLISHERS 1995, ISSN 1381-5067

Epstein-Barr virus (VCA)
ANTIBODY DETECTION (IgM)

Manufacturer: Diesse
Cat. No./Trade name: 91056/ENZYWELL Epstein Barr VCA IgM

SUMMARY

[Well-Ag]–**Ab**–[AHIgM-POD]–[TMB]–A_{450}

Assay type: EIA (non-competitive)
Detection: Colorimetric A_{450}
Format: Microtitre well, Ag coated
Sample type:
Sample pre-treatment:
  None
Sample volume: 50 µl of 1:26 dilution x 2*
Number of tests: 96
Controls - standards run in assay:
  Controls: Neg (2), cut-off (3), Pos (2)
Incubation:
  45 min (37°C) + 45 min (37°C) + 15 min (RT)
Washes: 2

CONTENTS

Antibodies, antigens, labelled components:
  EBV-VCA Ag bound to well
  Anti-human IgM MAb POD conjugated
Substrate: TMB
Controls - standards supplied:
  Controls: Neg, cut-off and Pos (human serum)
Additional reagents:
  None
Special equipment required:
  None

INTERPRETATION

Comments on interpretation:
  Equivocal: within cut-off ±10%; retest to confirm
  Repeatably equivocal: repeat with a fresh sample
No. of references: None

NOTES

131500.0

*Sorbent M solution (provided) is also added to sample wells to neutralize non-specific reactions

---

Epstein-Barr virus (VCA)
ANTIBODY DETECTION (IgM)

Manufacturer: Gull Laboratories Inc
Cat. No./Trade name: EBE150/EBV IgM ELISA Test

SUMMARY

[Well-Ag]–**Ab**–[AHIgM-AP]–[PNP]–A_{405}

Assay type: EIA (non-competitive)
Detection: Colorimetric A_{405}
Format: Microtitre well, Ag coated
Sample type: Serum
Sample pre-treatment:
  None
Sample volume: 100 µl of 1:11 dilution*
Number of tests: 96
Controls - standards run in assay:
  Controls: Neg (1), Pos (1)
  Reference Serum: 1 (3)
Incubation:
  30 min (37°C) + 30 min (37°C) + 30 min (37°C)
Washes: 2

CONTENTS

Antibodies, antigens, labelled components:
  EBV-VCA (gp 125 from P3HR-1 cells) bound to well
  Anti-human IgM Ab (caprine) AP conjugated
Substrate: PNP
Controls - standards supplied:
  Controls: Neg and Pos
  Reference serum: 1 (all human serum)
Additional reagents:
  None
Special equipment required:
  None

INTERPRETATION

Comments on interpretation:
  Equivocal: where sample OD is between mean OD of Reference Serum and mean OD of Reference Serum x 0.9; retest to confirm
  Repeatably equivocal: retest using an alternative method
No. of references: 11

NOTES

131701.0

*Diluent contains an absorbent to eliminate interference due to RF and IgG in serum

© KLUWER ACADEMIC PUBLISHERS 1995, ISSN 1381-5067

Epstein-Barr virus (VCA)
ANTIBODY DETECTION (IgM)

Manufacturer: Human GmbH
Cat. No./Trade name: 51104/Epstein-Barr Virus IgM Elisa

SUMMARY

[Well-Ag]–**Ab**–[AHIgM-HRP]–[TMB]–A$_{450}$

Assay type: EIA (non-competitive)
Detection: Colorimetric A$_{450}$
Format: Microtitre well, Ag coated
Sample type: Serum
Sample pre-treatment:
  Dilute sample with Dilution Buffer IgM and incubate 5 min (RT) to eliminate interference due to RF and IgG
Sample volume: 100 µl of prepared sample
Number of tests: 96
Controls - standards run in assay:
  Controls: Neg (2), Pos (2)
Incubation:
  30 min (RT) + 30 min (RT) + 15 min (RT)
Washes: 2

CONTENTS

Antibodies, antigens, labelled components:
  EBV-VCA Ag bound to well
  Anti-human IgM Ab (r) HRP conjugated
Substrate: TMB
Controls - standards supplied:
  Controls: Neg and Pos (human)
Additional reagents:
  None
Special equipment required:
  None

INTERPRETATION

Comments on interpretation:
  Equivocal: within cut-off ± 15%; retest in parallel with a fresh sample taken 7–14 days later
No. of references: 7

NOTES

131348.0

Epstein-Barr virus (VCA)
ANTIBODY DETECTION (IgM)

Manufacturer: IFCI Clone Systems
Cat. No./Trade name: EIAGEN/Epstein Barr VCA IgM

SUMMARY

[Well-Ag]–**Ab**–[AHIgM-POD]–[TMB]–A$_{450}$

Assay type: EIA (non-competitive)
Detection: Colorimetric A$_{450}$
Format: Microtitre well, Ag coated
Sample type: Serum
Sample pre-treatment:
  None
Sample volume: 40 µl (+ 1 ml diluent)*
Number of tests: 96
Controls - standards run in assay:
  Controls: Neg (2), cut-off (3), Pos (2)
Incubation:
  45 min (37°C) + 45 min (37°C) + 15 min (RT)
Washes: 2

CONTENTS

Antibodies, antigens, labelled components:
  EBV-VCA Ag bound to well
  Anti-human IgM MAb POD conjugated
Substrate: TMB
Controls - standards supplied:
  Controls: Neg, cut-off and Pos (human serum)
Additional reagents:
  None
Special equipment required:
  None

INTERPRETATION

Comments on interpretation:
  Equivocal: within cut-off ± 10%; retest to confirm
  Repeatably equivocal: retest a fresh sample
No. of references: None

NOTES

131676.0

*Samples are assayed in duplicate and Sorbitol M Reagent (provided) is added to diluted serum to eliminate interference due to RF and IgG

© *KLUWER ACADEMIC PUBLISHERS 1995, ISSN 1381-5067*

Epstein-Barr virus (VCA)
ANTIBODY DETECTION (IgM)

Manufacturer: Immuno Concepts Inc
Cat. No./Trade name: EBV-VCA IgM Antibody Test System

SUMMARY

[Slide-Ag]–**Ab**–[AHIgM-HRP]–[4CN]–purple stain

Assay type: EIA (non-competitive)
Detection: Light microscopy
Format: Slide well, Ag coated
Sample type: Serum
Sample pre-treatment:
  None
Sample volume: 100 µl of 1:10 dilution
Number of tests: 7, 13*
Controls - standards run in assay:
  Controls: Neg (1), Pos (1)
Incubation:
  1 hr 30 min (37°C) + 30 min (RT) + 30 min (RT)
Washes: 2

CONTENTS

Antibodies, antigens, labelled components:
  EBV-VCA infected and uninfected lymphoblastoid cells from Burkitt's lymphoma bound to slide well
  Anti-human IgM Ab (g) HRP conjugated
Substrate: 4-chloro-1-naphol (4CN)
Controls - standards supplied:
  Controls: Neg, Pos and titratable control (human serum)
Additional reagents:
  None
Special equipment required:
  None

INTERPRETATION

Comments on interpretation:
  Positive: light/dark blue purple staining of nucleus and cytoplasm of 10–20% of cells per field at ≥1:10 dilution
No. of references: 16

NOTES

131683.0

*7 or 13 wells per slide

Epstein-Barr virus (VCA)
ANTIBODY DETECTION (IgM)

Manufacturer: Immunobiological Laboratories
Cat. No./Trade name: VE 57101

SUMMARY

[Well-Ag]–**Ab**–[AHIgM-HRP]–[TMB]–A_{450}

Assay type: EIA (non-competitive)
Detection: Colorimetric A_{450}
Format: Microtitre well, Ag coated
Sample type: Serum, plasma
Sample pre-treatment:
  Mix diluted sample with RF-Absorbent and incubate 20 min (RT) to eliminate interference due to RF and IgG
Sample volume: 100 µl of treated sample x 2
Number of tests: 96
Controls - standards run in assay:
  Controls: Neg (2), low Pos (2), high Pos (2)
Incubation:
  20 min (RT) + 20 min (RT) + 10 min (RT)
Washes: 2

CONTENTS

Antibodies, antigens, labelled components:
  EBV-VCA Ag bound to well
  Anti-human IgM Ab HRP conjugated
Substrate: TMB
Controls - standards supplied:
  Controls: Neg, low Pos and high Pos (serum)
Additional reagents:
  None
Special equipment required:
  None

INTERPRETATION

Comments on interpretation:
  Classification of sample is according to cut-off; no further testing
No. of references: 9

NOTES

131257.0

Epstein-Barr virus (VCA)
ANTIBODY DETECTION (IgM)

Manufacturer: Incstar
Cat. No./Trade name: 8590 Epstein Barr VCA IgM

SUMMARY

[Well-AHIgM]–**Ab**–[Ag]–[MAb-HRP]–[TMB]–A_{450}

Assay type: EIA (non-competitive)
Detection: Colorimetric A_{450}
Format: Microtitre well, Ab coated
Sample type: Serum, plasma
Sample pre-treatment:
   None
Sample volume: 100 µl of 1:101 dilution x 2
Number of tests: 96
Controls - standards run in assay:
   Controls: Neg (2), cut-off (4), Pos (2)
Incubation:
   1 hr (37°C) + 1 hr (37°C) + 30 min (RT)
Washes: 2

CONTENTS

Antibodies, antigens, labelled components:
   Anti-human IgM Ab (r) bound to well
   EBV-VCA (p18 peptide) Ag
   Anti-EBV-VCA (p18 peptide) IgG MAb (m) HRP
      conjugated
Substrate: TMB
Controls - standards supplied:
   Controls: Neg, cut-off and Pos (human serum)
Additional reagents:
   None
Special equipment required:
   ETI-System Reader and ETI-System Washer (optional)

INTERPRETATION

Comments on interpretation:
   Equivocal: within cut-off ± 10%; retest to confirm
No. of references: None

NOTES

131217.0

Epstein-Barr virus (VCA)
ANTIBODY DETECTION (IgM)

Manufacturer: Ortho Diagnostic Systems
Cat. No./Trade name: 520015/ORTHO® Epstein-Barr
Virus VCA-IgM Antibody ELISA Test

SUMMARY

[Well-Ag]–**Ab**–[AHIgM-POD]–[OPD]–A_{490}

Assay type: EIA (non-competitive)
Detection: Colorimetric A_{490}
Format: Microtitre well, Ag coated
Sample type: Serum
Sample pre-treatment:
   Dilute samples and controls with Pretreatment Reagent
      provided and incubate 15-60 min (RT) to eliminate
      interference due to RF and IgG
Sample volume: 10 µl of treated sample (+ 100 µl diluent)
Number of tests: 192
Controls - standards run in assay:
   Controls: Neg (2), cut-off (3), Pos (2)
Incubation:
   30 min (37°C) + 1 hr (37°C) + 30 min (RT)
Washes: 2

CONTENTS

Antibodies, antigens, labelled components:
   EBV-VCA (protein 125 kD) Ag bound to well
   Anti-human IgM Ab (g) POD conjugated
Substrate: OPD
Controls - standards supplied:
   Controls: Neg, cut-off and Pos (human serum)
Additional reagents:
   H_2SO_4
Special equipment required:
   None

INTERPRETATION

Comments on interpretation:
   Classification of samples is according to cut-off; no
      further testing
No. of references: 16

NOTES

131219.0

© *KLUWER ACADEMIC PUBLISHERS 1995, ISSN 1381-5067*

Epstein-Barr virus (VCA)
ANTIBODY DETECTION (IgM)

Manufacturer: PanBio
Cat. No./Trade name: EBM-200/EBV Synthetic Peptide IgM Elisa Test

SUMMARY

[Well-Ag]–**Ab**–[AHIgM-HRP]–[TMB]–A_{450}

Assay type: EIA (non-competitive)
Detection: Colorimetric A_{450}
Format: Microtitre well, Ag coated
Sample type: Serum
Sample pre-treatment:
  Mix diluted serum with absorbent solution, incubate 15 min (RT) to eliminate interference due to RF and IgG
Sample volume: 100 μl of prepared sample
Number of tests: 96
Controls - standards run in assay:
  Controls: Neg (1), Pos (1)
  Calibrator: cut-off (2)
Incubation:
  20 min (RT) + 20 min (RT) + 10 min (RT)
Washes: 2

CONTENTS

Antibodies, antigens, labelled components:
  EBV-VCA (syn peptide) Ag bound to well
  Anti-human IgM Ab (sh) HRP conjugated
Substrate: TMB
Controls - standards supplied:
  Controls: Neg and Pos (human serum)
  Calibrator: cut-off (human serum)
Additional reagents:
  None
Special equipment required:
  None

INTERPRETATION

Comments on interpretation:
  Equivocal: where the ratio of cut-off value:sample OD is between 0.9-1.1; retest to confirm
No. of references: 9

NOTES

131402.0

Epstein-Barr virus (VCA)
ANTIBODY DETECTION (IgM)

Manufacturer: Savyon Diagnostics Ltd
Cat. No./Trade name: 014-01-096/IPAzyme® EBV/VCA TRUE IgM

SUMMARY

[Slide-Ag]–**Ab**–[AHIgM-HRP]–[TMB]–blue precipitate

Assay type: EIA (non-competitive)
Detection: Light microscopy
Format: Slide well, Ag coated
Sample type: Serum
Sample pre-treatment:
  Perform IgG/RF strip treatment of samples and controls using instructions and reagents provided
Sample volume: 20 μl of treated sample
Number of tests: 96
Controls - standards run in assay:
  Controls: Neg (1), Pos (1)
Incubation:
  2 hr (37°C) + 45 min (37°C) + 10 min (RT)
Washes: 2

CONTENTS

Antibodies, antigens, labelled components:
  EBV-VCA infected cells (P3HR-1 Burkitt's lymphoma cell line) bound to slide well
  Anti-human IgM Ab HRP conjugated
Substrate: TMB
Controls - standards supplied:
  Controls: Neg and Pos (human serum)
Additional reagents:
  None
Special equipment required:
  None

INTERPRETATION

Comments on interpretation:
  Positive: blue precipitate in cells
No. of references: 14

NOTES

131144.0

© KLUWER ACADEMIC PUBLISHERS 1995, ISSN 1381-5067

Epstein-Barr virus (VCA)
ANTIBODY DETECTION (IgM)

Manufacturer: Sigma Diagnostics
Cat. No./Trade name: SIA 131/SIA® Epstein-Barr VCA IgM

SUMMARY

[Well-Ag]–**Ab**–[AHIgM-AP]–[PNP]–A$_{405}$

Assay type: EIA (non-competitive)
Detection: Colorimetric A$_{405}$
Format: Microtitre well, Ag coated
Sample type: Serum (do not heat inactivate)
Sample pre-treatment:
   Add Pretreatment Solution to sample, incubate 30 min (RT) to eliminate interference due to RF and IgG
Sample volume: 200 µl of treated sample
Number of tests: 96
Controls - standards run in assay:
   Controls: Neg (1), Pos (1)
   Calibrators: 1 (2), 2 (2), 3 (2)
Incubation:
   30 min (RT) + 30 min (RT) + 45 min (RT)
Washes: 2

CONTENTS

Antibodies, antigens, labelled components:
   EBV-VCA Ag bound to well
   Anti-human IgM Ab (sh or g) AP conjugated
Substrate: PNP
Controls - standards supplied:
   Controls: Neg and Pos (human serum)
   Calibrators: 1, 2 and 3 (IgM Ab titre in AU/ml – human serum)
Additional reagents:
   None
Special equipment required:
   Sigma EIA Microwell Plate Reader (Cat. no. EQ104 – optional) or EIA Microwell Strip Reader (Cat. no. EQ103 – optional)

INTERPRETATION

Comments on interpretation:
   Equivocal: >90 but <110 AU/ml; retest in parallel with a fresh sample 7–14 days later
   Repeatably equivocal: report as such and retest by other methods
No. of references: 12

NOTES

131430.0

Epstein-Barr virus (VCA)
ANTIBODY DETECTION (IgG)

Manufacturer: Bion Enterprises Ltd
Cat. No./Trade name: /EBV-G (VCA) Antibody Test System

SUMMARY

[Slide-Ag]–**Ab**–[AHIgG-FITC]–fluorescence

Assay type: Immunofluorescence assay (indirect)
Detection: Fluorescence microscopy
Format: Slide well, Ag coated
Sample type: Serum
Sample pre-treatment:
   None
Sample volume: 50 µl (+450 µl diluent)
Number of tests: 60, 120
Controls - standards run in assay:
   Controls: Neg (1), Pos (1)
Incubation:
   30 min (RT) + 30 min (RT)
Washes: 2

CONTENTS

Antibodies, antigens, labelled components:
   EBV-VCA (P3HR1 strain) infected and uninfected lymphocytic cells bound to slide well
   Anti-human IgG Ab (g) FITC conjugated
Substrate:
Controls - standards supplied:
   Controls: Neg and Pos (human serum)
Additional reagents:
   None
Special equipment required:
   None

INTERPRETATION

Comments on interpretation:
   Positive: ≥(1+) fluorescence in 10–50% of cells at ≥1:10 dilution
No. of references: 22

NOTES

131290.0

© KLUWER ACADEMIC PUBLISHERS 1995, ISSN 1381-5067

Epstein-Barr virus (VCA)
ANTIBODY DETECTION (IgG)

Manufacturer: Gull Laboratories Inc
Cat. No./Trade name: EB100/EBV IgG Test

SUMMARY

[Slide-Ag]–**Ab**–[AHIgG-FITC]–fluorescence

Assay type: Immunofluorescence assay (indirect)
Detection: Fluorescence microscopy
Format: Slide well, Ag coated
Sample type: Serum
Sample pre-treatment:
    None
Sample volume: 15 µl of 1:10 dilution
Number of tests: 100
Controls - standards run in assay:
    Controls: Neg (1), Pos (1)
Incubation:
    30 min (RT) + 30 min (RT)
Washes: 2

CONTENTS

Antibodies, antigens, labelled components:
    EBV-VCA infected and uninfected HR1 Burkitt's
        lymphocytic cells bound to slide well
    Anti-human IgG Ab (caprine) FITC conjugated
Substrate:
Controls - standards supplied:
    Controls: Neg and Pos (human serum)
Additional reagents:
    None
Special equipment required:
    None

INTERPRETATION

Comments on interpretation:
    Positive: yellow-green fluorescence of 10–20% of cells
        per field at ≥1:10 dilution
No. of references: 15

NOTES

131704.0

Epstein-Barr virus (VCA)
ANTIBODY DETECTION (IgG)

Manufacturer: Hemagen Diagnostics Inc
Cat. No./Trade name: 2900/VIRGO® EBV-VCA IgG IFA

SUMMARY

[Slide-Ag]–**Ab**–[AHIgG-FITC]–fluorescence

Assay type: Immunofluorescence assay (indirect)
Detection: Fluorescence microscopy
Format: Slide well, Ag coated
Sample type: Serum
Sample pre-treatment:
    None
Sample volume: 10–20 µl of 1:10 dilution x2
Number of tests: 96
Controls - standards run in assay:
    Controls: Neg (1), Pos (1)
Incubation:
    30 min (37°C) + 30 min (37°C)
Washes: 2

CONTENTS

Antibodies, antigens, labelled components:
    EBV-VCA infected and uninfected cells bound to slide
        well
    Anti-human IgG Ab (g) FITC conjugated
Substrate:
Controls - standards supplied:
    Controls: Neg and Pos (human serum)
Additional reagents:
    None
Special equipment required:
    None

INTERPRETATION

Comments on interpretation:
    Positive: ≥(1+) fluorescence at 1:10 dilution
No. of references: 15

NOTES

131108.0

© *KLUWER ACADEMIC PUBLISHERS 1995, ISSN 1381-5067*

Epstein-Barr virus (VCA)
ANTIBODY DETECTION (IgG)

Manufacturer: Immuno Concepts Inc
Cat. No./Trade name: EBV-VCA IgG Antibody Test System

SUMMARY

[Slide-Ag]–**Ab**–[AHIgG-FITC]–fluorescence

Assay type: Immunofluorescence assay (indirect)
Detection: Fluorescence microscopy
Format: Slide well, Ag coated
Sample type: Serum
Sample pre-treatment:
 None
Sample volume: 100 µl of 1:10 dilution
Number of tests: 7 or 13*
Controls - standards run in assay:
 Controls: Neg (1), Pos (1)
Incubation:
 30 min (RT) + 30 min (RT)
Washes: 2

CONTENTS

Antibodies, antigens, labelled components:
 EBV-VCA infected and uninfected lymphoblastoid cells
  from Burkitt's lymphoma bound to slide well
 Anti-human IgG Ab (g) FITC conjugated
Substrate:
Controls - standards supplied:
 Controls: Neg, Pos and titrable control (human serum)
Additional reagents:
 None
Special equipment required:
 None

INTERPRETATION

Comments on interpretation:
 Positive: apple-green fluorescence of nucleus and
  cytoplasm of 10–20% of cells per field at ⩾1:10
  dilution
No. of references: 14

NOTES

131813.0

*7 or 13 wells per slide

Epstein-Barr virus (VCA)
ANTIBODY DETECTION (IgG)

Manufacturer: MRL Diagnostics
Cat. No./Trade name: IF0600G/Epstein-Barr Virus VCA IgG

SUMMARY

[Slide-Ag]–**Ab**–[AHIgG-FITC]–fluorescence

Assay type: Immunofluorescence assay (indirect)
Detection: Fluorescence microscopy
Format: Slide well, Ag coated
Sample type: Serum
Sample pre-treatment:
 None
Sample volume: 10 µl of 1:10 dilution
Number of tests: 120
Controls - standards run in assay:
 Controls: Neg (1), Pos (1)
Incubation:
 30 min (37°C) + 30 min (37°C)
Washes: 2

CONTENTS

Antibodies, antigens, labelled components:
 EBV-VCA infected lymphocytes and uninfected cells
  bound to slide well
 Anti-human IgG Ab (g) FITC conjugated
Substrate:
Controls - standards supplied:
 Controls: Neg and Pos (human serum)
Additional reagents:
 None
Special equipment required:
 None

INTERPRETATION

Comments on interpretation:
 Positive: ⩾(1+) apple-green fluorescence
No. of references: 9

NOTES

131116.0

Epstein-Barr virus (VCA)
ANTIBODY DETECTION (IgG)

Manufacturer: Stellar Bio Systems Inc
Cat. No./Trade name: Indirect Fluorescent Assay for EBV-VCA IgG Antibody

SUMMARY

[Slide-Ag]–**Ab**–[AHIgG-FITC]–fluorescence

Assay type: Immunofluorescence assay (indirect)
Detection: Fluorescence microscopy
Format: Slide well, Ag coated
Sample type: Serum
Sample pre-treatment:
 None
Sample volume: 20 µl of 1:10 dilution
Number of tests:
Controls - standards run in assay:
 Controls: Neg (1), Pos (1)
Incubation:
 30 min (37°C) + 30 min (37°C)
Washes: 2

CONTENTS

Antibodies, antigens, labelled components:
 EBV-VCA infected and uninfected cells (human B cell line) bound to slide
 Anti-human IgG Ab (g) FITC conjugated
Substrate:
Controls - standards supplied:
 Controls: Neg and Pos (human serum)
Additional reagents:
 None
Special equipment required:
 None

INTERPRETATION

Comments on interpretation:
 Positive: (1 +) – (4 +) bright green fluorescent staining of infected cells
No. of references: 8

NOTES

131630.0

Epstein-Barr virus (VCA)
ANTIBODY DETECTION (IgG)

Manufacturer: Zeus Scientific Inc
Cat. No./Trade name: 9150-10/EBV-VCA IFA Test System

SUMMARY

[Slide-Ag]–**Ab**–[AHIgG-FITC]–fluorescence

Assay type: Immunofluorescence assay (indirect)
Detection: Fluorescence microscopy
Format: Slide, Ag coated
Sample type: Serum
Sample pre-treatment:
 None
Sample volume: 20 µl of 1:10 dilution
Number of tests: 100
Controls - standards run in assay:
 Controls: Neg (1), Pos (1)
Incubation:
 30 min (RT) + 30 min (RT)
Washes: 2

CONTENTS

Antibodies, antigens, labelled components:
 EBV-VCA infected cells bound to slide
 Anti-human IgG Ab (g) FITC conjugated
Substrate:
Controls - standards supplied:
 Controls: Neg and Pos (human serum)
Additional reagents:
 None
Special equipment required:
 None

INTERPRETATION

Comments on interpretation:
 Positive: (1 +)–(4 +) apple-green fluorescence in cell membrane and cytoplasm of infected cells
No. of references: 17

NOTES

131296.0

© *KLUWER ACADEMIC PUBLISHERS 1995, ISSN 1381-5067*

Epstein-Barr virus (VCA)
ANTIBODY DETECTION (IgM)

Manufacturer: Bion Enterprises Ltd
Cat. No./Trade name: EBV-M (VCA) Antibody Test System

SUMMARY

[Slide-Ag]–**Ab**–[AHIgM-FITC]–fluorescence

Assay type: Immunofluorescence assay (indirect)
Detection: Fluorescence microscopy
Format: Slide well, Ag coated
Sample type: Serum
Sample pre-treatment:
Treat serum by ion exchange chromatography or IgG immunoprecipitation to eliminate interference due to RF and IgG
Sample volume: 20-30 μl of 1:10 and 1:40 dilution
Number of tests: 60, 120
Controls - standards run in assay:
Controls: Neg (1), Pos (1)
Incubation:
1 hr 30 min (37°C) + 1 hr (37°C)
Washes: 2

CONTENTS

Antibodies, antigens, labelled components:
EBV-VCA (P3HR1 strain) infected and uninfected lymphocytic cells bound to slide well
Anti-human IgM Ab (g) FITC conjugated
Substrate:
Controls - standards supplied:
Controls: Neg and Pos (human serum)
Additional reagents:
None
Special equipment required:
None

INTERPRETATION

Comments on interpretation:
Positive: ≥(1+) fluorescence in 10–50% of cells at ≥1:10 dilution
No. of references: 33

NOTES

131291.0

Epstein-Barr virus (VCA)
ANTIBODY DETECTION (IgM)

Manufacturer: Gull Laboratories Inc
Cat. No./Trade name: EB140/EBV IgM Test

SUMMARY

[Slide-Ag]–**Ab**–[AHIgM-FITC]–fluorescence

Assay type: Immunofluorescence assay (indirect)
Detection: Fluorescence microscopy
Format: Slide well, Ag coated
Sample type: Serum
Sample pre-treatment:
None
Sample volume: 15 μl of 1:10 dilution
Number of tests: 50, 100
Controls - standards run in assay:
Controls: Neg (1), Pos (1)
Incubation:
30 min (RT) + 30 min (RT)
Washes: 2

CONTENTS

Antibodies, antigens, labelled components:
EBV-VCA infected and uninfected HR1 Burkitt's lymphocytic cells bound to slide well
Anti-human IgM Ab (caprine) FITC conjugated
Substrate:
Controls - standards supplied:
Controls: Neg and Pos (human serum)
Additional reagents:
None
Special equipment required:
None

INTERPRETATION

Comments on interpretation:
Positive: yellow-green fluorescence of 10–20% of cells per field at ≥1:10 dilution
No. of references: 7

NOTES

131707.0

© KLUWER ACADEMIC PUBLISHERS 1995, ISSN 1381-5067

Epstein-Barr virus (VCA)
ANTIBODY DETECTION (IgM)

Manufacturer: Immuno Concepts Inc
Cat. No./Trade name: EBV-VCA IgM Antibody Test System

SUMMARY

[Slide-Ag]–**Ab**–[AHIgM-FITC]–fluorescence

Assay type: Immunofluorescence assay (indirect)
Detection: Fluorescence microscopy
Format: Slide well, Ag coated
Sample type: Serum
Sample pre-treatment:
    None
Sample volume: 100 µl of 1:10 dilution
Number of tests: 7 or 13
Controls - standards run in assay:
    Controls: Neg (1), Pos (1)
Incubation:
    30 min (RT) + 30 min (RT)
Washes: 2

CONTENTS

Antibodies, antigens, labelled components:
    EBV-VCA infected and uninfected lymphoblastoid cells from Burkitt's lymphoma bound to slide well
    Anti-human IgM Ab (g) FITC conjugated
Substrate:
Controls - standards supplied:
    Controls: Neg, Pos and titrable control (human serum)
Additional reagents:
    None
Special equipment required:
    None

INTERPRETATION

Comments on interpretation:
    Positive: apple-green fluorescence of nucleus and cytoplasm of 10–20% of cells per field at ≥1:10 dilution
No. of references: 16

NOTES

131814.0

*7 or 13 wells per slide

Epstein-Barr virus (VCA)
ANTIBODY DETECTION (IgM)

Manufacturer: MRL Diagnostics
Cat. No./Trade name: RIF0600M/EBV VCA RIFA® Test

SUMMARY

[Slide-Ag]–**Ab**–[AHIgM-FITC]–fluorescence

Assay type: Immunofluorescence assay (indirect)
Detection: Fluorescence microscopy
Format: Slide well, Ag coated
Sample type: Serum
Sample pre-treatment:
    Mix serum with IgM pretreatment diluent, stand 5 min and centrifuge to eliminate interference due to RF and IgG
Sample volume: 10 µl of treated sample
Number of tests: 120
Controls - standards run in assay:
    Controls: Neg (1), Pos (1)
Incubation:
    1 hr 30 min (37°C) + 30 min (RT)
Washes: 2

CONTENTS

Antibodies, antigens, labelled components:
    rec EBV-VCA infected mammalian cells and uninfected cells bound to slide well
    Anti-human IgM Ab (g) FITC conjugated
Substrate:
Controls - standards supplied:
    Controls: Neg and Pos (human serum)
Additional reagents:
    None
Special equipment required:
    None

INTERPRETATION

Comments on interpretation:
    Positive: ≥(1+) fluorescence
No. of references: 9

NOTES

131120.0

Epstein-Barr virus (VCA)
ANTIBODY DETECTION (IgM)

Manufacturer: Stellar Bio Systems Inc
Cat. No./Trade name: Indirect Fluorescent Assay for EBV-VCA IgM Antibody

SUMMARY

[Slide-Ag]–**Ab**–[AHIgM-FITC]–fluorescence

Assay type: Immunofluorescence assay (indirect)
Detection: Fluorescence microscopy
Format: Slide well, Ag coated
Sample type: Serum
Sample pre-treatment:
Removal of interference due to RF and IgG using commercially available pre-treatment devices (not provided)
Sample volume: 20 μl of 1:40 dilution (or 1:10 if pretreated sample)
Number of tests:
Controls - standards run in assay:
Controls: Neg (1), Pos (1)
Incubation:
1 hr (37°C) + 30 min (37°C)
Washes: 2

CONTENTS

Antibodies, antigens, labelled components:
EBV-VCA infected and uninfected cells (human B cell line) bound to slide
Anti-human IgM Ab (g) FITC conjugated
Substrate:
Controls - standards supplied:
Controls: Neg and Pos (human serum)
Additional reagents:
Pre-treatment device for removal of RF and IgG in sample
Special equipment required:
None

INTERPRETATION

Comments on interpretation:
Positive: (1 +) – (4 +) bright green fluorescent staining of infected cells
No. of references: 8

NOTES

131637.0

Heterophile antigen
ANTIBODY DETECTION

Manufacturer: BIOKIT SA
Cat. No./Trade name: Monolatex

SUMMARY

[Latex-Ag]–**Ab**–agglutination

Assay type: Particle agglutination assay
Detection: Visual
Format: Latex particles, Ag coated (test done on slide)
Sample type: Serum, plasma
Sample pre-treatment:
None
Sample volume: 50 μl
Number of tests:
Controls - standards run in assay:
Controls: Neg (1), Pos (1)
Incubation:
3 min (RT) with shaking
Washes:

CONTENTS

Antibodies, antigens, labelled components:
Heterophile Ag (from bovine cell membranes) bound to latex particles (Latex Reagent)
Substrate:
Controls - standards supplied:
Controls: Neg and Pos (human serum)
Additional reagents:
None
Special equipment required:
None

INTERPRETATION

Comments on interpretation:
Positive: agglutination of sample with Latex Reagent
No. of references: 11

NOTES

131728.0

© KLUWER ACADEMIC PUBLISHERS 1995, ISSN 1381-5067

Heterophile antigen
ANTIBODY DETECTION

Manufacturer: Diesse
Cat. No./Trade name: 024/IMN SLIDE

SUMMARY

[Latex-Ab]–**Ag**–agglutination

Assay type: Particle agglutination assay
Detection: Visual
Format: Latex particles, Ag coated (test done on slide)
Sample type: Serum, plasma
Sample pre-treatment:
  None
Sample volume: 40 µl
Number of tests: 50
Controls - standards run in assay:
  Controls: Neg (1), Pos (1)
Incubation:
  Read results 1 min after mixing sample and reagents
Washes:

CONTENTS

Antibodies, antigens, labelled components:
  Heterophile Ag (from bovine RBCs) bound to latex
    particles
Substrate:
Controls - standards supplied:
  Controls: Neg and Pos (human)
Additional reagents:
  None
Special equipment required:
  None

INTERPRETATION

Comments on interpretation:
  Positive: agglutination of sample with latex reagent
No. of references: 4

NOTES

131600.0

Heterophile antigen
ANTIBODY DETECTION

Manufacturer: Unipath Limited
Cat. No./Trade name: Infectious Mononucleosis Kit

SUMMARY

[Latex-Ag]–**Ab**–agglutination

Assay type: Particle agglutination assay
Detection: Visual
Format: Latex particles, Ag coated (test done on card)
Sample type: Serum, plasma
Sample pre-treatment:
  None
Sample volume: 1 drop
Number of tests: 30
Controls - standards run in assay:
  Controls: Neg (1), Pos (1)
Incubation:
  Read result within 2 min of mixing sample and reagent
Washes:

CONTENTS

Antibodies, antigens, labelled components:
  Heterophile Ag (from bovine red cell membranes) bound
    to latex particles (Test Latex)
Substrate:
Controls - standards supplied:
  Controls: Neg and Pos (rabbit serum)
Additional reagents:
  None
Special equipment required:
  None

INTERPRETATION

Comments on interpretation:
  Positive: agglutination of sample with Test Latex
No. of references: 7

NOTES

131618.0

© KLUWER ACADEMIC PUBLISHERS 1995, ISSN 1381-5067

Escherichia coli

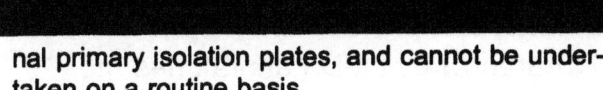

E.coli is a facultatively anaerobic, Gram-negative bacillus which may be motile or non-motile, and which is commonly found as part of the normal colonic flora of man and a variety of other animals. However *E.coli* is frequently the causal organism of a number of common bacterial infections including urinary tract infection, bacteraemia and various diarrhoeal illnesses, as well as being an important cause of neonatal meningitis and nosocomial pneumonia. Those strains which are responsible for enteric disease can be further subdivided into enterotoxigenic (ETEC), enteropathogenic (EPEC), enteroinvasive (EIEC) and enterohaemorrhagic (EHEC) groups according to their underling mechanism of pathogenesis and the nature of the resultant illness.

The organism grows well on a variety of routine laboratory media after overnight incubation at 37°C, and may be readily distinguished from other Enterobacteriaceae on the basis of biochemical testing. A number of distinct subtypes may be defined on the basis of serological identification of lipopolysaccharide somatic 'O' antigens, and protein flagellar 'H' antigens. It has been demonstrated that strains of a given O serogroup often share similar biochemical characteristics. Moreover, it is now quite apparent from extensive epidemiological studies that certain serotypes are often associated with particular types of infection. Perhaps most notable in this context has been the association of *E.coli* O157:H7 and other EHEC serotypes with the occurrence of haemorrhagic colitis and the haemolytic-uraemic syndrome. The ability of different serotypes to produce different types of disease is often associated with the elaboration of a variety of toxins, adhesins and other virulence factors, which are not infrequently encoded by extrachromosomal genetic elements such as plasmids and bacteriophages. In addition to the significant pathogenic role of *E.coli*, this organism is also of some importance in public health microbiology, where it is frequently used as a surrogate marker of faecal contamination in the examination of foods, milks and waters.

Diagnosis

E.coli is readily cultured from a wide range of clinical samples including blood, pus, wound swabs, cerebrospinal fluid and joint aspirates, as outlined above, to yield a definitive diagnosis. However the organism is invariably isolated from stool specimens on routine enteric laboratory media and this can result in difficulties in the diagnosis of enteric infections with *E.coli*. The identification of those serogroups associated with enteric disease generally requires painstaking serological testing of representative subcultured colonies from the origi-nal primary isolation plates, and cannot be undertaken on a routine basis.

Fortunately *E.coli* O157:H7 organisms rarely ferment the sugar D-sorbitol, which observation led to the development of sorbitol-MacConkey agar (SMAC) for the rapid screening of faeces for the organism. Commercial latex agglutination tests are now widely available for the presumptive identification of this serotype, and more recently immunomagnetic separation methods have been described to enhance the isolation of *E.coli* O157 from food and clinical samples. Faeces may also be examined for the presence of verocytotoxins produced by EHEC serotypes by means of cell-culture or by ELISA-based systems, and DNA probes tests for verocytotoxin genes are commercially available. Serological tests for the detection of specific antibodies to *E.coli* O157 are available, but are not in widespread use.

References

Eisenstein B. Enterobacteriaceae. In: Mandell GL, Bennett JE, Dolin R, eds. Principles and Practice of Infectious Diseases, 4th edn. New York, Edinburgh, London: Churchill Livingstone; 1995:1964–80.

Chapman PA. Isolation, identification and typing of vero cytotoxin-producing *Escherichia coli* O157. PHLS Microbiology Digest. 1994;11:13–17.

See also Multipathogen Assays section under: Gastrointestinal pathogens Meningitis pathogens

© *KLUWER ACADEMIC PUBLISHERS 1995, ISSN 1381-5067*

Escherichia coli
ANTIGEN DETECTION

Manufacturer: LMD Laboratories Inc
Cat. No./Trade name: ECO-63/E. coli 0157 Antigen Detection ELISA Kit

SUMMARY

[Well-Ab]–**Ag**–[Ab-HRP]–[TMB]–A$_{450}$

Assay type: EIA (non-competitive)
Detection: Colorimetric A$_{450}$
Format: Microtitre well, Ab coated
Sample type: Faecal specimen, rectal swab
Sample pre-treatment:
  Mix sample with diluted wash buffer and vortex
Sample volume: 100 µl of prepared sample
Number of tests: 96
Controls - standards run in assay:
  Controls: Neg (1), Pos (1)
Incubation:
  20 min (RT) + 10 min (RT) + 5 min (RT)
Washes: 2

CONTENTS

Antibodies, antigens, labelled components:
  Anti-*E. coli* (0157) Ab bound to well
  Anti-*E. coli* (0157) Ab HRP conjugated
Substrate: TMB
Controls - standards supplied:
  Controls: Neg and Pos
Additional reagents:
  None
Special equipment required:
  None

INTERPRETATION

Comments on interpretation:
  Equivocal: sample OD between 0.20–0.40; retest to confirm
No. of references: 14

NOTES

131181.0
For research use only

Escherichia coli
ANTIGEN DETECTION

Manufacturer: Tecra® Diagnostics
Cat. No./Trade name: ECOVIA 48/TECRA® E. coli 0.157

SUMMARY

[Well-Ab]–**Ag**–[Ab-enzyme]–[substrate]–green colour

Assay type: EIA (non-competitive)
Detection: Visual or colorimetric
Format: Microtitre well, Ab coated
Sample type: Food, food-related and environmental samples
Sample pre-treatment:
  Samples require culture in enrichment broth (18 hr) prior to assay
Sample volume: 0.2 ml of prepared sample
Number of tests: 48, 96
Controls - standards run in assay:
  Controls: Neg (1), Pos (1)
Incubation:
  30 min (37°C) + 30 min (37°C)
Washes: 2

CONTENTS

Antibodies, antigens, labelled components:
  Anti-*E. coli* (0157) Ab bound to well
  Anti-*E. coli* (0157) Ab enzyme conjugated
Substrate: ABTS
Controls - standards supplied:
  Controls: Neg (diluent) and Pos
Additional reagents:
  Modified Tryptic Soy Broth with acriflavin (dairy products), Modified EC medium with novobiocin (meat products)
Special equipment required:
  None

INTERPRETATION

Comments on interpretation:
  Positive: green colouration in well confirmed by standard tests
No. of references: None

NOTES

131726.0
This kit detects only *E. coli* 0157 serotype

Escherichia coli
ANTIGEN DETECTION

Manufacturer: Microgen Bioproducts
Cat. No./Trade name: M44/MicroScreen® E. coli 0157

SUMMARY

[Latex-Ab]–**Ag**–agglutination

Assay type: Particle agglutination assay
Detection: Visual
Format: Latex particles, Ab coated (test done on slide)
Sample type: Colourless colonies from Sorbitol MacConkey agar culture (isolated from faecal specimens)
Sample pre-treatment:
None
Sample volume: 2-3 colonies (+ 30 ml saline)
Number of tests: 50
Controls - standards run in assay:
Controls: Pos (2)
Incubation:
Read result 30 sec after mixing sample and reagent
Washes:

CONTENTS

Antibodies, antigens, labelled components:
Anti-*E. coli* (0157:H7) Ab bound to latex particles (Test Latex)
Non-immune Ig Ab bound to latex particles (Control Latex)
Substrate:
Controls - standards supplied:
None
Additional reagents:
Known E. coli 0157 culture for use as Positive control
Special equipment required:
None

INTERPRETATION

Comments on interpretation:
Positive: agglutination of sample with Test Latex and not with Control Latex
Samples that react with Control Latex should be retested after mixing with saline and boiling for 10 min
No. of references: 4

NOTES

131193.0

Escherichia coli
ANTIGEN DETECTION

Manufacturer: PRO-LAB Diagnostics
Cat. No./Trade name: PL070/E. coli 0157 Latex Test Reagent Kit

SUMMARY

[Latex-Ab]–**Ag**–agglutination

Assay type: Particle agglutination assay
Detection: Visual
Format: Latex particles, Ab coated (test done on card)
Sample type: Colonies from Sorbitol MacConkey agar
Sample pre-treatment:
Mix colonies with saline in culture tube
Sample volume: 1 drop
Number of tests: 50, 100
Controls - standards run in assay:
Controls: Pos (1)
Incubation:
Read result within 2 min of mixing sample and reagent
Washes:

CONTENTS

Antibodies, antigens, labelled components:
Anti-*E. coli* (serotype 0157) IgG Ab (r) bound to latex particles (Latex Reagent)
Non-immune rabbit IgG Ab bound to latex particles (Control Latex Reagent)
Substrate:
Controls - standards supplied:
Controls: Neg and Pos
Additional reagents:
None
Special equipment required:
None

INTERPRETATION

Comments on interpretation:
Positive: agglutination of sample with Latex Reagent and not with Control Latex Reagent
No. of references: 13

NOTES

131685.0

© *KLUWER ACADEMIC PUBLISHERS 1995, ISSN 1381-5067*

Escherichia coli
ANTIGEN DETECTION

Manufacturer: Unipath Limited
Cat. No./Trade name: E. coli 0157 Test

SUMMARY

[Latex-Ab]–**Ag**–agglutination

Assay type: Particle agglutination assay
Detection: Visual
Format: Latex particles, Ab coated (test done on card)
Sample type: Colonies from Sorbitol MacConkey Agar
Sample pre-treatment:
  None
Sample volume: Portion of colony (+ saline) x 2
Number of tests: 100
Controls - standards run in assay:
  Controls: Neg (1), Pos (1)
Incubation:
  Read result within 2 min of mixing sample and reagent
Washes:

CONTENTS

Antibodies, antigens, labelled components:
  Anti-*E. coli* (0157) Ab bound to latex particles (Test Latex)
  Non-immune rabbit Ig Ab bound to latex particles (Control Latex)
Substrate:
Controls - standards supplied:
  Controls: Neg and Pos
Additional reagents:
  Sorbitol MacConkey agar (Cat. no. Oxoid CM 813)
Special equipment required:
  None

INTERPRETATION

Comments on interpretation:
  Positive: agglutination of sample with Test Latex and no agglutination with Control Latex
No. of references: 9

NOTES

131613.0

Escherichia coli (verotoxins)
ANTIGEN DETECTION

Manufacturer: Meridian Diagnostics Inc
Cat. No./Trade name: 608096/Premier® EHEC

SUMMARY

[Well-MAb]–**Ag**–[Ab]–[ARIgG-AP]–[TMB]–A_{450}

Assay type: EIA (non-competitive)
Detection: Colorimetric A_{450}
Format: Microtitre well, Ab coated
Sample type: Faecal specimens, or culture isolates (from clinical specimens)
Sample pre-treatment:
  Mix with sample diluent
Sample volume: 100 µl of prepared sample
Number of tests: 96
Controls - standards run in assay:
  Controls: Neg (1), Pos (1)
Incubation:
  1 hr (RT) + 30 min (RT) + 30 min (RT) + 10 min (RT)
Washes: 3

CONTENTS

Antibodies, antigens, labelled components:
  Anti-*E. coli* (verotoxins 1 & 2) MAb bound to well
  Anti-*E. coli* (verotoxins 1 & 2) Ab (r)
  Anti-rabbit IgG Ab (g) HRP conjugated
Substrate: TMB
Controls - standards supplied:
  Controls: Neg and Pos
Additional reagents:
  None
Special equipment required:
  None

INTERPRETATION

Comments on interpretation:
  Low Positive: where OD of sample is between 0.130–0.180; retest to confirm or retest using plate or broth culture prior to testing
No. of references: 19

NOTES

131387.0

This kit can detect *E. coli* verotoxins 1 and 2 but does not differentiate between them

Escherichia coli (verotoxin 1 and verotoxin 2)
ANTIGEN DETECTION

Manufacturer: Denka Seiken Co. Ltd
Cat. No./Trade name: 342101/VEROTOX-F

SUMMARY

[Latex-Ab]–**Ag**–agglutination

Assay type: Particle agglutination assay
Detection: Visual
Format: Latex particles, Ab coated (test done in well)
Sample type: Culture fluid
Sample pre-treatment:
   Culture or Stationary Culture, then centrifuged
Sample volume: 25 µl of supernatant (serial dilutions)
Number of tests: 20
Controls - standards run in assay:
   Controls: Pos VT1 (1), Pos VT2 (1)
Incubation:
   18–20 hr (RT)
Washes:

CONTENTS

Antibodies, antigens, labelled components:
   Anti-*E. coli* verotoxin 1 (VT1) IgG Ab (r) bound to latex
      particles (Latex VT1 Reagent)
   Anti-*E. coli* verotoxin 2 (VT2) IgG Ab (r) bound to latex
      particles (Latex VT2 Reagent)
   Non-immune rabbit IgG Ab bound to latex particles
      (Control Latex)
Substrate:
Controls - standards supplied:
   Controls: Pos VT1 and Pos VT2
Additional reagents:
   CA-YE culture medium
Special equipment required:
   None

INTERPRETATION

Comments on interpretation:
   Positive: agglutination of the sample with the appropriate
      Latex Reagent and not with the Control Latex
      identifies the verotoxin present in the sample
No. of references: 3

NOTES

131815.0

This kit can detect *E. coli* verotoxin 1 and 2 and can
   differentiate between them

© *KLUWER ACADEMIC PUBLISHERS 1995, ISSN 1381-5067*

Giardia lamblia

The flagellated protozoan *Giardia lamblia* (also known as *G.intestinalis* or *G.duodenalis*) is the causal organism of one of the most common intestinal infections worldwide. The organism exists as a free-living trophozoite and as a thin-walled cyst form. The trophozoite is pear-shaped and dorsally convex with a concave ventral sucking disc, and a characteristic arrangement of nuclei and other organelles which results in a face-like appearance. Infection is acquired by ingesting the cysts, usually in faecally contaminated water or food, although person-to-person spread also occurs. The disease is more common in areas of poor sanitation. Numerous large waterborne outbreaks in the United States, Europe and elsewhere have been described. Although humans are the main reservoir of infection, a variety of animals may carry *Giardia* spp., and are considered by many to play a role in the epidemiology of the infection.

The organism adheres to the surface of the small intestine by means of its sucking disc and may interfere with the integrity of the brush border. *G.lamblia* does not elaborate an enterotoxin and the exact mechanism of pathogenesis is unclear. Some workers have suggested that symptoms may be related to the malabsorption of certain solutes within the gut which arises from damage to the mucosa, rather than the direct pathogenic effect of the organism. Cell-mediated and humoral responses, particularly secreted IgA and IgM, appear to be important in the eradication of the parasite. Infection with *G.lamblia* may result in a wide spectrum of disease ranging from asymptomatic cyst passage, through acute self-limited episodes of diarrhoea, to chronic diarrhoea with malabsorption and weight loss. Symptomatic patients usually have abdominal cramps, bloating, anorexia, malaise and nausea with foul smelling diarrhoea.

Diagnosis

Most cases of giardiasis can be diagnosed by the observation of trophozoites or cysts in faecal samples either by direct examination of a wet preparation, or following staining with a variety of simple stains. The organism may also be detected in stools which have been preserved, and a number of commercial preservation systems are available. Organisms are found in only about 50% of cases when a single stool specimen is examined. If three stools are examined the detection rate approaches 90%. If stools are negative, duodenal material may be sampled by the Enterotest technique, direct duodenal aspiration, or biopsy.

A number of systems for the detection of *Giardia* antigens have now been described. These have employed CIE, immunofluorescence and ELISA technologies. A number of commercial kits are available, mostly ELISA-based, and have been employed for stool and water testing. Serological tests for the diagnosis of giardiasis are not widely available.

Reference

Wolfe MF. Giardiasis. Clinical Microbiology Reviews. 1992;5:93–100.

See also Multipathogen Assays section under: Gastrointestinal pathogens

Giardia lamblia
ANTIGEN DETECTION

Manufacturer: Alexon Inc
Cat. No./Trade name: 585-96/ProSpecT® Giardia EZ Microplate Assay®

SUMMARY

[Well-Ab]–**Ag**–[MAb-HRP]–[TMB]–A$_{450}$

Assay type: EIA (non-competitive)
Detection: Colorimetric A$_{450}$
Format: Microtitre well, Ab coated
Sample type: Faecal specimens
Sample pre-treatment:
   Mix specimens with dilution buffer
Sample volume: 100 µl of diluted sample
Number of tests: 96
Controls - standards run in assay:
   Controls: Neg (1), Pos (1)
Incubation:
   1 hr (RT) + 10 min (RT)
Washes: 2

CONTENTS

Antibodies, antigens, labelled components:
   Anti-*Giardia lamblia* (GSA 65) Ab (r) bound to well
   Anti-*Giardia lamblia* (GSA 65) MAb HRP conjugated
Substrate: TMB
Controls - standards supplied:
   Controls: Neg and Pos (human faecal material)
Additional reagents:
   None
Special equipment required:
   None

INTERPRETATION

Comments on interpretation:
   Classification of sample is according to cut-off; no further testing
No. of references: 16

NOTES

131304.0

Giardia lamblia
ANTIGEN DETECTION

Manufacturer: Alexon Inc
Cat. No./Trade name: 380-20/ProSpecT® Giardia Rapid Assay

SUMMARY

[Membrane-Ab]–**Ag**–[MAb-HRP]–[TMB]–blue spot

Assay type: EIA (non-competitive)
Detection: Visual
Format: Test device, Ab coated membrane
Sample type: Faecal specimens
Sample pre-treatment:
   Mix specimens with dilution buffer
Sample volume: 4 drops of diluted sample
Number of tests: 20
Controls - standards run in assay:
   Controls: Pos (1) on membrane
Incubation:
   2 min (RT) + 2 min (RT) + 3 min (RT)
Washes: 2

CONTENTS

Antibodies, antigens, labelled components:
   Anti-*Giardia lamblia* (GSA 65) Ab bound to membrane (test area)
   Anti-*Giardia lamblia* (GSA 65) MAb HRP conjugated
Substrate: TMB
Controls - standards supplied:
   Controls: Integral Pos controls incorporated on membrane
Additional reagents:
   None
Special equipment required:
   None

INTERPRETATION

Comments on interpretation:
   Negative: circular blue spot in Pos control area and no blue spot in specimen test area
   Positive: circular blue spot in Pos control area and specimen test area
No. of references: 17

NOTES

131306.0

© KLUWER ACADEMIC PUBLISHERS 1995, ISSN 1381-5067

Giardia lamblia
ANTIGEN DETECTION

Manufacturer: Cellabs Pty Ltd
Cat. No./Trade name: KG2/GIARDIA CELISA®

SUMMARY

[Well-MAb]–**Ag**–[Ab]–[ARIgG-POD]–[TMB]–A_{450}

Assay type: EIA (non-competitive)
Detection: Colorimetric A_{450}
Format: Microtitre well, Ab coated
Sample type: Faecal specimens
Sample pre-treatment:
 Mix with formalin 10%, centrifuge and stand overnight at 2-8°C
Sample volume: 100 µl of prepared sample
Number of tests: 96
Controls - standards run in assay:
 Controls: Neg (1), Pos (1)
Incubation:
 10 min (37°C) + 1 hr (37°C) + 1 hr (37°C) + 20 min (RT)
Washes: 3

CONTENTS

Antibodies, antigens, labelled components:
 Anti-Giardia lamblia (cysts and trophozoites) MAb bound to well
 Anti-Giardia lamblia Ab (r)
 Anti-rabbit IgG Ab POD conjugated
Substrate: TMB
Controls - standards supplied:
 Controls: Neg (diluent) and Pos
Additional reagents:
 None
Special equipment required:
 None

INTERPRETATION

Comments on interpretation:
 Classification of samples is according to cut-off; no further testing
No. of references: 20

NOTES
131413.0

Giardia lamblia
ANTIGEN DETECTION

Manufacturer: LMD Laboratories Inc
Cat. No./Trade name: GL-33/Giardia lamblia Antigen Detection ELISA Kit

SUMMARY

[Well-Ab]–**Ag**–[Ab]–[AGIg-biotin]–[strept-HRP]–[TMB]–A_{450}

Assay type: EIA (non-competitive)
Detection: Colorimetric A_{450}
Format: Microtitre well, Ab coated
Sample type: Faecal specimens, rectal swab
Sample pre-treatment:
 Mix sample with diluted wash buffer and vortex
Sample volume: 100 µl of prepared sample
Number of tests: 48, 96
Controls - standards run in assay:
 Controls: Neg (1), Pos (1)
Incubation:
 20 min (RT) + 10 min (RT) + 5 min (RT) + 5 min (RT)
Washes: 3

CONTENTS

Antibodies, antigens, labelled components:
 Anti-Giardia lamblia Ab bound to well
 Anti-Giardia lamblia Ab (g)
 Anti-mouse Ig Ab biotinylated
 Streptavidin HRP conjugated
Substrate: TMB
Controls - standards supplied:
 Controls: Neg and Pos
Additional reagents:
 None
Special equipment required:
 None

INTERPRETATION

Comments on interpretation:
 Classification of samples is according to cut-off; no further testing
No. of references: 12

NOTES
131182.0

Giardia lamblia
ANTIGEN DETECTION

Manufacturer: Melotec S.A.
Cat. No./Trade name: Melotest Giardiasis Ag

SUMMARY

[Well-Ab]–**Ag**–[Ab-biotin]–[strept-HRP]–[TMB]–A_{450}

Assay type: EIA (non-competitive)
Detection: Colorimetric A_{450}
Format: Microtitre well, Ab coated
Sample type: Faecal sample
Sample pre-treatment:
 Mix sample with diluent
Sample volume: 100 µl of prepared sample x 2
Number of tests: 96
Controls - standards run in assay:
 Controls: Neg (2), Pos (1)
Incubation:
 20 min (RT) + 10 min (RT) + 5 min (RT) + 10 min (RT)
Washes: 3

CONTENTS

Antibodies, antigens, labelled components:
 Anti-*Giardia lamblia* Ab bound to well
 Anti-*Giardia lamblia* Ab (m) biotinylated
 Streptavidin HRP conjugated
Substrate: TMB
Controls - standards supplied:
 Controls: Neg and Pos
Additional reagents:
 None
Special equipment required:
 None

INTERPRETATION

Comments on interpretation:
 Equivocal: ratio of sample OD:cut-off value between 0.9
  and 1.1; retest a fresh sample
 Repeatably equivocal: retest a fresh sample taken in 2-4
  weeks
No. of references: 12

NOTES

131139.0

Giardia lamblia
ANTIGEN DETECTION

Manufacturer: Seradyn
Cat. No./Trade name: 70163/Color Vue® Giardia

SUMMARY

[Well-Ab]–**Ag**–[Ab-biotin]–[strept-HRP]–[TMB]–A_{450}

Assay type: EIA (non-competitive)
Detection: Colorimetric A_{450}
Format: Microtitre well, Ab coated
Sample type: Faecal specimen
Sample pre-treatment:
 Mix sample with diluted wash buffer
Sample volume: 100 µl of prepared sample
Number of tests: 96
Controls - standards run in assay:
 Controls: Neg (1), Pos (1)
Incubation:
 20 min (RT) + 10 min (RT) + 5 min (RT) + 5 min (RT)
Washes: 3

CONTENTS

Antibodies, antigens, labelled components:
 Anti-*Giardia lamblia* Ab bound to well
 Anti-*Giardia lamblia* Ab biotinylated
 Streptavidin HRP conjugated
Substrate: TMB
Controls - standards supplied:
 Controls: Neg and Pos
Additional reagents:
 None
Special equipment required:
 None

INTERPRETATION

Comments on interpretation:
 Classification of samples is according to cut-off; no
  further testing
No. of references: 13

NOTES

131652.0

© *KLUWER ACADEMIC PUBLISHERS 1995, ISSN 1381-5067*

Giardia lamblia
ANTIGEN DETECTION

Manufacturer: Cellabs Pty Ltd
Cat. No./Trade name: KG1/Giardia - CEL® IF Test

SUMMARY

[Slide]–**Ag**–[MAb-FITC]–fluorescence

Assay type: Immunofluorescence assay (direct)
Detection: Fluorescence microscopy
Format: Slide, specimen coated
Sample type: Faecal specimens
Sample pre-treatment:
Specimens may be concentrated prior to use by filtration or with Faecal Parasite Concentration Kit*. Otherwise dilute samples with 10% formalin or PBS, stand before use, apply to slide, dry, fix
Sample volume: 20 µl of prepared specimen
Number of tests: 50
Controls - standards run in assay:
Controls: Pos (1)
Incubation:
30 min (37°C)
Washes: 1

CONTENTS

Antibodies, antigens, labelled components:
Anti-*Giardia lamblia* (cysts) MAb FITC conjugated
Substrate:
Controls - standards supplied:
Controls: Pos (slide)
Additional reagents:
None
Special equipment required:
None

INTERPRETATION

Comments on interpretation:
Positive: green fluorescence of ≥one cyst
No. of references: 16

NOTES

131410.0

This kit detects Giardia cysts in rats, dogs, cattle and sheep specimens in addition to human specimens
*Kit purchased from Evergreen Scientific, CA 90058, USA (Cat. no. 240-3074-030)

© *KLUWER ACADEMIC PUBLISHERS 1995, ISSN 1381-5067*

Hantaviruses

Natural history

Hantaviruses belong to the genus Hantavirus of the family Bunyaviridae. All members are rodent-borne viruses, generally associated with haemorrhagic fever. The type species is Hantaan virus, the agent of Korean haemorrhagic fever with renal syndrome (HFRS) described in US soldiers in the Korean war and more recently recognised in other parts of Asia and Europe. Hantaan virus is a natural infection of mice and rats, with both sylvan and urban animals being persistently infected. Infection of humans results in a severe generalised fever, with haemorrhage and shock, with a 10% fatality rate. A milder form of infection termed nephropathica epidemica occurs in Scandinavia and is caused by the closely related Puumala virus. Both viruses were first isolated in the early 1980s. Infection occurs by aerosol from infected dust or fomites, contamination of food or by bite.

In 1992/93, an outbreak of acute respiratory disease in adults occurred in the Four Corners region of the south-western USA and was shown to be due to a new hantavirus spread by the American deer mouse. This virus is currently called Four Corners virus or Sin Nombre virus and is associated with Hanta pulmonary syndrome (HPS). Each virus tends to be particularly associated with a single species of rodent but may be isolated from several types.

The main symptoms of HFRS include abrupt onset of fever, malaise, headache, anorexia and back pain, followed by marked proteinuria and petechial skin rash. Shock develops and 1/3 of fatalities occur at this stage. Increased oliguria and bleeding follow, with severe vomiting and CNS involvement. The mortality is reduced by dialysis and intense supportive treatment. In HPS, there is a sudden onset of fever, headache, severe myalgia and cough leading to acute respiratory failure in 75% of cases.

Diagnosis

Isolation is difficult and rarely successful. Demonstration of viral antigen in clinical specimens has also been unsuccessful. Viral nucleic acid can be detected in tissue but most diagnoses are made serologically with detection of specific IgM by ELISA the most useful test.

Comment

Cross-reactivity between different hantavirus species makes it difficult to determine the serotype and consideration must be given to the clinical picture and location. It seems likely that other hantaviruses will be found in native rodent species and may or may not prove to be associated with disease.

Reference

Lloyd G. Hantavirus. In: Morgan-Capner P, ed. Current Topics in Clinical Virology. Public Health Laboratory Service; 1991.

Swanepoel R. Bunvaviridae.In: Zuckerman AJ, Banatvala JE, Pattison JR, eds. Principles and Practice of Clinical Virology, 3rd edn. Chichester: John Wiley; 1994.

Hantaan virus
ANTIBODY DETECTION (IgG)

Manufacturer: Progen
Cat. No./Trade name: PR 59065/Hantaan Virus IgG Elisa

SUMMARY

[Well-Ag]–**Ab**–[AHIgG-POD]–[TMB]–A_{450}

Assay type: EIA (non-competitive)
Detection: Colorimetric A_{450}
Format: Microtitre well, Ag coated
Sample type: Serum
Sample pre-treatment:
　None
Sample volume: 100 μl of diluted sample
Number of tests: 48
Controls - standards run in assay:
　Controls: Neg (1), Pos (1)
　Calibrators: 1 (1)
Incubation:
　45 min (37°C) + 45 min (37°C) + 10 min (RT)
Washes: 2

CONTENTS

Antibodies, antigens, labelled components:
　rec Hantaan virus (strain 78–118) Ag bound to well
　Anti-human IgG Ab POD conjugated
Substrate: TMB
Controls - standards supplied:
　Controls: Neg and Pos (human)
　Calibrator: 1 (human)
Additional reagents:
　None
Special equipment required:
　None

INTERPRETATION

Comments on interpretation:
　Equivocal: where the ratio of sample OD:calibrator OD is between 1 and 1.5; no clear interpretation is possible. If hantavirus is suspected, then it is recommended to test with Puumala virus IgG Elisa (Cat. no. PR59056)
No. of references: 5

NOTES

131075.0

Hantaan virus
ANTIBODY DETECTION (IgM)

Manufacturer: Progen
Cat. No./Trade name: PR59066/Hantaan Virus IgM Elisa

SUMMARY

[Well-Ag]–**Ab**–[AHIgM-POD]–[TMB]–A_{450}

Assay type: EIA (non-competitive)
Detection: Colorimetric A_{450}
Format: Microtitre well, Ag coated
Sample type: Serum
Sample pre-treatment:
　Incubate diluted serum 30 min (RT) with IgG absorbent provided to eliminate RF and IgG
Sample volume: 100 μl of treated sample
Number of tests: 48
Controls - standards run in assay:
　Controls: Neg (1), Pos (1)
　Calibrator: 1 (1)
Incubation:
　45 min (37°C) + 45 min (37°C) + 10 min (RT)
Washes: 2

CONTENTS

Antibodies, antigens, labelled components:
　rec Hantaan virus (strain 78-118) Ag bound to well
　Anti-human IgM Ab POD conjugated
Substrate: TMB
Controls - standards supplied:
　Controls: Neg and Pos (human)
　Calibrator: 1 (human)
Additional reagents:
　None
Special equipment required:
　None

INTERPRETATION

Comments on interpretation:
　Equivocal: where the ratio of sample OD:calibrator OD is between 1 and 1.5; no clear interpretation is possible
　If hantavirus infection is suspected, it is recommended to test with Puumala virus IgM Elisa also (Cat. no. PR 59057)
No. of references: 5

NOTES

131819.0

Hantaan virus
ANTIBODY DETECTION

Manufacturer: Progen
Cat. No./Trade name: PR 77065/Hantaan Virus Antibody IF Test

SUMMARY

[Slide-Ag]–**Ab**–[AHIgG&IgM-FITC]–fluorescence

Assay type: Immunofluorescence assay (indirect)
Detection: Fluorescence microscopy
Format: Slide, Ag coated
Sample type: Serum
Sample pre-treatment:
   None
Sample volume: 20 μl of 1:16 and 1:32 dilution
Number of tests: 60
Controls - standards run in assay:
   Controls: Neg (1), Pos (1)
Incubation:
   30 min (RT) + 30 min (RT)
Washes: 2

CONTENTS

Antibodies, antigens, labelled components:
   Hantaan virus infected and uninfected Vero E6 cells
      bound to slide
   Anti-human IgG & IgM Abs FITC conjugated
Substrate:
Controls - standards supplied:
   Controls: Neg and Pos (human)
Additional reagents:
   None
Special equipment required:
   None

INTERPRETATION

Comments on interpretation:
   Positive: (1+) – (3+) yellow-green fluorescence of virus
      particles in cytoplasm with red counterstain at 1:16
      titre
No. of references: 5

NOTES

131078.0

Puumala virus
ANTIBODY DETECTION (IgG)

Manufacturer: Progen
Cat. No./Trade name: PR 59056/Puumala Virus IgG Elisa

SUMMARY

[Well-Ag]–**Ab**–[AHIgG-POD]–[TMB]–A_{450}

Assay type: EIA (non-competitive)
Detection: Colorimetric A_{450}
Format: Microtitre well, Ag coated
Sample type: Serum
Sample pre-treatment:
   None
Sample volume: 100 μl of diluted sample
Number of tests: 48
Controls - standards run in assay:
   Controls: Neg (1), Pos (1)
   Calibrators: 1 (1)
Incubation:
   45 min (37°C) + 45 min (37°C) + 10 min (RT)
Washes: 2

CONTENTS

Antibodies, antigens, labelled components:
   rec Puumala virus (strain CG 18-12) Ag bound to well
   Anti-human IgG Ab POD conjugated
Substrate: TMB
Controls - standards supplied:
   Controls: Neg and Pos
Additional reagents:
   None
Special equipment required:
   None

INTERPRETATION

Comments on interpretation:
   Equivocal: where the ratio of sample OD:calibrator OD is
      between 1 and 1.5; no clear interpretation is possible.
      If hantavirus infection is suspected, then it is
      recommended to test with Hantaan Virus IgG Elisa
      (Cat. no. PR59065)
No. of references: 5

NOTES

131074.0

© KLUWER ACADEMIC PUBLISHERS 1995, ISSN 1381-5067

Puumala virus
ANTIBODY DETECTION (IgM)

Manufacturer: Progen
Cat. No./Trade name: PR59057/Puumala Virus IgM Elisa

SUMMARY

[Well-Ag]–**Ab**–[AHIgM-POD]–[TMB]–A_{450}

Assay type: EIA (non-competitive)
Detection: Colorimetric A_{450}
Format: Microtitre well, Ag coated
Sample type: Serum
Sample pre-treatment:
    Incubate diluted serum 30 min (RT) with IgG absorbent provided to eliminate RF and IgG
Sample volume: 100 µl of treated sample
Number of tests: 48
Controls - standards run in assay:
    Controls: Neg (1), Pos (1)
    Calibrator: 1 (1)
Incubation:
    45 min (37°C) + 45 min (37°C) + 10 min (RT)
Washes: 2

CONTENTS

Antibodies, antigens, labelled components:
    rec Puumala virus (strain CG 18-20) Ag bound to well
    Anti-human IgM Ab POD conjugated
Substrate: TMB
Controls - standards supplied:
    Controls: Neg and Pos (human)
    Calibrator: 1 (human)
Additional reagents:
    None
Special equipment required:
    None

INTERPRETATION

Comments on interpretation:
    Equivocal: where the ratio of sample OD:calibrator OD is between 1 and 1.5; no clear interpretation is possible
    If hantavirus infection is suspected, it is recommended to test with Hantaan virus IgM Elisa also (Cat. no. PR 59066)
No. of references: 5

NOTES

131818.0

Puumala virus
ANTIBODY DETECTION

Manufacturer: Progen
Cat. No./Trade name: PR 77065/Puumala Virus Antibody IF Test

SUMMARY

[Slide-Ag]–**Ab**–[AHIgG&IgM-FITC]–fluorescence

Assay type: Immunofluorescence assay (indirect)
Detection: Fluorescence microscopy
Format: Slide, Ag coated
Sample type: Serum
Sample pre-treatment:
    None
Sample volume: 20 µl of 1:16 and 1:32 dilution
Number of tests: 60
Controls - standards run in assay:
    Controls: Neg (1), Pos (1)
Incubation:
    30 min (RT) + 30 min (RT)
Washes: 2

CONTENTS

Antibodies, antigens, labelled components:
    Puumala virus infected and uninfected Vero E6 cells bound to slide
    Anti-human IgG & IgM Abs FITC conjugated
Substrate:
Controls - standards supplied:
    Controls: Neg and Pos (human)
Additional reagents:
    None
Special equipment required:
    None

INTERPRETATION

Comments on interpretation:
    Positive: (1+) – (3+) yellow-green fluorescence of virus particles in cytoplasm with red counterstain at 1:16 titre
No. of references: 5

NOTES

131077.0

© KLUWER ACADEMIC PUBLISHERS 1995, ISSN 1381-5067

Seoul virus
ANTIBODY DETECTION

Manufacturer: Progen
Cat. No./Trade name: PR 77060/Seoul Virus Antibody IF Test

SUMMARY

[Slide-Ag]–**Ab**–[AHIgG&IgM-FITC]–fluorescence

Assay type: Immunofluorescence assay (indirect)
Detection: Fluorescence microscopy
Format: Slide, Ag coated
Sample type: Serum
Sample pre-treatment:
    None
Sample volume: 20 μl of 1:16 and 1:32 dilution
Number of tests: 60
Controls - standards run in assay:
    Controls: Neg (1), Pos (1)
Incubation:
    30 min (RT) + 30 min (RT)
Washes: 2

CONTENTS

Antibodies, antigens, labelled components:
    Seoul virus infected and uninfected Vero E6 cells bound
        to slide
    Anti-human IgG & IgM Abs FITC conjugated
Substrate:
Controls - standards supplied:
    Controls: Neg and Pos (human)
Additional reagents:
    None
Special equipment required:
    None

INTERPRETATION

Comments on interpretation:
    Positive: (1+) – (3+) yellow-green fluorescence of virus
        particles in cytoplasm with red counterstain at 1:16
        titre
No. of references: 5

NOTES

131079.0

© KLUWER ACADEMIC PUBLISHERS 1995, ISSN 1381-5067

Helicobacter pylori

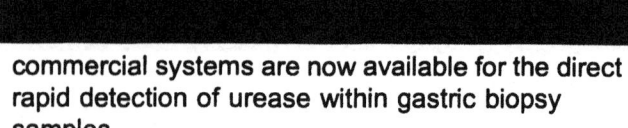

H.pylori is a small, curved, microaerophilic Gram-negative bacillus which is highly motile by means of numerous polar sheathed flagellae. Although first identified in man in 1982, it is already well established that this organism is responsible for the majority of peptic ulcer disease. In addition there is increasing evidence that chronic gastritis and gastric cancer may also be associated with infection by this organism. More recently it has also been suggested that *H.pylori* may be involved in the pathogenesis of certain types of gastric MALT lymphomas, and possibly even of coronary artery disease in some individuals. The organism has a world-wide distribution, although rates of infection appear higher in poorer socio-economic conditions.

Following infection, the organism colonises the mucous layer overlying the gastric mucosal epithelium. The major biochemical activity of the organism is the production of high levels of the urease enzyme. This is thought to play a major role in the ability of the organism to survive in the acid milieu of the stomach. Colonisation results in inflammation in the underlying mucosa. The precise pathogenic mechanisms are unclear, but a number of colonisation and virulence antigens have now been described. Infected subjects develop hypergastrinaemia as a result of decreased gastric somatostatin levels which are thought to be the result of cytokine production within the inflamed mucosa. This is thought to account for the increased parietal cell mass observed in peptic ulcer patients. Ultimately, chronic colonisation may diminish the number of parietal cells and result in atrophic changes which paradoxically favours colonisation and overgrowth by other bacterial species. This process may be important in the development of subsequent malignancy.

The organism can be isolated by culture on chocolate blood agar, or on antibiotic-containing media such as those described by Skirrow and Marshall which inhibit overgrowth by other species. Prolonged incubation for 2–5 days in a microaerophilic (5% oxygen) environment at 35–37°C is required. Typical colonies yielding organisms with characteristic microscopic morphology and motility which are catalase-, oxidase- and urease-positive may be presumptively identified as *H.pylori*.

Diagnosis

Definitive diagnosis of infection may be established by culture of the organism from gastric biopsy specimens as outlined above. The organism may also be demonstrated by histological examination by a variety of methods including Giemsa and silver stains, or fluorescent staining with acridine orange or specific immunofluorescence. A number of commercial systems are now available for the direct rapid detection of urease within gastric biopsy samples.

Non-invasive tests comprise urea breath tests and serological investigation. In the former tests fasted subjects are given a meal containing ^{13}C- or ^{14}C-labelled urea, and isotopically labelled CO_2 produced by the action of bacterial urease is subsequently detected in the patient's breath. These tests are particularly useful in establishing that eradication therapy has been successful. Serological tests are predominantly ELISA-based and detect specific antibodies to *H.pylori*. Infected persons almost invariably develop high-titre antibodies. The rate of decline of antibody levels after successful therapy is variable, and as yet serology is less useful in confirming eradication.

References

Blaser MJ. *Helicobacter pylori* and related organisms. In: Mandell GL, Bennett JE, Dolin R, eds. Principles and Practice of Infectious Diseases, 4th edn. New York, Edinburgh, London: Churchill Livingstone. 1995:1956–64.

Glupczynski Y. The diagnosis of *Helicobacter pylori* infection: a microbiologist's perspective. Reviews in Medical Microbiology. 1994;5:199–208.

Helicobacter pylori
ANTIBODY DETECTION (IgA)

Manufacturer: Bio-Rad
Cat. No./Trade name: 404 3002/G.A.P. H. pylori IgA Test

SUMMARY

[Well-Ag]–**Ab**–[AHIgA-HRP]–TMB–A_{450}

Assay type: EIA (non-competitive)
Detection: Colorimetric A_{450}
Format: Microtitre well, Ag coated
Sample type: Serum, plasma
Sample pre-treatment:
  None
Sample volume: 25 µl (+5 ml diluent)*
Number of tests: 96
Controls - standards run in assay:
  Controls: Neg (1), Pos (1)
  Standards: 0 (2), 1 (2), 2 (2), 3 (2), 4 (2)
Incubation:
  1 hr (RT) + 30 min (RT) + 10 min (RT)
Washes: 2

CONTENTS

Antibodies, antigens, labelled components:
  H. pylori Ag bound to well
  Anti-human IgA Ab (g) HRP conjugated
Substrate: TMB
Controls - standards supplied:
  Controls: Neg (< 10 units/ml) and Pos (25-50 units/ml)
  Standards: 0–4 (0-100 units/ml) all human serum
Additional reagents:
  None
Special equipment required:
  None

INTERPRETATION

Comments on interpretation:
  Equivocal: between 12.5 and 20 units/ml; retest fresh sample in 2 weeks in parallel with initial sample
No. of references: None

NOTES

131175.0

*Samples assayed in duplicate

Helicobacter pylori
ANTIBODY DETECTION (IgA)

Manufacturer: Biomerica
Cat. No./Trade name: 7008/GAP® IgA

SUMMARY

[Well-Ag]–**Ab**–[AHIgA-HRP]–[TMB]–A_{450}

Assay type: EIA (non-competitive)
Detection: Colorimetric A_{450}
Format: Microtitre well, Ag coated
Sample type: Serum
Sample pre-treatment:
  None
Sample volume: 25 µl of 1:200 dilution
Number of tests: 96
Controls - standards run in assay:
  Controls: Neg (1), Pos (1)
  Calibrators: 1 (1), 2 (1), 3 (1), 4 (1)
Incubation:
  1 hr (RT) + 30 min (RT) + 10 min (RT)
Washes: 2

CONTENTS

Antibodies, antigens, labelled components:
  H. pylori Ag bound to well
  Anti-human IgA Ab HRP conjugated
Substrate: TMB
Controls - standards supplied:
  Controls: Neg and Pos
  Calibrators: 0–4 (human serum)
Additional reagents:
  None
Special equipment required:
  None

INTERPRETATION

Comments on interpretation:
  Equivocal: between 12.5–25 units/ml; retest a second sample in 2 weeks in parallel with initial sample
No. of references: 23

NOTES

131032.0

© KLUWER ACADEMIC PUBLISHERS 1995, ISSN 1381-5067

Helicobacter pylori
ANTIBODY DETECTION (IgA)

Manufacturer: BioWhittaker Inc
Cat. No./Trade name: 30-677U/Pylori Elisa II IgA

SUMMARY

[Well-Ag]–**Ab**–[AHIgA-AP]–[PMP]–A_{550}

Assay type: EIA (non-competitive)
Detection: Colorimetric A_{550}
Format: Microtitre well, Ag coated
Sample type: Serum (do not heat inactivate)
Sample pre-treatment:
 None
Sample volume: 10 µl (+200 µl diluent)
Number of tests: 192
Controls - standards run in assay:
 Controls: A (1), B (1)
 Calibrators: Neg (1), Pos 1 (1), Pos 2 (1)
Incubation:
 45 min (RT) + 45 min (RT) + 45 min (RT)
Washes: 2 (+preliminary plate wash)

CONTENTS

Antibodies, antigens, labelled components:
 H. pylori Ag bound to well
 Anti-human IgA Ab (g) AP conjugated
Substrate: PMP
Controls - standards supplied:
 Controls: A and B
 Calibrators: Neg, Pos 1, Pos 2 (all human serum)
Additional reagents:
 None
Special equipment required:
 BioWhittaker automated system and software (optional)

INTERPRETATION

Comments on interpretation:
 Equivocal: ELISA value between 0.18–0.19; retest to
  confirm
 Repeatably equivocal: retest a fresh sample in 3 weeks
  or use an alternative method
No. of references: 14

NOTES

131758.0

Helicobacter pylori
ANTIBODY DETECTION (IgA)

Manufacturer: Chimica Diagnostica
Cat. No./Trade name: 42860/Helicobacter Pylori IgA

SUMMARY

[Well-Ag]–**Ab**–[AHIgA-POD]–[TMB]–A_{450}

Assay type: EIA (non-competitive)
Detection: Colorimetric A_{450}
Format: Microtitre well, Ag coated
Sample type: Serum
Sample pre-treatment:
 None
Sample volume: 100 µl of 1:101 dilution x 2*
Number of tests: 96
Controls - standards run in assay:
 Controls: Neg (2), Pos (2)
Incubation:
 1 hr (37°C) + 30 min (37°C) + 15 min (RT)
Washes: 2

CONTENTS

Antibodies, antigens, labelled components:
 H. pylori Ag bound to well
 Anti-human IgA Ab (r) POD conjugated
Substrate: TMB
Controls - standards supplied:
 Controls: Neg and Pos
Additional reagents:
 None
Special equipment required:
 None

INTERPRETATION

Comments on interpretation:
 Equivocal: between 85% and 115% of cut-off value;
  report as equivocal
No. of references: 2

NOTES

131369.0

*Diluent contains AHIgG to eliminate interference due to RF
 and IgG in sample

© KLUWER ACADEMIC PUBLISHERS 1995, ISSN 1381-5067

Helicobacter pylori
ANTIBODY DETECTION (IgA)

Manufacturer: Diesse
Cat. No./Trade name: ENZYWELL Helicobacter pylori IgA

SUMMARY

[Well-Ag]–**Ab**–[AHIgA-POD]–[TMB]–A_{450}

Assay type: EIA (non-competitive)
Detection: Colorimetric A_{450}
Format: Microtitre well, Ag coated
Sample type: Serum (do not heat inactivate)
Sample pre-treatment:
   None
Sample volume: 100 µl of 1:101 dilution x 2
Number of tests: 96
Controls - standards run in assay:
   Controls: cut-off (2)
Incubation:
   45 min (37°C) + 45 min (37°C) + 15 min (RT)
Washes: 2

CONTENTS

Antibodies, antigens, labelled components:
   H. pylori Ag bound to well
   Anti-human IgA MAb POD conjugated
Substrate: TMB

Helicobacter pylori
ANTIBODY DETECTION (IgA)

Manufacturer: Hycor Biomedical Inc
Cat. No./Trade name: 80525/PYLORAGEN® H. pylori IgA Test Kit

SUMMARY

[Well-Ag]–**Ab**–[AHIgA-HRP]–[TMB]–A_{450}

Assay type: EIA (non-competitive)
Detection: Colorimetric A_{450}
Format: Microtitre well, Ag coated
Sample type: Serum
Sample pre-treatment:
   None
Sample volume: 20 µl (+2 ml diluent)*
Number of tests: 96
Controls - standards run in assay:
   Controls: Neg (1), low Pos (1), mid Pos (1), high Pos (1)
Incubation:
   20 min (RT) + 20 min (RT) + 4 min (RT)
Washes: 2

CONTENTS

Antibodies, antigens, labelled components:
   H. pylori Ag bound to well
   Anti-human IgA Ab HRP conjugated
Substrate: TMB
Controls - standards supplied:

Helicobacter pylori
ANTIBODY DETECTION (IgA)

Manufacturer: IFCI Clone Systems
Cat. No./Trade name: EIAGEN Helicobacter pylori IgA

SUMMARY

[Well-Ag]–**Ab**–[AHIgA-POD]–[TMB]–A_{450}

Assay type: EIA (non-competitive)
Detection: Colorimetric A_{450}
Format: Microtitre well, Ag coated
Sample type: Serum
Sample pre-treatment:
　None
Sample volume: 10 μl (+1 ml diluent)*
Number of tests: 96
Controls - standards run in assay:
　Controls: Neg (2), cut-off (3), Pos (2)
Incubation:
　45 min (37°C) + 45 min (37°C) + 15 min (RT)
Washes: 2

CONTENTS

Antibodies, antigens, labelled components:
　H. pylori Ag bound to well
　Anti-human IgA Ab POD conjugated
Substrate: TMB
Controls - standards supplied:
　Controls: Neg, cut-off and Pos (human serum)
Additional reagents:
　None
Special equipment required:
　None

INTERPRETATION

Comments on interpretation:
　Equivocal: within cut-off ±10%; retest to confirm
　Repeatably equivocal: retest a fresh sample
No. of references: 6

NOTES

131679.0

*Samples are assayed in duplicate

Helicobacter pylori
ANTIBODY DETECTION (IgA)

Manufacturer: Immuno Pharmacology Research
Cat. No./Trade name: HELICOBACTER PYLORI IgA ELISA

SUMMARY

[Well-Ag]–**Ab**–[AHIgA-HRP]–[TMB]–A_{405}

Assay type: EIA (non-competitive)
Detection: Colorimetric A_{450}
Format: Microtitre well, Ag coated
Sample type: Serum
Sample pre-treatment:
　None
Sample volume: 100 μl of 1:100 dilution x 2*
Number of tests: 96
Controls - standards run in assay:
　Controls: Neg (1), Pos (1)
Incubation:
　1 hr (37°C) + 30 min (37°C) + 15 min (RT)
Washes: 2

CONTENTS

Antibodies, antigens, labelled components:
　H. pylori Ag bound to well
　Anti-human IgA Ab HRP conjugated
Substrate: TMB
Controls - standards supplied:
　Controls: Neg and Pos (human serum)
Additional reagents:
　None
Special equipment required:
　None

INTERPRETATION

Comments on interpretation:
　Classification of samples is according to cut-off; no
　further testing
No. of references: 8

NOTES

131326.0

*Diluent contains AHIgG and blocking agents to prevent interference due to RF and IgG in sample

© KLUWER ACADEMIC PUBLISHERS 1995, ISSN 1381-5067

Helicobacter pylori
ANTIBODY DETECTION (IgA)

Manufacturer: Immunobiological Laboratories
Cat. No./Trade name: VE 57321

SUMMARY

$$[Well\text{-}Ag]–\mathbf{Ab}–[AHIgA\text{-}HRP]–[TMB]–A_{450}$$

Assay type: EIA (non-competitive)
Detection: Colorimetric A_{450}
Format: Microtitre well, Ag coated
Sample type: Serum, plasma
Sample pre-treatment:
  None
Sample volume: 100 µl of 1:100 dilution x 2
Number of tests: 96
Controls - standards run in assay:
  Controls: Neg (2), low Pos (2), high Pos (2)
Incubation:
  1 hr (RT) + 30 min (RT) + 15 min (RT)
Washes: 2

CONTENTS

Antibodies, antigens, labelled components:
  H. pylori Ag bound to well
  Anti-human IgA Ab HRP conjugated
Substrate: TMB
Controls - standards supplied:
  Controls: Neg, low Pos and high Pos (serum)
Additional reagents:
  None
Special equipment required:
  None

INTERPRETATION

Comments on interpretation:
  Classification of sample is according to cut-off; no further testing
No. of references: 5

NOTES

131258.0

Helicobacter pylori
ANTIBODY DETECTION (IgG)

Manufacturer: Alexon Inc
Cat. No./Trade name: 780-96 H. pylori EIA (HM-CAP)

SUMMARY

$$[Well\text{-}Ag]–\mathbf{Ab}–[AHIgG\text{-}POD]–[TMB]–A_{450}$$

Assay type: EIA (non-competitive)
Detection: Colorimetric A_{450}
Format: Microtitre well, Ag coated
Sample type: Serum
Sample pre-treatment:
  None
Sample volume: 5 µl (+500 µl diluent)
Number of tests: 96
Controls - standards run in assay:
  Controls: Neg (1), low Pos (1), high Pos (1)
Incubation:
  20 min (RT) + 20 min (RT) + 10 min (RT)
Washes: 2

CONTENTS

Antibodies, antigens, labelled components:
  H. pylori (HM-CAP) Ag bound to well
  Anti-human IgG Ab (g) POD conjugated
Substrate: TMB
Controls - standards supplied:
  Calibrators: Neg, low Pos and high Pos (assigned titres – human serum)
Additional reagents:
  None
Special equipment required:
  None

INTERPRETATION

Comments on interpretation:
  Equivocal: where sample result is between 1.8–2.2 unit values; retest a fresh sample at a later date
No. of references: 31

NOTES

131298.0

© KLUWER ACADEMIC PUBLISHERS 1995, ISSN 1381-5067

Helicobacter pylori
ANTIBODY DETECTION (IgG)

Manufacturer: Amrad Corporation Ltd
Cat. No./Trade name: /HEL-pTEST®

SUMMARY

[Well-Ag]–**Ab**–[AHIgG-HRP]–TMB–A_{450}

Assay type: EIA (non-competitive)
Detection: Colorimetric A_{450}
Format: Microtitre well, Ab coated
Sample type: Serum, plasma
Sample pre-treatment:
  None
Sample volume: 100 µl of 1:200 dilution
Number of tests: 96, 192, 480
Controls - standards run in assay:
  Qualitative: Controls: Neg (3), Pos (3)
  Quantitative: Controls: Neg (1), Pos (1); Standards: 1 (2), 2(2), 3 (2), 4 (2)
Incubation:
  30 min (37°C) + 30 min (37°C) + 30 min (RT)
Washes: 2

CONTENTS

Antibodies, antigens, labelled components:
  H. pylori Ag bound to well
  Anti-human IgG Ab (sh) HRP conjugated
Substrate: TMB
Controls - standards supplied:
  Controls: Neg and Pos (serum)
  Standards: 4
Additional reagents:
  None
Special equipment required:
  None

INTERPRETATION

Comments on interpretation:
  Grey zone: between upper cut-off value and lower cut-off value (*qualitative*) or between 15–60 units/ml (*quantitative*); retest to confirm
  Repeatably equivocal: retest using an alternative procedure
No. of references: 24

NOTES

131017.0

Helicobacter pylori
ANTIBODY DETECTION (IgG)

Manufacturer: Bard Diagnostic Sciences, Inc
Cat. No./Trade name: 30070/ColorVue® pylori

SUMMARY

[Well-Ag]–**Ab**–[AHIgG-HRP]–[TMB]–A_{450}

Assay type: EIA (non-competitive)
Detection: Colorimetric A_{450}
Format: Microtitre well, Ag coated
Sample type: Serum
Sample pre-treatment:
  None
Sample volume: 10 µl (+1 ml diluent)*
Number of tests: 96
Controls - standards run in assay:
  Controls: Neg (3), Pos (2)
Incubation:
  30 min (37°C) + 30 min (37°C) + 15 min (RT)
Washes: 2

CONTENTS

Antibodies, antigens, labelled components:
  H. pylori Ag bound to well
  Anti-human IgG Ab (g) HRP conjugated
Substrate: TMB
Controls - standards supplied:
  Controls: Neg and Pos (human serum)
Additional reagents:
  None
Special equipment required:
  None

INTERPRETATION

Comments on interpretation:
  Equivocal: 0.72–0.99; retest to confirm
  Repeatably equivocal: report as such and retest using an alternative procedure
No. of references: 21

NOTES

131019.0

*Samples are assayed in duplicate

Helicobacter pylori
ANTIBODY DETECTION (IgG)

Manufacturer: Behringwerke AG
Cat. No./Trade name: /Enzygnost® Anti-Helicobacter pylori/IgG

SUMMARY

[Well-Ag]–**Ab**–[AHIgG-POD]–[TMB]–A$_{450}$

Assay type: EIA (non-competitive)
Detection: Colorimetric A$_{450}$
Format: Microtitre well, Ag coated
Sample type: Serum
Sample pre-treatment:
   None
Sample volume: 100 µl of 1:200 dilution
Number of tests: 96
Controls - standards run in assay:
   Qualitative: Neg (2), borderline (2)
   Quantitative: Neg (2), borderline (2), high Pos undiluted
      (2), high Pos dilutions 1.2–1.32 (x2)
Incubation:
   1 hr (37°C) + 30 min (37°C) + 10 min (RT)
Washes: 2

CONTENTS

Antibodies, antigens, labelled components:
   H. pylori Ag bound to well
   Anti-human IgG Ab POD conjugated
Substrate: TMB
Controls - standards supplied:
   Controls: Neg (wash solution), borderline and high Pos
      (human-Ab levels in U/ml)
Additional reagents:
   None
Special equipment required:
   Behring ELISA Processor II (optional)
   Behring ELISA Processor III (optional)

INTERPRETATION

Comments on interpretation:
   Qualitative: classification of samples is according to cut-
      off; no further testing
   Quantitative: Positive: Ab levels ⩾10 U/ml
No. of references: 12

NOTES

131021.0

Helicobacter pylori
ANTIBODY DETECTION (IgG)

Manufacturer: Bio-Rad
Cat. No./Trade name: 404 1002/G.A.P. H. pylori IgG Test

SUMMARY

[Well-Ag]–**Ab**–[AHIgG-HRP]–TMB–A$_{450}$

Assay type: EIA (non-competitive)
Detection: Colorimetric A$_{450}$
Format: Microtitre well, Ag coated
Sample type: Serum, plasma
Sample pre-treatment:
   None
Sample volume: 25 µl (+5 ml diluent)*
Number of tests: 96
Controls - standards run in assay:
   Controls: Neg (1), Pos (1)
   Standards: 0 (2), 1 (2), 2 (2), 3 (2), 4 (2)
Incubation:
   1 hr (RT) + 30 min (RT) + 10 min (RT)
Washes: 2

CONTENTS

Antibodies, antigens, labelled components:
   H. pylori Ag bound to well
   Anti-human IgG Ab (g) HRP conjugated
Substrate: TMB
Controls - standards supplied:
   Controls: Neg (<10 units/ml) and Pos (25-50 units/ml)
   Standards: 0-4 (0-100 units/ml) all human serum
Additional reagents:
   None
Special equipment required:
   None

INTERPRETATION

Comments on interpretation:
   Equivocal: between 12.5 and 20 units/ml; retest fresh
      sample in 2 weeks in parallel with initial sample
No. of references: 28

NOTES

131173.0

*Samples assayed in duplicate

© KLUWER ACADEMIC PUBLISHERS 1995, ISSN 1381-5067

Helicobacter pylori
ANTIBODY DETECTION (IgG)

Manufacturer: Biomerica
Cat. No./Trade name: 7004/GAP® IgG

SUMMARY

[Well-Ag]–**Ab**–[AHIgG-HRP]–[TMB]–A_{450}

Assay type: EIA (non-competitive)
Detection: Colorimetric A_{450}
Format: Microtitre well, Ag coated
Sample type: Serum
Sample pre-treatment:
    None
Sample volume: 25 µl of 1:200 dilution
Number of tests: 96
Controls - standards run in assay:
    Controls: Neg (1), Pos (1)
    Calibrators: 1 (1), 2 (1), 3 (1), 4 (1)
Incubation:
    1 hr (RT) + 30 min (RT) + 10 min (RT)
Washes: 2

CONTENTS

Antibodies, antigens, labelled components:
    H. pylori Ag bound to well
    Anti-human IgG Ab HRP conjugated
Substrate: TMB
Controls - standards supplied:
    Controls: Neg and Pos
    Calibrators: 0–4 (human serum)
Additional reagents:
    None
Special equipment required:
    None

INTERPRETATION

Comments on interpretation:
    Equivocal: between 12.5–25 units/ml; retest a second
    sample in 2 weeks in parallel with initial sample
No. of references: 27

NOTES

131033.0

Helicobacter pylori
ANTIBODY DETECTION (IgG)

Manufacturer: BioWhittaker Inc
Cat. No./Trade name: 30-678U/Pylori Elisa II Test Kit

SUMMARY

[Well-Ag]–**Ab**–[AHIgG-AP]–[PMP]–A_{550}

Assay type: EIA (non-competitive)
Detection: Colorimetric A_{550}
Format: Microtitre well, Ag coated
Sample type: Serum (do not heat inactivate)
Sample pre-treatment:
    None
Sample volume: 10 µl (+200 µl diluent)
Number of tests: 192
Controls - standards run in assay:
    Controls: A (1), B (1)
    Calibrators: Neg (1), Pos 1 (1), Pos 2 (1)
Incubation:
    45 min (RT) + 45 min (RT) + 45 min (RT)
Washes: 2 (+preliminary plate wash)

CONTENTS

Antibodies, antigens, labelled components:
    H. pylori Ag bound to well
    Anti-human IgG Ab (g) AP conjugated
Substrate: PMP
Controls - standards supplied:
    Controls: A and B
    Calibrators: Neg, Pos 1, Pos 2 (all human serum)
Additional reagents:
    None
Special equipment required:
    BioWhittaker automated system and software (optional)

INTERPRETATION

Comments on interpretation:
    Equivocal: Elisa value between 0.15–0.16 or Index value
        between 0.88–0.89; retest to confirm
    Repeatably equivocal: retest a fresh sample or use an
        alternative method
No. of references: 3

NOTES

131744.0

© KLUWER ACADEMIC PUBLISHERS 1995, ISSN 1381-5067

Helicobacter pylori
ANTIBODY DETECTION (IgG)

Manufacturer: BioWhittaker Inc
Cat. No./Trade name: 30-378U/Pyloristat

SUMMARY

[Well-Ag]–**Ab**–[AHIgG-AP]–[PMP]–A_{550}

Assay type: EIA (non-competitive)
Detection: Colorimetric A_{550}
Format: Microtitre well, Ag coated
Sample type: Serum (do not heat inactivate)
Sample pre-treatment:
  None
Sample volume: 10 μl (+200 μl diluent)
Number of tests: 192
Controls - standards run in assay:
  Controls: Neg (1), Pos (1)
  Standards: Neg (1), low Pos (1), high Pos (1)
Incubation:
  15 min (RT) + 15 min (RT) + 15 min (RT)
Washes: 2 (+preliminary plate wash)

CONTENTS

Antibodies, antigens, labelled components:
  H. pylori Ag bound to well
  Anti-human IgG Ab (g) AP conjugated
Substrate: PMP
Controls - standards supplied:
  Controls: Neg and Pos
  Standards: Neg, low Pos and high Pos (all human serum)
Additional reagents:
  None
Special equipment required:
  BioWhittaker automated system and software (optional)

INTERPRETATION

Comments on interpretation:
  Equivocal: Predicted Index value (PIV) between 0.80–0.90; retest to confirm
  Repeatably equivocal: retest a fresh sample or use alternative method
  Low Positive: PIV between 1.00–1.74
  Mid Positive: PIV between 1.75–3.89
  High Positive: PIV ≥3.90
No. of references: 16

NOTES

131751.0

Helicobacter pylori
ANTIBODY DETECTION (IgG)

Manufacturer: Chimica Diagnostica
Cat. No./Trade name: 42800/Helicobacter Pylori IgG

SUMMARY

[Well-Ag]–**Ab**–[AHIgG-POD]–[TMB]–A_{450}

Assay type: EIA (non-competitive)
Detection: Colorimetric A_{450}
Format: Microtitre well, Ag coated
Sample type: Serum
Sample pre-treatment:
  None
Sample volume: 100 μl of 1:101 dilution x 2
Number of tests: 96
Controls - standards run in assay:
  Standards: 1 (2), 2 (2), 3 (2), 4 (2), 5 (2), 6 (2)
Incubation:
  1 hr (37°C) + 1 hr (37°C) + 20 min (RT)
Washes: 2

CONTENTS

Antibodies, antigens, labelled components:
  H. pylori Ag bound to well
  Anti-human IgG Ab (r) POD conjugated
Substrate: TMB
Controls - standards supplied:
  Standards: 1–6 (titres 0, 8, 16, 32, 64 and 128 IU/ml)
Additional reagents:
  None
Special equipment required:
  None

INTERPRETATION

Comments on interpretation:
  Results are expressed quantitatively in IU/ml with reference to the Standards provided
No. of references: 2

NOTES

131359.0

© KLUWER ACADEMIC PUBLISHERS 1995, ISSN 1381-5067

Helicobacter pylori
ANTIBODY DETECTION (IgG)

Manufacturer: Cortecs Diagnostics
Cat. No./Trade name: 902040H H/Helisal®

SUMMARY

[Well-Ag]–**Ab**–[AHIgG-biotin]–[strept-HRP]–[TMB]–A$_{450}$

Assay type: EIA (non-competitive)
Detection: Colorimetric A$_{450}$
Format: Microtitre well, Ag coated
Sample type: Saliva*
Sample pre-treatment:
  Collect with device provided, add to buffer and filter using filter samplers/buffer tubes provided
Sample volume: 100 µl of prepared sample x 2
Number of tests: 96
Controls - standards run in assay:
  Controls: Neg (2), borderline (2)
  Standards: high (2), cut-off (2)
Incubation:
  30 min (RT) + 30 min (RT) + 15 min (RT) + 30 min (RT)
Washes: 3

CONTENTS

Antibodies, antigens, labelled components:
  H. pylori Ag bound to well
  Anti-human IgG Ab (r) biotinylated
  Streptavidin HRP conjugated
Substrate: TMB
Controls - standards supplied:
  Controls: Neg and borderline (1.25–2.25 U/ml)
  Standards: High (8 U/ml – dilute to produce cut-off standard and/or standard curve for quantitative analysis)
Additional reagents:
  None
Special equipment required:
  None

INTERPRETATION

Comments on interpretation:
  Equivocal: where Elisa Unit value is between 0.8–0.99; retest to confirm
  Repeatably equivocal: retest fresh saliva sample after 2–4 weeks or test serum using Cortecs HELISAL® (serum) test kit
No. of references: 7

NOTES

131037.0

*Use Omni-SAL® Saliva Collection Devices included in kit

Helicobacter pylori
ANTIBODY DETECTION (IgG)

Manufacturer: Cortecs Diagnostics
Cat. No./Trade name: 902042D/Helisal® (serum)

SUMMARY

[Well-Ag]–**Ab**–[AHIgG-HRP]–[TMB]–A$_{450}$

Assay type: EIA (non-competitive)
Detection: Colorimetric A$_{450}$
Format: Microtitre well, Ag coated
Sample type: Serum, plasma
Sample pre-treatment:
  Dilute in buffer provided
Sample volume: 100 µl of diluted sample x 2
Number of tests: 96
Controls - standards run in assay:
  Controls: Neg (2), borderline (2)
Incubation:
  45 min (RT) + 15 min (RT) + 15 min (RT)
Washes: 3

CONTENTS

Antibodies, antigens, labelled components:
  H. pylori Ag bound to well
  Anti-human IgG Ab (r) HRP conjugated
Substrate: TMB
Controls - standards supplied:
  Controls: Neg and borderline (1.25–2.25 U/ml)
  Standard curve for quantitative analysis provided
Additional reagents:
  None
Special equipment required:
  None

INTERPRETATION

Comments on interpretation:
  Equivocal: where sample volume is between 0.80–0.99; retest to confirm
  Repeatably equivocal: retest a fresh sample after 2–4 weeks
No. of references: 7

NOTES

131553.0

Helicobacter pylori
ANTIBODY DETECTION (IgG)

Manufacturer: Dako A/S
Cat. No./Trade name: K3015/DAKO® Helicobacter Pylori Test Kit

SUMMARY

[Well-Ag]–**Ab**–[AHIgG–HRP]–[TMB]–A_{450}

Assay type: EIA (non-competitive)
Detection: Colorimetric A_{450}
Format: Microtitre well, Ag coated
Sample type: Serum
Sample pre-treatment:
    None
Sample volume: 10 µl (+2.5 ml diluent)*
Number of tests: 96
Controls - standards run in assay:
    Controls: Neg (1), Borderline (1), Pos (1)
Incubation:
    20 min (RT) + 20 min (RT) + 20 min (RT)
Washes: 2

CONTENTS

Antibodies, antigens, labelled components:
    H. pylori Ag bound to well
    Anti-human IgG Ab HRP conjugated
Substrate: TMB
Controls - standards supplied:
    Controls: Neg, Borderline, Pos (human serum)
Additional reagents:
    None
Special equipment required:
    None

INTERPRETATION

Comments on interpretation:
    Equivocal: within designated range; retest to confirm
No. of references: 24

NOTES

131226.0

*Samples are assayed in duplicate

Helicobacter pylori
ANTIBODY DETECTION (IgG)

Manufacturer: Diesse
Cat. No./Trade name: 91060/ENZYWELL Helicobacter IgG

SUMMARY

[Well-Ag]–**Ab**–[AHIgG-POD]–[TMB]–A_{450}

Assay type: EIA (non-competitive)
Detection: Colorimetric A_{450}
Format: Microtitre well, Ag coated
Sample type: Serum (do not heat inactivate)
Sample pre-treatment:
    None
Sample volume: 100 µl of 1:101 dilution x 2
Number of tests: 96
Controls - standards run in assay:
    Calibrators: 1 (2), 2 (2), 3 (2), 4 (2), 5 (2)
Incubation:
    45 min (37°C) + 45 min (37°C) + 15 min (RT)
Washes: 2

CONTENTS

Antibodies, antigens, labelled components:
    H. pylori Ag bound to well
    Anti-human IgG MAb POD conjugated
Substrate:
Controls - standards supplied:
    Calibrators: 5 (human serum)
Additional reagents:
    None
Special equipment required:
    None

INTERPRETATION

Comments on interpretation:
    Equivocal: within cut-off ±10%; retest to confirm
No. of references: 5

NOTES

131442.0

© KLUWER ACADEMIC PUBLISHERS 1995, ISSN 1381-5067

Helicobacter pylori
ANTIBODY DETECTION (IgG)

Manufacturer: Enteric Products Inc
Cat. No./Trade name: 86677/HM CAP®

SUMMARY

[Well-Ag]–**Ab**–[AHIgG-HRP]–[substrate]–A_{450}

Assay type: EIA (non-competitive)
Detection: Colorimetric A_{450}
Format: Microtitre well, Ag coated
Sample type: Serum
Sample pre-treatment:
  None
Sample volume: 5 µl (+500 µl diluent)
Number of tests: 96
Controls - standards run in assay:
  Calibrators: Neg (1), low Pos (1), high Pos (1)
Incubation:
  20 min (RT) + 20 min (RT) + 10 min (RT)
Washes: 2

CONTENTS

Antibodies, antigens, labelled components:
  H. pylori Ag bound to well
  Anti-human IgG Ab HRP conjugated
Substrate: Not specified
Controls - standards supplied:
  Calibrators: Neg, low Pos and high Pos
Additional reagents:
  None
Special equipment required:
  None

INTERPRETATION

Comments on interpretation:
  Equivocal: where value is between 1.8 and 2.2; retest to
  confirm
No. of references: None

NOTES

131680.0

Helicobacter pylori
ANTIBODY DETECTION (IgG)

Manufacturer: Hycor Biomedical Inc
Cat. No./Trade name: 80500-80525/PYLORAGEN® H.
Pylori IgG Test Kit

SUMMARY

[Well-Ag]–**Ab**–[AHIgG-HRP]–[substrate]–A_{450}

Assay type: EIA (non-competitive)
Detection: Colorimetric A_{450}
Format: Microtitre well, Ag coated
Sample type: Serum
Sample pre-treatment:
  None
Sample volume: 10 µl (+2.5 ml diluent)*
Number of tests: 96
Controls - standards run in assay:
  Controls: Neg (1), Pos (2)
  Calibrators: (1)
Incubation:
  20 min (RT) + 20 min (RT) + 3 min (RT)
Washes:

CONTENTS

Antibodies, antigens, labelled components:
  H. pylori Ag bound to well
  Anti-human IgG Ab HRP conjugated
Substrate: TMB
Controls - standards supplied:
  Controls: Neg and Pos
  Calibrators: 1
Additional reagents:
  None
Special equipment required:
  None

INTERPRETATION

Comments on interpretation:
  Equivocal: where value is between 0.88–0.99; retest to
  confirm
No. of references: 24

NOTES

131720.0

*Samples are assayed in duplicate

© KLUWER ACADEMIC PUBLISHERS 1995, ISSN 1381-5067

Helicobacter pylori
ANTIBODY DETECTION (IgG)

Manufacturer: IFCI Clone Systems
Cat. No./Trade name: EIAGEN Helicobacter pylori IgG

SUMMARY

[Well-Ag]–**Ab**–[AHIgG-POD]–[TMB]–A$_{450}$

Assay type: EIA (non-competitive)
Detection: Colorimetric A$_{450}$
Format: Microtitre well, Ag coated
Sample type: Serum
Sample pre-treatment:
    None
Sample volume: 10 µl (+1 ml diluent)*
Number of tests: 96
Controls - standards run in assay:
    Calibrators: 1 (2), 2 (2), 3 (2), 4 (2), 5 (2)
Incubation:
    45 min (37°C) + 45 min (37°C) + 15 min (RT)
Washes: 2

CONTENTS

Antibodies, antigens, labelled components:
    H. pylori Ag bound to well
    Anti-human IgG MAb POD conjugated
Substrate: TMB
Controls - standards supplied:
    Calibrators: 5
Additional reagents:
    None
Special equipment required:
    None

INTERPRETATION

Comments on interpretation:
    Equivocal: within cut-off ±10%; retest to confirm
    Repeatably equivocal: retest a fresh sample
No. of references: 5

NOTES

131673.0

*Samples are assayed in duplicate

Helicobacter pylori
ANTIBODY DETECTION (IgG)

Manufacturer: Immuno Pharmacology Research
Cat. No./Trade name: HELICOBACTER PYLORI IgG ELISA

SUMMARY

[Well-Ag]–**Ab**–[AHIgG-HRP]–[TMB]–A$_{405}$

Assay type: EIA (non-competitive)
Detection: Colorimetric A$_{450}$
Format: Microtitre well, Ag coated
Sample type: Serum
Sample pre-treatment:
    None
Sample volume: 100 µl of 1:50 dilution x 2
Number of tests: 96
Controls - standards run in assay:
    Standards: 0 (1), 1 (1), 2 (1), 3 (1), 4 (1), 5 (1)
Incubation:
    1 hr (37°C) + 1 hr (37°C) + 20 min (RT)
Washes: 2

CONTENTS

Antibodies, antigens, labelled components:
    H. pylori Ag bound to well
    Anti-human IgG MAb (g) HRP conjugated
Substrate: TMB
Controls - standards supplied:
    Standards: 0–5 (0, 8, 16, 32, 64 and 128 arbitrary units/ml)
Additional reagents:
    None
Special equipment required:
    None

INTERPRETATION

Comments on interpretation:
    Positive: ≥8 U/ml
No. of references: 5

NOTES

131321.0

Helicobacter pylori
ANTIBODY DETECTION (IgG)

Manufacturer: Immunobiological Laboratories
Cat. No./Trade name: VE 57311/Helicobacter pylori IgG Elisa Test (HRP)

SUMMARY

[Well-Ag]–**Ab**–[AHIgG-HRP]–[TMB]–A$_{450}$

Assay type: EIA (non-competitive)
Detection: Colorimetric A$_{450}$
Format: Microtitre well, Ag coated
Sample type: Serum, plasma
Sample pre-treatment:
  None
Sample volume: 100 μl of 1:100 dilution x 2
Number of tests: 96
Controls - standards run in assay:
  Controls: Neg (2), low Pos (2), high Pos (2)
Incubation:
  1 hr (RT) + 30 min (RT) + 15 min (RT)
Washes: 2

CONTENTS

Antibodies, antigens, labelled components:
  H. pylori Ag bound to well
  Anti-human IgG Ab HRP conjugated
Substrate: TMB
Controls - standards supplied:
  Controls: Neg, low Pos and high Pos (serum)
Additional reagents:
  None
Special equipment required:
  None

INTERPRETATION

Comments on interpretation:
  Equivocal: within cut-off ±20%; no data is provided for further classification of samples within this range
No. of references: 5

NOTES

131250.0

Helicobacter pylori
ANTIBODY DETECTION (IgG)

Manufacturer: Meridian Diagnostics Inc
Cat. No./Trade name: 606096/Premier® H. pylori

SUMMARY

[Well-Ag]–**Ab**–[AHIgG-POD]–[TMB]–A$_{450}$

Assay type: EIA (non-competitive)
Detection: Colorimetric A$_{450}$
Format: Microtitre well, Ag coated
Sample type: Serum, plasma
Sample pre-treatment:
  None
Sample volume: 100 μl of 1:50 dilution
Number of tests: 96
Controls - standards run in assay:
  Controls: Neg (1), Pos (1)
Incubation:
  20 min (RT) + 20 min (RT) + 10 min (RT)
Washes: 2

CONTENTS

Antibodies, antigens, labelled components:
  H. pylori Ag bound to well
  Anti-human IgG MAb POD conjugated
Substrate: TMB
Controls - standards supplied:
  Controls: Neg and Pos (human serum)
Additional reagents:
  None
Special equipment required:
  None

INTERPRETATION

Comments on interpretation:
  Classification of samples is according to cut-off; no further testing
No. of references: 22

NOTES

131384.0

© KLUWER ACADEMIC PUBLISHERS 1995, ISSN 1381-5067

Helicobacter pylori
ANTIBODY DETECTION (IgG)

Manufacturer: Orion Diagnostica
Cat. No./Trade name: 68926/New Pyloriset® EIA-G

SUMMARY

[Well-Ag]–**Ab**–[AHIgG-POD]–[PNP]–A_{405}

Assay type: EIA (non-competitive)
Detection: Colorimetric A_{405}
Format: Microtitre well, Ag coated
Sample type: Serum
Sample pre-treatment:
    None
Sample volume: 100 μl of 1:201 dilution x 2
Number of tests: 96
Controls - standards run in assay:
    Calibrators: 1 (2), 2 (2), 3 (2), 4 (2)
Incubation:
    1 hr (RT) + 1 hr (RT) + 30 min (RT)
Washes:

CONTENTS

Antibodies, antigens, labelled components:
    H. pylori Ag bound to well
    Anti-human IgG Ab (p) HRP conjugated
Substrate: PNP
Controls - standards supplied:
    Calibrators: 4 (human - assigned values)
Additional reagents:
    None
Special equipment required:
    None

INTERPRETATION

Comments on interpretation:
    Classification of samples is according to cut-off; no
    further testing
No. of references: 34

NOTES

131253.0

Helicobacter pylori
ANTIBODY DETECTION (IgG)

Manufacturer: Quidel Corporation
Cat. No./Trade name: C007/Quick-Vue H. Pylori Test

SUMMARY

[Membrane-Ag]–**Ab**–[AHIgG-AP]–[3IP]–blue circle

Assay type: EIA (non-competitive)
Detection: Visual
Format: Test cartridge, Ag coated membrane
Sample type: Serum, plasma
Sample pre-treatment:
    None
Sample volume: 30 μl
Number of tests: 20
Controls - standards run in assay:
    Controls: Neg (1), Pos (1)
    Integral controls: Neg (1), Pos (1) on membrane
Incubation:
    Read results after final wash
Washes: 2

CONTENTS

Antibodies, antigens, labelled components:
    H. pylori Ag bound to membrane
    Anti-human IgG Ab (r) AP conjugated
Substrate: 3IP
Controls - standards supplied:
    Controls: Neg and Pos
    Integral controls incorporated in membrane
Additional reagents:
    None
Special equipment required:
    None

INTERPRETATION

Comments on interpretation:
    Negative: blue negative bar across centre of membrane
    Positive: blue circle or oval at centre of membrane with or
        without blue negative bar
No. of references: 23

NOTES

131081.0

© *KLUWER ACADEMIC PUBLISHERS 1995, ISSN 1381-5067*

Helicobacter pylori
ANTIBODY DETECTION (IgG)

Manufacturer: Radim
Cat. No./Trade name: K5HPG/Helicobacter Pylori IgG EIA Well

SUMMARY

[Well-Ag]–**Ab**–[AHIgG-HRP]–[TMB]–A$_{450}$

Assay type: EIA (non-competitive)
Detection: Colorimetric A$_{450}$
Format: Microtitre well, Ag coated
Sample type: Serum, plasma
Sample pre-treatment:
None
Sample volume: 100 µl of 1:300 dilution
Number of tests: 96
Controls - standards run in assay:
Qualitative: Controls: Neg (1): Standards: 1 (1)
Quantitative: Controls: Neg (1): Standards: 1 (1), 2 (1), 3 (1), 4 (1)
Incubation:
1 hr (37°C) + 30 min (37°C) + 10 min (37°C) or 15 min (RT)
Washes: 2

CONTENTS

Antibodies, antigens, labelled components:
H. pylori Ag bound to well
Anti-human IgG Ab (g) HRP conjugated
Substrate: TMB
Controls - standards supplied:
Controls: Neg (0, IU/ml - serum)
Standards: 4 (15, 30, 60, 120 IU/ml - serum)
Additional reagents:
None
Special equipment required:
None

INTERPRETATION

Comments on interpretation:
Qualitative: Equivocal: within cut-off ±10%; retest to confirm
Quantitative: Negative: < 15 IU/ml
Weak Positive: between 15 and 30 IU/ml
Positive: > 30 IU/ml
No. of references: 3

NOTES

131272.0

Helicobacter pylori
ANTIBODY DETECTION (IgG)

Manufacturer: Roche Diagnostic Systems
Cat. No./Trade name: 07 3497 7/COBAS® CORE Anti-H. Pylori EIA

SUMMARY

[Bead-Ag]–**Ab**–[AHIgG-POD]–[TMB]–A$_{450}$

Assay type: EIA (non-competitive)
Detection: Colorimetric A$_{450}$
Format: Tube, Ag coated bead
Sample type: Serum
Sample pre-treatment:
None
Sample volume: 10 µl (+400 µl diluent)
Number of tests: 50
Controls - standards run in assay:
Qualitative: Controls: Neg (3), Pos (1)
Quantitative: Standards: 1 (2), 2 (2), 3 (2), 4 (2), 5 (2), 6 (2)
Incubation:
18 min (37°C) + 18 min (37°C) + 15 min (37°C) (all with shaking)
Washes: 2

CONTENTS

Antibodies, antigens, labelled components:
H. pylori Ag bound to bead
Anti-human IgG Ab (g) POD conjugated
Substrate: TMB (bought separately)
Controls - standards supplied:
Controls: Neg and Pos (human serum; titre 160 Units/ml)
Standards: 1 (Neg control) 2–6 (dilutions of Pos control giving 10, 20, 40, 80 and 160 Units/ml)
Additional reagents:
Cobas® Core TMB kit
H$_2$SO$_4$ (for manual assay only)
Special equipment required:
Cobas® Core immunoassay analyser (for automated method - optional)
Cobas® Core EIA shaking incubator, tube washer and photometer (optional)
Cobas® Core TORC negative control bead dispenser
Cobas® Core reaction tubes

INTERPRETATION

Comments on interpretation:
Qualitative: Equivocal: within cut-off ±10%; retest to confirm
Repeatably equivocal: retest using another method e.g. Western Blot
Quantitative: result obtained with reference to Standards provided
No. of references: None

NOTES

131156.0

© KLUWER ACADEMIC PUBLISHERS 1995, ISSN 1381-5067

Helicobacter pylori
ANTIBODY DETECTION (IgG)

Manufacturer: Shield Diagnostics
Cat. No./Trade name: FHEL 100/Helico-G®

SUMMARY

[Well-Ag]–**Ab**–[AHIgG-HRP]–[TMB]–A$_{450}$

Assay type: EIA (non-competitive)
Detection: Colorimetric A$_{450}$
Format: Microtitre well, Ag coated
Sample type: Serum
Sample pre-treatment:
   None
Sample volume: 100 µl of 1:200 dilution x 2
Number of tests: 96
Controls - standards run in assay:
   Calibrators: low level (2), high level (2)*
Incubation:
   1 hr (37°C) + 30 min (37°C) + 10 min (RT) with shaking
Washes: 2

CONTENTS

Antibodies, antigens, labelled components:
   H. pylori Ag bound to well
   Anti-human IgG Ab (g) HRP conjugated
Substrate: TMB
Controls - standards supplied:
   Calibrators: low level (10 units/ml) and high level (200 units/ml)
Additional reagents:
   None
Special equipment required:
   None

INTERPRETATION

Comments on interpretation:
   Classification of samples is according to cut-off; no further testing
No. of references: 15

NOTES

131715.0

*For a quantitative procedure use dilutions of high level calibrator to give values of 10–200 units/ml

Helicobacter pylori
ANTIBODY DETECTION (IgG)

Manufacturer: United Biotech Inc
Cat. No./Trade name: 1A-601/MAGIWEL® Helicobacter pylori IgG

SUMMARY

[Well-Ag]–**Ab**–[AHIgG-HRP]–[TMB]–A$_{450}$

Assay type: EIA (non-competitive)
Detection: Colorimetric A$_{450}$
Format: Microtitre well, Ag coated
Sample type: Serum
Sample pre-treatment:
   None
Sample volume: 100 µl of 1:101 dilution
Number of tests: 96
Controls - standards run in assay:
   Reference Standard Set: Neg (1), Pos (1)
   Calibrators: 1 (1)
Incubation:
   30 min (RT) + 30 min (RT) + 15 min (RT)
Washes: 2

CONTENTS

Antibodies, antigens, labelled components:
   H. pylori Ag bound to well
   Anti-human IgG Ab (g) HRP conjugated
Substrate: TMB
Controls - standards supplied:
   Reference Standard Set: Neg, Pos
   Calibrator: 1 (100 EU/ml)
Additional reagents:
   None
Special equipment required:
   None

INTERPRETATION

Comments on interpretation:
   Equivocal: between 30–40 EU/ml; retest a fresh sample 3 weeks later
No. of references: 8

NOTES

131147.0

© KLUWER ACADEMIC PUBLISHERS 1995, ISSN 1381-5067

Helicobacter pylori
ANTIBODY DETECTION (IgM)

Manufacturer: Bio-Rad
Cat. No./Trade name: 404 2002/G.A.P. H. pylori IgM Test

SUMMARY

[Well-Ag]–**Ab**–[AHIgM-HRP]–TMB–A$_{450}$

Assay type: EIA (non-competitive)
Detection: Colorimetric A$_{450}$
Format: Microtitre well, Ag coated
Sample type: Serum, plasma
Sample pre-treatment:
    None
Sample volume: 25 µl (+5 ml diluent)*
Number of tests: 96
Controls - standards run in assay:
    Controls: Neg (1), Pos (1)
    Standards: 0 (2), 1 (2), 2 (2), 3 (2), 4 (2)
Incubation:
    1 hr (RT) + 30 min (RT) + 10 min (RT)
Washes: 2

CONTENTS

Antibodies, antigens, labelled components:
    H. pylori Ag bound to well
    Anti-human IgM Ab (g) HRP conjugated
Substrate: TMB
Controls - standards supplied:
    Controls: Neg (< 10 units/ml) and Pos (25-50 units/ml)
    Standards: 0-4 (0-100 units/ml) all human serum
Additional reagents:
    None
Special equipment required:
    None

INTERPRETATION

Comments on interpretation:
    Equivocal: between 12.5 and 20 units/ml; retest fresh
    sample in 2 weeks in parallel with initial sample
No. of references: None

NOTES

131174.0

*Samples assayed in duplicate

Helicobacter pylori
ANTIBODY DETECTION (IgM)

Manufacturer: Biomerica
Cat. No./Trade name: 7006/GAP® IgM

SUMMARY

[Well-Ag]–**Ab**–[AHIgM-HRP]–[TMB]–A$_{450}$

Assay type: EIA (non-competitive)
Detection: Colorimetric A$_{450}$
Format: Microtitre well, Ag coated
Sample type: Serum
Sample pre-treatment:
    None
Sample volume: 25 µl of 1:200 dilution
Number of tests: 96
Controls - standards run in assay:
    Controls: Neg (1), Pos (1)
    Calibrators: 1 (1), 2 (1), 3 (1), 4 (1)
Incubation:
    1 hr (RT) + 30 min (RT) + 10 min (RT)
Washes: 2

CONTENTS

Antibodies, antigens, labelled components:
    H. pylori Ag bound to well
    Anti-human IgM Ab HRP conjugated
Substrate: TMB
Controls - standards supplied:
    Controls: Neg and Pos
    Calibrators: 0–4 (human serum)
Additional reagents:
    None
Special equipment required:
    None

INTERPRETATION

Comments on interpretation:
    Equivocal: between 12.5–25 units/ml; retest a second
    sample in 2 weeks in parallel with initial sample
No. of references: 17

NOTES

131184.0

© KLUWER ACADEMIC PUBLISHERS 1995, ISSN 1381-5067

Helicobacter pylori
ANTIBODY DETECTION (IgM)

Manufacturer: Chimica Diagnostica
Cat. No./Trade name: 42850/Helicobacter Pylori IgM

SUMMARY

[Well-Ag]–**Ab**–[AHIgM-POD]–[TMB]–A$_{450}$

Assay type: EIA (non-competitive)
Detection: Colorimetric A$_{450}$
Format: Microtitre well, Ag coated
Sample type: Serum
Sample pre-treatment:
 None
Sample volume: 100 µl of 1:101 dilution x 2*
Number of tests: 96
Controls - standards run in assay:
 Controls: Neg (2), Pos (2)
Incubation:
 30 min (37°C) + 30 min (37°C) + 15 min (RT)
Washes: 2

CONTENTS

Antibodies, antigens, labelled components:
 H. pylori Ag bound to well
 Anti-human IgM Ab (r) POD conjugated
Substrate: TMB
Controls - standards supplied:
 Controls: Neg and Pos
Additional reagents:
 None
Special equipment required:
 None

INTERPRETATION

Comments on interpretation:
 Equivocal: between 85% and 115% of cut-off value;
 report as equivocal
No. of references: 2

NOTES

131364.0

*Diluent contains AHIgG to eliminate interference due to RF
 and IgG in sample

Helicobacter pylori
ANTIBODY DETECTION (IgM)

**Manufacturer: Immuno Pharmacology
Research**
Cat. No./Trade name: HELICOBACTER PYLORI IgM
ELISA

SUMMARY

[Well-Ag]–**Ab**–[AHIgM-POD]–[TMB]–A$_{405}$

Assay type: EIA (non-competitive)
Detection: Colorimetric A$_{450}$
Format: Microtitre well, Ag coated
Sample type: Serum
Sample pre-treatment:
 None
Sample volume: 100 µl of 1:100 dilution x 2*
Number of tests: 96
Controls - standards run in assay:
 Controls: Neg (1), cut-off (1), Pos (1)
Incubation:
 30 min (RT) + 30 min (RT) + 10 min (RT)
Washes: 2

CONTENTS

Antibodies, antigens, labelled components:
 H. pylori Ag bound to well
 Anti-human IgM Ab (r) POD conjugated
Substrate: TMB
Controls - standards supplied:
 Controls: Neg, cut-off and Pos (human serum)
Additional reagents:
 None
Special equipment required:
 None

INTERPRETATION

Comments on interpretation:
 Equivocal: within cut-off ±10%; retest to confirm
No. of references: 9

NOTES

131325.0

*Diluent contains AHIgG and blocking agents to prevent
 interference due to RF and IgG in sample

© KLUWER ACADEMIC PUBLISHERS 1995, ISSN 1381-5067

Helicobacter pylori
ANTIBODY DETECTION (IgG)

Manufacturer: Enteric Products Inc
Cat. No./Trade name: 2944/FLEXSURE® HP

SUMMARY

[Membrane-Ag]–**Ab**–[AHIgG-gold]–[substrate]–pink band

Assay type: Immunochromatographic assay
Detection: Visual
Format: Test device, Ag coated membrane
Sample type: Serum
Sample pre-treatment:
   None
Sample volume: 1 drop
Number of tests:
Controls - standards run in assay:
   Controls: Neg (1), Pos (1)
Incubation:
   Read results 4 min after adding reagents to Test device
Washes:

CONTENTS

Antibodies, antigens, labelled components:
   H. pylori Ag bound to membrane
   Anti-human IgG Ab colloidal gold conjugated bound to
      membrane
Substrate: Not specified
Controls - standards supplied:
   Controls: Neg and Pos
Additional reagents:
   None
Special equipment required:
   None

INTERPRETATION

Comments on interpretation:
   Positive: pink bands in Test and Control windows
No. of references: None

NOTES

131681.0

Helicobacter pylori
ANTIBODY DETECTION (IgG)

Manufacturer: Medical Instruments Corporation
Cat. No./Trade name: CLOser® H. pylori Test

SUMMARY

[Membrane-Ag]–**Ab**–[AHIgG-dye]–band

Assay type: Immunochromatographic assay
Detection: Visual
Format: Test device, Ag coated membrane
Sample type: Serum, plasma, whole blood
Sample pre-treatment:
   None
Sample volume: 10 µl serum, plasma; 25 µl whole blood
Number of tests: 35
Controls - standards run in assay:
   Integral controls in Test Device
Incubation:
   Read result as soon as band appears in Test window
Washes:

CONTENTS

Antibodies, antigens, labelled components:
   H. pylori Ag bound to membrane (Test window)
   Anti-human IgG Ab dye conjugated (in absorbent pad)
Substrate:
Controls - standards supplied:
   Integral control in test device
Additional reagents:
   None
Special equipment required:
   None

INTERPRETATION

Comments on interpretation:
   Negative: one coloured band in Control region of device
   Positive: two coloured bands of varying intensity in the
      Control region of the device
   Invalid: no coloured bands in Control region of device
No. of references: 18

NOTES

131551.0

One developer solution is provided with the kit

© KLUWER ACADEMIC PUBLISHERS 1995, ISSN 1381-5067

Helicobacter pylori
ANTIBODY DETECTION (IgG)

Manufacturer: Meridian Diagnostics Inc
Cat. No./Trade name: 710005/ImmunoCard® H. pylori

SUMMARY

[Membrane-Ag]–**Ab**–[AHIgG-AP]–[BCIP]–blue dot

Assay type: Immunochromatographic assay
Detection: Visual
Format: Reaction unit, Ag coated membrane
Sample type: Serum, plasma
Sample pre-treatment:
   None
Sample volume: 2 drops
Number of tests: 5, 30
Controls - standards run in assay:
   Controls: Neg (1), Pos (1)
Incubation:
   1 min (RT) + 5 min (RT)
Washes: 5

CONTENTS

Antibodies, antigens, labelled components:
   H. pylori Ag bound to membrane (Test port)
   Anti-human IgG MAb AP conjugated
Substrate: BCIP
Controls - standards supplied:
   Controls: Neg and Pos (human serum)
   Integral procedural controls incorporated in membrane
      (Control port)
Additional reagents:
   None
Special equipment required:
   None

INTERPRETATION

Comments on interpretation:
   Negative: blue colour in Control port but no colour in Test
      port
   Positive: blue colour in Control and Test ports
   Invalid: no colour in Control port; repeat test
No. of references: 22

NOTES

131385.0

Helicobacter pylori
ANTIBODY DETECTION

Manufacturer: Orion Diagnostica
Cat. No./Trade name: 68194/Pyloriset® Dry

SUMMARY

[Latex spot-Ag]–**Ab**–agglutination

Assay type: Particle agglutination assay
Detection: Visual
Format: Dry latex spot on card, Ag coated
Sample type: Serum
Sample pre-treatment:
   None
Sample volume: 40 µl of 1:4 dilution
Number of tests: 24
Controls - standards run in assay:
   Controls: Neg (1), Pos (1)
Incubation:
   Read results within 3 min of mixing sample and reagent
Washes:

CONTENTS

Antibodies, antigens, labelled components:
   H. pylori Ag bound to dry latex spot on test card
Substrate:
Controls - standards supplied:
   Controls: Neg and Pos
Additional reagents:
   None
Special equipment required:
   None

INTERPRETATION

Comments on interpretation:
   Positive: partial or complete agglutination of sample in
      Latex Reagent spot on test card
No. of references: None

NOTES

131261.0

© KLUWER ACADEMIC PUBLISHERS 1995, ISSN 1381-5067

Helicobacter pylori
ANTIBODY DETECTION

Manufacturer: Unipath Limited
Cat. No./Trade name: Helicobacter pylori Test

SUMMARY

[Latex-Ag]–**Ab**–agglutination

Assay type: Particle agglutination assay
Detection: Visual
Format: Latex particles, Ag coated (test done on card)
Sample type: Serum
Sample pre-treatment:
   None
Sample volume: 1 drop
Number of tests: 50
Controls - standards run in assay:
   Controls: Neg (1), Pos (1)
Incubation:
   Read result within 3 min of mixing sample and reagent
Washes:

CONTENTS

Antibodies, antigens, labelled components:
   H. pylori Ag bound to latex particles (Test Latex)
Substrate:
Controls - standards supplied:
   Controls: Neg and Pos (human serum)
Additional reagents:
   None
Special equipment required:
   None

INTERPRETATION

Comments on interpretation:
   Positive: agglutination of sample and Test Latex
No. of references: 5

NOTES

131614.0

© KLUWER ACADEMIC PUBLISHERS 1995, ISSN 1381-5067

Human Herpes virus 6 (HHV6)

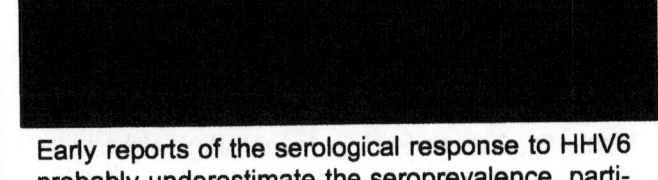

Natural history

A new lymphotropic human herpes virus was isolated in peripheral blood mononuclear cells in patients with lymphoproliferative disorders in 1986 and soon afterwards from peripheral blood of immunosuppressed patients. The isolates were called HHV6. A similar but antigenically different virus was later isolated from the blood of a healthy individual and termed HHV7. Recently a herpes virus from Kaposi's sarcoma has been referred to as HHV8.

HHV6 is an uniquitous virus transmitted in saliva. It is a typical herpes virus classified as a β-herpes virus because of its similar genetic arrangement and considerable sequence similarity to CMV. There are two recognised types, variants A and B, distinguished by the restriction endonuclease digestion pattern. There is no significant serological cross-reaction with other herpes viruses.

HHV6 was quickly shown to be associated with roseola infantum, sometimes called exanthem subitum or 6th disease. This is a mild childhood infection with sudden onset of fever in an otherwise well child. The fever lasts 3–5 days and subsides rapidly, leaving a rosy pink macular rash. Ninety-five percent of cases occur between the ages of 6 months and 3 years and most are associated with variant B. The majority of infants have been infected by the age of 2 years but many infections are subclinical. Occasional cases of severe illness following HHV6 infection have been documented but in children there is no evidence that HHV6 is the cause. HHV6 has however been shown to be pathogenic for immunosuppressed patients with both primary and reactivated infections being described, and associated with pneumonitis in bone marrow transplant patients and hepatitis, fever and neurological problems in liver transplant recipients. Other suggested associations include involvement in post-viral syndrome, B cell lymphoma and as a co-factor in acceleration of HIV disease but a causal relationship remains to be proved.

Diagnosis

Several techniques have been used to examine the serological response to HHV6 including indirect fluorescence antibody (IFA), anticomplement immunofluorescence, neutralisation and ELISA and immunoblotting. Both IgG and IgM antibodies can be detected by IFA and ELISA, with the latter assay showing greater sensitivity. Genome can be detected by *in situ* hybridisation and PCR but isolation is impractical for a diagnostic laboratory.

Comment

Early reports of the serological response to HHV6 probably underestimate the seroprevalence, particularly in detecting waning antibody in older subjects and it is now accepted that > 90% of adults will have evidence of HHV6 exposure.

References

Thomson BJ, Martin MED, Nicholas J. The molecular and cellular biology of human herpes-virus-6. Rev Med Virol. 1991;1:89–100.

Fox JD, Tedder RS, Briggs EM. Human herpes-viruses 6 and 7. In: Zuckerman AJ, Banatvala JE, Pattison JR, eds. Principles and Practice of Clinical Virology, 3rd edn. Chichester: John Wiley. 1994:135–52.

© KLUWER ACADEMIC PUBLISHERS 1995, ISSN 1381-5067

Human Herpes virus 6
ANTIBODY DETECTION (IgG)

Manufacturer: Biotrin International Ltd
Cat. No./Trade name: Human Herpes Virus 6 IgG EIA

SUMMARY

[Well-Ag]–**Ab**–[AHIgG-HRP]–[TMB]–A_{450}

Assay type: EIA (non-competitive)
Detection: Colorimetric A_{450}
Format: Microtitre well, Ag coated
Sample type: Serum
Sample pre-treatment:
　None
Sample volume: 100 µl of 1:101 dilution
Number of tests: 96
Controls - standards run in assay:
　Controls: Neg (1), cut-off (2), Pos (1)
Incubation:
　20 min (37°C) + 20 min (37°C) + 10 min (RT)
Washes: 2

CONTENTS

Antibodies, antigens, labelled components:
　HHV-6 Ag bound to well
　Anti-human IgG Ab (sh) HRP conjugated
Substrate: TMB
Controls - standards supplied:
　Controls: Neg, cut-off and Pos (human serum)
Additional reagents:
　None
Special equipment required:
　None

INTERPRETATION

Comments on interpretation:
　Equivocal: where ratio of sample OD:cut-off value is
　　beween 0.9–1.1; retest to confirm
No. of references: 11

NOTES

131558.0

Human Herpes virus 6
ANTIBODY DETECTION (IgG)

Manufacturer: PanBio
Cat. No./Trade name: H6G-100/Human Herpesvirus-6 IgG Elisa

SUMMARY

[Well-Ag]–**Ab**–[AHIgG-HRP]–[TMB]–A_{450}

Assay type: EIA (non-competitive)
Detection: Colorimetric A_{450}
Format: Microtitre well, Ag coated
Sample type: Serum
Sample pre-treatment:
　None
Sample volume: 100 µl of 1:100 dilution
Number of tests: 96
Controls - standards run in assay:
　Controls: Neg (1), Pos (1)
　Calibrator: cut-off (2)
Incubation:
　20 min (37°C) + 20 min (37°C) + 10 min (RT)
Washes: 2

CONTENTS

Antibodies, antigens, labelled components:
　HHV-6 Ag bound to well
　Anti-human IgG Ab (sh) HRP conjugated
Substrate: TMB
Controls - standards supplied:
　Controls: Neg and Pos (human serum)
　Calibrator: cut-off (human serum)
Additional reagents:
　None
Special equipment required:
　None

INTERPRETATION

Comments on interpretation:
　Equivocal: where the ratio of cut-off value:sample OD is
　　between 0.9-1.1; retest to confirm
No. of references: 11

NOTES

131395.0

© KLUWER ACADEMIC PUBLISHERS 1995, ISSN 1381-5067

Human Herpes virus 6
ANTIBODY DETECTION (IgG)

Manufacturer: Biotrin International Ltd
Cat. No./Trade name: Human Herpes Virus 6 IgG IFA

SUMMARY

[Slide-Ag]–**Ab**–[AHIgG-FITC]–fluorescence

Assay type: Immunofluorescence assay (indirect)
Detection: Fluorescence microscopy
Format: Slide, Ag coated
Sample type: Serum
Sample pre-treatment:
　None
Sample volume: 20 µl of 1:20 dilution
Number of tests: 40
Controls - standards run in assay:
　Controls: Neg (1), Pos (1)
Incubation:
　30 min (37°C) + 30 min (37°C)
Washes: 2

CONTENTS

Antibodies, antigens, labelled components:
　HHV-6 Ag from infected human lymphocytes bound to
　　slide
　Anti-human IgG Ab (g) FITC conjugated
Substrate:
Controls - standards supplied:
　Controls: Neg and Pos (human serum)
Additional reagents:
　None
Special equipment required:
　None

INTERPRETATION

Comments on interpretation:
　Positive: bright green fluorescence
No. of references: 18

NOTES

131082.0

Human Herpes virus 6
ANTIBODY DETECTION (IgG)

Manufacturer: Stellar Bio Systems Inc
Cat. No./Trade name: Indirect Fluorescent Assay for
HHV 6 IgG Antibody

SUMMARY

[Slide-Ag]–**Ab**–[AHIgG-FITC]–fluorescence

Assay type: Immunofluorescence assay (indirect)
Detection: Fluorescence microscopy
Format: Slide well, Ag coated
Sample type: Serum
Sample pre-treatment:
　None
Sample volume: 20 µl of 1:20 dilution
Number of tests:
Controls - standards run in assay:
　Controls: Neg (1), Pos (1)
Incubation:
　30 min (37°C) + 30 min (37°C)
Washes: 2

CONTENTS

Antibodies, antigens, labelled components:
　HHV-6 infected and uninfected cells (human
　　lymphocytes) bound to slide
　Anti-human IgG Ab (g) FITC conjugated
Substrate:
Controls - standards supplied:
　Controls: Neg and Pos (human serum)
Additional reagents:
　None
Special equipment required:
　None

INTERPRETATION

Comments on interpretation:
　Positive: (1 +) – (4 +) bright green fluorescent staining of
　　infected cells
No. of references: 26

NOTES

131631.0

For research use only

© KLUWER ACADEMIC PUBLISHERS 1995, ISSN 1381-5067

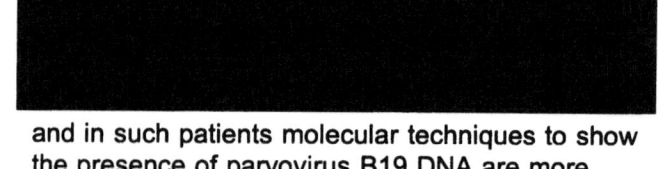

Human parvovirus B19

Natural history

Human parvovirus B19 belongs to the family Parvoviridae and is the only recognised human member of the family. It was first discovered in 1982 when serum specimens were being examined in the electron microscope for the presence of hepatitis B surface antigen. At first no disease could be associated with the 20 nm icosahedral particles, but it was soon realised that they were the cause of the acute childhood infection called 'fifth disease' or erythema infectiosum or, in lay terms 'slapped cheek syndrome'. This is a mild infection characterised by a transient lacey rash which spreads centrifugally from the face to the arms and legs. The rash is preceded by a prodrome of sore throat and fever and is frequently followed by arthralgia, particularly of the small joints of the hands and feet and most painful and prolonged in middle-aged women.

As more became known of the structure and biology of the virus, further associations with diseases were found. The virus has a single stranded DNA genome of around 5000 base pairs in length, sufficient only to code for its own protective coat protein. It therefore replicates only in cells which are rapidly dividing and can supply all other molecules for the replicative process. The only cells in vivo known to support the replication of human parvovirus B19 are erythroid progenitor cells. Lytic infection can therefore lead to a transient fall in haemoglobin of little consequence in otherwise healthy individuals. In those with underlying haematological abnormalities, parvovirus B19 infection can precipitate an aplastic crisis and in the immunocompromised, failure to clear the virus can lead to persistent anaemia. In the fetus, maternal infection can cross the placenta and lead to fatal hydrops fetalis in about 10% of cases.

Diagnosis

By the time the rash appears in acute infection a brisk antibody response is already present and indeed, the rash is the result of deposition of antigen–antibody complexes in the epithelial capillaries. Serological detection of parvovirus B19 specific IgM antibodies is the most useful diagnostic test and IgG antibodies can be sought when evidence of past infection is required. The first assays depended on a source of native antigen as the virus could not readily be grown in cell culture, but recombinant technology has allowed the development of a new range of EIAs and immunofluorescence tests based on recombinant antigens expressed in *E.coli* and increasingly from a baculovirus vector. In the immunocompromised, the serological response may be abnormal or absent and in such patients molecular techniques to show the presence of parvovirus B19 DNA are more useful. These include dot blot hybridisation assays, *in situ* hybridisation and polymerase chain reaction.

Comment

B19 specific IgM tests are the tests of choice for acute infection in immunocompetent people. False positive results can sometimes be obtained with some recombinant antigens and confirmatory assays using different methodologies may occasionally be necessary.

Reference

Pattison JR. In: Human parvoviruses. Zuckerman AJ, Banatvala JE, Pattison JR, eds. Principles and Practice of Clinical Virology, 3rd edn. Chichester: John Wiley. 1994:653–66.

© KLUWER ACADEMIC PUBLISHERS 1995, ISSN 1381-5067

Human parvovirus B19
ANTIBODY DETECTION (IgG)

Manufacturer: Biotrin International Ltd
Cat. No./Trade name: V519IG/Parvovirus B19 IgG (3rd Generation) EIA

SUMMARY

[Well-Ag]–**Ab**–[AHIgG-HRP]–[TMB]–A$_{450}$

Assay type: EIA (non-competitive)
Detection: Colorimetric A$_{450}$
Format: Microtitre well, Ag coated
Sample type: Serum
Sample pre-treatment:
   None
Sample volume: 10 μl (+1 ml diluent)
Number of tests: 96
Controls - standards run in assay:
   Controls: Neg (1), Pos (1)
Incubation:
   1 hr (RT) + 30 min (RT) + 10 min (RT)
Washes: 2

CONTENTS

Antibodies, antigens, labelled components:
   rec. Human parvovirus B19 (VP 2 protein) Ag bound to well
   Anti-human IgG Ab (r) HRP conjugated
Substrate: TMB
Controls - standards supplied:
   Controls: Neg and Pos (serum)
Additional reagents:
   None
Special equipment required:
   None

INTERPRETATION

Comments on interpretation:
   Equivocal: where sample OD is between cut-off value x 0.9 and cut-off value x 1.1; retest to confirm
   Repeatably equivocal: retest a fresh sample 1 week later
No. of references: 17

NOTES

131554.0

Human parvovirus B19
ANTIBODY DETECTION (IgG)

Manufacturer: Dako A/S
Cat. No./Trade name: K039/IDIEA® Parvovirus B19 IgG

SUMMARY

[Well-strept]–**Ab**–[Ag-biotin]–[AHIgG-POD]–[TMB]–A$_{450}$

Assay type: EIA (non-competitive)
Detection: Colorimetric A$_{450}$
Format: Microtitre well, streptavidin coated
Sample type: Serum
Sample pre-treatment:
   None
Sample volume: 10 μl (+1 ml diluent)*
Number of tests: 96
Controls - standards run in assay:
   Controls: Neg (1), cut-off (3), Pos (1)
Incubation:
   30 min (RT) + 30 min (RT) + 10 min (RT) (first two incubations with shaking)
Washes: 2

CONTENTS

Antibodies, antigens, labelled components:
   Streptavidin bound to well
   rec. Human parvovirus B19 (capsid protein VP2) Ag biotinylated
   Anti-human IgG (r) Ab POD conjugated
Substrate: TMB
Controls - standards supplied:
   Controls: Neg (sample diluent), cut-off and Pos (human serum)
Additional reagents:
   None
Special equipment required:
   None

INTERPRETATION

Comments on interpretation:
   Equivocal: retest a fresh specimen within 2 weeks
   Repeatably equivocal: report as negative
No. of references: 11

NOTES

131225.0

*Samples are assayed in duplicate

© KLUWER ACADEMIC PUBLISHERS 1995, ISSN 1381-5067

Human parvovirus B19
ANTIBODY DETECTION (IgG)

Manufacturer: Euro-Diagnostica B.V.
Cat. No./Trade name: 41.30.93/Parvovirus B19® IgG

SUMMARY

[Well-Ag]–**Ab**–[AHIgG-AP]–[PNP]–A_{405}

Assay type: EIA (non-competitive)
Detection: Colorimetric A_{405}
Format: Microtitre well, Ag coated
Sample type: Serum, plasma
Sample pre-treatment:
   None
Sample volume: 100 µl of 1:40 dilution x 2
Number of tests: 96
Controls - standards run in assay:
   Controls: Neg (2), Pos (2)
Incubation:
   45 min (37°C) + 45 min (37°C) + 30 min (37°C)
Washes: 2 (+ preliminary platewash)

CONTENTS

Antibodies, antigens, labelled components:
   Human parvovirus B19 (syn peptide and rec protein) Ags
      bound to well
   Anti-human IgG Ab (g) AP conjugated
Substrate: PNP
Controls - standards supplied:
   Controls: Neg and Pos (serum)
Additional reagents:
   NaOH
Special equipment required:
   None

INTERPRETATION

Comments on interpretation:
   Classification of samples is according to cut-off; no
      further testing
No. of references: 7

NOTES

131292.0

Human parvovirus B19
ANTIBODY DETECTION (IgG)

Manufacturer: Immunobiological
Laboratories
Cat. No./Trade name: RE57121/Parvovirus IgG ELISA

SUMMARY

[Well-Ag]–**Ab**–[AHIgG-POD]–[TMB]–A_{450}

Assay type: EIA (non-competitive)
Detection: Colorimetric A_{450}
Format: Microtitre well, Ag coated
Sample type: Serum, plasma
Sample pre-treatment:
   None
Sample volume: 100 µl of 1:101 dilution x 2
Number of tests: 96
Controls - standards run in assay:
   Standards:
   Qualitative: A (2), D (2)
   Quantitative: A (2), B (2), C (2), D (2)
Incubation:
   1 hr (37°C) + 30 min (37°C) + 10 min (RT)
Washes: 2

CONTENTS

Antibodies, antigens, labelled components:
   rec. Human parvovirus B19 (VP1/VP2) Ag bound to well
   Anti-human IgG Ab POD conjugated
Substrate: TMB
Controls - standards supplied:
   Standards: A–D (12, 40, 100 and 200 U/ml)
Additional reagents:
   None
Special equipment required:
   None

INTERPRETATION

Comments on interpretation:
   Equivocal: within cut-off ±0.05; retest to confirm
No. of references: 13

NOTES

131068.0

© *KLUWER ACADEMIC PUBLISHERS 1995, ISSN 1381-5067*

Human parvovirus B19
ANTIBODY DETECTION (IgG)

Manufacturer: Laboratoire Eurobio
Cat. No./Trade name: 900291/Parvovirus IgG ELIT®

SUMMARY

[Well-Ag]–**Ab**–[AHIgG-POD]–[TMB]–A_{450}

Assay type: EIA (non-competitive)
Detection: Colorimetric A_{450}
Format: Microtitre well, Ag coated
Sample type: Serum, plasma
Sample pre-treatment:
  None
Sample volume: 100 µl of 1:101 dilution
Number of tests: 96
Controls - standards run in assay:
  Standards: I (1), II (2), III (1), IV (1)
Incubation:
  1 hr (RT) + 30 min (RT) + 10 min (RT)
Washes: 2

CONTENTS

Antibodies, antigens, labelled components:
  Human parvovirus B19 Ag bound to well
  Anti-human IgG Ab (g) POD conjugated
Substrate: TMB
Controls - standards supplied:
  Standards: 4 (human serum)
Additional reagents:
  None
Special equipment required:
  None

INTERPRETATION

Comments on interpretation:
  Equivocal: where ratio of sample OD:standard OD is
    between 0.8 and 1.20; retest a fresh sample in 1–2
    weeks
No. of references: 0

NOTES

131096.0

Human parvovirus B19
ANTIBODY DETECTION (IgG)

Manufacturer: MRL Diagnostics
Cat. No./Trade name: EL0100G/Human Parvovirus B19
IgG Elisa

SUMMARY

[Well-Ag]–**Ab**–[AHIgG-POD]–[TMB]–A_{450}

Assay type: EIA (non-competitive)
Detection: Colorimetric A_{450}
Format: Microtitre well, Ag coated
Sample type: Serum (do not heat inactivate)
Sample pre-treatment:
  None
Sample volume: 10 µl of 1:101 dilution x 2
Number of tests: 96
Controls - standards run in assay:
  Controls: Neg (2), low Pos (2), high Pos (2)
  Calibrator: Cut-off (4)
Incubation:
  1 hr (RT) + 30 min (RT) + 10 min (RT)
Washes: 2 (+ preliminary plate wash)

CONTENTS

Antibodies, antigens, labelled components:
  Human parvovirus B19 Ag bound to well
  Anti-human IgG Ab (g) POD conjugated
Substrate: TMB
Controls - standards supplied:
  Controls: Neg, low Pos and high Pos (human serum)
  Calibrator: Cut-off (human serum)
Additional reagents:
  None
Special equipment required:
  None

INTERPRETATION

Comments on interpretation:
  Equivocal: when the ratio of sample OD:cut-off calibrator
    OD (Index value) is between ⩾0.80 and ⩽1.20;
    retest to confirm
  Repeatably equivocal: report as negative and retest a
    fresh sample taken several weeks later to identify any
    rise in Ab titre. If still equivocal, report as negative
No. of references: 8

NOTES

131111.0

© KLUWER ACADEMIC PUBLISHERS 1995, ISSN 1381-5067

Human parvovirus B19
ANTIBODY DETECTION (IgG)

Manufacturer: r-biopharm GmbH
Cat. No./Trade name: RIDASCREEN® Parvovirus IgG EIA

SUMMARY

[Well-Ag]–**Ab**–[AHIgG-HRP]–[TMB]–A_{450}

Assay type: EIA (non-competitive)
Detection: Colorimetric A_{450}
Format: Microtitre well, Ag coated
Sample type: Serum
Sample pre-treatment:
  None
Sample volume: 10 µl (+1 ml diluent)
Number of tests: 48, 96
Controls - standards run in assay:
  Controls: Neg (2), Pos (1)
Incubation:
  30 min (RT) + 30 min (RT) + 30 min (RT)
Washes: 2

CONTENTS

Antibodies, antigens, labelled components:
  rec. Human parvovirus B19 (capsid proteins) Ag bound to well
  Anti-human IgG Ab (g) HRP conjugated
Substrate: TMB
Controls - standards supplied:
  Controls: Neg and Pos (human serum
Additional reagents:
  None
Special equipment required:
  None

INTERPRETATION

Comments on interpretation:
  Equivocal: where OD of sample lies between the Neg cut-off value and the Pos cut-off value; retest a fresh sample taken 7–14 days later
No. of references: 18

NOTES

131149.0

Human parvovirus B19
ANTIBODY DETECTION (IgM)

Manufacturer: Biotrin International Ltd
Cat. No./Trade name: V619IM/Parvovirus B19 IgM (3rd Generation) EIA

SUMMARY

[Well-AHIgM]–**Ab**–[Ag-biotin]–[strept-HRP]–[TMB]–A_{450}

Assay type: EIA (non-competitive)
Detection: Colorimetric A_{450}
Format: Microtitre well, Ab coated
Sample type: Serum
Sample pre-treatment:
  None
Sample volume: 10 µl (+1 ml diluent)
Number of tests: 96
Controls - standards run in assay:
  Controls: Neg (1), Pos (1)
Incubation:
  1 hr (RT) + 30 min (RT) + 30 min (RT) + 10 min (RT)
Washes: 3

CONTENTS

Antibodies, antigens, labelled components:
  Anti-human IgM Ab (r) bound to well
  rec. Human parvovirus B19 (VP2 protein) Ag biotinylated
  Streptavidin HRP conjugated
Substrate: TMB
Controls - standards supplied:
  Controls: Neg and Pos (serum)
Additional reagents:
  None
Special equipment required:
  None

INTERPRETATION

Comments on interpretation:
  Equivocal: where sample OD is between cut-off value x 0.9 and cut-off value x 1.1; retest to confirm
  Repeatably equivocal: retest a fresh sample 1 week later
No. of references: 17

NOTES

131555.0

Human parvovirus B19
ANTIBODY DETECTION (IgM)

Manufacturer: Dako A/S
Cat. No./Trade name: K041/IDEIA® Parvovirus B19 IgM

SUMMARY

[Well-AHIgM]–**Ab**–[Ag-biotin]–[strept-POD]–[TMB]–A_{450}

Assay type: EIA (non-competitive)
Detection: Colorimetric A_{450}
Format: Microtitre well, Ab coated
Sample type: Serum
Sample pre-treatment:
　None
Sample volume: 10 μl (+2 ml diluent)*
Number of tests: 96
Controls - standards run in assay:
　Controls: Neg (1), cut-off (3), Pos (1)
Incubation:
　1 hr (RT) + 1 hr (RT) + 1 hr (RT) + 10 min (RT) (first
　　two incubations with shaking)
Washes: 3

CONTENTS

Antibodies, antigens, labelled components:
　Anti-human IgM Ab bound to well
　rec. Human parvovirus B19 (capsid protein VP2) Ag
　　biotinylated
　Streptavidin POD conjugated
Substrate: TMB
Controls - standards supplied:
　Controls: Neg (sample diluent), cut-off and Pos (human
　　serum)
Additional reagents:
　None
Special equipment required:
　None

INTERPRETATION

Comments on interpretation:
　Equivocal: retest a fresh specimen within 2 weeks
　Repeatably equivocal: report as negative
No. of references: 11

NOTES

131038.0

*Samples are assayed in duplicate

Human parvovirus B19
ANTIBODY DETECTION (IgM)

Manufacturer: Euro-Diagnostica B.V.
Cat. No./Trade name: 41.30.82/Parvoscan-B19® IgM

SUMMARY

[Well-Ag]–**Ab**–[AHIgM-AP]–[PNP]–A_{405}

Assay type: EIA (non-competitive)
Detection: Colorimetric A_{405}
Format: Microtitre well, Ag coated
Sample type: Serum, plasma
Sample pre-treatment:
　Absorption of sample to eliminate interference due to RF
　　and IgG is recommended
Sample volume: 100 μl of 1:40 dilution x 2
Number of tests: 96
Controls - standards run in assay:
　Controls: Neg (3), Pos (2)
Incubation:
　45 min (37°C) + 45 min (37°C) + 30 min (37°C)
Washes: 2 (+ preliminary platewash)

CONTENTS

Antibodies, antigens, labelled components:
　Human parvovirus B19 (syn peptide) Ag bound to well
　Anti-human IgM Ab (g) AP conjugated
Substrate: PNP
Controls - standards supplied:
　Controls: Neg and Pos (serum)
Additional reagents:
　NaOH
Special equipment required:
　None

INTERPRETATION

Comments on interpretation:
　Classification of samples is according to cut-off; no
　　further testing
No. of references: 8

NOTES

131293.0

© KLUWER ACADEMIC PUBLISHERS 1995, ISSN 1381-5067

Human parvovirus B19
ANTIBODY DETECTION (IgM)

Manufacturer: Immunobiological Laboratories
Cat. No./Trade name: RE 57101/Parvovirus IgM ELISA

SUMMARY

[Well-Ag]–**Ab**–[AHIgM-enzyme]–[TMB]–A_{450}

Assay type: EIA (non-competitive)
Detection: Colorimetric A_{450}
Format: Microtitre well, Ag coated
Sample type: Serum, plasma
Sample pre-treatment:
　Mix diluted serum with RF-Absorbent to eliminate interference due to RF and IgG in sample
Sample volume: 100 μl of treated sample x 2
Number of tests: 96
Controls - standards run in assay:
　Standards:
　Qualitative: A (2), D (2)
　Quantitative: A (2), B (2), C (2), D (2)
Incubation:
　1 hr (37°C) + 30 min (37°C) + 10 min (RT)
Washes: 2

CONTENTS

Antibodies, antigens, labelled components:
　rec. Human parvovirus B19 (VP1/VP2) Ag bound to well
　Anti-human IgM Ab enzyme conjugated
Substrate: TMB
Controls - standards supplied:
　Standards: A–D (10, 25, 50 and 150 U/ml)
Additional reagents:
　None
Special equipment required:
　None

INTERPRETATION

Comments on interpretation:
　Equivocal: within cut-off ±0.05; retest to confirm
No. of references: 13

NOTES

131069.0

Human parvovirus B19
ANTIBODY DETECTION (IgM)

Manufacturer: Laboratoire Eurobio
Cat. No./Trade name: 900292/Parvovirus IgM ELIT®

SUMMARY

[Well-Ag]–**Ab**–[AHIgM-POD]–[OPD]–A_{450}

Assay type: EIA (non-competitive)
Detection: Colorimetric A_{450}
Format: Microtitre well, Ag coated
Sample type: Serum, plasma
Sample pre-treatment:
　Treat sample with anti-human IgG reagent provided to reduce interference due to RF and IgG in sample
Sample volume: 100 μl of 1:101 dilution
Number of tests: 96
Controls - standards run in assay:
　Standards: I (1), II (2), III (1), IV (1)
Incubation:
　1 hr (RT) + 30 min (RT) + 10 min (RT)
Washes: 2

CONTENTS

Antibodies, antigens, labelled components:
　Human parvovirus B19 Ag bound to well
　Anti-human IgM Ab (g) POD conjugated
Substrate: OPD
Controls - standards supplied:
　Standards: 4 (human serum with defined Ab values)
Additional reagents:
　None
Special equipment required:
　None

INTERPRETATION

Comments on interpretation:
　Equivocal: where ratio of sample OD:standard OD is between 0.8 and 1.20; retest a fresh sample in 1–2 weeks
No. of references: 0

NOTES

131097.0

© KLUWER ACADEMIC PUBLISHERS 1995, ISSN 1381-5067

Human parvovirus B19
ANTIBODY DETECTION (IgM)

Manufacturer: MRL Diagnostics
Cat. No./Trade name: EL0100M/Human Parvovirus B19 IgM Elisa

SUMMARY

[Well-Ag]–**Ab**–[AHIgM-POD]–[TMB]–A$_{450}$

Assay type: EIA (non-competitive)
Detection: Colorimetric A$_{450}$
Format: Microtitre well, Ag coated
Sample type: Serum (do not heat inactivate)
Sample pre-treatment:
   None
Sample volume: 10 μl of 1:101 dilution*
Number of tests: 96
Controls - standards run in assay:
   Controls: Neg (2), low Pos (2), high Pos (2)
   Calibrator: Cut-off (4)
Incubation:
   1 hr (RT) + 30 min (RT) + 10 min (RT)
Washes: 2 (+ preliminary plate wash)

CONTENTS

Antibodies, antigens, labelled components:
   Human parvovirus B19 Ag bound to well
   Anti-human IgM Ab (g) POD conjugated
Substrate: TMB
Controls - standards supplied:
   Controls: Neg, low Pos and high Pos (human serum)
   Calibrator: Cut-off (human serum)
Additional reagents:
   None
Special equipment required:
   None

INTERPRETATION

Comments on interpretation:
   Equivocal: when the ratio of sample OD:cut-off calibrator OD (Index value) is between ≥0.80 and ≤1.20; retest to confirm
   Repeatably equivocal: report as negative and retest a fresh sample taken several weeks later to identify any rise in Ab titre. If still equivocal, report as negative
No. of references: 10

NOTES

131112.0

*MRL IgM sample diluent contains a hyper-immune anti-human IgG precipitating Ig to eliminate interference due to RF & IgG in sample (10 min waiting time after dilution)

Human parvovirus B19
ANTIBODY DETECTION (IgM)

Manufacturer: r-biopharm GmbH
Cat. No./Trade name: RIDASCREEN® Parvovirus IgM EIA

SUMMARY

[Well-Ag]–**Ab**–[AHIgM-HRP]–[TMB]–A$_{450}$

Assay type: EIA (non-competitive)
Detection: Colorimetric A$_{450}$
Format: Microtitre well, Ag coated
Sample type: Serum
Sample pre-treatment:
   Absorption of serum to eliminate interference due to RF and IgG is recommended
Sample volume: 10 μl (+1 ml diluent)
Number of tests: 48, 96
Controls - standards run in assay:
   Controls: Neg (2), Pos (1)
Incubation:
   30 min (RT) + 30 min (RT) + 30 min (RT)
Washes: 2

CONTENTS

Antibodies, antigens, labelled components:
   rec. Human parvovirus B19 (capsid proteins) Ag bound to well
   Anti-human IgM Ab (g) HRP conjugated
Substrate: TMB
Controls - standards supplied:
   Controls: Neg and Pos (human serum
Additional reagents:
   None
Special equipment required:
   None

INTERPRETATION

Comments on interpretation:
   Equivocal: where OD of sample lies between the Neg cut-off value and the Pos cut-off value; retest a fresh sample taken 7–14 days later
No. of references: 18

NOTES

131150.0

© KLUWER ACADEMIC PUBLISHERS 1995, ISSN 1381-5067

Human parvovirus B19
ANTIBODY DETECTION (IgG AND/OR IgM)

Manufacturer: Biotrin International Ltd
Cat. No./Trade name: V219WB/ParvoBlot

SUMMARY

[Strip-Ags]–**Ab**–[AHIgG/IgM-POD]–[DAB]–visible band

Assay type: Immunoblot assay
Detection: Visual
Format: Reaction tray, Ag coated nitrocellulose strip
Sample type: Serum
Sample pre-treatment:
  For IgM assay: mix serum with IgG Absorbent provided, incubate 15 min (RT), centrifuge and further dilute with Assay Diluent to eliminate interference due to RF and IgG
Sample volume: IgG assay: 1 ml of 1:101 dilution; IgM assay: 1 ml of prepared sample
Number of tests: 20
Controls - standards run in assay:
  Controls: Neg (1), IgG Pos (1), IgM Pos (1)
Incubation:
  3 hr (37°C) + 1 hr (37°C) + 15 min (RT) (first 2 incubations with shaking)
Washes: 2

CONTENTS

Antibodies, antigens, labelled components:
  rec. Human parvovirus B19 (VP1 and VP2 proteins) Ag bound to nitrocellulose strip as discrete bands
  Anti-human IgG Ab HRP conjugated
  Anti-human IgM Ab HRP conjugated
Substrate: DAB
Controls - standards supplied:
  Controls: Neg, IgG Pos and IgM Pos (serum)
Additional reagents:
  None
Special equipment required:
  None

INTERPRETATION

Comments on interpretation:
  Positive: appearance of coloured bands on strips. For interpretation of band patterns, see manufacturer's instructions
No. of references: 11

NOTES

131557.0

This kit can detect IgG and IgM antibodies and can differentiate between them

Human parvovirus B19
ANTIBODY DETECTION (IgG AND/OR IgM)

Manufacturer: r-biopharm GmbH
Cat. No./Trade name: RIDA® Blot Parvovirus B19

SUMMARY

[Strip-Ags]–**Ab**–[AHIgG/IgM-POD]–[substrate]– visible band

Assay type: Immunoblot assay
Detection: Visual
Format: Incubation dish, Ag coated nitrocellulose strip
Sample type: Serum, plasma
Sample pre-treatment:
  For IgM determination: preabsorption of serum with protein A-beads or RF absorbent to eliminate RF and IgG
Sample volume: 20 µl of 1:100 dilution
Number of tests: 20
Controls - standards run in assay:
  Controls: Neg (1), weak Pos (1) and Pos (1)
  Control strip: (1)
Incubation:
  Overnight (RT) + 1 h (RT) + 2–10 min (RT) (first two incubations on shaker)
Washes: 2

CONTENTS

Antibodies, antigens, labelled components:
  rec. Human parvovirus B19 (3 protein segments from VP1 and VP2) Ags bound to nitrocellulose strip as discrete bands
  Anti-human IgG Ab POD conjugated
  Anti-human IgM Ab POD conjugated
Substrate: Diaminobenzidine
Controls - standards supplied:
  Controls: Neg, weak Pos and Pos
  Control strip: (Ag strip incubated with Pos control serum)
Additional reagents:
  Protein A – beads or RF absorbent
Special equipment required:
  None

INTERPRETATION

Comments on interpretation:
  Positive: appearance of coloured bands on strips. For interpretation of band patterns, see manufacturer's instructions
No. of references: 18

NOTES

131151.0

This kit can detect IgG and IgM antibodies and can differentiate between them

© KLUWER ACADEMIC PUBLISHERS 1995, ISSN 1381-5067

Human parvovirus B19
ANTIBODY DETECTION (IgM)

Manufacturer: Immunetics
Cat. No./Trade name: Human Parvo B19 Western Blot Kit

SUMMARY

[Strip-Ag]–**Ab**–[AHIgM-AP]–BCIP/NBT]–bands

Assay type: Immunoblot assay
Detection: Visual
Format: Reaction tray, Ag coated nitrocellulose strip
Sample type: Serum, plasma
Sample pre-treatment:
   None
Sample volume: 1 ml of 1:100 dilution
Number of tests: 24
Controls - standards run in assay:
   Controls: Neg (1), Pos (1)
Incubation:
   1 hr (RT) + 30 min (RT)
Washes: 2

CONTENTS

Antibodies, antigens, labelled components:
   Human parvovirus B19 (purified viral capsid) Ags bound
      to nitrocellulose strip as discrete bands
   Anti-human IgM Ab AP conjugated
Substrate: BCIP/NBT
Controls - standards supplied:
   Controls: Neg and Pos (human serum)
Additional reagents:
   None
Special equipment required:
   None

INTERPRETATION

Comments on interpretation:
   Positive: coloured bands on membrane. For
      interpretation of band pattern, see manufacturer's
      instructions
No. of references: 3

NOTES

131821.0

For research use only

Human parvovirus B19
ANTIBODY DETECTION (IgG)

Manufacturer: Stellar Bio Systems Inc
Cat. No./Trade name: Indirect Fluorescent Assay for Parvovirus B19 IgG Antibody

SUMMARY

[Slide-Ag]–**Ab**–[AHIgG-FITC]–fluorescence

Assay type: Immunofluorescence assay (indirect)
Detection: Fluorescence microscopy
Format: Slide well, Ag coated
Sample type: Serum
Sample pre-treatment:
   None
Sample volume: 20 µl of 1:20 dilution
Number of tests:
Controls - standards run in assay:
   Controls: Neg (1), Pos (1)
Incubation:
   30 min (37°C) + 30 min (37°C)
Washes: 2

CONTENTS

Antibodies, antigens, labelled components:
   rec. Human parvovirus B19 infected and uninfected cells
      (from insects) bound to slide
   Anti-human IgG Ab (g) FITC conjugated
Substrate:
Controls - standards supplied:
   Controls: Neg and Pos (human serum)
Additional reagents:
   None
Special equipment required:
   None

INTERPRETATION

Comments on interpretation:
   Positive: (1+) – (4+) bright green fluorescent staining of
      infected cells
No. of references: 8

NOTES

131633.0

© KLUWER ACADEMIC PUBLISHERS 1995, ISSN 1381-5067

Human parvovirus B19
ANTIBODY DETECTION (IgG AND/OR IgM)

Manufacturer: Biotrin International Ltd
Cat. No./Trade name: Parvovirus B19 IgG and IgM IFA

SUMMARY

[Slide-Ag]–**Ab**–[AHIgG/IgM-FITC]–fluorescence

Assay type: Immunofluorescence assay (indirect)
Detection: Fluorescence microscopy
Format: Slide well, Ag coated
Sample type: Serum
Sample pre-treatment:
   For IgA and IgM assays:* pretreat sample with a suitable absorbent reagent to eliminate interference due to RF andn IgG
Sample volume: IgG assay: 20 µl of 1:64 dilution; IgM assay: 20 µl of 1:16 dilution
Number of tests: 60
Controls - standards run in assay:
   Controls: Neg (1), IgG Pos (1), IgM Pos (1)
Incubation:
   3 hr (37°C) + 30 min (37°C)
Washes: 2

CONTENTS

Antibodies, antigens, labelled components:
   rec. Human parvovirus B19 Ag from insect cells bound to slide well
   Anti-human IgG Ab (r) FITC conjugated
   Anti-human IgM Ab (r) FITC conjugated*
Substrate:
Controls - standards supplied:
   Controls: Neg, IgG Pos and IgM Pos (serum)
Additional reagents:
   RF and IgG absorbent reagent for treating samples prior to IgM assay
Special equipment required:
   None

INTERPRETATION

Comments on interpretation:
   Equivocal: weak (+/–) fluorescence at any titre; confirm by alternative method and/or clinical information
   Positive: green fluorescence of cell surface aggregates at titre of ≥ 1:64 (for IgG assay) and at titre of ≥ 1:16 (for IgM assay)
No. of references: 18

NOTES

131556.0

This kit can detect IgG and IgM antibodies and can differentiate between them
*Separate Human parvovirus B19 IgA IFA reagents are available for performing IgA assay if required (Cat. no. V119IA)

Human parvovirus B19
ANTIBODY DETECTION (IgM)

Manufacturer: Stellar Bio Systems Inc
Cat. No./Trade name: Indirect Fluorescent Assay for Parvovirus B19 IgM Antibody

SUMMARY

[Slide-Ag]–**Ab**–[AHIgM-FITC]–fluorescence

Assay type: Immunofluorescence assay (indirect)
Detection: Fluorescence microscopy
Format: Slide well, Ag coated
Sample type: Serum
Sample pre-treatment:
   None
Sample volume: 20 µl of 1:20 dilution
Number of tests:
Controls - standards run in assay:
   Controls: Neg (1), Pos (1)
Incubation:
   30 min (37°C) + 30 min (37°C)
Washes: 2

CONTENTS

Antibodies, antigens, labelled components:
   rec. Human parvovirus B19 infected and uninfected cells (from insects) bound to slide
   Anti-human IgM Ab (r) FITC conjugated
Substrate:
Controls - standards supplied:
   Controls: Neg and Pos (human serum)
Additional reagents:
   None
Special equipment required:
   None

INTERPRETATION

Comments on interpretation:
   Positive: (1+) – (4+) bright green fluorescent staining of infected cells
No. of references: 8

NOTES

131636.0

Leishmania species

Leishmaniasis is a zoonosis caused by protozoa of the genus *Leishmania*. The animal reservoirs are rodents, canines and other mammals. The organisms live as intracellular amastigotes within host macrophages and are ingested when sandfly vectors feed on the infected animal. Within the gut of the fly promastigote forms develop and multiply prior to inoculation into another host when the insect feeds.

Leishmaniasis is clinically divided into three main groups, visceral, cutaneous and mucosal. A large number of species of *Leishmania* have now been defined on the basis of isoenzyme and molecular analyses. Although a single species may give rise to a variety of clinical presentations, and any given clinical manifestation may be the result of infection with a number of different species, a number of associations are commonly recognised. Visceral leishmaniasis (Kala-azar) is caused when *L.donovani* infects the entire reticuloendothelial system resulting in fever, weight loss and hepatosplenomegaly following an incubation period which may last for up to a year. Ultimately there may be severe wasting, anaemia, cachexia and pigmentation of the skin with gross splenomegaly and other organ involvement. Visceral leishmaniasis is increasingly recognised amongst the immunocompromised and AIDS patients. The other species of *Leishmania* have mainly dermal or mucosal manifestations. In the Old World three species of *Leishmania*: *L.tropica*, *L.major* and *L.aethiopica* predominate and in the New World there are two main species groups: *L.mexicana* and *L.braziliensis*. Cutaneous disease manifestations include a wide spectrum of acute and chronic ulcerating and non-ulcerating lesions,which may be destructive, particularly if there is mucosal involvement, and are frequently secondarily infected. Old World cutaneous leishmaniasis is prevalent in Mediterranean countries, the Middle East through to North Eastern India, in the Congo basin and in China, whilst New World disease is widespread in South and Central America.

Diagnosis

The definitive diagnosis of leishmaniasis depends upon the demonstration of amastigotes in smears or aspirates taken from infected lesions and tissue aspirates, or the isolation of the organism in culture. Bone-marrow, spleen or liver are commonly examined in suspected visceral disease, and scrapings and biopsies from within and around cutaneous lesions, but the organisms have been successfully demonstrated in a number of other infected tissues. Wright or Giemsa staining is most commonly employed to demonstrate the organism. Culture may be performed by inoculation into Novy, McNeal and Nicolle (NNN) medium and incubation at 25°C. Within a few days to several weeks, motile promastigotes may be demonstrated in the culture.

Intradermal inoculation of *Leishmania* antigens in the leishmanin skin test may be helpful in the diagnosis of cutaneous disease, but is not particularly useful in visceral disease often remaining negative in invasive disease until successful chemotherapy has been completed. A number of serodiagnostic systems based predominantly upon immunofluorescence and ELISA technologies are now available and are of some value in the diagnosis of visceral disease, although specificity varies and they may be falsely negative in the HIV-infected individual. Serodiagnosis may also be of some value in suspected mucosal disease, but is somewhat less useful in cutaneous disease, where the antibody response may be variable, and titres correspondingly low.

References

Pearson RD, De Queiroz Sousa A. *Leishmania* species: Visceral (Kala-azar), cutaneous, and mucosal leishmaniasis. In: Mandell GL, Bennett JE, Dolin R, eds. Principles and Practice of Infectious Diseases, 4th edn. New York, Edinburgh, London: Churchill Livingstone; 1995:2428–42.

Palma G, Gutierrez Y. Laboratory diagnosis of *Leishmania*. Clinics in Laboratory Medicine. 1991;11:909–22.

Leishmania species
ANTIBODY DETECTION (IgG)

Manufacturer: Amico Laboratories Inc
Cat. No./Trade name: 5700G/AMIZYME® Leishmania Species IgG Assay

SUMMARY

[Well-Ag]–**Ab**–[AHIgG-HRP]–[ABTS]–A_{405}

Assay type: EIA (non-competitive)
Detection: Colorimetric A_{405}
Format: Microtitre well, Ag coated
Sample type: Serum
Sample pre-treatment:
  None
Sample volume: 5 µl (+245 µl diluent)
Number of tests: 96
Controls - standards run in assay:
  Controls: Neg (1)
  Calibrators: I (1)
Incubation:
  25 min (RT) + 25 min (RT) + 25 min (RT)
Washes: 2

CONTENTS

Antibodies, antigens, labelled components:
  Leishmania tropica Ag bound to well
  Anti-human IgG Ab (g) HRP conjugated
Substrate: ABTS
Controls - standards supplied:
  Controls: Neg
  Calibrator: 1
Additional reagents:
  None
Special equipment required:
  None

INTERPRETATION

Comments on interpretation:
  Equivocal: where sample OD is between 0.375–0.399 (95–99 U/ml); retest to confirm
No. of references: 7

NOTES

131785.0

Leishmania species
ANTIBODY DETECTION (IgG)

Manufacturer: Immuno Pharmacology Research
Cat. No./Trade name: ELISA LEISHMANIA

SUMMARY

[Well-Ag]–**Ab**–[AHIgG-AP]–[PNP]–A_{405}

Assay type: EIA (non-competitive)
Detection: Colorimetric A_{405}
Format: Microtitre well, Ag coated
Sample type: Serum
Sample pre-treatment:
  None
Sample volume: 100 µl of 1:100 dilution x 2
Number of tests: 48, 96
Controls - standards run in assay:
  Controls: Neg (1), Pos (1)
Incubation:
  1 hr (37°C) + 1 hr (37°C) + 30 min (37°C)
Washes: 2

CONTENTS

Antibodies, antigens, labelled components:
  Leishmania species Ag bound to well
  Anti-human IgG Ab AP conjugated
Substrate: PNP
Controls - standards supplied:
  Controls: Neg and Pos
Additional reagents:
  None
Special equipment required:
  None

INTERPRETATION

Comments on interpretation:
  Classification of samples is according to cut-off; no further testing
No. of references: None

NOTES

131327.0

© *KLUWER ACADEMIC PUBLISHERS 1995, ISSN 1381-5067*

Leishmania species
ANTIBODY DETECTION (IgG)

Manufacturer: Melotec S.A.
Cat. No./Trade name: Melotest Leishmania Ab

SUMMARY

[Well-Ag]–**Ab**–[AHIgG-HRP]–[TMB]–A_{450}

Assay type: EIA (non-competitive)
Detection: Colorimetric A_{450}
Format: Microtitre well, Ag coated
Sample type: Serum, plasma
Sample pre-treatment:
　None
Sample volume: 100 μl of 1:20 dilution x 2
Number of tests: 96
Controls - standards run in assay:
　Controls: Neg (1), low Pos (2), high Pos (1)
Incubation:
　10 min (RT) + 5 min (RT) + 10 min (RT)
Washes: 2

CONTENTS

Antibodies, antigens, labelled components:
　Leishmania species Ags bound to well
　Anti-human IgG Ab HRP conjugated
Substrate: TMB
Controls - standards supplied:
　Controls: Neg, low Pos and high Pos (sera)
Additional reagents:
　None
Special equipment required:
　None

INTERPRETATION

Comments on interpretation:
　Equivocal: ratio of sample OD:cut-off value between 0.9
　　and 1.1; retest a fresh sample
　Repeatably equivocal: retest a fresh sample in 2-4 weeks
No. of references: 8

NOTES

131137.0

There may be some cross-reactivity between Leishmania
　and *Trypanosoma cruzi*

Leishmania species
ANTIBODY DETECTION (IgM)

Manufacturer: Amico Laboratories Inc
Cat. No./Trade name: 5700M/AMIZYME® Leishmania
species IgM Assay

SUMMARY

[Well-Ag]–**Ab**–[AHIgM-HRP]–[ABTS]–A_{405}

Assay type: EIA (non-competitive)
Detection: Colorimetric A_{405}
Format: Microtitre well, Ag coated
Sample type: Serum
Sample pre-treatment:
　None
Sample volume: 5 μl (+245 μl diluent)*
Number of tests: 96
Controls - standards run in assay:
　Controls: Neg (1)
　Calibrators: I (1)
Incubation:
　25 min (RT) + 25 min (RT) + 25 min (RT)
Washes: 2

CONTENTS

Antibodies, antigens, labelled components:
　Leishmania tropica Ag bound to well
　Anti-human IgM Ab (g) HRP conjugated
Substrate: ABTS
Controls - standards supplied:
　Controls: Neg
　Calibrator: 1
Additional reagents:
　None
Special equipment required:
　None

INTERPRETATION

Comments on interpretation:
　Equivocal: where sample OD is between 0.275–0.299
　　(92–99 U/ml); retest to confirm
No. of references: 7

NOTES

131793.0

*Sample diluted in IgM serum diluent provided to eliminate
　interference due to RF and IgG

© *KLUWER ACADEMIC PUBLISHERS 1995, ISSN 1381-5067*

Leishmania species
ANTIBODY DETECTION (IgM)

Manufacturer: Amico Laboratories Inc
Cat. No./Trade name: 5700MC/AMIZYME® Leishmania species IgM Capture Assay

SUMMARY

[Well-Ag]–**Ab**–[AHIgM]–[AMIg-HRP]–[ABTS]–A_{405}

Assay type: EIA (non-competitive)
Detection: Colorimetric A_{405}
Format: Microtitre well, Ag coated
Sample type: Serum
Sample pre-treatment:
   None
Sample volume: 5 µl (+245 µl diluent)
Number of tests: 96
Controls - standards run in assay:
   Controls: Neg (1)
   Calibrators: I (1)
Incubation:
   25 min (RT) + 25 min (RT) + 25 min (RT) + 25 min (RT)
Washes: 3

CONTENTS

Antibodies, antigens, labelled components:
   Leishmania tropica Ag bound to well
   Anti-human IgM MAb (m)
   Anti-mouse Ig Ab (g) HRP conjugated
Substrate: ABTS
Controls - standards supplied:
   Controls: Neg
   Calibrator: 1
Additional reagents:
   None
Special equipment required:
   None

INTERPRETATION

Comments on interpretation:
   Equivocal: where sample OD is between 0.285–0.299; retest to confirm
No. of references: 2

NOTES

131803.0

Leishmania species
ANTIBODY DETECTION

Manufacturer: bioMerieux
Cat. No./Trade name: 75931/Leishmania-Spot IF

SUMMARY

[Slide-Ag]–**Ab**–[AHIg-FITC]–fluorescence

Assay type: Immunofluorescence assay (indirect)
Detection: Fluorescence microscopy
Format: Slide, Ag coated
Sample type: Serum
Sample pre-treatment:
   None
Sample volume: 10 µl of serial dilutions (1:40 initial dilution)
Number of tests: 100
Controls - standards run in assay:
   Controls: Neg (1)
Incubation:
   30 min (37°C) + 30 min (37°C)
Washes: 2

CONTENTS

Antibodies, antigens, labelled components:
   Leishmania infantum Ag (from culture) bound to slide
   Anti-human Ig Ab (g) FITC conjugated
Substrate:
Controls - standards supplied:
   Controls: Neg (PBS)
Additional reagents:
   None
Special equipment required:
   None

INTERPRETATION

Comments on interpretation:
   Positive: clearly defined fluorescence, mainly of membrane and flagellium of organisms
No. of references: 4

NOTES

131283.0

This kit detects antibodies to Mediterranean visceral leishmaniasis (none or few for cutaneous leishmaniasis). Cross reactivity may occur with other forms of leishmaniasis, malaria or trypanosomiasis
A kit is available for detection of canine leishmaniasis (Cat. No. 75701)

Leishmania species
ANTIBODY DETECTION

Manufacturer: Immuno Pharmacology Research
Cat. No./Trade name: IFI-LEISHMANIA

SUMMARY

[Slide-Ag]–**Ab**–[AHIg-FITC]–fluorescence

Assay type: Immunofluorescence assay (indirect)
Detection: Fluorescence microscopy
Format: Slide well, Ag coated
Sample type: Serum
Sample pre-treatment:
  None
Sample volume: 10 µl of 1:100 dilution
Number of tests: 50, 100
Controls - standards run in assay:
  Controls: Neg (1), Pos (1)
Incubation:
  30 min (37°C) + 30 min (37°C)
Washes: 2

CONTENTS

Antibodies, antigens, labelled components:
  Leishmania species Ag bound to slide well
  Anti-human Ig Ab (g or sh) FITC conjugated
Substrate:
Controls - standards supplied:
  Controls: Neg and Pos (serum)
Additional reagents:
  None
Special equipment required:
  None

INTERPRETATION

Comments on interpretation:
  Positive: fluorescence of membrane and flagellum or
    organism
No. of references: 6

NOTES

131334.0

Cross-reactivity may occur with other parasites such as
  plasmodium species, toxoplasma species and
  trypanosoma species

© *KLUWER ACADEMIC PUBLISHERS 1995, ISSN 1381-5067*

Leptospira species

Leptospirosis is caused by the organism *Leptospira interrogans*. Leptospires are finely coiled, thread-like spirochaetes, 6–20 mm long with typically hooked or bent ends. Two species are now recognised: the free-living saprophyte *Leptospira biflexa* and the pathogenic *L.interrogans*. Nineteen serogroups and almost 200 serotypes of this latter organism are recognised, and were in former years referred to as separate species. The common pathological serogroups causing human disease are *L.interrogans* serovars *icterohaemorrhagiae*, *hebdomadis*, *hardjo*, *canicola*, *pomona* and *grippo-typhosa*.

Leptospirosis is a zoonosis of worldwide distribution and infection is acquired by direct contact with the reservoir animal or, more commonly, by indirect contact with water or soil contaminated with infected urine. Leptospires gain access through ingestion, mucous membranes or skin abrasions. Occupational and recreational exposure are particularly important. Although rats are the most important reservoir globally, particularly for *icterohaemorrhagiae* infection, dogs, cattle, voles and a variety of other wild and domestic animals may harbour the organism. After an initial incubation period of 1–2 weeks, there is a bacteraemic phase which gives rise to nonspecific 'flu-like' symptoms and during which time the organism is widely distributed throughout the body. This phase may result in progressive fulminant disease, with hepatorenal failure. More commonly a secondary 'immune' phase occurs, during which time organisms are no longer found in the blood or CSF but may be present at other sites. Clinical features at this stage are of rash, meningitis, uveitis and in severe cases jaundice, renal failure and haemorrhage (Weil's disease). Subclinical infection is common in those exposed to infected animals and of those who are ill, only about 10% will develop severe disease with jaundice. Some cases do not appear to progress to the secondary immune phase of the disease.

The organism is unusual amongst the spirochaetes in that it may be cultured *in vitro*. Blood and other samples are inoculated into serum-containing media such as Fletcher's or Tween 80-albumin, and are incubated in the dark at 30°C for up to 6 weeks. The organisms grow as a linear disc below the surface of semi-solid media, or give rise to turbidity of liquid media. Cultures are examined by dark-field microscopy for characteristic morphology and motility of the leptospires. Serotyping may be performed by microscopic agglutination tests. The organisms may also be isolated by means of animal inoculation.

Diagnosis

Definitive diagnosis of the infection is by culture of the organism from blood, CSF, and othe samples as outlined above. Isolation from blood and CSF is only possible during the first 10 days of illness. Direct microscopy of body fluids and tissues by dark-field examination, fluorescent antibody techniques and silver impregnation have all been employed with varying degrees of success for the diagnosis of infection.

Direct microscopy and culture for this organism are not widely available and serology is more commonly used for diagnosis, particularly in the second phase of the disease. A number of genus-specific macroscopic agglutination, haemagglutination and complement fixation tests have been described. This type of test has been used as a screening procedure, with more specific microscopic agglutination tests being employed for subsequent examination or for seroepidemiological studies. Agglutinins appear on day 6–12 and reach a peak at 3–4 weeks. Antibiotic therapy may modify the immune response, and low positive titres may persist for many years. Recently, sensitive and specific ELISA-based assays have increasingly been used in the diagnosis of leptospirosis.

Reference

Farrar WE. *Leptospira* species. In: Mandell GL, Bennett JE, Dolin R, eds. Principles and Practice of Infectious Diseases, 4th edn. New York, Edinburgh, London: Churchill Livingstone. 1995:2137–41.

© KLUWER ACADEMIC PUBLISHERS 1995, ISSN 1381-5067

Leptospira species
ANTIBODY DETECTION (IgG AND/OR IgM))

Manufacturer: Immuno Pharmacology Research
Cat. No./Trade name: IgG/IgM LEPTOELISA

SUMMARY

[Well-Ag]–**Ab**–[AHIgG/IgM-AP]–[PNP]–A_{405}

Assay type: EIA (non-competitive)
Detection: Colorimetric A_{405}
Format: Microtitre well, Ag coated
Sample type: Serum
Sample pre-treatment:
    None
Sample volume: 100 µl of 1:100 dilution x 2
Number of tests: 48, 96
Controls - standards run in assay:
    Controls: Neg (1), Pos (1)
Incubation:
    1 hr (37°C) + 1 hr (37°C) + 30 min (37°C)
Washes: 2

CONTENTS

Antibodies, antigens, labelled components:
    Leptospira (cracked Ag, serovar Copenhagen) Ag bound to well
    Anti-human IgG Ab AP conjugated
    Anti-human IgM Ab AP conjugated
Substrate: PNP
Controls - standards supplied:
    Controls: Neg and Pos
Additional reagents:
    None
Special equipment required:
    None

INTERPRETATION

Comments on interpretation:
    Classification of samples is according to cut-off; no further testing
No. of references: 4

NOTES

131332.0

This kit can detect IgG and IgM antibodies and can differentiate between them

Leptospira species
ANTIBODY DETECTION (IgM)

Manufacturer: PanBio
Cat. No./Trade name: LPM-200/Leptospira IgM Elisa Test

SUMMARY

[Well-Ag]–**Ab**–[AHIgM-HRP]–[TMB]–A_{450}

Assay type: EIA (non-competitive)
Detection: Colorimetric A_{450}
Format: Microtitre well, Ag coated
Sample type: Serum
Sample pre-treatment:
    Mix diluted serum with absorbent solution, incubate 15 min (RT) to eliminate interference due to RF and IgG
Sample volume: 100 µl of prepared sample
Number of tests: 96
Controls - standards run in assay:
    Controls: Neg (1), Pos (1)
    Calibrator: cut-off (2)
Incubation:
    20 min (RT) + 20 min (RT) + 10 min (RT)
Washes: 2

CONTENTS

Antibodies, antigens, labelled components:
    Leptospira species Ag bound to well
    Anti-human IgM Ab (sh) HRP conjugated
Substrate: TMB
Controls - standards supplied:
    Controls: Neg and Pos (human serum)
    Calibrator: cut-off (human serum)
Additional reagents:
    None
Special equipment required:
    None

INTERPRETATION

Comments on interpretation:
    Equivocal: where the ratio of cut-off value:sample OD is between 0.9-1.1; retest to confirm
No. of references: 9

NOTES

131401.0

Listeria species

Listeria monocytogenes is a Gram-positive, non-sporing, aerobic bacillus and is the causal organism of listeriosis. Although commonly regarded as one of the zoonoses, the epidemiology of this infection is still relatively poorly understood, and there remains much controversy with regard to the role played by various animal reservoirs. The organism has been isolated from numerous domestic and wild birds and animals, and is also found in vegetation, silage, water and sewage. Although outbreaks of infection associated with the consumption of contaminated meat and dairy products are well documented, the significance of the isolation from foodstuffs of *L.monocytogenes* serotypes other than 1a, 1b and IVb, which are associated with more than 90% of human disease, is less certain. *L.ivanovii* has occasionally been implicated in human disease.

Listeriosis is primarily an opportunistic infection of the immunocompromised, the extremes of age, and pregnant women. The disease may manifest as a febrile illness in pregnancy which may result in premature labour or miscarriage; overwhelming granulomatous sepsis of the neonate as a result of transplacental infection (granulomatosus infanti-septica); pyrexia of unknown origin; meningoence-phalitis/cerebritis; focal infections as a result of direct contact and inoculation or secondary to metastatic dissemination of infection during the bacteraemic phase.

Clinical samples from sterile sites are normally cultured directly on blood agar at 35–37°C for 5–7 days. Clinical samples from non-sterile sites and food and environmental samples are often enriched prior to subculture, commonly by the incorporation of selective antibiotics and other agents in broth media, or occasionally by 'cold-enrichment', which exploits the ability of the organism to multiply at 4°C. The organism yields small, round, smooth translucent colonies surrounded by a narrow zone of beta-haemolysis, which may be enhanced in the CAMP test. In view of the coryneform appearance on Gram stain it is important to beware of lightly dismissing the organism as a 'diphtheroid'. A variety of novel media for the differential and selective isolation of *Listeria* have recently become available. The organism exhibits a characteristic tumbling motility, is catalase-positive, and can be definitively identified by biochemical testing. Serotyping schemes based predominantly on the flagellar and somatic antigens are of importance for epidemiological purposes.

Diagnosis

Direct microscopic examination of clinical material is of limited value, unless from a normally sterile site, due to the diphtheroid-like morphology of the organism. A direct immunofluorescence test has been described, but is not in routine use. Definitive diagnosis of infection is obtained by isolation of the organism from blood, CSF and other tissue samples as outlined above.

There has been increased interest in recent years in methodology for the rapid, sensitive detection of *Listeria* species in food and environmental samples, where the speed and/or sensitivity of conventional culture has been felt to be inadequate. A number of commercial systems are now available, largely ELISA-based, for the rapid detection of the organisms in broth enrichment cultures of food-stuffs.

Reference

Bille J, Doyle MP. *Listeria* and *Erysipelothrix*. In: Balows A, Hausler WJ, Hermann KL, Isenberg HD, Shadomy HJ, eds. Manual of Clinical Microbiology, 5th edn. Washington: ASM; 1991:287–295.

© *KLUWER ACADEMIC PUBLISHERS 1995, ISSN 1381-5067*

Listeria species
ANTIGEN DETECTION

Manufacturer: Tecra® Diagnostics
Cat. No./Trade name: LISVIA48/TECRA® Listeria Visual Immunoassay

SUMMARY

[Well-Ab]–**Ag**–[Ab-enzyme]–[substrate]–green colour

Assay type: EIA (non-competitive)
Detection: Visual or colorimetric
Format: Microtitre well, Ab coated
Sample type: Food, food related or environemental samples
Sample pre-treatment:
Samples require culture in enrichment broth (42–50 hr) prior to assay
Sample volume: 0.2 ml of prepared sample
Number of tests: 48, 96
Controls - standards run in assay:
Controls: Neg (1), Pos (1)
Incubation:
30 min (37°C) + 30 min (37°C)
Washes: 2

CONTENTS

Antibodies, antigens, labelled components:
Anti-Listeria species Ab bound to well
Anti-Listeria species Ab enzyme conjugated
Substrate: ABTS
Controls - standards supplied:
Controls: Neg (diluent) and Pos
Additional reagents:
Listeria enrichment broth (LEB), Fraser broth, University of Vermont Media (UVM)
Special equipment required:
None

INTERPRETATION

Comments on interpretation:
Positive: green colouration in well confirmed by standard tests
No. of references: 4

NOTES

131722.0

An independent evaluation report of this kit and a list of international organisation and professional groups who have given an approval to it are available – please contact manufacturer for details

Listeria species
ANTIGEN DETECTION

Manufacturer: Tecra® Diagnostics
Cat. No./Trade name: LISUNQ10/TECRA® UNIQUE® Listeria

SUMMARY

{Dipstick-Ab}–**Ag**–[Ab-enzyme]–[substrate]–purple colour

Assay type: EIA (non-competitive)
Detection: Visual
Format: Dipstick, Ab coated
Sample type: Food, food related and environmental samples
Sample pre-treatment:
Samples require culture in enrichment broth (24 hr) prior to assay
Sample volume: Not specified
Number of tests:
Controls - standards run in assay:
Controls: Integral Neg (1), Pos (1)
Incubation:
1 hr (28°C) + 6 hr (28°C) + 30 min (37°C)
Washes: 2

CONTENTS

Antibodies, antigens, labelled components:
Anti-Listeria species Ab bound to dipstick
Anti-Listeria species Ab enzyme conjugated
Substrate: Not specified
Controls - standards supplied:
Controls: Integral Neg and Pos incorporated in dipstick
Additional reagents:
Pre-enrichment broth
Special equipment required:
None

INTERPRETATION

Comments on interpretation:
Positive: purple colour on dipstick with white negative band across top of dipstick
No. of references: 0

NOTES

131810.0

© *KLUWER ACADEMIC PUBLISHERS 1995, ISSN 1381-5067*

Listeria species
ANTIGEN DETECTION

Manufacturer: Unipath Limited
Cat. No./Trade name: OXOID CLEARVIEW Listeria

SUMMARY

[Latex-Ab]–**Ag**–[Ab detector]–blue line

Assay type: Immunochromatographic assay
Detection: Visual
Format: Test device, Ab coated latex particles
Sample type: Culture isolates in Selective Enrichment Broth, from food or environmental samples
Sample pre-treatment:
 None
Sample volume: 135 µl of broth
Number of tests: 50
Controls - standards run in assay:
 Controls: Pos (1)
Incubation:
 Read result 20 min after adding sample
Washes:

CONTENTS

Antibodies, antigens, labelled components:
 Anti-Listeria species (flagella, antigen B) Ab bound to blue latex (Sample window)
 Anti-Listeria species (flagella, antigen B) Ab detector conjugated (Result window)
Substrate:
Controls - standards supplied:
 Controls: Pos
 Integral controls incorporated in Test device
Additional reagents:
 Oxoid Fraser Broth (FB) (Cat. no. CM 895)
 Oxoid Buffered Listeria Enrichment Broth (BLEB) (Cat. no. CM 897)
 BLEB (uninoculated) used as Negative control if required
 BLEB Selective Enrichment Supplement (Cat. no. SR 141E)
Special equipment required:
 None

INTERPRETATION

Comments on interpretation:
 Negative: blue line in Control window and no blue line in Result window
 Positive: blue line in both Control and Result window
No. of references: 4

NOTES

131617.0

Listeria species
ANTIGEN DETECTION

Manufacturer: Microgen Bioproducts
Cat. No./Trade name: M48/Microscreen Listeria

SUMMARY

[Latex-Ab]–**Ag**–agglutination

Assay type: Particle agglutination assay
Detection: Visual
Format: Latex particles, Ab coated (test done on slide)
Sample type: Broth culture or colonies from selective solid media culture (isolated from faecal/food specmens)*
Sample pre-treatment:
 None
Sample volume: 1 drop of broth or 1 colony (+1 drop saline)
Number of tests: 100
Controls - standards run in assay:
 Controls: Reagent control (1), Pos (1)
Incubation:
 Read result within 1 min of mixing sample and reagent
Washes:

CONTENTS

Antibodies, antigens, labelled components:
 Anti-Listeria monocytogenes (A,B and C) Ab and anti-Listeria grayi (E) Ab bound to latex particles (Test Latex)
Substrate:
Controls - standards supplied:
 None
Additional reagents:
 Pos control (Cat. no. M48b)
Special equipment required:
 None

INTERPRETATION

Comments on interpretation:
 Positive: agglutination of sample with Test Latex
No. of references: 5

NOTES

131228.0

*Confirm colonies are smooth strain, oxidase-negative cultures to avoid nonspecific reactions

Mumps virus

Natural history

Mumps is an acute fever of childhood caused by a single serotype in the Paramyxovirus genus of the Paramyxoviridae family. Efficient infection occurs by direct contact or through aerosols containing virus and most members of a community will be infected at some stage. Peak incidence of infection is in the 5–9 year age band, but many infections are subclinical (approximately 30%). Immunity is sufficiently durable to prevent reinfection and to dampen the rate of virus spread. The virus is a spherical particle of 150–200 nm with a phospholipid membrane bearing protein spikes (V antigen) and an internal helical nucleoprotein (S antigen) containing single-strand negative sense RNA. Antigenic differences between isolates do not appear to have epidemiological significance.

The most common presenting symptom is parotitis, often one-sided. Prodromal symptoms include fever, headache, myalgia and sometimes earache. The primary focus of infection can be at several possible sites with spread to local lymphoid tissue and thence to the parotid gland, but also to more distant sites such as the testes, ovaries, pancreas and central nervous system. Complications can therefore include meningoencephalitis (about 10% of cases, generally mild), orchitis, hearing impairment and pancreatitis. Myocarditis, monoarthritis and renal dysfunction can also occasionally occur.

Diagnosis

Virus can be isolated from respiratory specimens and from saliva, CSF and urine, in monkey kidney cells with subsequent identification by haemadsorption inhibition or neutralisation. However, serological diagnosis is also possible. A wide range of techniques have been used including complement fixation tests, using S and V antigens to detect antibodies early in infection, haemagglutination inhibition and haemolysis in gel. More recently, RIA and ELISA have been used both for diagnosis of acute infection (IgM) and for screening to determine past exposure (IgG).

Comment

Mumps vaccine has been available for some time, but is now usually incorporated into a triple vaccine for mumps, measles and rubella (MMR) given at age 12–15 months. The protection rate is >95% and immunity lasts for more than 10 years. Screening assays are becoming increasingly important in the assessment of vaccine-induced immunity and trials are being undertaken using saliva specimens rather than serum for antibody estimations.

References

Black FL. Epidemiology of paramyxoviridae. In: Kingsbury DW, ed. The Paramyxoviruses. Plenum Press. 1991:509–36.

Leinikki P. Mumps. In: Zuckerman AJ, Banatvala JE, Pattison JR, eds. Principles and Practice of Clinical Virology, 3rd edn. Chichester: John Wiley. 1994:401–16.

Mumps virus
ANTIBODY DETECTION (IgG)

Manufacturer: Amico Laboratories Inc
Cat. No./Trade name: 6300G/AMIZYME® Mumps Virus
IgG Assay

SUMMARY

[Well-Ag]–**Ab**–[AHIgG-HRP]–[ABTS]–A_{405}

Assay type: EIA (non-competitive)
Detection: Colorimetric A_{405}
Format: Microtitre well, Ag coated
Sample type: Serum
Sample pre-treatment:
  None
Sample volume: 5 µl (+245 µl diluent)
Number of tests: 96
Controls - standards run in assay:
  Controls: Neg (1)
  Calibrators: I (1)
Incubation:
  25 min (RT) + 25 min (RT) + 25 min (RT)
Washes: 2

CONTENTS

Antibodies, antigens, labelled components:
  Mumps virus Ag bound to well
  Anti-human IgG Ab (g) HRP conjugated
Substrate: ABTS
Controls - standards supplied:
  Controls: Neg
  Calibrator: 1
Additional reagents:
  None
Special equipment required:
  None

INTERPRETATION

Comments on interpretation:
  Equivocal: where sample OD is between 0.385–0.399
  (95–99 U/ml); retest to confirm
No. of references: 3

NOTES

131786.0

Mumps virus
ANTIBODY DETECTION (IgG)

**Manufacturer: BAG-Biologische
Analysensystem GmbH**
Cat. No./Trade name: 5227 BAG - Mumps - EIA - G

SUMMARY

[Well-Ag]–**Ab**–[AHIgG-HRP]–[TMB]–A_{450}

Assay type: EIA (non-competitive)
Detection: Colorimetric A_{450}
Format: Microtitre well, Ag coated
Sample type: Serum
Sample pre-treatment:
  None
Sample volume: 100 µl of 1:101 dilution
Number of tests: 96
Controls - standards run in assay:
  Controls: Neg (1), cut-off (2), Pos (1)
Incubation:
  30 min (RT) + 30 min (RT) + 10 min (RT)
Washes: 2

CONTENTS

Antibodies, antigens, labelled components:
  Mumps virus (Enders strain) Ag bound to well
  Anti-human IgG Ab (sh) HRP conjugated
Substrate: TMB
Controls - standards supplied:
  Controls: Neg, cut-off and Pos (human serum)
Additional reagents:
  None
Special equipment required:
  None

INTERPRETATION

Comments on interpretation:
  Equivocal: within cut-off ±10%; data is not provided for
  further assessment of samples within this range
No. of references: None

NOTES

131605.0

Mumps virus
ANTIBODY DETECTION (IgG)

Manufacturer: Behringwerke AG
Cat. No./Trade name: /Enzygnost® Anti-Parotitis Virus/
IgG

SUMMARY

[Well-Ag]–**Ab**–[AHIgG-POD]–[TMB]–A$_{450}$

Assay type: EIA (non-competitive)
Detection: Colorimetric A$_{450}$
Format: Microtitre well, Ag coated
Sample type: Serum, plasma
Sample pre-treatment:
    None
Sample volume: 20 µl of 1:21 dilution x 2*
Number of tests: 48
Controls - standards run in assay:
    Reference reagent: P/N (2)*
Incubation:
    1 hr (37°C) + 1 hr (37°C) + 30 min (RT)
Washes: 2

CONTENTS

Antibodies, antigens, labelled components:
    Mumps virus Ag from simian kidney cells bound to well
    Control Ag from uninfected simian kidney cells bound to well
    Anti-human IgG Ab (r) POD conjugated
Substrate: TMB (bought separately)
Controls - standards supplied:
    Reference reagent: P/N (Anti-Parotitis Virus Reference Reagent)
Additional reagents:
    Supplementary Reagents for Enzygnost®/TMB (Code No. OUVP)
Special equipment required:
    Behring ELISA Processor II (optional)
    Behring ELISA Processor III (optional)

INTERPRETATION

Comments on interpretation:
    Equivocal: where sample OD is between 0.10 and 0.20; retest to confirm
    Repeatably equivocal: report as equivocal
    For quantitative procedure see manufacturer's instructions
No. of references: 12

NOTES

131023.0

*Samples and controls assayed simultaneously in Ag coated and Control Ag coated wells
If high Ab titres are expected, a further serum dilution is required

Mumps virus
ANTIBODY DETECTION (IgG)

Manufacturer: bioMerieux
Cat. No./Trade name: 30 218/VIDAS Mumps IgG

SUMMARY

[Solid phase-Ag]–**Ab**–[AHIgG-AP]–[4MP]–fluorescence

Assay type: EIA (non-competitive)
Detection: Fluorometric
Format: Solid phase receptacle (SPR), Ag coated
Sample type: Serum (do not heat inactivate)
Sample pre-treatment:
    None
Sample volume: 100 µl
Number of tests: 60
Controls - standards run in assay:
    Controls: Neg (1), Pos (1)
    Standards: Pos (1)
Incubation:
    Automated - total time 40 min
Washes: Automated

CONTENTS

Antibodies, antigens, labelled components:
    Mumps virus (Enders strain) Ag bound to solid phase receptacle
    Anti-human IgG MAb (m) AP conjugated
Substrate: 4-MP
Controls - standards supplied:
    Controls: Neg and Pos (human plasma)
    Standards: Pos (human plasma)
Additional reagents:
    None
Special equipment required:
    Vitek immunodiagnostic assay system (VIDAS)

INTERPRETATION

Comments on interpretation:
    This is an automated assay
    Results designated equivocal; retest to confirm
No. of references: 7

NOTES

131308.0

© KLUWER ACADEMIC PUBLISHERS 1995, ISSN 1381-5067

Mumps virus
ANTIBODY DETECTION (IgG)

Manufacturer: BioWhittaker Inc
Cat. No./Trade name: 30-645U/MUMPS ELISA II

SUMMARY

[Well-Ag]–**Ab**–[AHIgG-AP]–[PMP]–A_{550}

Assay type: EIA (non-competitive)
Detection: Colorimetric A_{550}
Format: Microtitre well, Ag coated
Sample type: Serum (do not heat inactivate)
Sample pre-treatment:
  None
Sample volume: 10 µl (+200 µl diluent)
Number of tests: 192
Controls - standards run in assay:
  Controls: A (1), B (1)
  Calibrators: Neg (1), Pos 1 (1), Pos 2 (1)
Incubation:
  45 min (RT) + 45 min (RT) + 45 min (RT)
Washes: 2 (+preliminary plate wash)

CONTENTS

Antibodies, antigens, labelled components:
  Mumps virus Ag bound to well
  Anti-human IgG Ab (g) AP conjugated
Substrate: PMP
Controls - standards supplied:
  Controls: A and B
  Calibrators: Neg, Pos 1, Pos 2 (all human serum)
Additional reagents:
  None
Special equipment required:
  BioWhittaker automated system and software (optional)

INTERPRETATION

Comments on interpretation:
  Equivocal: Elisa value between 0.15–0.16 or Index value
    between 0.88–0.89; retest to confirm
  Repeatably equivocal: retest a fresh sample or use an
    alternative method
No. of references: 3

NOTES

131740.0

Mumps virus
ANTIBODY DETECTION (IgG)

Manufacturer: BioWhittaker Inc
Cat. No./Trade name: 30-345U/Mumps Stat

SUMMARY

[Well-Ag]–**Ab**–[AHIgG-AP]–[PMP]–A_{550}

Assay type: EIA (non-competitive)
Detection: Colorimetric A_{550}
Format: Microtitre well, Ag coated
Sample type: Serum (do not heat inactivate)
Sample pre-treatment:
  None
Sample volume: 10 µl (+200 µl diluent)
Number of tests: 192
Controls - standards run in assay:
  Controls: Neg (1), Pos (1)
  Standards: Neg (1), low Pos (1), high Pos (1)
Incubation:
  15 min (RT) + 15 min (RT) + 15 min (RT)
Washes: 2 (+preliminary plate wash)

CONTENTS

Antibodies, antigens, labelled components:
  Mumps virus Ag bound to well
  Anti-human IgG Ab (g) AP conjugated
Substrate: PMP
Controls - standards supplied:
  Controls: Neg and Pos
  Standards: Neg, low Pos and high Pos (all human serum)
Additional reagents:
  None
Special equipment required:
  BioWhittaker automated system and software (optional)

INTERPRETATION

Comments on interpretation:
  Equivocal: Predicted Index value (PIV) between 0.80–
    0.90; retest to confirm
  Repeatably equivocal: retest a fresh sample or use
    alternative method
  Low Positive: PIV between 1.00–1.74
  Mid Positive: PIV between 1.75–3.89
  High Positive: PIV $\geqslant 3.90$
No. of references: 16

NOTES

131747.0

Mumps virus
ANTIBODY DETECTION (IgG)

Manufacturer: Chimica Diagnostica
Cat. No./Trade name: 42900/Mumps IgG

SUMMARY

[Well-Ag]–**Ab**–[AHIgG-POD]–[TMB]–A_{450}

Assay type: EIA (non-competitive)
Detection: Colorimetric A_{450}
Format: Microtitre well, Ag coated
Sample type: Serum
Sample pre-treatment:
   None
Sample volume: 100 μl of 1:101 dilution x 2
Number of tests: 96
Controls - standards run in assay:
   Controls: Neg (2), cut-off (2), Pos (2)
Incubation:
   30 min (RT) + 30 min (RT) + 15 min (RT)
Washes: 2

CONTENTS

Antibodies, antigens, labelled components:
   Mumps virus Ag bound to well
   Anti-human IgG Ab (r) POD conjugated
Substrate: TMB
Controls - standards supplied:
   Controls: Neg, cut-off and Pos
Additional reagents:
   None
Special equipment required:
   None

INTERPRETATION

Comments on interpretation:
   Equivocal: where Index (ratio of OD sample:OD cut-off
      control) is between >0.9 and <1.1; report as
      equivocal
No. of references: 2

NOTES

131362.0

Mumps virus
ANTIBODY DETECTION (IgG)

Manufacturer: Clark Laboratories
Cat. No./Trade name: Mumps Virus IgG Elisa Test

SUMMARY

[Well-Ag]–**Ab**–[AHIgG-HRP]–[OPD]–A_{490}

Assay type: EIA (non-competitive)
Detection: Colorimetric A_{490}
Format: Microtitre well, Ag coated
Sample type: Serum (do not heat inactivate)
Sample pre-treatment:
   None
Sample volume: 10 μl (+200 μl diluent)
Number of tests: 96
Controls - standards run in assay:
   Controls: Neg (1), Pos (1)
   Calibrators: 1 (2)
Incubation:
   20 min (RT) + 20 min (RT) + 10 min (RT)
Washes: 2

CONTENTS

Antibodies, antigens, labelled components:
   Mumps virus Ag bound to well
   Anti-human IgG Ab (g) HRP conjugated
Substrate: TMB*
Controls - standards supplied:
   Controls: Neg and Pos (human serum or plasma)
   Calibrator: 1 (human serum or plasma)
Additional reagents:
   H_2SO_4
Special equipment required:
   None

INTERPRETATION

Comments on interpretation:
   Classification of samples is according to cut-off; no
      further testing
No. of references: None

NOTES

131660.0

*OPD can be supplied as an alternative

© KLUWER ACADEMIC PUBLISHERS 1995, ISSN 1381-5067

Mumps virus
ANTIBODY DETECTION (IgG)

Manufacturer: Denka Seiken Co. Ltd
Cat. No./Trade name: 321725/Mumps IgG (II) – EIA "SEIKEN"

SUMMARY

[Well-Ag]–**Ab**–[AHIgG-POD]–[TMB]–A_{450}

Assay type: EIA (non-competitive)
Detection: Colorimetric A_{450}
Format: Microtitre well, Ag coated
Sample type: Serum, plasma
Sample pre-treatment:
  None
Sample volume: 10 µl (+2 ml diluent)
Number of tests: 96
Controls - standards run in assay:
  Standards: I (1), II (2), III (1), IV (1)
Incubation:
  1 hr (RT) + 1 hr (RT) + 30 min (RT)
Washes: 2

CONTENTS

Antibodies, antigens, labelled components:
  Mumps virus Ag bound to well
  Anti-human IgG Ab (g) POD conjugated
Substrate: TMB
Controls - standards supplied:
  Standards: I–IV (EIA values 0–3200, human serum)
Additional reagents:
  None
Special equipment required:
  None

INTERPRETATION

Comments on interpretation:
  Equivocal: where EIA value is between 200 and 400;
    repeat test with a fresh specimen in 1–2 weeks or use
    an alternative method
No. of references: None

NOTES

131687.0

Mumps virus
ANTIBODY DETECTION (IgG)

Manufacturer: Gamma SA
Cat. No./Trade name: Mumps Virus IgG Elisa

SUMMARY

[Well-Ag]–**Ab**–[AHIgG-POD]–[TMB]–A_{450}

Assay type: EIA (non-competitive)
Detection: Colorimetric A_{450}
Format: Microtitre well, Ag coated
Sample type: Serum
Sample pre-treatment:
  None
Sample volume: 10 µl (+1 ml diluent)
Number of tests: 96
Controls - standards run in assay:
  Controls: Neg (2), Pos (2)
Incubation:
  30 min (RT) + 30 min (RT) + 15 min (RT)
Washes: 2

CONTENTS

Antibodies, antigens, labelled components:
  Mumps virus Ag bound to well
  Anti-human IgG Ab (r) POD conjugated
Substrate: TMB
Controls - standards supplied:
  Controls: Neg and Pos
Additional reagents:
  None
Special equipment required:
  None

INTERPRETATION

Comments on interpretation:
  Classification of samples is according to cut-off; no
    further testing
No. of references: 0

NOTES

131177.0

© KLUWER ACADEMIC PUBLISHERS 1995, ISSN 1381-5067

Mumps virus
ANTIBODY DETECTION (IgG)

Manufacturer: Human GmbH
Cat. No./Trade name: 51207/Mumps-Virus - IgG - Elisa

SUMMARY

[well-Ag]–**Ab**–[AHIgG-HRP]–[TMB]–A_{450}

Assay type: EIA (non-competitive)
Detection: Colorimetric A_{450}
Format: Microtitre well, Ag coated
Sample type: Serum
Sample pre-treatment:
  None
Sample volume: 100 µl of 1:100 dilution
Number of tests: 96
Controls - standards run in assay:
  Controls: Neg (2), Pos (2)
Incubation:
  30 min (RT) + 30 min (RT) + 15 min (RT)
Washes: 2

CONTENTS

Antibodies, antigens, labelled components:
  Mumps virus Ag bound to well
  Anti-human IgG Ab (r) HRP conjugated
Substrate: TMB
Controls - standards supplied:
  Controls: Neg and Pos (human)
Additional reagents:
  None
Special equipment required:
  None

INTERPRETATION

Comments on interpretation:
  Equivocal: within cut-off ±15%; retest in parallel with a
  fresh sample taken 7–14 days later
No. of references: 5

NOTES

131343.0

Mumps virus
ANTIBODY DETECTION (IgG)

Manufacturer: Immunobiological
Laboratories
Cat. No./Trade name: VE 57141/Mumps IgG

SUMMARY

[Well-Ag]–**Ab**–[AHIgG-HRP]–[TMB]–A_{450}

Assay type: EIA (non-competitive)
Detection: Colorimetric A_{450}
Format: Microtitre well, Ag coated
Sample type: Serum, plasma
Sample pre-treatment:
  None
Sample volume: 100 µl of 1:100 dilution x 2
Number of tests: 96
Controls - standards run in assay:
  Controls: Neg (2), low Pos (2), high Pos (2)
Incubation:
  1 hr (RT) + 30 min (RT) + 15 min (RT)
Washes: 2

CONTENTS

Antibodies, antigens, labelled components:
  Mumps virus Ag bound to well
  Anti-human IgG Ab HRP conjugated
Substrate: TMB
Controls - standards supplied:
  Controls: Neg, low Pos and high Pos (serum)
Additional reagents:
  None
Special equipment required:
  None

INTERPRETATION

Comments on interpretation:
  Equivocal: within cut-off ±20%; no data is provided for
  further classification of samples within this range
No. of references: 7

NOTES

131251.0

© KLUWER ACADEMIC PUBLISHERS 1995, ISSN 1381-5067

Mumps virus
ANTIBODY DETECTION (IgG)

Manufacturer: Laboratoire Eurobio
Cat. No./Trade name: 900273/Mumps IgG ELIT®

SUMMARY

[Well-Ag]–**Ab**–[AHIgG-POD]–[TMB]–A_{450}

Assay type: EIA (non-competitive)
Detection: Colorimetric A_{450}
Format: Microtitre well, Ag coated
Sample type: Serum, plasma
Sample pre-treatment:
   None
Sample volume: 100 µl of 1:201 dilution
Number of tests: 96
Controls - standards run in assay:
   Standards: I (1), II (2), III (1), IV (1)
Incubation:
   1 hr (RT) + 1 hr (RT) + 1 hr (RT) + 30 min (RT)
Washes: 2

CONTENTS

Antibodies, antigens, labelled components:
   Mumps virus Ag bound to well
   Anti-human IgG Ab (g) POD conjugated
Substrate: TMB
Controls - standards supplied:
   Standards: 4 (human serum)
Additional reagents:
   None
Special equipment required:
   None

INTERPRETATION

Comments on interpretation:
   Equivocal: where EIA value of sample is between 200
      and 400; retest a fresh sample in 1–2 weeks
No. of references: None

NOTES

131729.0

Mumps virus
ANTIBODY DETECTION (IgG)

Manufacturer: Medix Biotech Inc
Cat. No./Trade name: KBF 2067/Mumps IgG EIA Kit

SUMMARY

[Well-Ag]–**Ab**–[AHIgG-HRP]–[OPD]–A_{490}

Assay type: EIA (non-competitive)
Detection: Colorimetric A_{490}
Format: Microtitre well, Ag coated
Sample type: Serum (do not heat inactivate)
Sample pre-treatment:
   None
Sample volume: 10 µl (+200 µl diluent)
Number of tests: 96
Controls - standards run in assay:
   Controls: Neg (1), high Pos (1)
   Calibrator: low Pos (4)
Incubation:
   20 min (RT) + 20 min (RT) + 10 min (RT)
Washes: 2

CONTENTS

Antibodies, antigens, labelled components:
   Mumps virus Ag bound to well
   Anti-human IgG Ab (g) HRP conjugated
Substrate: OPD
Controls - standards supplied:
   Controls: Neg and high Pos (human serum or plasma)
   Calibrator: low Pos (human serum or plasma)
Additional reagents:
   H_2SO_4
Special equipment required:
   None

INTERPRETATION

Comments on interpretation:
   Equivocal: where ratio of sample OD:cut-off value is
      between 0.9-1.10; report as equivocal
No. of references: None

NOTES

131171.0

© KLUWER ACADEMIC PUBLISHERS 1995, ISSN 1381-5067

Mumps virus
ANTIBODY DETECTION (IgG)

Manufacturer: Melotec S.A.
Cat. No./Trade name: /Melotest Mumps IgG

SUMMARY

$$[\text{Well-Ag}]-\textbf{Ab}-[\text{AHIgG-HRP}]-[\text{TMB}]-\text{A}_{450}$$

Assay type: EIA (non-competitive)
Detection: Colorimetric A_{450}
Format: Microtitre well, Ag coated
Sample type: Serum, plasma
Sample pre-treatment:
  None
Sample volume: 100 μl of 1:20 dilution x 2
Number of tests: 96
Controls - standards run in assay:
  Controls: Neg (1), low Pos (2), high Pos (1)
Incubation:
  20 min (RT) + 20 min (RT) + 10 min (RT)
Washes: 2

CONTENTS

Antibodies, antigens, labelled components:
  Mumps virus Ag bound to well
  Anti-human IgG Ab HRP conjugated
Substrate: TMB
Controls - standards supplied:
  Controls: Neg, low Pos and high Pos (sera)
Additional reagents:
  None
Special equipment required:
  None

INTERPRETATION

Comments on interpretation:
  Equivocal: ratio of sample OD:cut-off value between 0.9
    and 1.1; retest a fresh sample
  Repeatably equivocal: retest a fresh sample taken in 2-4
    weeks
No. of references: 7

NOTES

131125.0

Mumps virus
ANTIBODY DETECTION (IgG)

Manufacturer: Radim
Cat. No./Trade name: K8PG/Mumps IgG EIA Well

SUMMARY

$$[\text{Well-Ag}]-\textbf{Ab}-[\text{AHIgG-HRP}]-[\text{TMB}]-\text{A}_{450}$$

Assay type: EIA (non-competitive)
Detection: Colorimetric A_{450}
Format: Microtitre well, Ag coated
Sample type: Serum, plasma
Sample pre-treatment:
  None
Sample volume: 100 μl of 1:300 dilution
Number of tests: 96
Controls - standards run in assay:
  Controls: Neg (1), cut-off (1), Pos (1)
Incubation:
  1 hr (37°C) + 30 min (37°C) + 10 min (37°C) or 15 min
    (RT)
Washes: 2

CONTENTS

Antibodies, antigens, labelled components:
  Mumps virus Ag bound to well
  Anti-human IgG Ab (g) HRP conjugated
Substrate: TMB
Controls - standards supplied:
  Controls: Neg, cut-off and Pos (human serum)
Additional reagents:
  None
Special equipment required:
  None

INTERPRETATION

Comments on interpretation:
  Negative: < cut-off
  Grey area: within cut-off ±10%; retest to confirm
  Positive: > cut-off
No. of references: 7

NOTES

131273.0

© KLUWER ACADEMIC PUBLISHERS 1995, ISSN 1381-5067

Mumps virus
ANTIBODY DETECTION (IgM)

Manufacturer: Amico Laboratories Inc
Cat. No./Trade name: 6300M/AMIZYME® Mumps virus IgM Assay

SUMMARY

[Well-Ag]–**Ab**–[AHIgM-HRP]–[ABTS]–A_{405}

Assay type: EIA (non-competitive)
Detection: Colorimetric A_{405}
Format: Microtitre well, Ag coated
Sample type: Serum
Sample pre-treatment:
   None
Sample volume: 5 µl (+245 µl diluent)*
Number of tests: 96
Controls - standards run in assay:
   Controls: Neg (1)
   Calibrators: I (1)
Incubation:
   25 min (RT) + 25 min (RT) + 25 min (RT)
Washes: 2

CONTENTS

Antibodies, antigens, labelled components:
   Mumps virus Ag bound to well
   Anti-human IgM Ab (g) HRP conjugated
Substrate: ABTS
Controls - standards supplied:
   Controls: Neg
   Calibrator: 1
Additional reagents:
   None
Special equipment required:
   None

INTERPRETATION

Comments on interpretation:
   Equivocal: where sample OD is between 0.275–0.299
      (92–99 U/ml); retest to confirm
No. of references: 3

NOTES

131794.0

*Sample diluted in IgM serum diluent provided to eliminate interference due to RF and IgG

Mumps virus
ANTIBODY DETECTION (IgM)

Manufacturer: Amico Laboratories Inc
Cat. No./Trade name: 6300MC/AMIZYME® Mumps virus IgM Capture Assay

SUMMARY

[Well-Ag]–**Ab**–[AHIgM]–[AMIg-HRP]–[ABTS]–A_{405}

Assay type: EIA (non-competitive)
Detection: Colorimetric A_{405}
Format: Microtitre well, Ag coated
Sample type: Serum
Sample pre-treatment:
   None
Sample volume: 5 µl (+245 µl diluent)
Number of tests: 96
Controls - standards run in assay:
   Controls: Neg (1)
   Calibrators: I (1)
Incubation:
   25 min (RT) + 25 min (RT) + 25 min (RT) + 25 min
      (RT)
Washes: 3

CONTENTS

Antibodies, antigens, labelled components:
   Mumps virus Ag bound to well
   Anti-human IgM MAb (m)
   Anti-mouse Ig Ab (g) HRP conjugated
Substrate: ABTS
Controls - standards supplied:
   Controls: Neg
   Calibrator: 1
Additional reagents:
   None
Special equipment required:
   None

INTERPRETATION

Comments on interpretation:
   Equivocal: where sample OD is between 0.285–0.299;
      retest to confirm
No. of references: 2

NOTES

131804.0

© *KLUWER ACADEMIC PUBLISHERS 1995, ISSN 1381-5067*

Mumps virus
ANTIBODY DETECTION (IgM)

Manufacturer: BAG-Biologische Analysensystem GmbH
Cat. No./Trade name: 5228 BAG - Mumps - EIA - M

SUMMARY

[Well-Ag]–**Ab**–[AHIgM-HRP]–[TMB]–A_{450}

Assay type: EIA (non-competitive)
Detection: Colorimetric A_{450}
Format: Microtitre well, Ag coated
Sample type: Serum (do not heat inactivate)
Sample pre-treatment:
   Treat diluted serum with BAG - IgM - Sep - System (Cat. no. 5540) to eliminate interference due to RF and IgG
Sample volume: 100 μl of treated sample
Number of tests: 96
Controls - standards run in assay:
   Controls: Neg (1), cut-off (2), Pos (1)
Incubation:
   30 min (RT) + 30 min (RT) + 10 min (RT)
Washes: 2

CONTENTS

Antibodies, antigens, labelled components:
   Mumps virus (Enders strain) Ag bound to well
   Anti-human IgM Ab (sh) HRP conjugated
Substrate: TMB
Controls - standards supplied:
   Controls: Neg, cut-off and Pos (human serum)
Additional reagents:
   None
Special equipment required:
   None

INTERPRETATION

Comments on interpretation:
   Equivocal: within cut-off ± 10%; data is not provided for further assessment of samples within this range
No. of references: None

NOTES

131610.0

Mumps virus
ANTIBODY DETECTION (IgM)

Manufacturer: Behringwerke AG
Cat. No./Trade name: Enzygnost® Anti-Parotitis virus/ IgM

SUMMARY

[Well-Ag]–**Ab**–[AHIgM-POD]–[TMB]–A_{450}

Assay type: EIA (non-competitive)
Detection: Colorimetric A_{450}
Format: Microtitre well, Ag coated
Sample type: Serum, plasma
Sample pre-treatment:
   Add RF absorbent provided to diluted sample, incubate 15 min (RT) or overnight (2–8°C) to eliminate interference due to RF and IgG
Sample volume: 150 μl of treated sample x 2*
Number of tests: 96
Controls - standards run in assay:
   Reference reagents: P/N (2), PP (2)*
Incubation:
   1 hr (37°C) + 1 hr (37°C) + 30 min (RT)
Washes: 2

CONTENTS

Antibodies, antigens, labelled components:
   Anti-mumps virus from simian kidney cells bound to well
   Control Ag from uninfected simian kidney cells bound to well
   Anti-human IgM Ab (g) POD conjugated
Substrate: TMB (bought separately)
Controls - standards supplied:
   Reference reagent: P/P and P/N (Anti-Parotitis Virus Reference Reagents)
Additional reagents:
   Supplementary reagents for Enzygnost/TMB (Cat. no. OUVP)
Special equipment required:
   Behring ELISA Processor II (optional)
   Behring ELISA Processor III (optional)

INTERPRETATION

Comments on interpretation:
   Equivocal: where sample OD is between 0.10 and 0.20; retest to confirm
   Repeatably equivocal: report as equivocal and retest a fresh specimen at a later date
No. of references: 12

NOTES

131027.0

*Samples and controls are assayed simultaneously in Ag coated and control Ag coated wells

© *KLUWER ACADEMIC PUBLISHERS 1995, ISSN 1381-5067*

Mumps virus
ANTIBODY DETECTION (IgM)

Manufacturer: Chimica Diagnostica
Cat. No./Trade name: 42950/Mumps IgM

SUMMARY

[Well-Ag]–**Ab**–[AHIgM-POD]–[TMB]–A_{450}

Assay type: EIA (non-competitive)
Detection: Colorimetric A_{450}
Format: Microtitre well, Ag coated
Sample type: Serum
Sample pre-treatment:
  None
Sample volume: 100 µl of 1:101 dilution x 2*
Number of tests: 96
Controls - standards run in assay:
  Controls: Neg (2), cut-off (2), Pos (2)
Incubation:
  30 min (RT) + 30 min (RT) + 10 min (RT)
Washes: 2

CONTENTS

Antibodies, antigens, labelled components:
  Mumps virus Ag bound to well
  Anti-human IgM Ab (r) POD conjugated
Substrate: TMB
Controls - standards supplied:
  Controls: Neg, cut-off and Pos
Additional reagents:
  None
Special equipment required:
  None

INTERPRETATION

Comments on interpretation:
  Equivocal: where Index (ratio of OD sample:OD cut-off control) is between > 0.9 and < 1.1; report as equivocal
No. of references: 2

NOTES

131365.0

*Diluent contains AHIgG to eliminate interference due to RF and IgG in sample

Mumps virus
ANTIBODY DETECTION (IgM)

Manufacturer: Clark Laboratories
Cat. No./Trade name: Mumps Virus IgM Elisa Test

SUMMARY

[Well-Ag]–**Ab**–[AHIgM-HRP]–[OPD]–A_{490}

Assay type: EIA (non-competitive)
Detection: Colorimetric A_{490}
Format: Microtitre well, Ag coated
Sample type: Serum (do not heat inactivate)
Sample pre-treatment:
  Add Absorbent Solution provided to diluted serum and incubate 20 min (RT) to eliminate interference due to RF and IgG
Sample volume: 100 µl of treated sample
Number of tests: 96
Controls - standards run in assay:
  Controls: Neg (1), Pos (1)
  Calibrators: 1 (2)
Incubation:
  20 min (RT) + 20 min (RT) + 10 min (RT)
Washes: 2

CONTENTS

Antibodies, antigens, labelled components:
  Mumps virus Ag bound to well
  Anti-human IgM Ab (g) HRP conjugated
Substrate: TMB*
Controls - standards supplied:
  Controls: Neg and Pos (human serum or plasma)
  Calibrator: 1 (human serum or plasma)
Additional reagents:
  H_2SO_4
Special equipment required:
  None

INTERPRETATION

Comments on interpretation:
  Classification of samples is according to cut-off; no further testing
No. of references: None

NOTES

131666.0

*OPD can be supplied as an alternative

© KLUWER ACADEMIC PUBLISHERS 1995, ISSN 1381-5067

Mumps virus
ANTIBODY DETECTION (IgM)

Manufacturer: Denka Seiken Co. Ltd
Cat. No./Trade name: 321787/Mumps IgM (II) – EIA "SEIKEN"

SUMMARY

[Well-AHIgM]–**Ab**–[Ag]–[MAb-POD]–[TMB]–A$_{450}$

Assay type: EIA (non-competitive)
Detection: Colorimetric A$_{450}$
Format: Microtitre well, Ab coated
Sample type: Serum, plasma
Sample pre-treatment:
  None
Sample volume: 10 μl (+2 ml diluent)
Number of tests: 96
Controls - standards run in assay:
  Standards: I (1), II (2), III (1)
Incubation:
  1 hr (RT) + 1 hr (RT) + 1 hr (RT) + 30 min (RT)
Washes: 3

CONTENTS

Antibodies, antigens, labelled components:
  Anti-human IgM MAb (m) bound to well
  Mumps virus Ag
  Anti-mumps virus MAb (m) POD conjugated
Substrate: TMB
Controls - standards supplied:
  Standards: I (Neg), II (low titre) and III (high titre) (all
    human serum)
Additional reagents:
  None
Special equipment required:
  None

INTERPRETATION

Comments on interpretation:
  Equivocal: where the calculated IgM Antibody Index is
    between 0.8 and 1.2; repeat test with a fresh
    specimen in 1–2 weeks or use an alternative method
No. of references: None

NOTES

131688.0

Mumps virus
ANTIBODY DETECTION (IgM)

Manufacturer: Gamma SA
Cat. No./Trade name: Mumps Virus IgM Elisa

SUMMARY

[Well-Ag]–**Ab**–[AHIgM-POD]–[TMB]–A$_{450}$

Assay type: EIA (non-competitive)
Detection: Colorimetric A$_{450}$
Format: Microtitre well, Ag coated
Sample type: Serum
Sample pre-treatment:
  Dilute serum with dilution buffer to eliminate interference
    due to RF and IgG
Sample volume: 100 μl of prepared sample
Number of tests: 96
Controls - standards run in assay:
  Controls: Neg (2), Pos (2)
Incubation:
  30 min (RT) + 30 min (RT) + 15 min (RT)
Washes: 2

CONTENTS

Antibodies, antigens, labelled components:
  Mumps virus Ag bound to well
  Anti-human IgM Ab (r) POD conjugated
Substrate: TMB
Controls - standards supplied:
  Controls: Neg and Pos
Additional reagents:
  None
Special equipment required:
  None

INTERPRETATION

Comments on interpretation:
  Classification of samples is according to cut-off; no
    further testing
No. of references: 0

NOTES

131176.0

© *KLUWER ACADEMIC PUBLISHERS 1995, ISSN 1381-5067*

Mumps virus
ANTIBODY DETECTION (IgM)

Manufacturer: Human GmbH
Cat. No./Trade name: 51107/Mumps Virus IgM Elisa

SUMMARY

[Well-Ag]–**Ab**–[AHIgM-HRP]–[TMB]–A$_{450}$

Assay type: EIA (non-competitive)
Detection: Colorimetric A$_{450}$
Format: Microtitre well, Ag coated
Sample type: Serum
Sample pre-treatment:
Dilute sample with Dilution Buffer IgM and incubate 5 min (RT) to eliminate interference due to RF and IgG
Sample volume: 100 µl of prepared sample
Number of tests: 96
Controls - standards run in assay:
Controls: Neg (2), Pos (2)
Incubation:
30 min (RT) + 30 min (RT) + 15 min (RT)
Washes: 2

CONTENTS

Antibodies, antigens, labelled components:
Mumps virus Ag bound to well
Anti-human IgM Ab (r) HRP conjugated
Substrate: TMB
Controls - standards supplied:
Controls: Neg and Pos (human)
Additional reagents:
None
Special equipment required:
None

INTERPRETATION

Comments on interpretation:
Equivocal: within cut-off ± 15%; retest in parallel with a fresh sample taken 7–14 days later
No. of references: 5

NOTES

131347.0

Mumps virus
ANTIBODY DETECTION (IgM)

Manufacturer: Immunobiological Laboratories
Cat. No./Trade name: VE 57151/Mumps IgM Elisa Test (HRP)

SUMMARY

[Well-Ag]–**Ab**–[AHIgM-HRP]–[TMB]–A$_{450}$

Assay type: EIA (non-competitive)
Detection: Colorimetric A$_{450}$
Format: Microtitre well, Ag coated
Sample type: Serum
Sample pre-treatment:
It is advised to pre-incubate sample with an absorbent to remove interference due to RF, IgG and ANA (bought separately)
Sample volume: 100 µl of 1:100 dilution x 2
Number of tests: 96
Controls - standards run in assay:
Controls: Neg (2), low Pos (2), high Pos (2)
Incubation:
1 hr (RT) + 30 min (RT) + 15 min (RT)
Washes: 2

CONTENTS

Antibodies, antigens, labelled components:
Mumps virus Ag bound to well
Anti-human IgM Ab HRP conjugated
Substrate: TMB
Controls - standards supplied:
Controls: Neg, low Pos and high Pos (serum)
Additional reagents:
Absorbent for removal of RF, IgG and ANA (available from manufacturer)
Special equipment required:
None

INTERPRETATION

Comments on interpretation:
Equivocal: within cut-off ± 5%; no data is provided for further assessment of samples within this range
No. of references: 7

NOTES

131260.0

Mumps virus
ANTIBODY DETECTION (IgM)

Manufacturer: Laboratoire Eurobio
Cat. No./Trade name: 900274/Mumps IgM ELIT®

SUMMARY

[Well-AHIgM]–**Ab**–[Ag]–[Ab-POD]–[TMB]–A_{450}

Assay type: EIA (non-competitive)
Detection: Colorimetric A_{450}
Format: Microtitre well, Ab coated
Sample type: Serum, plasma
Sample pre-treatment:
    None
Sample volume: 100 µl of 1:201 dilution
Number of tests: 96
Controls - standards run in assay:
    Standards: I (1), II (2), III (1)
Incubation:
    1 hr (RT) + 1 hr (RT) + 1 hr (RT) + 30 min (RT)
Washes: 3

CONTENTS

Antibodies, antigens, labelled components:
    Anti-human IgM MAb bound to well
    Mumps virus Ag
    Anti-mumps virus Ab POD conjugated
Substrate: TMB
Controls - standards supplied:
    Standards: 3 (human serum)
Additional reagents:
    None
Special equipment required:
    None

INTERPRETATION

Comments on interpretation:
    Equivocal: where IgM index is beween 0.8 and 1.2; retest
    a fresh sample in 1–2 weeks or test using an
    alternative method
No. of references: None

NOTES

131731.0

Mumps virus
ANTIBODY DETECTION (IgM)

Manufacturer: Medix Biotech Inc
Cat. No./Trade name: KBF 2068-2/Mumps IgM EIA Kit

SUMMARY

[Well-Ag]–**Ab**–[AHIgM-HRP]–[OPD]–A_{490}

Assay type: EIA (non-competitive)
Detection: Colorimetric A_{490}
Format: Microtitre well, Ag coated
Sample type: Serum (do not heat inactivate)
Sample pre-treatment:
    Add absorbent solution provided to diluted controls,
    calibrator and samples, incubate 20 min (RT) to
    eliminate interference due to RF & IgG
Sample volume: 100 µl of treated sample x 2*
Number of tests: 96
Controls - standards run in assay:
    Controls: Neg (2), high Pos (2)*
    Calibrators: low Pos (4)
Incubation:
    20 min (RT) + 20 min (RT) + 10 min (RT)
Washes: 2

CONTENTS

Antibodies, antigens, labelled components:
    Mumps Ag bound to well
    Control Ag bound to well
    Anti-human IgM Ab (g) HRP conjugated
Substrate: OPD
Controls - standards supplied:
    Controls: Neg and high Pos (human serum or plasma)
    Calibrator: low Pos (human serum or plasma)
Additional reagents:
    H_2SO_4
Special equipment required:
    None

INTERPRETATION

Comments on interpretation:
    Equivocal: where ratio of sample OD:cut-off value is
    between 0.9-1.10; report as equivocal
No. of references: None

NOTES

131172.0

*Samples and controls are assayed simultaneously in Ag
coated and Control Ag coated wells

© KLUWER ACADEMIC PUBLISHERS 1995, ISSN 1381-5067

Mumps virus
ANTIBODY DETECTION (IgM)

Manufacturer: Radim
Cat. No./Trade name: K8PM/Mumps IgM EIA Well

SUMMARY

[Well-Ag]–**Ab**–[AHIgM-HRP]–[TMB]–A_{450}

Assay type: EIA (non-competitive)
Detection: Colorimetric A_{450}
Format: Microtitre well, Ag coated
Sample type: Serum, plasma
Sample pre-treatment:
 None
Sample volume: 100 µl of 1:300 dilution x 2
Number of tests: 96
Controls - standards run in assay:
 Controls: Neg (2), cut-off (2), Pos (2)
Incubation:
 1 hr (37°C) + 30 min (37°C) + 10 min (37°C) or 15 min
 (RT)
Washes: 2

CONTENTS

Antibodies, antigens, labelled components:
 Mumps virus Ag bound to well
 Anti-human IgM Ab (g) HRP conjugated
Substrate: TMB
Controls - standards supplied:
 Controls: Neg, cut-off and Pos (human serum)
Additional reagents:
 None
Special equipment required:
 None

INTERPRETATION

Comments on interpretation:
 Equivocal: within cut-off ±10%; retest to confirm
 Positive: > cut-off and retested to confirm after treatment
 of specimen with Protein A (Cat. no. M23)
No. of references: 7

NOTES

131276.0

Mumps virus
ANTIBODY DETECTION (IgG)

Manufacturer: Bion Enterprises Ltd
Cat. No./Trade name: /Mumps-G Antibody Test System

SUMMARY

[Slide-Ag]–**Ab**–[AHIgG-FITC]–fluorescence

Assay type: Immunofluorescence assay (indirect)
Detection: Fluorescence microscopy
Format: Slide well, Ag coated
Sample type: Serum
Sample pre-treatment:
 None
Sample volume: 50 µl (+450 µl diluent)
Number of tests: 60, 120
Controls - standards run in assay:
 Controls: Neg (1), Pos (1)
Incubation:
 30 min (RT) + 30 min (RT)
Washes: 2

CONTENTS

Antibodies, antigens, labelled components:
 Mumps virus (CDC strain) infected and uninfected HEp-2
 culture cells bound to slide well
 Anti-human IgG Ab (g) FITC conjugated
Substrate:
Controls - standards supplied:
 Controls: Neg and Pos (human serum)
Additional reagents:
 None
Special equipment required:
 None

INTERPRETATION

Comments on interpretation:
 Positive: ≥(1+) fluorescence in 10–50% of cells at
 ≥1:10 dilution
No. of references: 14

NOTES

131289.0

Mumps virus
ANTIBODY DETECTION (IgG)

Manufacturer: Hemagen Diagnostics Inc
Cat. No./Trade name: 3200/VIRGO® Mumps IgG IFA

SUMMARY

[Slide well-Ag]–**Ab**–[AHIgG-FITC]–fluorescence

Assay type: Immunofluorescence assay (indirect)
Detection: Fluorescence microscopy
Format: Slide well, Ag coated
Sample type: Serum
Sample pre-treatment:
  None
Sample volume: 10–20 μl of 1:8 dilution x2
Number of tests: 96
Controls - standards run in assay:
  Controls: Neg (1), Pos (1)
Incubation:
  30 min (RT) + 30 min (RT)
Washes: 2

CONTENTS

Antibodies, antigens, labelled components:
  Mumps virus infected and uninfected cells bound to slide
    well
  Anti-human IgG Ab (g) FITC conjugated
Substrate:
Controls - standards supplied:
  Controls: Neg and Pos (human serum)
Additional reagents:
  None
Special equipment required:
  None

INTERPRETATION

Comments on interpretation:
  Positive: ≥(1+) fluorescence at 1:8 dilution
No. of references: 9

NOTES

131601.0

Mumps virus
ANTIBODY DETECTION (IgG)

Manufacturer: Stellar Bio Systems Inc
Cat. No./Trade name: Indirect Fluorescent Assay for
Mumps Virus IgG Antibody

SUMMARY

[Slide-Ag]–**Ab**–[AHIgG-FITC]–fluorescence

Assay type: Immunofluorescence assay (indirect)
Detection: Fluorescence microscopy
Format: Slide well, Ag coated
Sample type: Serum
Sample pre-treatment:
  None
Sample volume: 20 μl of 1:10 dilution
Number of tests:
Controls - standards run in assay:
  Controls: Neg (1), Pos (1)
Incubation:
  30 min (37°C) + 30 min (37°C)
Washes: 2

CONTENTS

Antibodies, antigens, labelled components:
  Mumps virus infected and uninfected cells (fibroblasts)
    bound to slide
  Anti-human IgG Ab (g) FITC conjugated
Substrate:
Controls - standards supplied:
  Controls: Neg and Pos (human serum)
Additional reagents:
  None
Special equipment required:
  None

INTERPRETATION

Comments on interpretation:
  Positive: (1+) – (4+) bright green fluorescent staining of
    infected cells
No. of references: 9

NOTES

131632.0

Mumps virus
ANTIBODY DETECTION (IgM)

Manufacturer: Stellar Bio Systems Inc
Cat. No./Trade name: Indirect Fluorescent Assay for Mumps virus IgM Antibody

SUMMARY

[Slide-Ag]–**Ab**–[AHIgM-FITC]–fluorescence

Assay type: Immunofluorescence assay (indirect)
Detection: Fluorescence microscopy
Format: Slide well, Ag coated
Sample type: Serum
Sample pre-treatment:
    None
Sample volume: 20 µl of 1:5 dilution
Number of tests:
Controls - standards run in assay:
    Controls: Neg (1), Pos (1)
Incubation:
    1 hr (37°C) + 1 hr (37°C)
Washes: 2

CONTENTS

Antibodies, antigens, labelled components:
    Mumps virus infected and uninfected cells (fibroblasts) bound to slide
    Anti-human IgM Ab (g) FITC conjugated
Substrate:
Controls - standards supplied:
    Controls: Neg and Pos (human serum)
Additional reagents:
    None
Special equipment required:
    None

INTERPRETATION

Comments on interpretation:
    Positive: (1+) – (4+) bright green fluorescent staining of infected cells
No. of references: 9

NOTES

131638.0

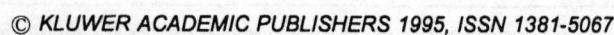

Neisseria meningitidis

Neisseria meningitidis is a bean-shaped, oxidase-positive, Gram-negative diplococcus and is the causal organism of meningococcal meningitis which occurs most commonly in children and young adults. The organism colonises the mucosa of the nasopharynx from where it may be transmitted by respiratory secretions; asymptomatic carriage is much more frequent than the occurrence of clinical disease. Infection is worldwide in distribution and the majority of cases occur in the winter and spring months.

Meningococcal meningitis presents with headache, confusion, neck stiffness and photophobia. Although focal cerebral lesions are less commonly found than in other acute bacterial meningitides, the disease is often complicated by septicaemia which may be fulminant and can be accompanied by the development of widespread petechiae and a characteristic purpuric rash. Septic shock may supervene and severe haemorrhage and necrosis of a number of organs may occur, with the development of haemorrhagic infarction of the suprarenal glands, the renal cortex, and intravascular coagulation being particularly ominous prognostic signs.

Meningococci are nutritionally fastidious, and may be cultured on chocolate blood agar, or the more selective NYC agar. The organism grows well in commercial blood culture media. Aerobic incubation for 18 hours at 36°C yields round, smooth, moist, colourless, convex colonies of 1 mm diameter, which may appear mucoid. Presumptive oxidase-positive colonies yielding organisms of characteristic morphology are further identified on the basis of inability to grow at room temperature, or on nutrient agar, and ability to ferment glucose and maltose, but not lactose, sucrose or fructose. Biochemical confirmation is now largely performed by rapid carbohydrate utilisation tests.

Thirteen serogroups of meningococci (A, B, C, D, 29E, H, I, K, L, W135, X, Y and Z) are recognised on the basis of polysaccharide capsular or outer membrane protein antigens, of which A, B, C, Y and W135 are most commonly associated with systemic disease. A variety of other conventional and molecular typing strategies have been employed to study the epidemiology of meningococcal disease.

Diagnosis

Definitive diagnosis of invasive meningococcal infection is established by the isolation of the organism, as outlined above, from a normally sterile body site such as CSF, blood, synovial, pleural or pericardial fluids. The organism may also be cultured from petechiae and other skin lesions. The examination of a gram-stained smear may be useful in the presumptive diagnosis of the disease. The organism may be coincidentally cultured from the nasopharynx in healthy individuals, and the demonstration of such carriage may be important in epidemic situations and in close contacts of cases.

Recently techniques have been developed for the rapid detection of meningococci or their products in various body fluids, particularly in those cases where direct microscopic examination and/or culture has been negative. Countercurrent-immunoelectrophoresis, and various agglutination and coagglutination tests for the detection of bacterial antigen have been described, and a number of commercial kits are available. Although these techniques have proven to be very useful, care is required in their interpretation as false-negative reactions are not uncommon, and cross-reactions with other bacteria (particularly type B and *E.coli* K1) occur. Blood, CSF and other fluids may be examined, although antigen may be more persistent in concentrated urine samples. More recently, PCR-based methodology for detection in blood and CSF has demonstrated considerable promise.

Reference

Morello JA, Janda WM, Doern GV. *Neisseria* and *Branhamella*. In: Balows A, Hausler WJ, Hermann KL, Isenberg HD, Shadomy HJ, eds. Manual of Clinical Microbiology, 5th edn. Washington: ASM. 1991:258–76.

See also Multipathogen Assays section under: Meningitis pathogens

Neisseria meningitidis
ANTIGEN DETECTION

Manufacturer: Becton Dickinson
Cat. No./Trade name: 8501-60/Directigen® N. meningitidis Groups A,C,Y and WI35 Kit

SUMMARY

[Latex-Ab]–**Ag**–agglutination

Assay type: Particle agglutination assay
Detection: Visual
Format: Latex particles, Ab coated (test done on slide)
Sample type: CSF, serum, urine, blood culture
Sample pre-treatment:
Serum, urine, blood culture media: dilute, heat (100°C) and centrifuge
CSF: heat (100°C) unless N. meningitidis (group B) is suspected
Sample volume: 50 µl of treated sample x 2
Number of tests: 90
Controls - standards run in assay:
Controls: Neg (2), Pos (2)**
Incubation:
Read result 10 min after mixing sample and reagents
Washes:

CONTENTS

Antibodies, antigens, labelled components:
Anti-N. meningitidis (groups A,C,Y and WI35) Ab bound to latex particles (Test Latex)
Non-immune rabbit Ig Ab bound to latex particles (Control Latex)
Substrate:
Controls - standards supplied:
Controls: Neg and Pos (polyvalent)
Additional reagents:
Directigen® Specimen Buffer (Cat. no. 8563-91)
Special equipment required:
Directigen® Test Slides (Cat. no. 8507-79)

INTERPRETATION

Comments on interpretation:
Positive: agglutination with the Test Latex and no agglutination with the Control Latex
No. of references: 15

NOTES

131232.0

© KLUWER ACADEMIC PUBLISHERS 1995, ISSN 1381-5067

Plasmodium species

Four species of the protozoan genus *Plasmodium* (*P.vivax*, *P.ovale*, *P.malariae* and *P.falciparum*) infect humans, and give rise to malaria. One hundred million people are infected annually, and the disease is very widespread in tropical and warm temperate regions where the appropriate anophiline mosquito vectors are found. Malaria is transmitted to humans by sporozoites from mosquito salivary glands. After inoculation, sporozoites enter liver parenchymal cells and undergo asexual multiplication (tissue schizogony) to form merozoite-filled tissue schizonts. Schizont rupture releases merozoites which then invade red cells by means of attachment to specific binding sites. The merozoites divide asexually forming trophozoites which subdivide to form red cell schizonts. The asexual replication cycle then recurs at regular intervals until treatment, acquisition of immunity or death supervenes. With the relapsing malarias, due to *P.vivax* and *P.ovale* merozoites may also be released from long-term persisting liver forms called hypnozoites (which are dormant sporozoite forms); these are responsible for late relapses. Some intraerythrocytic parasites differentiate to form gametocytes, which are taken up into the mosquito stomach when the insect feeds, and the sexual part of the cycle takes place (sporogony). Ultimately sporozoites are formed which migrate through the mosquito body cavity and enter the salivary glands, to complete the parasite life cycle. Blood transfusion and transplacental spread may result in non-vector borne transmission.

Malaria presents most commonly as fever, which is initially irregular in pattern, but which frequently becomes characteristically regular as the host immune response results in synchronisation of the later cycles of multiplication. Febrile episodes may be accompanied by rigors, hypotension, gastrointestinal disturbance and altered consciousness. Patients are often asymptomatic between paroxysms, although hepatosplenomegaly is frequently detectable. Haemolytic anaemia is an important complication, and often results in mild jaundice. Secondary tissue hypoxia may result both from anaemia, but also by blockage of small vessels and may result in seizures and coma, renal failure and ARDS-like responses. The majority of these complications are associated with *P.falciparum* infections.

Diagnosis

Diagnosis of malaria is still most commonly performed by detection and identification of the organisms in a Giemsa-stained thick or thin blood film. Thick films are more sensitive due to the much larger volume of blood per unit area of the slide, but are more difficult to interpret for the inexperienced laboratory. According to the parasite morphology observed, it is possible to diagnose the infecting species. Wright's stain has also been used for the examination of thin films, and some workers have described the use of fluorescent microscopy of acridine orange-stained preparations.

PCR or DNA probe-based methodologies have been described for the specific diagnosis of malaria, but are as yet not widely available, particularly in endemic area. Various ELISA-based systems to detect parasite antigens, such as the histidine-rich protein (HRP2), have been described and are commercially available. Detection of serum antibodies is generally unhelpful diagnostically in the acute situation due to the 3-4 week delay before seroconversion.

References

Krogstad DJ. *Plasmodium* species. In: Mandell GL, Bennett JE, Dolin R, eds. Principles and Practice of Infectious Diseases, 4th edn. New York, Edinburgh, London: Churchill Livingstone. 1995:2415–27.

Makler MT, Gibbins B. Laboratory diagnosis of malaria. Clin Lab Med. 1991;11:941–56.

Plasmodium falciparum
ANTIGEN DETECTION

Manufacturer: Cellabs Pty Ltd
Cat. No./Trade name: KM2/MALARIA CELISA®

SUMMARY

[Well-MAb]–**Ag**–[MAb-HRP]–[TMB]–A_{450}

Assay type: EIA (non-competitive)
Detection: Colorimetric A_{450}
Format: Microtitre well, Ab coated
Sample type: Lysed blood
Sample pre-treatment:
  None
Sample volume: 100 µl
Number of tests: 192
Controls - standards run in assay:
  Controls: Neg (1), high Pos (1)
Incubation:
  1 hr (RT) + 1 hr (RT) + 15 min (RT)
Washes: 2

CONTENTS

Antibodies, antigens, labelled components:
  Anti-*P. falciparum* MAb bound to well
  Anti-*P. falciparum* MAb HRP conjugated
Substrate: TMB
Controls - standards supplied:
  Controls: Neg (1), high Pos (1)
Additional reagents:
  None
Special equipment required:
  None

INTERPRETATION

Comments on interpretation:
  Classification of samples is according to cut-off; no
    further testing
  A positive result suggests a current or very recent
    infection. Test may remain Pos several days after
    parasites are no longer detectable in blood films
No. of references: None

NOTES
131414.0

For Research use only and for use as a confirmatory test for diagnostic problems, screening blood transfusion products and travel-related infection

Plasmodium falciparum
ANTIBODY DETECTION

Manufacturer: Cellabs Pty Ltd
Cat. No./Trade name: KM3/MALARIA Ab CELISA®

SUMMARY

[Well-Ag]–**Ab**–[AHIg-HRP]–[TMB]–A_{450}

Assay type: EIA (non-competitive)
Detection: Colorimetric A_{450}
Format: Microtitre well, Ag coated
Sample type: Serum, plasma
Sample pre-treatment:
  None
Sample volume: 100 µl
Number of tests: 192
Controls - standards run in assay:
  Controls: Neg (1), high Pos (1)
Incubation:
  1 hr (37°C) + 1 hr (37°C) + 15 min (RT)
Washes: 2

CONTENTS

Antibodies, antigens, labelled components:
  P. falciparum Ag bound to well
  Anti-human Ig Ab HRP conjugated
Substrate: TMB
Controls - standards supplied:
  Controls: Neg and high Pos
Additional reagents:
  None
Special equipment required:
  None

INTERPRETATION

Comments on interpretation:
  Classification of samples is according to cut-off; no
    further testing
  Serum samples which give values above the cut-off point
    should be considered as positive for malaria
    antibody. This suggests that the donor has or has had
    malaria. It does not imply that the donor is carrying
    malaria parasites at this particular time
No. of references: 7

NOTES
131781.0

This test system is not designed as a diagnostic method for active clinical malaria and must not be used for this purpose

© KLUWER ACADEMIC PUBLISHERS 1995, ISSN 1381-5067

Plasmodium falciparum
ANTIBODY DETECTION

Manufacturer: bioMerieux
Cat. No./Trade name: 72751/Falciparum-Spot IF

SUMMARY

[Slide-Ag]–**Ab**–[AHIg-FITC]–fluorescence

Assay type: Immunofluorescence assay (indirect)
Detection: Fluorescence microscopy
Format: Slide, Ag coated
Sample type: Serum
Sample pre-treatment:
   Absorption of serum with group A1 RBCs to avoid
      nonspecific fluorescence with RBCs in substrate
      preparation
Sample volume: 10 µl of 1:20 and 1:40 dilutions
Number of tests: 100
Controls - standards run in assay:
   Controls: Neg (1)
Incubation:
   30 min (37°C) + 30 min (37°C)
Washes: 2

CONTENTS

Antibodies, antigens, labelled components:
   P. falciparum Ag (from human RBCs, group A1) bound to
      slide
   Anti-human Ig Ab (g) FITC conjugated
Substrate:
Controls - standards supplied:
   Controls: Neg (PBS)
Additional reagents:
   None
Special equipment required:
   None

INTERPRETATION

Comments on interpretation:
   Positive: intense fluorescence of organisms situated
      inside infected RBCs
No. of references: 5

NOTES

131281.0

© *KLUWER ACADEMIC PUBLISHERS 1995, ISSN 1381-5067*

Rickettsiae

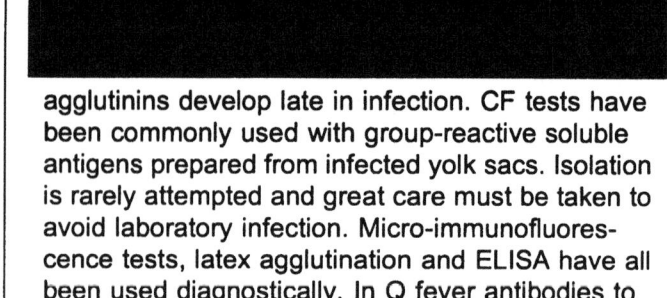

Natural history

Rickettsiae are pleomorphic obligate intracellular parasites. They are true bacteria by virtue of their 5-layered peptidoglycan cell wall containing muramic acid and diaminopimelic acid, they contain both RNA in ribosomes and DNA and they divide by binary fission. There are three genera (*Rickettsia*, *Rochalimaea* and *Coxiella*) within the family *Rickettsiaceae* and all except *Coxiella* are transmitted to humans by arthropods. Their endemicity is maintained by cyclic transmission from infected to uninfected arthropod through a vertebrate intermediary. Infection in humans is incidental, occurring as the result of a bite from an infected arthropod or inhalation of infectious aerosols in the case of *Coxiella*. The genus *Rickettsia* has two main subgroups, the Typhus fever group and the Spotted fever group, based on biological activity. The most important members of the family and their vectors and reservoirs are summarised in the table below.

The best known rickettsial disease is typhus, in epidemic form frequently found in conditions of poverty and malnutrition and frequently fatal in the absence of treatment. Endemic typhus is less severe and rarely fatal. Rickettsial diseases are characterised by a sudden onset of fever, severe headache, prostration and in all except *Coxiella*, a skin rash with prominent vascular lesions. Haemorrhagic complications and lymphadenopathy are common. Q fever has a quite different presentation with fever alone in approximately half of cases and an atypical pneumonia in the other half. Chronic infection can lead to endocarditis with involvement of the heart valves. This is a rare but serious complication. All rickettsial infections are suppressed by treatment with tetracyclines or chloramphenicol.

Diagnosis

Traditionally, the Weil-Felix reaction, an agglutination reaction based on common antigens between *Proteus* and rickettsial species, was used in diagnosis but lacks specificity and peak titres of agglutinins develop late in infection. CF tests have been commonly used with group-reactive soluble antigens prepared from infected yolk sacs. Isolation is rarely attempted and great care must be taken to avoid laboratory infection. Micro-immunofluorescence tests, latex agglutination and ELISA have all been used diagnostically. In Q fever antibodies to Phase I antigens (naturally infected source) are produced only late in infection, while antibodies to Phase II (yolk sac grown organisms) are produced in the acute phase. This can be of use in diagnosing Q fever endocarditis.

Comment

Differential testing using normal yolk sac antigen is necessary in any serological test which uses yolk sac derived rickettsial antigens. Control of infection is dependent on breaking the chain of infection by eradication of vectors, and immunisation where possible. Transmission from humans to human does not occur.

Reference

Ascher MS. In: Schmidt NJ, Emmons RW. Rickttsial Diseases inDiagnostic Procedures for Viral, Rickettsial and Chlamydial Infections, 6th edn. Am Pub Hlth Assoc. 1989:1141–64.

Genus	Species	Disease	Vector	Vertebrate reservoir
Rickettsia	*R. prowazekii*	Epidemic typhus	Louse	Humans
	R. typhi	Endemic typhus	Flea	Rodents
	R. tsutsugamushi	Scrub typhus	Mite	Rodents
	R. rickettsii	Rocky Mountain spotted fever	Tick	Rodents, dogs
	R. conorii	Boutonneuse fever	Tick	Rodents, dogs
	R. akari	Rickettsialpox	Mite	Mice
Coxiella	*C. burnetti*	Q fever	–	Cattle, sheep, goats
Rochalimaea	*R. quintana*	Trench fever	Louse	Humans

Rickettsia conorii
ANTIBODY DETECTION (IgG AND/OR IgM))

Manufacturer: Immuno Pharmacology Research
Cat. No./Trade name: ELISA RICKETTSIA

SUMMARY

[Well-Ag]–**Ab**–[AHIgG/IgM-AP]–[PNP]–A_{405}

Assay type: EIA (non-competitive)
Detection: Colorimetric A_{405}
Format: Microtitre well, Ag coated
Sample type: Serum
Sample pre-treatment:
  None
Sample volume: 100 µl of 1:200 dilution x 2
Number of tests: 48, 96
Controls - standards run in assay:
  Controls: Neg (1), Pos (1)
Incubation:
  1 hr (37°C) + 1 hr (37°C) + 30 min (37°C)
Washes: 2

CONTENTS

Antibodies, antigens, labelled components:
  Rickettsia conorii (isolated in Sicily) Ag bound to well
  Anti-human IgG Ab AP conjugated
  Anti-human IgM Ab AP conjugated
Substrate: PNP
Controls - standards supplied:
  Controls: Neg and Pos
Additional reagents:
  None
Special equipment required:
  None

INTERPRETATION

Comments on interpretation:
  Classification of samples is according to cut-off; no
    further testing
No. of references: 3

NOTES

131330.0

This kit can detect IgG and IgM antibodies and can
  differentiate between them

Rickettsia conorii
ANTIBODY DETECTION

Manufacturer: bioMerieux
Cat. No./Trade name: 75901/Rickettsia conori-Spot IF

SUMMARY

[Slide-Ag]–**Ab**–[AHIg-FITC]–fluorescence

Assay type: Immunofluorescence assay (indirect)
Detection: Fluorescence microscopy
Format: Slide, Ag coated
Sample type: Serum
Sample pre-treatment:
  None
Sample volume: 10 µl of 1:40 dilution
Number of tests: 100
Controls - standards run in assay:
  Controls: Neg (1)
Incubation:
  30 min (37°C) + 30 min (37°C)
Washes: 2

CONTENTS

Antibodies, antigens, labelled components:
  Rickettsia conorii Ag (from Vero cell line) bound to slide
  Anti-human Ig Ab FITC conjugated
Substrate:
Controls - standards supplied:
  Controls: Neg (PBS)
Additional reagents:
  None
Special equipment required:
  None

INTERPRETATION

Comments on interpretation:
  Positive: fluorescence of organisms either in or outside
    the cells
No. of references: 4

NOTES

131286.0

© KLUWER ACADEMIC PUBLISHERS 1995, ISSN 1381-5067

Rickettsia conorii
ANTIBODY DETECTION

Manufacturer: Immuno Pharmacology Research
Cat. No./Trade name: IFI-RICKETTSIA

SUMMARY

[Slide-Ag]–**Ab**–[AHIg-FITC]–fluorescence

Assay type: Immunofluorescence assay (indirect)
Detection: Fluorescence microscopy
Format: Slide well, Ag coated
Sample type: Serum
Sample pre-treatment:
None
Sample volume: 10 μl of 1:40 dilution
Number of tests: 50, 100
Controls - standards run in assay:
Controls: Neg (1), Pos (1)
Incubation:
30 min (37°C) + 30 min (37°C)
Washes: 2

CONTENTS

Antibodies, antigens, labelled components:
Rickettsia conorii Ag bound to slide well
Anti-human Ig Ab (g or sh) FITC conjugated
Substrate:
Controls - standards supplied:
Controls: Neg and Pos (serum)
Additional reagents:
None
Special equipment required:
None

INTERPRETATION

Comments on interpretation:
Positive: fluorescence of organism
No. of references: 7

NOTES

131333.0

Spotted Fever bio-group and Typhus Fever bio-group
ANTIBODY DETECTION (IgG)

Manufacturer: MRL Diagnostics
Cat. No./Trade name: IF100G/Rickettsia IFA IgG

SUMMARY

[Slide-Ag]–**Ab**–[AHIgG-FITC]–fluorescence

Assay type: Immunofluorescence assay (indirect)
Detection: Fluorescence microscopy
Format: Slide well, Ag coated
Sample type: Serum
Sample pre-treatment:
Dilute serum with Yolk Sac Diluent and incubate for 15 min (RT) prior to use
Sample volume: 25 μl of prepared sample
Number of tests: 80
Controls - standards run in assay:
Controls: Neg (1), R. rickettsii Pos (1), R. typhii Pos (1)
Incubation:
30 min (37°C) + 30 min (37°C)
Washes: 2

CONTENTS

Antibodies, antigens, labelled components:
R. rickettsia Ag and R. typhii Ag bound to slide well (2 spots per well: one spot for each antigen-yolk sac suspensiqon)
Anti-human IgG Ab (g) FITC conjugated
Substrate:
Controls - standards supplied:
Controls: Neg, R. rickettsia Pos and R. typhii (Pos (human serum)
Additional reagents:
None
Special equipment required:
None

INTERPRETATION

Comments on interpretation:
Negative: fluorescence ≤ Neg control well
Positive: ≥ (1+) fluorescence in appropriate Ag spot in slide well identifies the relevant group antibodies present
No. of references: 7

NOTES

131115.0

This kit can detect Abs to Spotted Fever bio-group and Typhus Fever bio-group and can differentiate between the groups. The Spotted Fever group include infection with R. rickettsia, R. akari, R. conorii, R. australis and R. sibirica. The Typhus Fever group include infection with R. typhii and R. prowazekii

© KLUWER ACADEMIC PUBLISHERS 1995, ISSN 1381-5067

Spotted Fever bio-group and Typhus Fever bio-group
ANTIBODY DETECTION (IgM)

Manufacturer: MRL Diagnostics
Cat. No./Trade name: IF100M/Rickettsia IFA IgM

SUMMARY

[Slide-Ag]–**Ab**–[AHIgG-FITC]–fluorescence

Assay type: Immunofluorescence assay (indirect)
Detection: Fluorescence microscopy
Format: Slide well, Ag coated
Sample type: Serum
Sample pre-treatment:
Dilute serum with IgM sample diluent, incubate 15 min (RT) and centrifuge to eliminate interference due to RF and IgG
Sample volume: 25 μl of prepared sample
Number of tests: 80
Controls - standards run in assay:
Controls: Neg (1), *R. rickettsii* Pos (1), *R. typhii* Pos (1)
Incubation:
1 hr 30 min (37°C) + 30 min (37°C)
Washes: 2

CONTENTS

Antibodies, antigens, labelled components:
R. rickettsia Ag and *R. typhii* Ag bound to slide well (2 spots per slide well: one spot for each antigen-yolk sac preparation)
Anti-human IgM Ab (g) FITC conjugated
Substrate:
Controls - standards supplied:
Controls: Neg, *R. rickettsii* Pos and *R. typhii* Pos (human serum)
Additional reagents:
None
Special equipment required:
None

INTERPRETATION

Comments on interpretation:
Positive: ≥(1+) fluorescence in appropriate Ag spot in slide well identifies the relevant group antibodies present
No. of references: 7

NOTES

131121.0

This kit can detect Abs to Spotted Fever bio-group and Typhus Fever bio-group and can differentiate between the groups. The Spotted Fever group include infection with *R. rickettsii*, *R. akari*, *R. conorii*, *R. australis* and *R. sibirica*. The Typhus Fever group include infection with *R. typhii* and *R. prowazekii*

© KLUWER ACADEMIC PUBLISHERS 1995, ISSN 1381-5067

Rotavirus

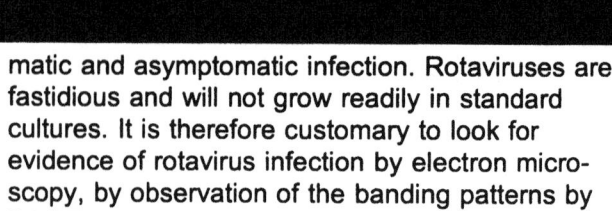

Natural history

Rotaviruses form one genus in the family Reoviridae and are unique in possessing a genome of 11 segments of double-stranded RNA. They are a major viral cause of endemic acute gastroenteritis in young children and are only occasionally associated with outbreaks. Rotavirus infection frequently leads to hospitalisation of infants because of the severe dehydration which can follow the vomiting and diarrhoea. Hospital-acquired infection is not uncommon and should be considered if a new case develops more than 2 days after admission. Rotaviruses are a major cause of death in children worldwide, with approximately 800,000 deaths per annum, mostly but not entirely in the developing world.

Diarrhoea and vomiting are sometimes accompanied by fever and there may be some cases with diarrhoea alone and even of asymptomatic excretion. The main site of infection is the upper small bowel particularly in the columnar cells of the intestinal villi. These become damaged and are sloughed off, thereby reducing the absorptive area of the mucosa. However as the crypts remain undamaged, recovery is rapid after the infected cells have been lost. Respiratory symptoms sometimes accompany rotavirus infection but there is no good evidence of infection within the respiratory tract. Infections in temperate climates are more frequent in late autumn and winter as the weather becomes colder.

Rotaviruses are spherical viruses with cubic symmetry and the characteristic double-shelled nature of the particles in the electron microscope gave them their name. The most important proteins serologically in the outer layer are VP4, which forms 60 dimeric spikes and VP7, the major outer capsid protein. Both carry neutralising epitopes. The most important protein of the inner capsid is VP6 which specifies both group and subgroup specific antigens. The 11 segments of the genome can be separated electrophoretically on a polyacrylamide gel (PAGE) with characteristic banding patterns for each antigenic group (A–E). Most strains causing disease in humans belong to group A which has at least 4 subgroups and 14 serotypes. Group B human strains have been implicated in a large epidemic in China.

Diagnosis

Serological studies are difficult because most people over the age of 3 years have some antibody and because of antigenic cross-reactions between different strains. Humoral local IgA and systemic IgM and IgG responses occur after both symptomatic and asymptomatic infection. Rotaviruses are fastidious and will not grow readily in standard cultures. It is therefore customary to look for evidence of rotavirus infection by electron microscopy, by observation of the banding patterns by PAGE or by antigen detection using latex agglutination or ELISA.

Comment

There is a great need for a vaccine to combat the morbidity of rotavirus infection, particularly in the developing world. However, the epidemiology of rotaviruses is very complex, with different serotypes co-circulating in any one area and antigenic drift occurring within a serotype, leading to reinfection. Live attenuated vaccines have given protection only against the serotype of the vaccine. New avenues using subunit or recombinant particle vaccines have still to be explored.

References

Desselberger U. Towards rotavirus vaccines. Rev Med Virol. 1993;3:15–21.

Madeley C. Viruses associated with acute diarrhoeal disease. In: Zuckerman AJ, Banatvala JE, Pattison JR, eds. Principles and Practice of Clinical Virology, 3rd edn. 1994:189–227.

See also Multipathogen Assays section under: Gastrointestinal pathogens

Rotavirus
ANTIGEN DETECTION

Manufacturer: Abbott Laboratories
Cat. No./Trade name: 7181/ROTAZYME® II Kit

SUMMARY

[Bead-Ab]–**Ag**–[Ab-HRP]–[OPD]–A_{492}

Assay type: EIA (non-competitive)
Detection: Colorimetric A_{492}
Format: Reaction wells, Ab coated beads
Sample type: Faecal specimens
Sample pre-treatment:
    Prepare 10% dilution of sample with sample diluent
Sample volume: 200 µl of prepared sample
Number of tests: 50
Controls - standards run in assay:
    Controls: Neg (1), Pos (1)
Incubation:
    1 hr (37°C) + 1 hr (37°C) + 30 min (RT)
Washes: 2

CONTENTS

Antibodies, antigens, labelled components:
    Anti-Rotavirus Ab (gp) bound to bead
    Anti-Rotavirus Ab (r) HRP conjugated
Substrate:
Controls - standards supplied:
    Controls: Neg (sample diluent) and Pos (simian)
Additional reagents:
    H_2SO_4
Special equipment required:
    Pentawash® II and Gast® Vacuum Pump (or similar)
    Quantum Analyser (optional)

INTERPRETATION

Comments on interpretation:
    Classification of samples is according to cut-off; no
    further testing
No. of references: 14

NOTES

131355.0

Rotavirus
ANTIGEN DETECTION

Manufacturer: Abbott Laboratories
Cat. No./Trade name: 6892-16/ABBOTT TESTPACK®
ROTAVIRUS

SUMMARY

[Particles-Ab]–**Ag**–[MAb-AP]–[reaction disc-Ag]–
[substrate]–purple (+) sign

Assay type: EIA (non-competitive)
Detection: Visual
Format: Microparticles, Ab coated; reaction disc Ag coated
Sample type: Faecal specimens
Sample pre-treatment:
    Add 0.1 gm undiluted sample to Specimen Dilution Cup
    with Filter tube, dilute with PBS and clarify
Sample volume: 0.1 ml/0.1 g (one level scoop)
Number of tests: 20
Controls - standards run in assay:
    Not specified
Incubation:
    5 min (RT) + 2 min (RT)
Washes: 2

CONTENTS

Antibodies, antigens, labelled components:
    Anti-Rotavirus Ab (gp) bound to microparticles
    Rotavirus Ag bound to reaction disc
    Anti-Rotavirus MAb (m) AP conjugated
Substrate: Not specified
Controls - standards supplied:
    Integral procedural controls incorporated in test
Additional reagents:
    Abbott Testpack® Rotavirus Positive Control (Cat. no.
    3186-02)
Special equipment required:
    None

INTERPRETATION

Comments on interpretation:
    Negative: grey/purple (–) negative sign on reaction disc
    Positive: grey/purple (+) positive sign on reaction disc
    Unsatisfactory: if no (–) or (+) signs appear; repeat test
No. of references: 5

NOTES

131356.0

© *KLUWER ACADEMIC PUBLISHERS 1995, ISSN 1381-5067*

Rotavirus
ANTIGEN DETECTION

Manufacturer: Alexon Inc
Cat. No./Trade name: 920-48 ProSpecT® Rotavirus EZ Microplate Assay®

SUMMARY

[Well-Ab]–**Ag**–[Ab-HRP]–[TMB]–A_{450}

Assay type: EIA (non-competitive)
Detection: Colorimetric A_{450}
Format: Microtitre well, Ab coated
Sample type: Faecal specimens
Sample pre-treatment:
    Mix specimen with dilution buffer
Sample volume: 100 µl of sample
Number of tests: 48, 96
Controls - standards run in assay:
    Controls: Neg (1), Pos (1)
Incubation:
    1 hr (RT) + 10 min (RT)
Washes: 1

CONTENTS

Antibodies, antigens, labelled components:
    Anti-Rotavirus (group A) Ab (r) bound to well
    Anti-Rotavirus (group A) Ab (r) HRP conjugated
Substrate: TMB
Controls - standards supplied:
    Controls: Neg (dilution buffer) and Pos (bovine)
Additional reagents:
    None
Special equipment required:
    None

INTERPRETATION

Comments on interpretation:
    Equivocal: within cut-off ±0.010; interpret in conjunction with clinical findings, or retest to confirm or repeat with a fresh specimen in 2–4 weeks
No. of references: 22

NOTES

131300.0

Rotavirus
ANTIGEN DETECTION

Manufacturer: Cambridge Biotech Corporation
Cat. No./Trade name: 6004/Rotaclone® - Rotavirus Diagnostic Kit

SUMMARY

[Well-MAb]–**Ag**–[MAb-HRP]–[TMB]–A_{450}

Assay type: EIA (non-competitive)
Detection: Colorimetric A_{450}
Format: Microtitre well, Ab coated
Sample type: Faecal specimens, rectal swab
Sample pre-treatment:
    Mix stool or swab with 1 ml diluent using sample transfer pipette provided
Sample volume: 100 µl of prepared sample
Number of tests: 48
Controls - standards run in assay:
    Controls: Neg (1), Pos (1)
Incubation:
    1 hr (RT) + 10 min (RT)
Washes: 1

CONTENTS

Antibodies, antigens, labelled components:
    Anti-rotovirus MAb bound to well
    Anti-rotovirus MAb HRP conjugated
Substrate: TMB
Controls - standards supplied:
    Controls: Neg (sample diluent) and Pos (simian rotovirus SA-11)
Additional reagents:
    None
Special equipment required:
    None

INTERPRETATION

Comments on interpretation:
    Classification of samples is according to cut-off; no further testing
No. of references: 14

NOTES

131179.0

Rotavirus
ANTIGEN DETECTION

Manufacturer: Dako A/S
Cat. No./Trade name: K6020/IDEIA® Rotavirus

SUMMARY

[Well-Ab]–**Ag**–[Ab-HRP]–[TMB]–A_{450}

Assay type: EIA (non-competitive)
Detection: Colorimetric A_{450}
Format: Microtitre well, Ab coated
Sample type: Faecal samples, rectal swab
Sample pre-treatment:
   Prepare 10% suspension of faecal sample or rectal swab in diluent
Sample volume: 100 µl of prepared sample
Number of tests: 96
Controls - standards run in assay:
   Controls: Neg (1), Pos (1)
Incubation:
   1 hr (RT) + 10 min (RT)
Washes: 1

CONTENTS

Antibodies, antigens, labelled components:
   Anti-Rotavirus PAb bound to well
   Anti-Rotavirus PAb HRP conjugated
Substrate: TMB
Controls - standards supplied:
   Controls: Neg (sample diluent) and Pos (bovine)
Additional reagents:
   None
Special equipment required:
   None

INTERPRETATION

Comments on interpretation:
   Equivocal: within cut-off ±0.01; retest to confirm or repeat using a fresh sample
No. of references: 23

NOTES

131041.0

Rotavirus
ANTIGEN DETECTION

Manufacturer: LMD Laboratories Inc
Cat. No./Trade name: Rota-2/Rotavirus Stool Antigen Microwell ELISA Kit

SUMMARY

[Well-Ab]–**Ag**–[Ab]–[AGIg-biotin]–[strept-HRP]–
[TMB]–A_{450}

Assay type: EIA (non-competitive)
Detection: Colorimetric A_{450}
Format: Microtitre well, Ab coated
Sample type: Faecal specimen
Sample pre-treatment:
   Mix sample with wash buffer and vortex
Sample volume: 100 µl of prepared sample
Number of tests: 48, 96
Controls - standards run in assay:
   Controls: Neg (1), Pos (1)
Incubation:
   45 min (37°C) + 20 min (37°C) + 10 min (37°C) + 10 min (37°C) + 10 min (RT)
Washes: 4

CONTENTS

Antibodies, antigens, labelled components:
   Anti-Rotavirus Ab bound to well
   Anti-Rotavirus Ab (g)
   Anti-mouse Ig Ab biotinylated
   Streptavidin HRP conjugated
Substrate: TMB
Controls - standards supplied:
   Controls: Neg and Pos
Additional reagents:
   None
Special equipment required:
   None

INTERPRETATION

Comments on interpretation:
   Classification of samples is according to cut-off; no further testing
No. of references: 5

NOTES

131647.0

© KLUWER ACADEMIC PUBLISHERS 1995, ISSN 1381-5067

Rotavirus
ANTIGEN DETECTION

Manufacturer: Melotec S.A.
Cat. No./Trade name: Melotest Rotavirus Ag

SUMMARY

[Well-Ab]–**Ag**–[Ab]–[AMIg-biotin]–[strept-HRP]–
[TMB]–A_{450}

Assay type: EIA (non-competitive)
Detection: Colorimetric A_{450}
Format: Microtitre well, Ab coated
Sample type: Faecal sample
Sample pre-treatment:
   Mix sample with diluent (1:15) and further dilute
      supernatant (1:3) before use
Sample volume: 100 μl of prepared sample x 2
Number of tests: 96
Controls - standards run in assay:
   Controls: Neg (2), Pos (1)
Incubation:
   45 min (37°C) + 20 min (37°C) + 10 min (37°C) + 10
      min (37°C) + 10 min (RT)
Washes: 4

CONTENTS

Antibodies, antigens, labelled components:
   Anti-Rotavirus Ab bound to well
   Anti-Rotavirus Ab (m)
   Anti-mouse Ig Ab biotinylated
   Streptavidin HRP conjugated
Substrate: TMB
Controls - standards supplied:
   Controls: Neg and Pos
Additional reagents:
   None
Special equipment required:
   None

INTERPRETATION

Comments on interpretation:
   Equivocal: ratio of sample OD:cut-off value between 0.9
      and 1.1; retest a fresh sample
   Repeatably equivocal: retest a fresh sample taken in 2-4
      weeks
No. of references: 5

NOTES

131140.0

Rotavirus
ANTIGEN DETECTION

Manufacturer: Microgen Bioproducts
Cat. No./Trade name: M430/RotaScreen® EIA

SUMMARY

[Well-Ab]–**Ag**–[Ab-biotin]–[strept-HRP]–[TMB]–A_{450}

Assay type: EIA (non-competitive)
Detection: Colorimetric A_{450}
Format: Microtitre well, Ab coated
Sample type: Faecal specimens*
Sample pre-treatment:
   Mix faecal material with sample diluent, centrifuge
Sample volume: 100 μl of prepared sample
Number of tests: 96
Controls - standards run in assay:
   Controls: Neg (2), Pos (1)
Incubation:
   1 hr (37°C) + 1 hr (37°C) + 30 min (RT)
Washes: 2

CONTENTS

Antibodies, antigens, labelled components:
   Anti-Rotavirus (subgroup I, II; serotypes 1, 2, 3) Ab (r)
      bound to well
   Anti-Rotavirus Ab (r) biotinylated
   Streptavidin HRP conjugated
Substrate: TMB
Controls - standards supplied:
   Controls: Neg (sample diluent) and Pos (bovine
      rotavirus)
Additional reagents:
   H_2SO_4
Special equipment required:
   None

INTERPRETATION

Comments on interpretation:
   Equivocal: within cut-off ±10%; retest to confirm
No. of references: 10

NOTES

131191.0

*Swab specimens are only advised if there is no alternative
   as there may be insufficient material to perform test

Rotavirus
ANTIGEN DETECTION

Manufacturer: r-biopharm GmbH
Cat. No./Trade name: RIDASCREEN® Rotavirus

SUMMARY

[Well-MAb]–**Ag**–[MAb-HRP]–[TMB]–A_{450}

Assay type: EIA (non-competitive)
Detection: Colorimetric A_{450}
Format: Microtitre well, Ab coated
Sample type: Faecal specimens
Sample pre-treatment:
Add sample diluent to specimen and mix
Sample volume: 100 μl of diluted sample
Number of tests: 96
Controls - standards run in assay:
Controls: Neg (1), Pos (1)
Incubation:
1 hr (RT) + 10 min (RT)
Washes: 2

CONTENTS

Antibodies, antigens, labelled components:
Anti-Rotavirus (types 2, 13–15) MAb bound to well
Anti-Rotavirus MAb HRP conjugated
Substrate: TMB
Controls - standards supplied:
Controls: Neg (sample diluent) and Pos (simian rotavirus SA-11)
Additional reagents:
None
Special equipment required:
None

INTERPRETATION

Comments on interpretation:
Classification of sample is according to cut-off; no further testing
No. of references: 15

NOTES

131152.0

Rotavirus
ANTIGEN DETECTION

Manufacturer: Sanofi Diagnostics Pasteur
Cat. No./Trade name: Pathfinder® Rotavirus

SUMMARY

[Tube-Ab]–**Ag**–[MAb-HRP]–[TMB]–A_{450}

Assay type: EIA (non-competitive)
Detection: Colorimetric A_{450}
Format: Tube, Ab coated
Sample type: Faecal specimen, or swab
Sample pre-treatment:
Mix with diluent
Sample volume: 300 μl of prepared specimen
Number of tests: 50
Controls - standards run in assay:
Controls: Neg (1), Pos (1)
Incubation:
1 hr (RT) + 15 min (RT)
Washes: 1

CONTENTS

Antibodies, antigens, labelled components:
Anti-Rotavirus (simian SA-11) IgG Ab bound to tube
Anti-Rotavirus (EDIM strain from mice) MAb HRP conjugated
Substrate: TMB
Controls - standards supplied:
Controls: Neg (sample diluent) and Pos (simian Ag)
Additional reagents:
H_2SO_4
Special equipment required:
None

INTERPRETATION

Comments on interpretation:
Equivocal: cut-off ± 10%; retest a fresh sample
Repeatably equivocal: report as indeterminant result
No. of references: 25

NOTES

131767.0

© KLUWER ACADEMIC PUBLISHERS 1995, ISSN 1381-5067

Rotavirus
ANTIGEN DETECTION

Manufacturer: Meridian Diagnostics Inc
Cat. No./Trade name: 708030/ImmunoCard® Rotavirus

SUMMARY

[Membrane-Ab]–**Ag**–[Ab-AP]–[BCIP]–blue dot

Assay type: Immunochromatographic assay
Detection: Visual
Format: Reaction unit, Ab coated membrane
Sample type: Faecal specimens
Sample pre-treatment:
  Mix with diluent in test tube and vortex prior to use
Sample volume: 3 drops of prepared sample
Number of tests: 30
Controls - standards run in assay:
  Controls: Neg (1), Pos (1)
Incubation:
  3 min (RT) + 3 min (RT)
Washes: 1

CONTENTS

Antibodies, antigens, labelled components:
  Anti-Rotavirus Ab (r) bound to membrane (Test port)
  Anti-Rotavirus Ab (r) AP conjugated (incorporated in
    Sample ports)
Substrate: BCIP
Controls - standards supplied:
  Controls: Neg (buffer) and Pos
  Integral procedural controls incorporated in membrane
    (Control port)
Additional reagents:
  None
Special equipment required:
  None

INTERPRETATION

Comments on interpretation:
  Negative: blue colour in Control port but no colour in Test
    port
  Positive: blue colour in Control and Test ports
  Invalid: no colour in Control port; repeat test
No. of references: 17

NOTES

131381.0

Rotavirus
ANTIGEN DETECTION

Manufacturer: BIOKIT SA
Cat. No./Trade name: rotagen

SUMMARY

[Latex-Ab]–**Ag**–agglutination

Assay type: Particle agglutination assay
Detection: Visual
Format: Latex particles, Ab coated (test done on slide)
Sample type: Faecal specimens, faecal swabs
Sample pre-treatment:
  Dilute with buffer, centrifuge
Sample volume: 50 μl of supernatant
Number of tests:
Controls - standards run in assay:
  Controls: Pos (1)
Incubation:
  2 min (RT) on rotary shaker
Washes:

CONTENTS

Antibodies, antigens, labelled components:
  Anti-Rotavirus Ab bound to latex particles (Latex
    Reagent)
  Non-immune rabbit Ig Ab bound to latex particles (Latex
    Control)
Substrate:
Controls - standards supplied:
  Controls: Pos
Additional reagents:
  None
Special equipment required:
  None

INTERPRETATION

Comments on interpretation:
  Equivocal: equal agglutination with Latex Reagent and
    Latex Control; dilute sample 1:2 and retest
  Repeatably equivocal: consider as unsatisfactory and
    retest sample using another method
No. of references: 7

NOTES

131377.0

Rotavirus
ANTIGEN DETECTION

Manufacturer: bioMerieux
Cat. No./Trade name: 58842/Slidex Rota-Kit 2

SUMMARY

[Latex-Ab]–**Ag**–agglutination

Assay type: Particle agglutination assay
Detection: Visual
Format: Latex particles, Ab coated (test done on card)
Sample type: Faecal specimens
Sample pre-treatment:
   Dilute sample, vortex and centrifuge
Sample volume: 50 µl of prepared sample x2
Number of tests: 30
Controls - standards run in assay:
   Controls: Pos (1)
Incubation:
   Read result within 2 min of mixing sample and reagent
Washes:

CONTENTS

Antibodies, antigens, labelled components:
   Anti-Rotavirus MAb (m) bound to latex particles (Test Latex)
   Non-immune mouse Ig Ab bound to latex particles (Control Latex)
Substrate:
Controls - standards supplied:
   Controls: Pos
Additional reagents:
   None
Special equipment required:
   None

INTERPRETATION

Comments on interpretation:
   Positive: agglutination of sample with Test Latex and no agglutination with Control Latex
No. of references: 8

NOTES

131279.0

Rotavirus
ANTIGEN DETECTION

Manufacturer: Meridian Diagnostics Inc
Cat. No./Trade name: 201030/Meritec® - Rotavirus

SUMMARY

[Latex-Ab]–**Ag**–agglutination

Assay type: Particle agglutination assay
Detection: Visual
Format: Latex particles, Ab coated (test done on card)
Sample type: Faecal specimens, rectal swabs, meconium*
Sample pre-treatment:
   Mix with sample buffer and vortex
Sample volume: 1 drop of prepared sample x 2
Number of tests: 20
Controls - standards run in assay:
   Controls: Pos (1)
Incubation:
   5 min (RT) with rotation
Washes:

CONTENTS

Antibodies, antigens, labelled components:
   Anti-Rotavirus Ab bound to latex particles (Detection Latex)
   Non-immune rabbit Ig Ab bound to latex (Control Latex)
Substrate:
Controls - standards supplied:
   Controls: Pos
Additional reagents:
   None
Special equipment required:
   None

INTERPRETATION

Comments on interpretation:
   Positive: agglutination of sample with Detection Latex and no agglutination with Control Latex
No. of references: 23

NOTES

131390.0

*Interpret test results from meconium with caution

© KLUWER ACADEMIC PUBLISHERS 1995, ISSN 1381-5067

Rotavirus
ANTIGEN DETECTION

Manufacturer: Microgen Bioproducts
Cat. No./Trade name: M80/RotaScreen®

SUMMARY

[Latex-Ab]–**Ag**–agglutination

Assay type: Particle agglutination assay
Detection: Visual
Format: Latex particles, Ab coated (test done on slide)
Sample type: Faecal specimens
Sample pre-treatment:
Mix faecal material with Extraction Buffer, centrifuge or filter (using Pack Cat. no. 802)
Sample volume: 50 µl (supernatant) or 1 drop (filtrate) x 2
Number of tests: 50
Controls - standards run in assay:
Controls: Pos (2)
Incubation:
Read result within 2 min of mixing sample and reagent
Washes:

CONTENTS

Antibodies, antigens, labelled components:
Anti-Rotavirus Ab (r) bound to latex particles (Test Latex)
Non-immune rabbit Ig Ab bound to latex particles (Control Latex)
Substrate:
Controls - standards supplied:
Controls: Pos (bovine rotavirus)
Additional reagents:
None
Special equipment required:
None

INTERPRETATION

Comments on interpretation:
Positive: agglutination of sample with Test Latex and not with Control Latex
No. of references: 9

NOTES

131192.0

Rotavirus
ANTIGEN DETECTION

Manufacturer: Murex Diagnostics Limited
Cat. No./Trade name: ZL40/Murex Rotavirus Latex

SUMMARY

[Latex-Ab]–**Ag**–agglutination

Assay type: Particle agglutination assay
Detection: Visual
Format: Latex particles, Ab coated (test done on slide)
Sample type: Faecal specimens
Sample pre-treatment:
Mix 0.1 ml specimen with 1 ml of Extraction Buffer, centrifuge
Sample volume: 1 drop of supernatant
Number of tests: 20
Controls - standards run in assay:
Controls: Neg (1), Pos (1)
Incubation:
Read results within 2 min of mixing sample and reagent
Washes:

CONTENTS

Antibodies, antigens, labelled components:
Anti-Rotavirus (Nebraska calf diarrhoea) Ab (r) bound to latex particles (Test Latex)
Non-immune rabbit Ig Ab bound to latex particles (Control Latex)
Substrate:
Controls - standards supplied:
Controls: Pos
Additional reagents:
Faecal specimens negative for Rotavirus for use as quality control
Special equipment required:
None

INTERPRETATION

Comments on interpretation:
Positive: agglutination of sample with Test Latex and not with Control Latex
No. of references: 18

NOTES

131357.0

Rotavirus
ANTIGEN DETECTION

Manufacturer: Omega Diagnostics Ltd
Cat. No./Trade name: ODO38/Virotect Rota

SUMMARY

[Latex-Ab]–**Ag**–agglutination

Assay type: Particle agglutination assay
Detection: Visual
Format: Latex particles, Ab coated (test done on slide)
Sample type: Faecal samples
Sample pre-treatment:
    Prepare a 10% suspension of sample using Extraction Buffer provided
Sample volume: 1 drop of prepared sample x 2
Number of tests: 50
Controls - standards run in assay:
    Controls: Pos (1)
Incubation:
    Read result 2 min after mixing reagent with sample
Washes:

CONTENTS

Antibodies, antigens, labelled components:
    Anti-Rotavirus (human & animal) Ab (r) bound to latex particles
    Non-immune rabbit Ig Ab bound to latex particles (control reagent)
Substrate:
Controls - standards supplied:
    Controls: Pos
Additional reagents:
    None
Special equipment required:
    None

INTERPRETATION

Comments on interpretation:
    Positive: agglutination in test latex circle and no agglutination in control latex circle
No. of references: 9

NOTES

131160.0

Rotavirus
ANTIGEN DETECTION

Manufacturer: Orion Diagnostica
Cat. No./Trade name: 67201/Rotalex®

SUMMARY

[Latex-Ab]–**Ag**–agglutination

Assay type: Particle agglutination assay
Detection: Visual
Format: Latex particles, Ab coated (test done on card)
Sample type: Faecal specimens
Sample pre-treatment:
    Filter, using Faecal Specimen Filtration Vial, or centrifuge diluted samples
Sample volume: 50 µl of prepared sample x 2
Number of tests: 20, 100
Controls - standards run in assay:
    Controls: Pos (1)
Incubation:
    Read result within 2 min of mixing sample and reagents
Washes:

CONTENTS

Antibodies, antigens, labelled components:
    Anti-Rotavirus Ab bound to latex particles (Test Latex)
    Non-immune Ig Ab bound to latex particles (Control Latex)
Substrate:
Controls - standards supplied:
    Controls: Pos
Additional reagents:
    None
Special equipment required:
    Faecal Specimen Filtration Vials Kit (Cat. no. 68312) (optional)

INTERPRETATION

Comments on interpretation:
    Positive: agglutination of sample with Test Latex and not with Control Latex
    Non-specific: agglutination of sample only with Control Latex; retest a fresh sample
No. of references: None

NOTES

131262.0

© KLUWER ACADEMIC PUBLISHERS 1995, ISSN 1381-5067

Rotavirus
ANTIGEN DETECTION

Manufacturer: Orion Diagnostica
Cat. No./Trade name: 67487/Diarlex® Rota

SUMMARY

[Latex spot-Ab]–**Ag**–agglutination

Assay type: Particle agglutination assay
Detection: Visual
Format: Dry latex spot on card, Ab coated
Sample type: Faecal specimens
Sample pre-treatment:
Filter, using Faecal Specimen Filtration Vial, or centrifuge diluted samples
Sample volume: 50 µl of prepared sample x 2
Number of tests: 20
Controls - standards run in assay:
Controls: Pos (1)
Incubation:
Read result within 2 min of mixing sample and reagents
Washes:

CONTENTS

Antibodies, antigens, labelled components:
Anti-Rotavirus Ab bound to dry latex spot on test card (Rota Latex)
Non-immune Ig Ab bound to dry latex spot on test card (Control Latex)
Substrate:
Controls - standards supplied:
Controls: Pos
Additional reagents:
None
Special equipment required:
Faecal Specimen Filtration Vials Kit (Cat. no. 68312 – optional)

INTERPRETATION

Comments on interpretation:
Positive: agglutination of sample in dry Rota Latex spot and no agglutination in dry Control Latex spot on test card
Non-specific: agglutination of sample with Control Latex only; retest a fresh sample
No. of references: None

NOTES

131265.0

Rotavirus
ANTIBODY DETECTION (IgG)

Manufacturer: Amico Laboratories Inc
Cat. No./Trade name: 6500G/AMIZYME® Rotavirus IgG Assay

SUMMARY

[Well-Ag]–**Ab**–[AHIgG-HRP]–[ABTS]–A_{405}

Assay type: EIA (non-competitive)
Detection: Colorimetric A_{405}
Format: Microtitre well, Ag coated
Sample type: Serum
Sample pre-treatment:
None
Sample volume: 5 µl (+245 µl diluent)
Number of tests: 96
Controls - standards run in assay:
Controls: Neg (1)
Calibrators: I (1)
Incubation:
25 min (RT) + 25 min (RT) + 25 min (RT)
Washes: 2

CONTENTS

Antibodies, antigens, labelled components:
Rotavirus Ag bound to well
Anti-human IgG Ab (g) HRP conjugated
Substrate: ABTS
Controls - standards supplied:
Controls: Neg
Calibrator: 1
Additional reagents:
None
Special equipment required:
None

INTERPRETATION

Comments on interpretation:
Equivocal: where sample OD is between 0.375–0.399 (95–99 U/ml); retest to confirm
No. of references: 8

NOTES

131787.0

© KLUWER ACADEMIC PUBLISHERS 1995, ISSN 1381-5067

Rotavirus
ANTIBODY DETECTION (IgG)

Manufacturer: Immuno Pharmacology Research
Cat. No./Trade name: ROTAVIRUS IgG

SUMMARY

[Well-Ag]–**Ab**–[AHIgG-AP]–[PNP]–A_{405}

Assay type: EIA (non-competitive)
Detection: Colorimetric A_{405}
Format: Microtitre well, Ag coated
Sample type: Serum
Sample pre-treatment:
  None
Sample volume: 200 µl of 1:100 dilution x 2
Number of tests: 48, 96
Controls - standards run in assay:
  Controls: Neg (1), Pos (1)
Incubation:
  1 hr (37°C) + 1 hr (37°C) + 30 min (37°C)
Washes: 2

CONTENTS

Antibodies, antigens, labelled components:
  Rotavirus Ag bound to well
  Anti-human IgG Ab AP conjugated
Substrate: PNP
Controls - standards supplied:
  Controls: Neg and Pos
Additional reagents:
  None
Special equipment required:
  None

INTERPRETATION

Comments on interpretation:
  Equivocal: where sample OD is between 0.150 and
    0.300; retest to confirm
  Repeatably equivocal: retest a fresh specimen in 15 days
No. of references: None

NOTES

131316.0

Rotavirus
ANTIBODY DETECTION (IgM)

Manufacturer: Amico Laboratories Inc
Cat. No./Trade name: 6500MC/AMIZYME® Rotavirus IgM Capture Assay

SUMMARY

[Well-Ag]–**Ab**–[AHIgM]–[AMIg-HRP]–[ABTS]–A_{405}

Assay type: EIA (non-competitive)
Detection: Colorimetric A_{405}
Format: Microtitre well, Ag coated
Sample type: Serum
Sample pre-treatment:
  None
Sample volume: 5 µl (+245 µl diluent)
Number of tests: 96
Controls - standards run in assay:
  Controls: Neg (1)
  Calibrators: I (1)
Incubation:
  25 min (RT) + 25 min (RT) + 25 min (RT) + 25 min
    (RT)
Washes: 3

CONTENTS

Antibodies, antigens, labelled components:
  Rotavirus Ag bound to well
  Anti-human IgM MAb (m)
  Anti-mouse Ig Ab (g) HRP conjugated
Substrate: ABTS
Controls - standards supplied:
  Controls: Neg
  Calibrator: 1
Additional reagents:
  None
Special equipment required:
  None

INTERPRETATION

Comments on interpretation:
  Equivocal: where sample OD is between 0.285–0.299;
    retest to confirm
No. of references: 2

NOTES

131808.0

© KLUWER ACADEMIC PUBLISHERS 1995, ISSN 1381-5067

Rotavirus
ANTIBODY DETECTION (IgM)

Manufacturer: Amico Laboratories Inc
Cat. No./Trade name: 6500M/AMIZYME® Rotavirus IgM Assay

SUMMARY

[Well-Ag]–**Ab**–[AHIgM-HRP]–[ABTS]–A_{405}

Assay type: EIA (non-competitive)
Detection: Colorimetric A_{405}
Format: Microtitre well, Ag coated
Sample type: Serum
Sample pre-treatment:
    None
Sample volume: 5 µl (+245 µl diluent)*
Number of tests: 96
Controls - standards run in assay:
    Controls: Neg (1)
    Calibrators: I (1)
Incubation:
    25 min (RT) + 25 min (RT) + 25 min (RT)
Washes: 2

CONTENTS

Antibodies, antigens, labelled components:
    Rotavirus Ag bound to well
    Anti-human IgM Ab (g) HRP conjugated
Substrate: ABTS
Controls - standards supplied:
    Controls: Neg
    Calibrator: 1
Additional reagents:
    None
Special equipment required:
    None

INTERPRETATION

Comments on interpretation:
    Equivocal: where sample OD is between 0.275–0.299
        (92–99 U/ml); retest to confirm
No. of references: 3

NOTES

131809.0

*Sample diluted in IgM serum diluent provided to eliminate interference due to RF and IgG

© KLUWER ACADEMIC PUBLISHERS 1995, ISSN 1381-5067

Rubella virus

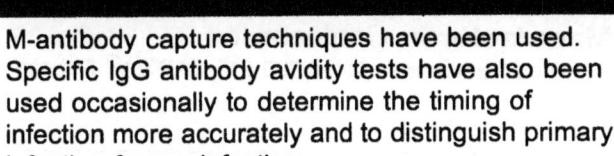

Natural history

Rubella or German measles is caused by the rubella virus belonging to the Rubivirus genus within the family Togaviridae. It is very similar in structure to the alphaviruses but it has no antigenic similarity to them and is not arthropod-borne. It is therefore an RNA virus of about 60 nm in diameter with a central core and a fragile envelope.

Rubella is a generally mild infection particularly of school aged children, with worldwide distribution and increased incidence every 3–4 years. The short-lived macular rash appears first on the face and spreads rapidly to the trunk and limbs. There may be slight malaise in adults and some lymphadenopathy. The most common complication is joint involvement, particularly in adult women, and this can persist for up to one month with fingers, wrists and ankles most affected. Patients are infectious for a week before the rash appears and up to 7–10 days thereafter leading to significant opportunity for spread.

More than 50 years ago, specific congenital defects were observed following maternal rubella infection in the first trimester of pregnancy. These included cataract, heart defects and deafness. The recognition of the congenital rubella syndrome (CRS) led to an enormous amount of research to establish good diagnostic tests for both current infection and past exposure and to produce an effective vaccine. From the late 1960s, live attenuated vaccines have been available in developed countries and specific programmes instituted to prevent CRS. Thus in the United States, vaccine was offered to all preschool children, while in the UK vaccine was limited to prepubertal schoolgirls and sero-negative women of childbearing age in the hope that circulating wild-type virus would boost immunity and render re-vaccination unnecessary. Since the late 1980s, the triple vaccine, MMR, containing live attenuated measles, mumps and rubella viruses has been widely used in controlling infection and has further reduced the incidence of CRS.

Diagnosis

Acute infection can be shown serologically by several means. For a long time the haemagglutination inhibition (HAI) test was used on paired sera several days apart to show a rise in antibody titre. Specific IgM antibodies could also be demonstrated in a single sample by this test if the serum was first separated into IgG and IgM fractions on a sucrose density gradient. Recently, ELISA and RIA have taken over for the detection of rubella specific IgM antibodies. Both indirect immunoassays using specimens from which IgG has been removed and

M-antibody capture techniques have been used. Specific IgG antibody avidity tests have also been used occasionally to determine the timing of infection more accurately and to distinguish primary infection from reinfection.

Far more screening tests for pre-existing rubella antibody are carried out. There are several techniques available including single radial haemolysis (SRH), latex agglutination and ELISA. The latter are readily available commercially. The detection of IgG antibodies in saliva is possible using a G-antibody capture technique. In all cases a control containing an internationally determined amount of rubella antibody should be included for comparison. Sera containing > 15 IU/ml of rubella antibody are considered to contain protective levels.

Comment

The extensive nature of rubella vaccination programmes has greatly reduced CRS. However, greater travel and immigration from less developed countries mean that wild-type virus can still circulate and susceptible women will still be entering pregnancy. Furthermore the duration of vaccine-induced immunity in the absence of reglar exposure to wild-type virus is unknown and rubella infections are more likely. Persistence of virus with shedding occurs following congenital infection. Care must therefore continue to be taken to prevent spread from all sources and vigilance over the protective levels in the community is essential.

References

Best J. Rubella. In: Greenough A, Osborne J, Sutherland S, eds. Congenital, Perinatal and Neonatal Infections. 1992:171–84.

Best J, Banatvala JE. Rubella. In: Zuckerman AJ, Banatvala JE, Pattison JR, eds. Principles and Practice of Clinical Virology, 3rd edn. Wiley; 1994:363–400.

See also Multipathogen Assays section under: TORCH Screening Panel

Rubella virus
ANTIBODY DETECTION (IgG)

Manufacturer: Centocor Inc
Cat. No./Trade name: M416/800-915/CAPTIA® Rubella-G

SUMMARY

[Well-Ag]–**Ab**–[AHIgG-HRP]–[TMB]–A_{450}

Assay type: EIA (non-competitive)
Detection: Colorimetric A_{450}
Format: Microtitre well, Ag coated
Sample type: Serum (do not heat inactivate)
Sample pre-treatment:
 None
Sample volume: 100 μl of 1:101 dilution
Number of tests: 96
Controls - standards run in assay:
 Controls: Neg (1), low Pos (2), high Pos (1)*
Incubation:
 1 hr (37°C) + 1 hr (37°C) + 30 min (RT)
Washes: 2

CONTENTS

Antibodies, antigens, labelled components:
 Rubella virus Ag bound to well
 Anti-human IgG Ab HRP conjugated
Substrate: TMB
Controls - standards supplied:
 Controls: Neg, low Pos and high Pos (human serum - IU/ml - WHO reference standard)
Additional reagents:
 H_2SO_4
Special equipment required:
 None

INTERPRETATION

Comments on interpretation:
 Equivocal: where Antibody Index is between 0.9 and 1.1; retest to confirm
 Repeatably equivocal: repeat test using fresh sample
No. of references: 9

NOTES

131620.0

*If results are to be expressed in IU/ml, use high Pos control in duplicate

Rubella virus
ANTIBODY DETECTION (IgG)

Manufacturer: Centocor Inc
Cat. No./Trade name: 800-916/CAPTIA® Select Rub-G

SUMMARY

[Well-Ag]–**Ab**–[AHIgG-HRP]–[TMB]–A_{450}

Assay type: EIA (non-competitive)
Detection: Colorimetric A_{450}
Format: Microtitre well, Ag coated
Sample type: Serum (do not heat inactivate)
Sample pre-treatment:
 None
Sample volume: 10 μl (+200 μl diluent)*
Number of tests: 96, 480
Controls - standards run in assay:
 Controls: Neg (1), cut-off (2), Pos (1)**
Incubation:
 30 min (37°C) + 30 min (37°C) + 30 min (RT)
Washes: 2

· CONTENTS

Antibodies, antigens, labelled components:
 Rubella virus Ag bound to well
 Anti-human IgG MAb HRP conjugated
Substrate: TMB
Controls - standards supplied:
 Controls: Neg, cut-off and Pos (human serum - IU/ml - WHO reference standard)
Additional reagents:
 H_2SO_4
Special equipment required:
 None

INTERPRETATION

Comments on interpretation:
 Equivocal: within cut-off ±10%; retest to confirm
No. of references: 6

NOTES

131623.0

*A Specimen Delivery Indicator reagent in diluent changes to purple on addition of specimen (with in-well dilution)
**If results are to be expressed in IU/ml, use Pos control in duplicate

© *KLUWER ACADEMIC PUBLISHERS 1995, ISSN 1381-5067*

Rubella virus
ANTIBODY DETECTION (IgM)

Manufacturer: Centocor Inc
Cat. No./Trade name: M415/800-920/CAPTIA® Rubella-M

SUMMARY

[Well:AHIgM]–**Ab**–[Ag]–[MAb-biotin]–[strept-HRP]–
[TMB]–A_{450}

Assay type: EIA (non-competitive)
Detection: Colorimetric A_{450}
Format: Microtitre well, Ab coated
Sample type: Serum (do not heat inactivate)
Sample pre-treatment:
    None
Sample volume: 100 µl of 1:101 dilution
Number of tests: 96
Controls - standards run in assay:
    Controls: Neg (1), low Pos (2), high Pos (1)
Incubation:
    1 hr (37°C) + 1 hr (37°C) + 30 min (RT)
Washes: 2

CONTENTS

Antibodies, antigens, labelled components:
    Anti-human IgM Ab (g) bound to well
    Rubella virus Ag
    Anti-Rubella virus MAb biotinylated
    Streptavidin HRP conjugated
Substrate: TMB
Controls - standards supplied:
    Controls: Neg, low Pos and high Pos (human serum)
Additional reagents:
    H_2SO_4
Special equipment required:
    None

INTERPRETATION

Comments on interpretation:
    Equivocal: where Antibody Index is between 0.9 and 1.1;
    retest to confirm
No. of references: 10

NOTES

131621.0

Rubella virus
ANTIBODY DETECTION

Manufacturer: Diesse
Cat. No./Trade name: 91030/ENZYWELL Rubella IgG

SUMMARY

[Well-Ag]–**Ab**–[AHIgG-POD]–[TMB]–A_{450}

Assay type: EIA (non-competitive)
Detection: Colorimetric A_{450}
Format: Microtitre well, Ag coated
Sample type: Serum (do not heat inactivate)
Sample pre-treatment:
    None
Sample volume: 100 µl of 1:101 dilution x 2
Number of tests: 96
Controls - standards run in assay:
    Controls: Neg (2), cut-off (3), Pos (2)
Incubation:
    45 min (37°C) + 45 min (37°C) + 15 min (RT)
Washes: 2

CONTENTS

Antibodies, antigens, labelled components:
    Rubella virus Ag bound to well
    Anti-human IgG MAb POD conjugated
Substrate: TMB
Controls - standards supplied:
    Controls: Neg, cut-off and Pos (human serum)
Additional reagents:
    Rubella IgG Calibration Set (Cat. no. 91930 - for optional
    quantitative results)
Special equipment required:
    None

INTERPRETATION

Comments on interpretation:
    Equivocal: within cut-off ±10%; retest to confirm
    Repeatably equivocal: repeat with a fresh sample
    Results may be expressed quantitatively in IU/ml using 6
    calibrators bought separately
No. of references: 3

NOTES

131441.0

© KLUWER ACADEMIC PUBLISHERS 1995, ISSN 1381-5067

Rubella virus
ANTIBODY DETECTION

Manufacturer: GenBio
Cat. No./Trade name: 4525/ImmunoDot® Rubella Test

SUMMARY

[Dipstick-Ag]–**Ab**–[AHIg-AP]–[BCIP]–dot

Assay type: EIA (non-competitive)
Detection: Visual
Format: Reaction vessel, Ag coated dipstick
Sample type: Serum, heparinized whole blood
Sample pre-treatment:
   None
Sample volume: 10 µl serum or 20 µl whole blood (+ 2 ml diluent)
Number of tests: 25, 100
Controls - standards run in assay:
   Controls: Pos (1)
   Integral procedural: Neg (1), Pos (1)
Incubation:
   5 min (44–48°C) + 5 min (44–48°C) + 15 min (44–48°C) + 5 min (44–48°C)
Washes: 4

CONTENTS

Antibodies, antigens, labelled components:
   Rubella virus (HPV-77) Ag (at 4 concentration levels) bound to dipsticks as 4 discrete dots
   Anti-human Ig Ab (g) AP conjugated
Substrate: BCIP
Controls - standards supplied:
   Integral Procedural Controls: Neg and Pos (bound to dipstick as separate dots)
Additional reagents:
   Positive Control Reagent (Cat. no. 2215)
Special equipment required:
   GenBio Workstation (Cat. no. 4001 or 4990)

INTERPRETATION

Comments on interpretation:
   Qualitative:
   Negative: first Rubella Ag dot non reactive
   Positive: first Rubella Ag dot reactive
   Quantitative:
   Positive: antibody titre is determined by degree of reactivity in Rubella Ag dots 1–4 (of varying concentrations)
No. of references: 2

NOTES

131424.0

Rubella virus
ANTIBODY DETECTION (IgG)

Manufacturer: Abbott Laboratories
Cat. No./Trade name: 7870-24/RUBAZYME® Kit

SUMMARY

[Bead-Ag]–**Ab**–[AHIgG-HRP]–[OPD]–A_{492}

Assay type: EIA (non-competitive)
Detection: Colorimetric A_{492}
Format: Reaction wells, Ag coated beads
Sample type: Serum
Sample pre-treatment:
   None
Sample volume: 10 µl (+ 200 µl diluent)
Number of tests: 100, 500
Controls - standards run in assay:
   Controls: Neg (1), low Pos (2), high Pos (1)
Incubation:
   1 hr (37°C) + 1 hr (37°C) + 30 min (RT)
Washes: 2

CONTENTS

Antibodies, antigens, labelled components:
   Rubella virus Ag bound to beads
   Anti-human IgG Ab (g) HRP conjugated
Substrate: OPD
Controls - standards supplied:
   Controls: Neg, low Pos and high Pos (human plasma)
Additional reagents:
   H_2SO_4
Special equipment required:
   Pentawash® II and Gast® Vacuum Pump (or similar)
   Quantum Analyser (optional)

INTERPRETATION

Comments on interpretation:
   Assessment of immune status by calculation of Rubazyme Index
   Serodiagnosis of recent infection is by calculation of Rubazyme Diagnostic Ratio, confirmed when necessary by retesting a fresh specimen later or further testing with Abbott Rubazyme-M Assay Kit
No. of references: 12

NOTES

131351.0

© KLUWER ACADEMIC PUBLISHERS 1995, ISSN 1381-5067

Rubella virus
ANTIBODY DETECTION (IgG)

Manufacturer: Alfa Biotech
Cat. No./Trade name: 05779212/Rubella IgG Elisa System

SUMMARY

[Well-Ag]–**Ab**–[AHIgG-HRP]–[TMB]–A_{450}

Assay type: EIA (non-competitive)
Detection: Colorimetric A_{450}
Format: Microtitre well, Ag coated
Sample type: Serum
Sample pre-treatment:
   None
Sample volume: 100 µl of 1:100 dilution x 2
Number of tests: 96
Controls - standards run in assay:
   Calibrators: 1 (2), 2 (2), 3 (2), 4 (2), 5 (2), 6 (2)
Incubation:
   20 min (37°C) + 20 min (37°C) + 20 min (37°C)
Washes: 2

CONTENTS

Antibodies, antigens, labelled components:
   Rubella virus Ag bound to well
   Anti-human IgG Ab (g) HRP conjugated
Substrate: TMB
Controls - standards supplied:
   Calibrators: 6 (human serum; range 0–100 IU/ml)
Additional reagents:
   None
Special equipment required:
   None

INTERPRETATION

Comments on interpretation:
   Equivocal: where Ab level is between 10 and 12.5 IU/ml;
      retest to confirm
   Where value is > 100 IU/ml; report as high Pos or retest
      at a higher serum dilution
No. of references: 8

NOTES

131099.0

Rubella virus
ANTIBODY DETECTION (IgG)

Manufacturer: BAG-Biologische Analysensystem GmbH
Cat. No./Trade name: 5272/BAG - Rubella - EIA - G

SUMMARY

[Well-Ag]–**Ab**–[AHIgG-HRP]–[TMB]–A_{450}

Assay type: EIA (non-competitive)
Detection: Colorimetric A_{450}
Format: Microtitre well, Ag coated
Sample type: Serum
Sample pre-treatment:
   None
Sample volume: 100 µl of 1:101 dilution
Number of tests: 96
Controls - standards run in assay:
   Controls: Neg (1), cut-off (2), Pos (1)
Incubation:
   30 min (RT) + 30 min (RT) + 10 min (RT)
Washes: 2

CONTENTS

Antibodies, antigens, labelled components:
   Rubella virus (strain HPV-77) Ag bound to well
   Anti-human IgG Ab (sh) HRP conjugated
Substrate: TMB
Controls - standards supplied:
   Controls: Neg, cut-off and Pos (human serum)
Additional reagents:
   None
Special equipment required:
   None

INTERPRETATION

Comments on interpretation:
   Equivocal: within cut-off ±10%; data is not provided for
      further assessment of samples within this range
No. of references: None

NOTES

131606.0

© KLUWER ACADEMIC PUBLISHERS 1995, ISSN 1381-5067

Rubella virus
ANTIBODY DETECTION (IgG)

Manufacturer: Behringwerke AG
Cat. No./Trade name: /Enzygnost® Anti-Rubella Virus/ IgG

SUMMARY

[Well-Ag]–**Ab**–[AHIgG-POD]–[TMB]–A_{450}

Assay type: EIA (non-competitive)
Detection: Colorimetric A_{450}
Format: Microtitre well, Ag coated
Sample type: Serum, plasma
Sample pre-treatment:
None
Sample volume: 20 µl of 1:21 dilution x 2*
Number of tests: 48
Controls - standards run in assay:
Reference reagent: P/N (2)*
Incubation:
1 hr (37°C) + 1 hr (37°C) + 30 min (RT)
Washes: 2

CONTENTS

Antibodies, antigens, labelled components:
Rubella virus Ag from BHK cells bound to well
Control Ag from uninfected BHK cells bound to well
Anti-human IgG Ab (r) POD conjugated
Substrate: TMB (bought separately)
Controls - standards supplied:
Reference reagent: P/N (Anti-Rubella Reference Reagent)
Additional reagents:
Supplementary Reagents for Enzygnost®/TMB (Code No. OUVP)
Special equipment required:
Behring ELISA Processor II (optional)
Behring ELISA Processor III (optional)

INTERPRETATION

Comments on interpretation:
Equivocal: where sample OD is between 0.10 and 0.20; retest to confirm
Repeatably equivocal: report as equivocal
For quantitative procedure see manufacturer's instructions
No. of references: 21

NOTES

131024.0

*Samples and controls assayed simultaneously in Ag coated and Control Ag coated wells

Rubella virus
ANTIBODY DETECTION (IgG)

Manufacturer: BIOKIT SA
Cat. No./Trade name: 3000-1215/bioelisa RUBELLA IgG

SUMMARY

[Well-Ag]–**Ab**–[AHIgG-HRP]–[TMB]–A_{450}

Assay type: EIA (non-competitive)
Detection: Colorimetric A_{450}
Format: Microtitre well, Ag coated
Sample type: Serum (do not heat inactivate)
Sample pre-treatment:
None
Sample volume: 100 µl of 1:100 dilution
Number of tests: 96, 480
Controls - standards run in assay:
Calibrators: Neg (2), low Pos (2), high Pos (2)
Incubation:
1 hr (37°C) + 30 min (37°C) + 30 min (RT)
Washes: 2

CONTENTS

Antibodies, antigens, labelled components:
Rubella virus Ag bound to well
Anti-human IgG Ab (r) HRP conjugated
Substrate: TMB
Controls - standards supplied:
Calibrators: Neg, low Pos (10 IU/ml) and high Pos (200 IU/ml) – human serum
Additional reagents:
None
Special equipment required:
None

INTERPRETATION

Comments on interpretation:
Classification of samples is according to cut-off; no further testing. Positive samples are considered to have IgG Ab levels of ⩾10 IU/ml
No. of references: 5

NOTES

131373.0

Rubella virus
ANTIBODY DETECTION (IgG)

Manufacturer: bioMerieux
Cat. No./Trade name: 30 213/VIDAS RUB IgG

SUMMARY

[Solid phase-Ag]–**Ab**–[AHIgG-AP]–[4MP]–fluorescence

Assay type: EIA (non-competitive)
Detection: Fluorometric
Format: Solid phase receptacle (SPR), Ag coated
Sample type: Serum
Sample pre-treatment:
  None
Sample volume: 100 μl
Number of tests: 60
Controls - standards run in assay:
  Controls: Neg (1), Pos (1)
  Calibrators: Pos (1)
Incubation:
  Automated - total time 40 min
Washes: Automated

CONTENTS

Antibodies, antigens, labelled components:
  Rubella virus Ag bound to solid phase receptacle
  Anti-human IgG Ab (m) AP conjugated
Substrate: 4-MP
Controls - standards supplied:
  Controls: Neg and Pos (titre in IU/ml - human serum)
  Calibrators: Pos (titre in IU/ml - human serum)
Additional reagents:
  None
Special equipment required:
  Vitek immunodiagnostic assay system (VIDAS)

INTERPRETATION

Comments on interpretation:
  This is an automated assay
  Results designated equivocal; retest to confirm
No. of references: None

NOTES

131307.0

Rubella virus
ANTIBODY DETECTION (IgG)

Manufacturer: BioWhittaker Inc
Cat. No./Trade name: 30-625U/Rubelisa II

SUMMARY

[Well-Ag]–**Ab**–[AHIgG-AP]–[PMP]–A_{550}

Assay type: EIA (non-competitive)
Detection: Colorimetric A_{550}
Format: Microtitre well, Ag coated
Sample type: Serum (do not heat inactivate)
Sample pre-treatment:
  None
Sample volume: 10 μl (+200 μl diluent)
Number of tests: 192
Controls - standards run in assay:
  Controls: A (1), B (1)
  Calibrators: Neg (1), Pos 1 (1), Pos 2 (1)
Incubation:
  45 min (RT) + 45 min (RT) + 45 min (RT)
Washes: 2 (+preliminary plate wash)

CONTENTS

Antibodies, antigens, labelled components:
  Rubella virus Ag bound to well
  Anti-human IgG Ab (g) AP conjugated
Substrate: PMP
Controls - standards supplied:
  Controls: A and B
  Calibrators: Neg, Pos 1, Pos 2 (all human serum)
Additional reagents:
  None
Special equipment required:
  BioWhittaker automated system and software (optional)

INTERPRETATION

Comments on interpretation:
  Equivocal: Elisa value between 0.15–0.16 or Index value
    between 0.88–0.89; retest to confirm
  Repeatably equivocal: retest a fresh sample or use an
    alternative method
No. of references: 3

NOTES

131743.0

© KLUWER ACADEMIC PUBLISHERS 1995, ISSN 1381-5067

Rubella virus
ANTIBODY DETECTION (IgG)

Manufacturer: BioWhittaker Inc
Cat. No./Trade name: 30-336U/Rubestat

SUMMARY

[Well-Ag]–**Ab**–[AHIgG-AP]–[PMP]–A_{550}

Assay type: EIA (non-competitive)
Detection: Colorimetric A_{550}
Format: Microtitre well, Ag coated
Sample type: Serum (do not heat inactivate)
Sample pre-treatment:
   None
Sample volume: 10 µl (+200 µl diluent)
Number of tests: 192
Controls - standards run in assay:
   Controls: Neg (1), Pos (1)
   Standards: Neg (1), low Pos (1), high Pos (1)
Incubation:
   15 min (RT) + 15 min (RT) + 15 min (RT)
Washes: 2 (+preliminary plate wash)

CONTENTS

Antibodies, antigens, labelled components:
   Rubella virus Ag bound to well
   Anti-human IgG Ab (g) AP conjugated
Substrate: PMP
Controls - standards supplied:
   Controls: Neg and Pos
   Standards: Neg, low Pos and high Pos (all human serum)
Additional reagents:
   None
Special equipment required:
   BioWhittaker automated system and software (optional)

INTERPRETATION

Comments on interpretation:
   Equivocal: Predicted Index value (PIV) between 0.80–
   0.90; retest to confirm
   Repeatably equivocal: retest a fresh sample or use
     alternative method
   Low Positive: PIV between 1.00–1.74
   Mid Positive: PIV between 1.75–3.89
   High Positive: PIV ⩾3.90
No. of references: 16

NOTES

131750.0

Rubella virus
ANTIBODY DETECTION (IgG)

Manufacturer: Bouty SpA
Cat. No./Trade name: 22294/BEIA Rubella IgG Quant

SUMMARY

[Well-Ag]–**Ab**–[AHIgG-HRP]–[TMB]–A_{450}

Assay type: EIA (non-competitive)
Detection: Colorimetric A_{450}
Format: Microtitre well, Ag coated
Sample type: Serum
Sample pre-treatment:
   None
Sample volume: 10 µl (+800 µl diluent)*
Number of tests: 96, 192
Controls - standards run in assay:
   Controls: Neg (2)
   Calibrators: 1 (2), 2 (2), 3 (2), 4 (2)
Incubation:
   30 min (RT) + 30 min (RT) + 15 min (RT)
Washes: 2

CONTENTS

Antibodies, antigens, labelled components:
   Rubella virus Ag bound to well
   Anti-human IgG Ab (g) HRP conjugated
Substrate: TMB
Controls - standards supplied:
   Controls: Neg (human serum)
   Calibrators: 4 (8, 25, 75 and 250 IU/ml - human serum)
Additional reagents:
   None
Special equipment required:
   None

INTERPRETATION

Comments on interpretation:
   Negative: <8 IU/ml
   Low Positive: between 8 and 15 IU/ml
   Positive: ⩾15 IU/ml
No. of references: 6

NOTES

131239.0

*Samples can be assayed singly or in duplicate

Rubella virus
ANTIBODY DETECTION (IgG)

Manufacturer: Chimica Diagnostica
Cat. No./Trade name: 41400/Rubella Virus IgG

SUMMARY

[Well-Ag]–**Ab**–[AHIgG-POD]–[TMB]–A$_{450}$

Assay type: EIA (non-competitive)
Detection: Colorimetric A$_{450}$
Format: Microtitre well, Ag coated
Sample type: Serum
Sample pre-treatment:
  None
Sample volume: 100 µl of 1:101 dilution x 2
Number of tests: 96
Controls - standards run in assay:
  Standards: 1 (2), 2 (2), 3 (2), 4 (2), 5 (2), 6 (2)
Incubation:
  1 hr (37°C) + 1 hr (37°C) + 20 min (RT)
Washes: 2

CONTENTS

Antibodies, antigens, labelled components:
  Rubella virus Ag bound to well
  Anti-human IgG Ab (r) POD conjugated
Substrate: TMB
Controls - standards supplied:
  Standards: 1–6 (titres 0, 10, 20, 50, 100 and 250 IU/ml)
Additional reagents:
  None
Special equipment required:
  None

INTERPRETATION

Comments on interpretation:
  Results are expressed quantitatively in IU/ml with
    reference to the Standards provided
No. of references: 2

NOTES

131360.0

Rubella virus
ANTIBODY DETECTION (IgG)

Manufacturer: Clark Laboratories
Cat. No./Trade name: Rubella virus IgG Elisa Test

SUMMARY

[Well-Ag]–**Ab**–[AHIgG-HRP]–[OPD]–A$_{490}$

Assay type: EIA (non-competitive)
Detection: Colorimetric A$_{490}$
Format: Microtitre well, Ag coated
Sample type: Serum (do not heat inactivate)
Sample pre-treatment:
  None
Sample volume: 10 µl (+200 µl diluent)
Number of tests: 96
Controls - standards run in assay:
  Controls: Neg (1), Pos (1)
  Calibrators: 1 (2)
Incubation:
  20 min (RT) + 20 min (RT) + 10 min (RT)
Washes: 2

CONTENTS

Antibodies, antigens, labelled components:
  Rubella virus Ag bound to well
  Anti-human IgG Ab (g) HRP conjugated
Substrate: TMB*
Controls - standards supplied:
  Controls: Neg and Pos (human serum or plasma)
  Calibrator: 1 (human serum or plasma)
Additional reagents:
  H_2SO_4
Special equipment required:
  None

INTERPRETATION

Comments on interpretation:
  Classification of samples is according to cut-off; no
    further testing
No. of references: None

NOTES

131661.0

*OPD can be supplied as an alternative

© KLUWER ACADEMIC PUBLISHERS 1995, ISSN 1381-5067

Rubella virus
ANTIBODY DETECTION (IgG)

Manufacturer: Denka Seiken Co. Ltd
Cat. No./Trade name: 321718/Rubella IgG (II) – EIA "SEIKEN"

SUMMARY

[Well-Ag]–**Ab**–[AHIgG-POD]–[TMB]–A_{450}

Assay type: EIA (non-competitive)
Detection: Colorimetric A_{450}
Format: Microtitre well, Ag coated
Sample type: Serum, plasma
Sample pre-treatment:
  None
Sample volume: 10 µl (+2 ml diluent)
Number of tests: 96
Controls - standards run in assay:
  Standards: I (1), II (2), III (1), IV (1)
Incubation:
  1 hr (RT) + 1 hr (RT) + 30 min (RT)
Washes: 2

CONTENTS

Antibodies, antigens, labelled components:
  Rubella virus Ag bound to well
  Anti-human IgG Ab (g) POD conjugated
Substrate: TMB
Controls - standards supplied:
  Standards: I–IV (EIA values 0–3200, human serum)
Additional reagents:
  None
Special equipment required:
  None

INTERPRETATION

Comments on interpretation:
  Equivocal: where EIA value is between 200 and 400;
    repeat test with a fresh specimen in 1–2 weeks or use
    an alternative method
No. of references: None

NOTES

131686.0

Rubella virus
ANTIBODY DETECTION (IgG)

Manufacturer: Diamedix Corporation
Cat. No./Trade name: 783-360/Rubella IgG Microassay

SUMMARY

[Well-Ag]–**Ab**–[AHIgG-AP]–[PNP]–A_{405}

Assay type: EIA (non-competitive)
Detection: Colorimetric A_{405}
Format: Microtitre well, Ag coated
Sample type: Serum (do not heat inactivate)
Sample pre-treatment:
  None
Sample volume: 10 µl (+200 µl diluent)
Number of tests: 96
Controls - standards run in assay:
  Controls: Neg (1), Pos (1)
  Calibrator: 1 (1)
Incubation:
  20 min (RT) + 20 min (RT) + 20 min (RT)
Washes: 2

CONTENTS

Antibodies, antigens, labelled components:
  Rubella virus (Gilchrist strain) Ag bound to well
  Anti-human IgG Ab AP conjugated
Substrate: PNP
Controls - standards supplied:
  Controls: Neg and Pos
  Calibrator: 1 (assigned Ab titre)
Additional reagents:
  None
Special equipment required:
  None

INTERPRETATION

Comments on interpretation:
  Negative: <20 Elisa units/ml; report as non-immune
  Positive: ≥20 Elisa units/ml; report as immune
No. of references: 2

NOTES

131050.0

© KLUWER ACADEMIC PUBLISHERS 1995, ISSN 1381-5067

Rubella virus
ANTIBODY DETECTION (IgG)

Manufacturer: E. Merck
Cat. No./Trade name: 14082/Rubella IgG MAGIA®

SUMMARY

[Particles-Ag]–**Ab**–[AHIgG-AP]–[substrate]–A_{405}

Assay type: EIA (non-competitive)
Detection: Colorimetric A_{405}
Format: Cuvette, Ag coated particles (magnetisable)
Sample type: Serum
Sample pre-treatment:
   None
Sample volume: 10 µl
Number of tests: 100
Controls - standards run in assay:
   Calibrators: zero, sero G (automated)
Incubation:
   Automated
Washes: Automated

CONTENTS

Antibodies, antigens, labelled components:
   Rubella virus Ag bound to particles (magnetisable)
   Anti-human IgG Ab AP conjugated
Substrate: Bought separately
Controls - standards supplied:
   Bought separately
Additional reagents:
   Particle wash and system solutions (Cat. nos. 14097, 14096
   Substrate (Cat. no. 14095), calibrators (Cat. nos. 14090, 14092)
   Anti-foam (Cat. no. 14098)
Special equipment required:
   MAGIA® 7000 Immunoanalyzer (Cat. no. 14011) or
      MAGIA® 8000 (Cat. no. 114526)
   Reaction cuvettes (Cat. no. 14085)

INTERPRETATION

Comments on interpretation:
   This is an automated assay
No. of references: 2

NOTES

131065.0

Rubella virus
ANTIBODY DETECTION (IgG)

Manufacturer: Gull Laboratories Inc
Cat. No./Trade name: RUE101/Rubella IgG ELISA Test

SUMMARY

[Well-Ag]–**Ab**–[AHIgG-AP]–[PNP]–A_{405}

Assay type: EIA (non-competitive)
Detection: Colorimetric A_{405}
Format: Microtitre well, Ag coated
Sample type: Serum
Sample pre-treatment:
   None
Sample volume: 100 µl of 1:21 dilution
Number of tests: 96
Controls - standards run in assay:
   Controls: Neg (1), Pos (1)
   Calibrators: 1 (1), 2 (1), 3 (1)*
   Reference serum: 1 (3)
Incubation:
   30 min (37°C) + 30 min (37°C) + 30 min (37°C)
Washes: 2

CONTENTS

Antibodies, antigens, labelled components:
   Rubella virus Ag bound to well
   Anti-human IgG Ab (caprine) AP conjugated
Substrate: PNP
Controls - standards supplied:
   Controls: Neg and Pos
   Calibrators: 3
   Reference serum: 1 (all human serum)
Additional reagents:
   None
Special equipment required:
   None

INTERPRETATION

Comments on interpretation:
   Equivocal: where sample OD is between mean OD of
      Reference Serum x 0.91 and mean OD of Reference
      Serum x 0.99; retest to confirm
   Repeatably equivocal: retest a fresh specimen or use an
      alternative method
No. of references: 11

NOTES

131695.0

*Omit calibrators for qualitative assay

© *KLUWER ACADEMIC PUBLISHERS 1995, ISSN 1381-5067*

Rubella virus
ANTIBODY DETECTION (IgG)

Manufacturer: Human GmbH
Cat. No./Trade name: 51208/Rubella virus - IgG - Elisa

SUMMARY

[well-Ag]–**Ab**–[AHIgG-HRP]–[TMB]–A$_{450}$

Assay type: EIA (non-competitive)
Detection: Colorimetric A$_{450}$
Format: Microtitre well, Ag coated
Sample type: Serum
Sample pre-treatment:
  None
Sample volume: 100 µl of 1:100 dilution
Number of tests: 96
Controls - standards run in assay:
  Controls: Neg (2), cut-off (2), Pos (2)
Incubation:
  30 min (RT) + 30 min (RT) + 15 min (RT)
Washes: 2

CONTENTS

Antibodies, antigens, labelled components:
  Rubella virus Ag bound to well
  Anti-human IgG Ab (r) HRP conjugated
Substrate: TMB
Controls - standards supplied:
  Controls: Neg, cut-off and Pos (human)
Additional reagents:
  None
Special equipment required:
  None

INTERPRETATION

Comments on interpretation:
  Equivocal: within cut-off ± 15%; retest in parallel with a
  fresh sample taken 7–14 days later
No. of references: 5

NOTES

131342.0

Rubella virus
ANTIBODY DETECTION (IgG)

Manufacturer: IFCI Clone Systems
Cat. No./Trade name: 08.1002/EIAGEN Rubella IgG

SUMMARY

[Well-Ag]–**Ab**–[AHIgG-POD]–[TMB]–A$_{450}$

Assay type: EIA (non-competitive)
Detection: Colorimetric A$_{450}$
Format: Microtitre well, Ag coated
Sample type: Serum
Sample pre-treatment:
  None
Sample volume: 10 µl (+ 1 ml diluent)*
Number of tests: 96
Controls - standards run in assay:
  Controls: Neg (2), cut-off (3), Pos (2)
Incubation:
  45 min (37°C) + 45 min (37°C) + 15 min (RT)
Washes: 2

CONTENTS

Antibodies, antigens, labelled components:
  Rubella virus Ag bound to well
  Anti-human IgG MAb POD conjugated
Substrate: TMB
Controls - standards supplied:
  Controls: Neg, cut-off and Pos (human serum)
Additional reagents:
  Rubella Standard Set (6 calibrators) is available for a
  performance of a quantitative assay (optional)
Special equipment required:
  None

INTERPRETATION

Comments on interpretation:
  Equivocal: within cut-off ± 10%; retest to confirm
  Repeatably equivocal: retest a fresh sample
No. of references: 3

NOTES

131675.0

*Samples are assayed in duplicate

© KLUWER ACADEMIC PUBLISHERS 1995, ISSN 1381-5067

Rubella virus
ANTIBODY DETECTION (IgG)

Manufacturer: Immuno Pharmacology Research
Cat. No./Trade name: RUBELLA IgG ELISA

SUMMARY

[Well-Ag]–**Ab**–[AHIgG-AP]–[PNP]–A_{405}

Assay type: EIA (non-competitive)
Detection: Colorimetric A_{405}
Format: Microtitre well, Ag coated
Sample type: Serum
Sample pre-treatment:
 None
Sample volume: 200 μl of 1:100 dilution x 2
Number of tests: 48, 96
Controls - standards run in assay:
 Controls: Neg (1), Pos (1)
Incubation:
 1 hr (37°C) + 1 hr (37°C) + 30 min (37°C)
Washes: 2

CONTENTS

Antibodies, antigens, labelled components:
 Rubella virus Ag bound to well
 Anti-human IgG Ab AP conjugated
Substrate: PNP
Controls - standards supplied:
 Controls: Neg and Pos
Additional reagents:
 None
Special equipment required:
 None

INTERPRETATION

Comments on interpretation:
 Equivocal: where OD is between 0.150 and 0.300; retest to confirm
 Repeatably equivocal: retest a fresh specimen in 15 days
No. of references: None

NOTES

131317.0

Rubella virus
ANTIBODY DETECTION (IgG)

Manufacturer: Incstar
Cat. No./Trade name: 4540/Rubella IgG Clin-ELISA

SUMMARY

[Well-Ag]–**Ab**–[AHIgG-AP]–[PNP]–A_{450}

Assay type: EIA (non-competitive)
Detection: Colorimetric A_{450}
Format: Microtitre well, Ag coated
Sample type: Serum
Sample pre-treatment:
 None
Sample volume: 10 μl (+500 μl diluent)
Number of tests: 96
Controls - standards run in assay:
 Controls: low Pos (1), high Pos (1)
 Calibrators: I (1), II (1), III (1)
Incubation:
 30 min (RT) + 30 min (RT) + 45 min (RT)
Washes: 2

CONTENTS

Antibodies, antigens, labelled components:
 Rubella virus Ag bound to well
 Anti-human IgG Ab (g or sh) AP conjugated
Substrate: PNP
Controls - standards supplied:
 Controls: low Pos, high Pos (human serum)
 Calibrators: I-Neg, II-low Pos, III-high Pos
Additional reagents:
 IgG Diluent Colorizer (Cat. no. 4506 – optional)
Special equipment required:
 None

INTERPRETATION

Comments on interpretation:
 Classification of samples is according to cut-off; no further testing
No. of references: 10

NOTES

131717.0

This product is not available in the UK but is sold in the rest of Europe

Rubella virus
ANTIBODY DETECTION (IgG)

Manufacturer: Kreatech Diagnostics
Cat. No./Trade name: EL-3001-gRUB/Rubella IgG EIA

SUMMARY

[Well-Ag]–**Ab**–[AHIgG–HRP]–[TMB]–A_{450}

Assay type: EIA (non-competitive)
Detection: Colorimetric A_{450}
Format: Microtitre well, Ag coated
Sample type: Serum (do not heat inactivate)
Sample pre-treatment:
  None
Sample volume: 10 µl (+1 ml diluent)*
Number of tests: 96
Controls - standards run in assay:
  Controls: Neg (2), cut-off (4), Pos (2)*
Incubation:
  1 hr (37°C) + 1 hr (37°C) + 30 min (RT)
Washes: 2

CONTENTS

Antibodies, antigens, labelled components:
  Rubella virus Ag from infected Vero cells bound to well
  Control Ag from uninfected Vero cells bound to well
  Anti-human IgG Ab HRP conjugated
Substrate: TMB
Controls - standards supplied:
  Controls: Neg, cut-off and Pos (human serum)
Additional reagents:
  None
Special equipment required:
  None

INTERPRETATION

Comments on interpretation:
  Classification of samples is according to cut-off; no futher testing
No. of references: None

NOTES

131084.0

*Samples and controls are assayed simultaneously and in duplicate in Ag coated and control Ag coated wells to reduce nonspecific activity

Rubella virus
ANTIBODY DETECTION (IgG)

Manufacturer: Laboratoire Eurobio
Cat. No./Trade name: 900277/Rubella IgG ELIT®

SUMMARY

[Well-Ag]–**Ab**–[AHIgG-POD]–[TMB]–A_{450}

Assay type: EIA (non-competitive)
Detection: Colorimetric A_{450}
Format: Microtitre well, Ag coated
Sample type: Serum, plasma
Sample pre-treatment:
  None
Sample volume: 100 µl of 1:201 dilution
Number of tests: 96
Controls - standards run in assay:
  Standards: I (1), II (2), III (1), IV (1)
Incubation:
  1 hr (RT) + 30 min (RT) + 10 min (RT)
Washes: 2

CONTENTS

Antibodies, antigens, labelled components:
  Rubella virus Ag bound to well
  Anti-human IgG Ab (g) POD conjugated
Substrate: TMB
Controls - standards supplied:
  Standards: 4 (human serum with defined Ab values)
Additional reagents:
  None
Special equipment required:
  None

INTERPRETATION

Comments on interpretation:
  Equivocal: where EIA value of sample is between 200 and 400; retest a fresh sample in 1–2 weeks
No. of references: None

NOTES

131730.0

Rubella virus
ANTIBODY DETECTION (IgG)

Manufacturer: Labsystems Oy
Cat. No./Trade name: 61 07 201S/Rubella IgG EIA

SUMMARY

[Well-Ag]–**Ab**–[AHIgG-AP]–[PNP]–A_{405}

Assay type: EIA (non-competitive)
Detection: Colorimetric A_{405}
Format: Microtitre well, Ag coated
Sample type: Serum*
Sample pre-treatment:
  None
Sample volume: 100 µl of 1:100 dilution x 2
Number of tests: 96
Controls - standards run in assay:
  Controls: Pos (2)
  Calibrators: 1 (2), 2 (2), 3 (2), 4 (2), 5 (2), 6 (2)
Incubation:
  1 hr (37°C) + 1 hr (37°C) + 30 min (37°C)
Washes: 2

CONTENTS

Antibodies, antigens, labelled components:
  Rubella virus Ag bound to well
  Anti-human IgG Ab (sh) AP conjugated
Substrate: PNP
Controls - standards supplied:
  Controls: Pos (titre in IU/ml - human serum)
  Calibrators: 1–6 (titres in IU/ml - human serum)
Additional reagents:
  NaOH
Special equipment required:
  Auto-EIA II analyser (optional)

INTERPRETATION

Comments on interpretation:
  Results are expressed quantitatively in Enzyme
    Immunoassay Units (EIU) or International Unit per ml
    (IU/ml)
No. of references: 9

NOTES

131336.0

*Heat treatment may slightly change assay results

Rubella virus
ANTIBODY DETECTION (IgG) *

Manufacturer: Labsystems Oy
Cat. No./Trade name: 61 07 202/Rubella IgG Avidity

SUMMARY

[Well-Ag]–**Ab**–[AHIgG-AP]–[PNP]–A_{405}

Assay type: EIA (non-competitive)
Detection: Colorimetric A_{405}
Format: Microtitre well, Ag coated
Sample type: Serum**
Sample pre-treatment:
  None
Sample volume: 100 µl of serial dilutions (four-fold: 1:50–
  1:12 000)
Number of tests: 96
Controls - standards run in assay:
  Controls: low avidity (serial dilutions), high avidity (serial
    dilutions)
Incubation:
  1 hr (37°C) + 1 hr (37°C) + 30 min (37°C)
Washes: 2*

CONTENTS

Antibodies, antigens, labelled components:
  Rubella virus Ag bound to well
  Anti-human IgG Ab (sh) AP conjugated
Substrate: PNP
Controls - standards supplied:
  Controls: low avidity and high avidity (human serum)
Additional reagents:
  NaOH
Special equipment required:
  Auto-EIA II analyser (optional)

INTERPRETATION

Comments on interpretation:
  Results are expressed as a ratio of high affinity IgG Ab
    titre:total IgG Ab titre
  Low avidity: < 15%; acute primary injection within last 3
    months
  Borderline avidity: 15–25%; retest and report average
  High avidity: > 25%; no primary infection within last 3
    months
No. of references: 12

NOTES

131339.0

*This assay distinguishes low affinity IgG Abs (recent
  primary infection) from high affinity IgG Ab (pre-existing
  immunity) by means of washing steps for paired samples
  with and without protein denaturant
**Heat treatment may slightly change assay results

© *KLUWER ACADEMIC PUBLISHERS 1995, ISSN 1381-5067*

Rubella virus
ANTIBODY DETECTION (IgG)

Manufacturer: Medix Biotech Inc
Cat. No./Trade name: KBF 2078/Rubella IgG EIA Kit

SUMMARY

[Well-Ag]–**Ab**–[AHIgG-HRP]–[OPD]–A_{490}

Assay type: EIA (non-competitive)
Detection: Colorimetric A_{490}
Format: Microtitre well, Ag coated
Sample type: Serum (do not heat inactivate)
Sample pre-treatment:
  None
Sample volume: 10 μl (+200 μl diluent)
Number of tests: 96
Controls - standards run in assay:
  Controls: Neg (1), high Pos (1)
  Calibrators: low Pos (4)
Incubation:
  20 min (RT) + 20 min (RT) + 10 min (RT)
Washes: 2

CONTENTS

Antibodies, antigens, labelled components:
  Rubella virus Ag bound to well
  Anti-human IgG Ab (g) HRP conjugated
Substrate: OPD
Controls - standards supplied:
  Controls: Neg and high Pos (human serum or plasma)
  Calibrator: low Pos (human serum or plasma)
Additional reagents:
  H_2SO_4
Special equipment required:
  None

INTERPRETATION

Comments on interpretation:
  Equivocal: where ratio of sample OD:cut-off value is
    between 0.9-1.10; report as equivocal
No. of references: None

NOTES

131163.0

Rubella virus
ANTIBODY DETECTION (IgG)

Manufacturer: Melotec S.A.
Cat. No./Trade name: /Melotest Rubella IgG

SUMMARY

[Well-Ag]–**Ab**–[AHIgG-HRP]–[TMB]–A_{450}

Assay type: EIA (non-competitive)
Detection: Colorimetric A_{450}
Format: Microtitre well, Ag coated
Sample type: Serum, plasma
Sample pre-treatment:
  None
Sample volume: 100 μl of 1:20 dilution x 2
Number of tests: 96
Controls - standards run in assay:
  Controls: Neg (1), low Pos (2), high Pos (1)
Incubation:
  20 min (RT) + 20 min (RT) + 10 min (RT)
Washes: 2

CONTENTS

Antibodies, antigens, labelled components:
  Rubella virus Ag bound to well
  Anti-human IgG Ab HRP conjugated
Substrate: TMB
Controls - standards supplied:
  Controls: Neg, low Pos and high Pos (sera)
Additional reagents:
  None
Special equipment required:
  None

INTERPRETATION

Comments on interpretation:
  Equivocal: ratio of sample OD:cut-off value between 0.9
    and 1.1; retest a fresh sample
  Repeatably equivocal: retest a fresh sample taken in 2-4
    weeks
No. of references: 9

NOTES

131124.0

© *KLUWER ACADEMIC PUBLISHERS 1995, ISSN 1381-5067*

Rubella virus
ANTIBODY DETECTION (IgG)

Manufacturer: Menarini Diagnostics
Cat. No./Trade name: M6138/HF Rubella IgG

SUMMARY

[Well-Ag]–**Ab**–[AHIgG-POD]–[TMB]–A_{450}

Assay type: EIA (non-competitive)
Detection: Colorimetric A_{450}
Format: Microtitre well, Ag coated
Sample type: Serum
Sample pre-treatment:
  None
Sample volume: 100 µl of 1:101 dilution
Number of tests: 96
Controls - standards run in assay:
  Controls: Neg (1), cut-off (1), Pos (1)
Incubation:
  45 min (37°C) + 45 min (37°C) + 15 min (RT)
Washes: 2

CONTENTS

Antibodies, antigens, labelled components:
  Rubella virus Ag bound to well
  Anti-human IgG MAb POD conjugated
Substrate: TMB
Controls - standards supplied:
  Controls: Neg, cut-off and Pos (serum)
Additional reagents:
  None
Special equipment required:
  None

INTERPRETATION

Comments on interpretation:
  No data is provided for interpretation of results
No. of references: None

NOTES

131090.0

Rubella virus
ANTIBODY DETECTION (IgG)

Manufacturer: PanBio
Cat. No./Trade name: RUG-100/Rubella IgG Elisa Test

SUMMARY

[Well-Ag]–**Ab**–[AHIgG-HRP]–[TMB]–A_{450}

Assay type: EIA (non-competitive)
Detection: Colorimetric A_{450}
Format: Microtitre well, Ag coated
Sample type: Serum
Sample pre-treatment:
  None
Sample volume: 100 µl of 1:100 dilution
Number of tests: 96
Controls - standards run in assay:
  Standards: A (1), B (1), C (1), D (1), E (1)
Incubation:
  20 min (37°C) + 20 min (37°C) + 10 min (RT)
Washes: 2

CONTENTS

Antibodies, antigens, labelled components:
  Rubella virus Ag bound to well
  Anti-human IgG Ab (sh) HRP conjugated
Substrate: TMB
Controls - standards supplied:
  Standards A-E (titre - 0, 10, 25, 50 and 100 IU/ml) -
    human serum
Additional reagents:
  None
Special equipment required:
  None

INTERPRETATION

Comments on interpretation:
  Equivocal: where the ratio cut-off value:sample OD is
    between 0.9-1.1; retest to confirm
No. of references: 7

NOTES

131394.0

© *KLUWER ACADEMIC PUBLISHERS 1995, ISSN 1381-5067*

Rubella virus
ANTIBODY DETECTION (IgG)

Manufacturer: Radim
Cat. No./Trade name: K2RG/Rubella IgG EIA Well

SUMMARY

[Well-Ag]–**Ab**–[AHIgG-HRP]–[TMB]–A_{450}

Assay type: EIA (non-competitive)
Detection: Colorimetric A_{450}
Format: Microtitre well, Ag coated
Sample type: Serum, plasma
Sample pre-treatment:
  None
Sample volume: 100 μl of 1:300 dilution
Number of tests: 96, 192
Controls - standards run in assay:
  Qualitative: Control: Neg (1): Standards: 1 (1)
  Quantitative: Control: Neg (1): Standards: 1 (1), 2 (1), 3
    (1), 4 (1), 5 (1)
Incubation:
  1 hr (37°C) + 30 min (37°C) + 10 min (37°C) or 15 min
  (RT)
Washes: 2

CONTENTS

Antibodies, antigens, labelled components:
  Rubella virus Ag bound to well
  Anti-human IgG Ab (g) HRP conjugated
Substrate: TMB
Controls - standards supplied:
  Controls: Neg (0, IU/ml - serum)
  Standards: 5 (15, 30, 60, 120, 240 IU/ml - serum)
Additional reagents:
  None
Special equipment required:
  None

INTERPRETATION

Comments on interpretation:
  Qualitative: Equivocal: within cut-off ±10%; retest to
    confirm
  Quantitative: Negative: < 15 IU/ml
  Weak Positive: between 15 and 20 IU/ml
  Positive: > 20 IU/ml
No. of references: 6

NOTES

131270.0

Rubella virus
ANTIBODY DETECTION (IgG)

Manufacturer: Roche Diagnostic Systems
Cat. No./Trade name: 07 5477 3/COBAS® CORE
Rubella IgG EIA recomb

SUMMARY

[Bead-Ag]–**Ab**–[AHIgG-HRP]–[TMB]–A_{450}

Assay type: EIA (non-competitive)
Detection: Colorimetric A_{450}
Format: Tube, Ag coated bead
Sample type: Serum, plasma
Sample pre-treatment:
  None
Sample volume: 3 μl
Number of tests: 100
Controls - standards run in assay:
  Controls: high Pos 2 (1)*
  Calibrators: a (2), b (2), c (2), d (2)
Incubation:
  15 min (37°C) + 30 min (37°C) + 15 min (37°C) (all with
  shaking)
Washes: 2

CONTENTS

Antibodies, antigens, labelled components:
  rec Rubella virus Ag bound to bead
  Anti-human IgG MAb (m) HRP conjugated
Substrate: TMB (bought separately)
Controls - standards supplied:
  Controls: 1 (low Pos) and 2 (high Pos) (human serum)
  Calibrators: a, b, c, d (0-500 IU/ml, human serum)
Additional reagents:
  Cobas® Core TMB Kit
  H_2SO_4 (for manual assay only)
Special equipment required:
  Cobas® Core immunoassay analyser (for automated
    method - optional)
  Cobas® Core EIA shaking incubator, tube washer and
    photometer (optional)
  Cobas® Core TORC negative control bead dispenser
  Cobas® Core reaction tubes

INTERPRETATION

Comments on interpretation:
  Negative: < 15 IU/ml
  Equivocal: between 10 and 15 IU/ml; retest a fresh
    sample taken in 2-3 weeks time in parallel with initial
    sample
  Positive: > 15 IU/ml
No. of references: None

NOTES

131154.0

*Control 1 (low Pos) may be run optionally in the sample rack
  in manual assay

Rubella virus
ANTIBODY DETECTION (IgG)

Manufacturer: Sanofi Diagnostics Pasteur
Cat. No./Trade name: U72910/Platelia® Rubella IgG

SUMMARY

[Well-Ag]–**Ab**–[AHIgG-HRP]–[OPD]–A_{492}

Assay type: EIA (non-competitive)
Detection: Colorimetric A_{492}
Format: Microtitre well, Ag coated
Sample type: Serum
Sample pre-treatment:
　None
Sample volume: 10 µl (+1 ml diluent)
Number of tests: 96
Controls - standards run in assay:
　Standards: Neg (1), Pos I (2), Pos II (1), Pos III (1)
Incubation:
　1 hr (37–40°C) + 1 hr (37–40°C) + 30 min (RT)
Washes: 2 (+ preliminary plate wash)

CONTENTS

Antibodies, antigens, labelled components:
　Rubella virus Ag bound to well
　Anti-human IgG MAb (m) HRP conjugated
Substrate: OPD
Controls - standards supplied:
　Standards: Neg, Pos I, Pos II and Pos III (human serum –
　　calibrated against WHO reference sera)
Additional reagents:
　None
Special equipment required:
　None

INTERPRETATION

Comments on interpretation:
　Equivocal: Ab levels between > 10 IU/ml and < 15 IU/ml;
　　retest a fresh sample in 3 weeks in parallel with initial
　　specimen
No. of references: 12

NOTES

131760.0

Rubella virus
ANTIBODY DETECTION (IgG)

Manufacturer: Sigma Diagnostics
Cat. No./Trade name: SIA 406/SAI® Rubella IgG

SUMMARY

[Well-Ag]–**Ab**–[AHIgG-AP]–[PMP]–A_{550}

Assay type: EIA (non-competitive)
Detection: Colorimetric A_{550}
Format:
Sample type: Serum (do not heat inactivate)
Sample pre-treatment:
　None
Sample volume: 10 µl (+200 µl diluent)
Number of tests: 96
Controls - standards run in assay:
　Controls: A (1), B (1)
　Calibrators: 1 (2), 2 (2), 3 (2)
Incubation:
　45 min (RT) + 45 min (RT) + 45 min (RT)
Washes: 2 (+ preliminary plate wash)

CONTENTS

Antibodies, antigens, labelled components:
　Rubella virus Ag bound to well
　Anti-human IgG Ab (g) AP conjugated
Substrate: PMP
Controls - standards supplied:
　Controls: A & B (human serum)
　Calibrators: 1, 2 and 3 (IgG Ab titre in AU/ml – human
　　serum)
Additional reagents:
　None
Special equipment required:
　Sigma EIA Multiwell Plate Reader (Cat. no. EQ104 –
　　optional)

INTERPRETATION

Comments on interpretation:
　Equivocal: 0.15–0.16 SIA® Rubella IgG Values, or 0.88–
　　0.99 Index Value; retest to confirm
　Repeatably equivocal: report as such and retest using
　　alternative method or fresh sample
　For information on reporting in IU/ml, contact
　　manufacturers
No. of references: 3

NOTES

131434.0

Rubella virus
ANTIBODY DETECTION (IgG)

Manufacturer: Sorin Biomedica
Cat. No./Trade name: P2859/ETI-RUBEK-G

SUMMARY

[Well-Ag]–**Ab**–[AHIgG-HRP]–[TMB]–A_{450}

Assay type: EIA (non-competitive)
Detection: Colorimetric A_{450}
Format: Microtitre well, Ag coated
Sample type: Serum, plasma
Sample pre-treatment:
  None
Sample volume: 100 µl of 1:505 dilution x 2*
Number of tests:
Controls - standards run in assay:
  Controls: Neg (2)
  Calibrators: 1 (2), 2 (2), 3 (2), 4 (2)
Incubation:
  1 hr (37°C) + 1 hr (37°C) + 30 min (RT)
Washes: 2

CONTENTS

Antibodies, antigens, labelled components:
  Rubella virus (Putnam strain) Ag bound to well
  Anti-human IgG IgG Ab (g) HRP conjugated
Substrate: TMB
Controls - standards supplied:
  Controls: Neg (human serum)
  Calibrators: 4 (10, 25, 50 and 100 IU/ml)
Additional reagents:
  None
Special equipment required:
  ETI-System Reader and ETI-System Washer (optional)

INTERPRETATION

Comments on interpretation:
  Equivocal: within cut-off ± 10%; retest to confirm
No. of references: 10

NOTES

131212.0

*To assist the automation of this assay a sample dilution of
  1:100 has also been validated

Rubella virus
ANTIBODY DETECTION (IgG)

Manufacturer: United Biotech Inc
Cat. No./Trade name: 1A-301/MAGIWEL® Rubella IgG

SUMMARY

[Well-Ag]–**Ab**–[AHIgG-HRP]–[TMB]–A_{450}

Assay type: EIA (non-competitive)
Detection: Colorimetric A_{450}
Format: Microtitre well, Ag coated
Sample type: Serum
Sample pre-treatment:
  None
Sample volume: 100 µl of 1:101 dilution
Number of tests: 96
Controls - standards run in assay:
  Reference Standard Set: Neg (1), Pos (1)
  Calibrators: 1 (1)
Incubation:
  30 min (RT) + 30 min (RT) + 15 min (RT)
Washes: 2

CONTENTS

Antibodies, antigens, labelled components:
  Rubella virus Ag bound to well
  Anti-human IgG Ab (g) HRP conjugated
Substrate: TMB
Controls - standards supplied:
  Reference Standard Set: Neg, Pos
  Calibrator: 1 (100 EU/ml)
Additional reagents:
  None
Special equipment required:
  None

INTERPRETATION

Comments on interpretation:
  Positive: ≥ 15 EU/ml
No. of references: 4

NOTES

131146.0

Rubella virus
ANTIBODY DETECTION (IgG)

Manufacturer: Zeus Scientific Inc
Cat. No./Trade name: 980 1GL/Rubella IgG Elisa Test System

SUMMARY

[Well-Ag]–**Ab**–[AHIgG-POD]–[TMB]–A_{450}

Assay type: EIA (non-competitive)
Detection: Colorimetric A_{450}
Format: Microtitre well, Ag coated
Sample type: Serum
Sample pre-treatment:
　None
Sample volume: 10 µl (+200 µl diluent)
Number of tests: 96
Controls - standards run in assay:
　Controls: Neg (1), low Pos (3), high Pos (1)
Incubation:
　20 min (RT) + 20 min (RT) + 10 min (RT)
Washes: 2

CONTENTS

Antibodies, antigens, labelled components:
　Rubella virus Ag bound to well
　Anti-human IgG Ab (g) POD conjugated
Substrate: PNP
Controls - standards supplied:
　Controls: Neg, low Pos and high Pos (human serum)
Additional reagents:
　None
Special equipment required:
　None

INTERPRETATION

Comments on interpretation:
　Equivocal: 4.5-6.5 IU/ml; retest to confirm
　Repeatably equivocal: retest using alternative procedure
No. of references: 12

NOTES

131294.0

Rubella virus
ANTIBODY DETECTION (IgM)

Manufacturer: Abbott Laboratories
Cat. No./Trade name: 7205-22/RUBAZYME® - M

SUMMARY

[Bead-Ag]–**Ab**–[AHIgM-HRP]–[OPD]–A_{492}

Assay type: EIA (non-competitive)
Detection: Colorimetric A_{492}
Format: Reaction wells, Ag coated beads
Sample type: Serum (do not heat inactivate)
Sample pre-treatment:
　None
Sample volume: 10 µl (+200 µl diluent)
Number of tests: 50
Controls - standards run in assay:
　Controls: Neg (1), low Pos (3), high Pos (1)
Incubation:
　1 hr (45°C) + 1 hr 30 min (45°C) + 1 hr 30 min (45°C) +
　30 min (RT)
Washes: 3

CONTENTS

Antibodies, antigens, labelled components:
　Rubella virus Ag bound to beads
　Anti-human IgM Ab (g) HRP conjugated
Substrate: OPD
Controls - standards supplied:
　Controls: Neg, low Pos and high Pos (human plasma)
Additional reagents:
　H_2SO_4
Special equipment required:
　Pentawash® II and Gast® Vacuum Pump (or similar)
　Quantum Analyser (optional)

INTERPRETATION

Comments on interpretation:
　Equivocal: where Rubazyme-M Index is between 0.910
　and 1.090; if initial specimen is an acute specimen,
　retest a second early convalescent specimen. If initial
　specimen is late convalescent specimen, retest acute
　specimen if available
No. of references: 13

NOTES

131353.0

© *KLUWER ACADEMIC PUBLISHERS 1995, ISSN 1381-5067*

Rubella virus
ANTIBODY DETECTION (IgM)

Manufacturer: Alfa Biotech
Cat. No./Trade name: 05772954/Rubella IgM Elisa System

SUMMARY

[Well-AHIgM]–**Ab**–[Ag:MAb]–[AMIg-HRP]–[TMB]–A$_{450}$

Assay type: EIA (non-competitive)
Detection: Colorimetric A$_{450}$
Format: Microtitre well, Ab coated
Sample type: Serum
Sample pre-treatment:
    None
Sample volume: 100 μl of 1:100 dilution x2
Number of tests: 96
Controls - standards run in assay:
    Controls: Neg (2), cut-off (4), Pos (2)
Incubation:
    20 min (37°C) + 20 min (37°C) + 20 min (37°C)
Washes: 2

CONTENTS

Antibodies, antigens, labelled components:
    Anti-human IgM Ab bound to well
    Rubella virus Ag
    Anti-Rubella virus MAb
    Anti-mouse Ig Ab (g) HRP conjugated
Substrate: TMB
Controls - standards supplied:
    Controls: Neg, cut-off and Pos (human serum)
Additional reagents:
    None
Special equipment required:
    None

INTERPRETATION

Comments on interpretation:
    Equivocal: within cut-off ±15%; retest to confirm
    Repeatably equivocal: retest a fresh sample in 2 weeks
No. of references: 5

NOTES

131103.0

Rubella virus
ANTIBODY DETECTION (IgM)

Manufacturer: Amico Laboratories Inc
Cat. No./Trade name: 6000M/AMIZYME® Rubella IgM Assay

SUMMARY

[Well-Ag]–**Ab**–[AHIgM-HRP]–[ABTS]–A$_{405}$

Assay type: EIA (non-competitive)
Detection: Colorimetric A$_{405}$
Format: Microtitre well, Ag coated
Sample type: Serum
Sample pre-treatment:
    None
Sample volume: 5 μl (+245 μl diluent)*
Number of tests: 96
Controls - standards run in assay:
    Controls: Neg (1)
    Calibrators: I (1)
Incubation:
    25 min (RT) + 25 min (RT) + 25 min (RT)
Washes: 2

CONTENTS

Antibodies, antigens, labelled components:
    Rubella virus Ag bound to well
    Anti-human IgM Ab (g) HRP conjugated
Substrate: ABTS
Controls - standards supplied:
    Controls: Neg
    Calibrator: 1
Additional reagents:
    None
Special equipment required:
    None

INTERPRETATION

Comments on interpretation:
    Equivocal: where sample OD is between 0.275–0.299
    (92–99 U/ml); retest to confirm
No. of references: 2

NOTES

131795.0

*Sample diluted in IgM serum diluent provided to eliminate interference due to RF and IgG

© *KLUWER ACADEMIC PUBLISHERS 1995, ISSN 1381-5067*

Rubella virus
ANTIBODY DETECTION (IgM)

Manufacturer: Amico Laboratories Inc
Cat. No./Trade name: 6000MC/AMIZYME® Rubella IgM Capture Assay

SUMMARY

[Well-Ag]–**Ab**–[AHIgM]–[AMIg-HRP]–[ABTS]–A_{405}

Assay type: EIA (non-competitive)
Detection: Colorimetric A_{405}
Format: Microtitre well, Ag coated
Sample type: Serum
Sample pre-treatment:
 None
Sample volume: 5 μl (+245 μl diluent)
Number of tests: 96
Controls - standards run in assay:
 Controls: Neg (1)
 Calibrators: I (1)
Incubation:
 25 min (RT) + 25 min (RT) + 25 min (RT) + 25 min (RT)
Washes: 3

CONTENTS

Antibodies, antigens, labelled components:
 Rubella virus Ag bound to well
 Anti-human IgM MAb (m)
 Anti-mouse Ig Ab (g) HRP conjugated
Substrate: ABTS
Controls - standards supplied:
 Controls: Neg
 Calibrator: 1
Additional reagents:
 None
Special equipment required:
 None

INTERPRETATION

Comments on interpretation:
 Equivocal: where sample OD is between 0.285–0.299; retest to confirm
No. of references: 2

NOTES

131805.0

Rubella virus
ANTIBODY DETECTION (IgM)

Manufacturer: BAG-Biologische Analysensystem GmbH
Cat. No./Trade name: 5273 BAG - Rubella - EIA - M

SUMMARY

[Well-Ag]–**Ab**–[AHIgM-HRP]–[TMB]–A_{450}

Assay type: EIA (non-competitive)
Detection: Colorimetric A_{450}
Format: Microtitre well, Ag coated
Sample type: Serum (do not heat inactivate)
Sample pre-treatment:
 Treat diluted serum with BAG - IgM - Sep - System (Cat. no. 5540) to eliminate interference due to RF and IgG
Sample volume: 100 μl of treated sample
Number of tests: 96
Controls - standards run in assay:
 Controls: Neg (1), cut-off (2), Pos (1)
Incubation:
 30 min (RT) + 30 min (RT) + 10 min (RT)
Washes: 2

CONTENTS

Antibodies, antigens, labelled components:
 Rubella virus (strain HPV-77) Ag bound to well
 Anti-human IgM Ab (sh) HRP conjugated
Substrate: TMB
Controls - standards supplied:
 Controls: Neg, cut-off and Pos (human serum)
Additional reagents:
 None
Special equipment required:
 None

INTERPRETATION

Comments on interpretation:
 Equivocal: within cut-off ±10%; data is not provided for further assessment of samples within this range
No. of references: None

NOTES

131611.0

© KLUWER ACADEMIC PUBLISHERS 1995, ISSN 1381-5067

Rubella virus
ANTIBODY DETECTION (IgM)

Manufacturer: Behringwerke AG
Cat. No./Trade name: Enzygnost® Anti-Rubella virus/ IgM

SUMMARY

[Well-Ag]–**Ab**–[AHIgM-POD]–[TMB]–A$_{450}$

Assay type: EIA (non-competitive)
Detection: Colorimetric A$_{450}$
Format: Microtitre well, Ag coated
Sample type: Serum, plasma
Sample pre-treatment:
Add RF absorbent provided to diluted sample, incubate 15 min (RT) or overnight (2–8°C) to eliminate interference due to RF and IgG
Sample volume: 150 µl of treated sample x 2*
Number of tests: 96
Controls - standards run in assay:
Reference reagents: P/N (2), P/P (2)*
Incubation:
1 hr (37°C) + 1 hr (37°C) + 30 min (RT)
Washes: 2

CONTENTS

Antibodies, antigens, labelled components:
Rubella virus from BHK cells bound to well
Control Ag from uninfected BHK cells bound to well
Anti-human IgM Ab (g) POD conjugated
Substrate: TMB (bought separately)
Controls - standards supplied:
Reference reagent: P/P and P/N (Anti-Rubella Virus Reference Reagents)
Additional reagents:
Supplementary reagents for Enzygnost/TMB (Cat. no. OUVP)
Special equipment required:
Behring ELISA Processor II (optional)
Behring ELISA Processor III (optional)

INTERPRETATION

Comments on interpretation:
Equivocal: where sample OD is between 0.10 and 0.20; retest to confirm
Repeatably equivocal: report as equivocal and retest a fresh specimen at a later date
No. of references: 33

NOTES

131028.0

*Samples and controls are assayed simultaneously in Ag coated and control Ag coated wells

---

Rubella virus
ANTIBODY DETECTION (IgM)

Manufacturer: BIOKIT SA
Cat. No./Trade name: 3000-1231/bioelisa RUBELLA IgM (Immunocapture)

SUMMARY

[Well-AHIgM]–**Ab**–[Ag-HRP]–[TMB]–A$_{450}$

Assay type: EIA (non-competitive)
Detection: Colorimetric A$_{450}$
Format: Microtitre well, Ab coated
Sample type: Serum (do not heat inactivate)
Sample pre-treatment:
None
Sample volume: 100 µl of 1:101 dilution
Number of tests: 96
Controls - standards run in assay:
Controls: Neg (2), cut-off (4), Pos (2)
Incubation:
1 hr (37-40°C) + 1 hr (37-40°C) + 30 min (RT)
Washes: 2

CONTENTS

Antibodies, antigens, labelled components:
Anti-human IgM Ab (r) bound to well
Rubella virus (strain HPV77 from VERO cells) Ag HRP conjugated*
Substrate: TMB
Controls - standards supplied:
Neg, cut-off and Pos (human serum)
Additional reagents:
None
Special equipment required:
None

INTERPRETATION

Comments on interpretation:
Equivocal: within cut-off ±10%; retest to confirm
No. of references: 4

NOTES

131375.0

*Control Ag consisting of uninfected cellular components is added to conjugate to reduce non-specific activity

Rubella virus
ANTIBODY DETECTION (IgM)

Manufacturer: bioMerieux
Cat. No./Trade name: 30 214/VIDAS RUB IgM

SUMMARY

[Solid phase-AHIgM]–**Ab**–[Ag]–[MAb-AP]–[4MP]–
fluorescence

Assay type: EIA (non-competitive)
Detection: Fluorometric
Format: Solid phase receptacle (SPR), Ab coated
Sample type: Serum
Sample pre-treatment:
None
Sample volume: 100 µl
Number of tests: 30
Controls - standards run in assay:
Controls: Neg (1), Pos (1)
Standards: Pos (1)
Incubation:
Automated - total time 1 hr
Washes: Automated

CONTENTS

Antibodies, antigens, labelled components:
Anti-human IgM Ab (g) bound to solid phase receptacle
Rubella virus Ag
Anti-Rubella virus MAb (m) AP conjugated
Substrate:
Controls - standards supplied:
Controls: Neg and Pos (human serum)
Standards: Pos (human serum)
Additional reagents:
None
Special equipment required:
Vitek immunodiagnostic assay system (VIDAS)

INTERPRETATION

Comments on interpretation:
This is an automated assay
Results designated equivocal; retest to confirm
Repeatably equivocal: retest using a fresh sample in 2-3
weeks
No. of references: 4

NOTES

131310.0

Rubella virus
ANTIBODY DETECTION (IgM)

Manufacturer: BioWhittaker Inc
Cat. No./Trade name: 30-347U/Rubella Stat M

SUMMARY

[Well-Ag]–**Ab**–[AHIgM-AP]–[PMP]–A_{550}

Assay type: EIA (non-competitive)
Detection: Colorimetric A_{550}
Format: Microtitre well, Ag coated
Sample type: Serum (do not heat inactivate)
Sample pre-treatment:
Mix sample with Pretreatment Serum, incubate 30 min
(RT) and centrifuge to eliminate interference due to
IgG and RF
Sample volume: 25 µl of treated sample (+ 100 µl diluent)
Number of tests: 96
Controls - standards run in assay:
Calibrators: Neg (1), Pos 1 (1), Pos 2 (1)
Incubation:
45 min (RT) + 30 min (RT) + 30 min (RT) (all with
shaking)
Washes: 2 (+ preliminary plate wash)

CONTENTS

Antibodies, antigens, labelled components:
Rubella virus Ag bound to well
Anti-human IgM Ab (g) AP conjugated
Substrate: PMP
Controls - standards supplied:
Calibrators: Neg, Pos 1, Pos 2 (all human serum)
Rheumatoid Factor Control (human serum)
Additional reagents:
None
Special equipment required:
BioWhittaker automated system and software (optional)

INTERPRETATION

Comments on interpretation:
Equivocal: test value between 0.10–0.12; indicates
probable positive, retest in duplicate to confirm
No. of references: 11

NOTES

131754.0

© *KLUWER ACADEMIC PUBLISHERS 1995, ISSN 1381-5067*

Rubella virus
ANTIBODY DETECTION (IgM)

Manufacturer: Bouty SpA
Cat. No./Trade name: 22292/BEIA Rubella-IgM Capture

SUMMARY

[Well-AHIgM]–**Ab**–[Ag-biotin]–[strept–HRP]–[TMB]–A_{450}

Assay type: EIA (non-competitive)
Detection: Colorimetric A_{450}
Format: Microtitre well, Ab coated
Sample type: Serum
Sample pre-treatment:
   None
Sample volume: 10 µl (+800 µl diluent)*
Number of tests: 96
Controls - standards run in assay:
   Controls: Neg (2), cut-off (2), Pos (2)
Incubation:
   1 hr (RT) + 1 hr (RT) + 30 min (RT)
Washes: 2

CONTENTS

Antibodies, antigens, labelled components:
   Anti-human IgM IgG Ab bound to well
   Rubella virus Ag biotinylated
   Streptavidin HRP conjugated
Substrate: TMB
Controls - standards supplied:
   Controls: Neg, cut-off and Pos (human serum)
Additional reagents:
   None
Special equipment required:
   None

INTERPRETATION

Comments on interpretation:
   Equivocal: within cut-off ±5%; retest to confirm
   Repeatably equivocal: repeat test with a fresh sample
No. of references: None

NOTES

131240.0

*Samples can be assayed singly or in duplicate

Rubella virus
ANTIBODY DETECTION (IgM)

Manufacturer: Chimica Diagnostica
Cat. No./Trade name: 41450/Rubella Virus IgM

SUMMARY

[Well-AHIgM]–**Ab**–[Ag]–[MAb-POD]–[TMB]–A_{450}

Assay type: EIA (non-competitive)
Detection: Colorimetric A_{450}
Format: Microtitre well, Ab coated
Sample type: Serum
Sample pre-treatment:
   None
Sample volume: 100 µl of 1:50 dilution x 2*
Number of tests: 96
Controls - standards run in assay:
   Standards: 1 (2), 2 (2), 3 (2), 4 (2), 5 (2), 6 (2)
Incubation:
   1 hr (37°C) + 1 hr (37°C) + 20 min (RT)
Washes: 2

CONTENTS

Antibodies, antigens, labelled components:
   Anti-human IgM MAb bound to well
   Rubella virus Ag
   Anti-Rubella virus MAb POD conjugated
Substrate: TMB
Controls - standards supplied:
   Standards: 1–6 (titres 0, 5, 10, 40, 80 and 160 IU/ml)
Additional reagents:
   None
Special equipment required:
   None

INTERPRETATION

Comments on interpretation:
   Results are expressed quantitatively in IU/ml with
   reference to the standards provided
No. of references: 2

NOTES

131367.0

*Diluent contains AHIgG to eliminate interference due to RF
  and IgG in sample

© *KLUWER ACADEMIC PUBLISHERS 1995, ISSN 1381-5067*

Rubella virus
ANTIBODY DETECTION (IgM)

Manufacturer: Clark Laboratories
Cat. No./Trade name: Rubella Virus IgM Elisa Test

SUMMARY

[Well-Ag]–**Ab**–[AHIgM-HRP]–[OPD]–A_{490}

Assay type: EIA (non-competitive)
Detection: Colorimetric A_{490}
Format: Microtitre well, Ag coated
Sample type: Serum (do not heat inactivate)
Sample pre-treatment:
  Add Absorbent Solution provided to diluted serum and incubate 20 min (RT) to eliminate interference due to RF and IgG
Sample volume: 100 μl of treated sample
Number of tests: 96
Controls - standards run in assay:
  Controls: Neg (1), Pos (1)
  Calibrators: 1 (2)
Incubation:
  20 min (RT) + 20 min (RT) + 10 min (RT)
Washes: 2

CONTENTS

Antibodies, antigens, labelled components:
  Rubella virus Ag bound to well
  Anti-human IgM Ab (g) HRP conjugated
Substrate: TMB*
Controls - standards supplied:
  Controls: Neg and Pos (human serum or plasma)
  Calibrator: 1 (human serum or plasma)
Additional reagents:
  H_2SO_4
Special equipment required:
  None

INTERPRETATION

Comments on interpretation:
  Classification of samples is according to cut-off; no further testing
No. of references: None

NOTES

131667.0

*OPD can be supplied as an alternative

Rubella virus
ANTIBODY DETECTION (IgM)

Manufacturer: Dade International Inc (Bartels Division)
Cat. No./Trade name: B1029-325/Bartels PRIMA System® Rubella IgM EIA

SUMMARY

[Well-Ag]–**Ab**–[AHIgM-enzyme]–[PNP]–A_{405}

Assay type: EIA (non-competitive)
Detection: Colorimetric A_{405}
Format: Microtitre well, Ag coated
Sample type: Serum
Sample pre-treatment:
  None
Sample volume: 100 μl of 1:11 dilution*
Number of tests: 96
Controls - standards run in assay:
  Controls: Neg (1), Pos (1)
  Reference serum: (1)
Incubation:
  30 min (37°C) + 30 min (37°C) + 30 min (37°C)
Washes: 2

CONTENTS

Antibodies, antigens, labelled components:
  Rubella virus (HPV 777 strain) Ag bound to well
  Anti-human IgM Ab enzyme conjugated
Substrate: PNP
Controls - standards supplied:
  Controls: Neg and Pos
  Reference serum: 1
Additional reagents:
  None
Special equipment required:
  None

INTERPRETATION

Comments on interpretation:
  Classification of samples is according to cut-off; no further testing
No. of references: 10

NOTES

131046.0

*Specimen diluent contains an absorbent to eliminate interference due to RF and IgG

© KLUWER ACADEMIC PUBLISHERS 1995, ISSN 1381-5067

Rubella virus
ANTIBODY DETECTION (IgM)

Manufacturer: Denka Seiken Co. Ltd
Cat. No./Trade name: 321770/Rubella IgM (II) – EIA "SEIKEN"

SUMMARY

[Well-AHIgM]–**Ab**–[Ag]–[MAb-POD]–[TMB]–A$_{450}$

Assay type: EIA (non-competitive)
Detection: Colorimetric A$_{450}$
Format: Microtitre well, Ab coated
Sample type: Serum, plasma
Sample pre-treatment:
　None
Sample volume: 10 μl (+2 ml diluent)
Number of tests: 96
Controls - standards run in assay:
　Standards: I (1), II (2), III (1)
Incubation:
　1 hr (RT) + 1 hr (RT) + 1 hr (RT) + 30 min (RT)
Washes: 3

CONTENTS

Antibodies, antigens, labelled components:
　Anti-human IgM MAb (m) bound to well
　Rubella virus Ag
　Anti-Rubella virus MAb (m) POD conjugated
Substrate: TMB
Controls - standards supplied:
　Standards: I (Neg), II (low titre) and III (high titre) (all human serum)
Additional reagents:
　None
Special equipment required:
　None

INTERPRETATION

Comments on interpretation:
　Equivocal: where the calculated IgM Antibody Index is between 0.8 and 1.2; repeat test with a fresh specimen in 1–2 weeks or use an alternative method
No. of references: None

NOTES

131689.0

Rubella virus
ANTIBODY DETECTION (IgM)

Manufacturer: Diamedix Corporation
Cat. No./Trade name: 783-370/Rubella IgM Microassay

SUMMARY

[Well-Ag]–**Ab**–[AHIgM-AP]–[PNP]–A$_{405}$

Assay type: EIA (non-competitive)
Detection: Colorimetric A$_{405}$
Format: Microtitre well, Ag coated
Sample type: Serum (do not heat inactivate)
Sample pre-treatment:
　None
Sample volume: 5 μl (+200 μl diluent)*
Number of tests: 96
Controls - standards run in assay:
　Controls: Neg (1), Pos (1)
　Calibrator: 1 (1)
Incubation:
　30 min (RT) + 30 min (RT) + 30 min (RT)
Washes: 2

CONTENTS

Antibodies, antigens, labelled components:
　Rubella virus (HPV-77) Ag bound to well
　Anti-human IgM Ab AP conjugated
Substrate: PNP
Controls - standards supplied:
　Controls: Neg and Pos
　Calibrator: 1 (assigned Ab titre)
Additional reagents:
　None
Special equipment required:
　None

INTERPRETATION

Comments on interpretation:
　Positive: ⩾40 Elisa units/ml
No. of references: 2

NOTES

131052.0

*Specimen Diluent contains an aggregated human IgG to reduce interference due to RF and IgG in sample

Rubella virus
ANTIBODY DETECTION (IgM)

Manufacturer: Diesse
Cat. No./Trade name: 91031/ENZYWELL Rubella IgM*

SUMMARY

[Well-AHIgM]–**Ab**–[Ag]–[MAb-POD]–[TMB]–A_{450}

Assay type: EIA (non-competitive)
Detection: Colorimetric A_{450}
Format: Microtitre well, Ab coated
Sample type: Serum
Sample pre-treatment:
 None
Sample volume: 100 µl of 1:101 dilution x 2
Number of tests: 96
Controls - standards run in assay:
 Controls: Neg (2), cut-off (3), Pos (2)
Incubation:
 45 min (37°C) + 45 min (37°C) + 15 min (RT)
Washes: 2

CONTENTS

Antibodies, antigens, labelled components:
 Anti-human IgM MAb bound to well
 Rubella virus Ag
 Anti-Rubella virus MAb POD conjugated
Substrate: TMB
Controls - standards supplied:
 Controls: Neg, cut-off and Pos (human serum)
Additional reagents:
 None
Special equipment required:
 None

INTERPRETATION

Comments on interpretation:
 Equivocal: within cut-off ± 10%; retest to confirm
 Repeatably equivocal: repeat with a fresh sample
No. of references: 4

NOTES

131449.0

*In the ENZYWELL Rubella IgM Plus Kit (Cat. no. 91032) the
 contents and method are the same but the
 Immunocomplex remains stable for 2 weeks after
 reconstitution

Rubella virus
ANTIBODY DETECTION (IgM)

Manufacturer: Gull Laboratories Inc
Cat. No./Trade name: RUE150/Rubella IgM ELISA Test

SUMMARY

[Well-Ag]–**Ab**–[AHIgM-AP]–[PNP]–A_{405}

Assay type: EIA (non-competitive)
Detection: Colorimetric A_{405}
Format: Microtitre well, Ag coated
Sample type: Serum
Sample pre-treatment:
 None
Sample volume: 100 µl of 1:11 dilution*
Number of tests: 96
Controls - standards run in assay:
 Controls: Neg (1), Pos (1)
 Reference Serum: 1 (3)
Incubation:
 30 min (37°C) + 30 min (37°C) + 30 min (37°C)
Washes: 2

CONTENTS

Antibodies, antigens, labelled components:
 Rubella virus (HPV 77 strain) Ag bound to well
 Anti-human IgM Ab (caprine) AP conjugated
Substrate: PNP
Controls - standards supplied:
 Controls: Neg and Pos
 Reference serum: 1 (all human serum)
Additional reagents:
 None
Special equipment required:
 None

INTERPRETATION

Comments on interpretation:
 Equivocal: where sample OD is between mean OD of
  Reference Serum and mean OD of Reference Serum
  x 0.9; retest to confirm
 Repeatably equivocal: retest using an alternative method
No. of references: 13

NOTES

131700.0

*Diluent contains an absorbent to eliminate interference due
 to RF and IgG in serum

© KLUWER ACADEMIC PUBLISHERS 1995, ISSN 1381-5067

Rubella virus
ANTIBODY DETECTION (IgM)

Manufacturer: Human GmbH
Cat. No./Trade name: 51108/Rubella Virus IgM Elisa

SUMMARY

[Well-Ag]–**Ab**–[AHIgM-HRP]–[TMB]–A_{450}

Assay type: EIA (non-competitive)
Detection: Colorimetric A_{450}
Format: Microtitre well, Ag coated
Sample type: Serum
Sample pre-treatment:
  Dilute sample with Dilution Buffer IgM and incubate 5 min
    (RT) to eliminate interference due to RF and IgG
Sample volume: 100 µl of prepared sample
Number of tests: 96
Controls - standards run in assay:
  Controls: Neg (2), Pos (2)
Incubation:
  30 min (RT) + 30 min (RT) + 15 min (RT)
Washes: 2

CONTENTS

Antibodies, antigens, labelled components:
  Rubella virus Ag bound to well
  Anti-human IgM Ab (r) HRP conjugated
Substrate: TMB
Controls - standards supplied:
  Controls: Neg and Pos (human)
Additional reagents:
  None
Special equipment required:
  None

INTERPRETATION

Comments on interpretation:
  Equivocal: within cut-off ± 15%; retest in parallel with a
    fresh sample taken 7–14 days later
No. of references: 5

NOTES

131346.0

Rubella virus
ANTIBODY DETECTION (IgM)

Manufacturer: Human GmbH
Cat. No./Trade name: 51118/Rubella Virus direct IgM Elisa

SUMMARY

[Well-AHIgM]–**Ab**–[Ag:MAb-HRP]–[TMB]–A_{450}

Assay type: EIA (non-competitive)
Detection: Colorimetric A_{450}
Format: Microtitre well, Ab coated
Sample type: Serum
Sample pre-treatment:
  None
Sample volume: 10 µl
Number of tests: 96
Controls - standards run in assay:
  Controls: Neg (1), cut-off (2), Pos (1)
Incubation:
  30 min (37°C) + 30 min (37°C) + 15 min (RT)
Washes: 2

CONTENTS

Antibodies, antigens, labelled components:
  Anti-human IgM Ab (g) bound to well
  Rubella virus Ag
  Anti-Rubella virus MAb HRP conjugated
Substrate: TMB
Controls - standards supplied:
  Controls: Neg, cut-off, and Pos (human serum)
Additional reagents:
  None
Special equipment required:
  None

INTERPRETATION

Comments on interpretation:
  Equivocal: where ratio of mean OD of cut-off control:OD
    of sample (patient index) is ⩽1.0; retest in parallel
    with a fresh sample taken 7–14 days later
No. of references: 5

NOTES

131350.0

© KLUWER ACADEMIC PUBLISHERS 1995, ISSN 1381-5067

Rubella virus
ANTIBODY DETECTION (IgM)

Manufacturer: IFCI Clone Systems
Cat. No./Trade name: EIAGEN Rubella IgM

SUMMARY

[Well-AHIgM]–**Ab**–[Ag:MAb-POD]–[TMB]–A_{450}

Assay type: EIA (non-competitive)
Detection: Colorimetric A_{450}
Format: Microtitre well, Ab coated
Sample type: Serum
Sample pre-treatment:
    None
Sample volume: 10 μl (+1 ml diluent)*
Number of tests: 96
Controls - standards run in assay:
    Controls: Neg (2), cut-off (3), Pos (2)
Incubation:
    45 min (37°C) + 45 min (37°C) + 15 min (RT)
Washes: 2

CONTENTS

Antibodies, antigens, labelled components:
    Anti-human IgM MAb bound to well
    Rubella virus Ag
    Anti-Rubella virus MAb POD conjugated
Substrate: TMB
Controls - standards supplied:
    Controls: Neg, cut-off and Pos (human serum)
Additional reagents:
    None
Special equipment required:
    None

INTERPRETATION

Comments on interpretation:
    Equivocal: within cut-off ±10%; retest to confirm
    Repeatably equivocal: retest a fresh sample
No. of references: 8

NOTES

131678.0

*Samples are assayed in duplicate

Rubella virus
ANTIBODY DETECTION (IgM)

Manufacturer: Immuno Pharmacology Research
Cat. No./Trade name: RUBELLA IgM ELISA

SUMMARY

[Well-Ag]–**Ab**–[AHIgM-AP]–[PNP]–A_{405}

Assay type: EIA (non-competitive)
Detection: Colorimetric A_{450}
Format: Microtitre well, Ag coated
Sample type: Serum
Sample pre-treatment:
    None
Sample volume: 200 μl of 1:100 dilution x 2
Number of tests: 48
Controls - standards run in assay:
    Controls: Neg (1), Pos (1)
Incubation:
    1 hr (37°C) + 1 hr (37°C) + 30 min (37°C)
Washes: 2

CONTENTS

Antibodies, antigens, labelled components:
    Rubella virus Ag bound to well
    Anti-human IgM Ab AP conjugated
Substrate: PNP
Controls - standards supplied:
    Controls: Neg and Pos
Additional reagents:
    None
Special equipment required:
    None

INTERPRETATION

Comments on interpretation:
    Equivocal: where sample OD is between 0.150 and
        0.300; retest to confirm
    Repeatably equivocal: retest a fresh specimen in 15 days
No. of references: None

NOTES

131322.0

© *KLUWER ACADEMIC PUBLISHERS 1995, ISSN 1381-5067*

Rubella virus
ANTIBODY DETECTION (IgM)

Manufacturer: Incstar
Cat. No./Trade name: 5540/Rubella IgM Clin-ELISA® Kit

SUMMARY

[Well-Ag]–**Ab**–[AHIgM-AP]–[PNP]–A_{405}

Assay type: EIA (non-competitive)
Detection: Colorimetric A_{405}
Format: Microtitre well, Ag coated
Sample type: Serum
Sample pre-treatment:
  Dilute of serum with Serum and Control Pretreatment Reagent and incubate 30 min (RT) to eliminate interference due to RF and IgG
Sample volume: 200 μl of treated sample
Number of tests: 96
Controls - standards run in assay:
  Controls: Neg (1), Pos (1)
  Calibrators: I-Neg (2), II-low Pos (2), III-high Pos (2)
Incubation:
  30 min (RT) + 30 min (RT) + 45 min (RT)
Washes: 2

CONTENTS

Antibodies, antigens, labelled components:
  Rubella virus Ag bound to well
  Anti-human IgM Ab (g) AP conjugated
Substrate: PNP
Controls - standards supplied:
  Controls: Neg and Pos (human serum)
  Calibrators: I-Neg, II-low Pos, III-high Pos (human serum)
Additional reagents:
  NaOH
Special equipment required:
  None

INTERPRETATION

Comments on interpretation:
  Equivocal: ELISA value between 90-100; retest to confirm or retest in parallel with a fresh sample
  Repeatably equivocal: report as equivocal antibody status
No. of references: 5

NOTES

131083.0

Rubella virus
ANTIBODY DETECTION (IgM)

Manufacturer: Kreatech Diagnostics
Cat. No./Trade name: EL-3002-mRUB/Rubella IgM EIA

SUMMARY

[Well-AHIgM]–**Ab**–[Ag–HRP]–[TMB]–A_{450}

Assay type: EIA (non-competitive)
Detection: Colorimetric A_{450}
Format: Microtitre well, Ab coated
Sample type: Serum (do not heat inactivate)
Sample pre-treatment:
  None
Sample volume: 10 μl (+1 ml diluent)*
Number of tests: 96
Controls - standards run in assay:
  Controls: Neg (2), cut-off (4), Pos (2)
Incubation:
  1 hr (37°C) + 1 hr (37°C) + 30 min (RT)
Washes: 2

CONTENTS

Antibodies, antigens, labelled components:
  Anti-human IgM Ab bound to well
  Rubella virus Ag from infected Vero cells HRP conjugated**
Substrate: TMB
Controls - standards supplied:
  Controls: Neg, cut-off and Pos (human serum)
Additional reagents:
  None
Special equipment required:
  None

INTERPRETATION

Comments on interpretation:

No. of references: None

NOTES

131086.0

*Samples are assayed in duplicate
**Control antigen from uninfected Vero cells is added to conjugate to reduce non-specific activity

© KLUWER ACADEMIC PUBLISHERS 1995, ISSN 1381-5067

Rubella virus
ANTIBODY DETECTION (IgM)

Manufacturer: Laboratoire Eurobio
Cat. No./Trade name: 900276/Rubella IgM ELIT®

SUMMARY

[Well-AHIgM]–**Ab**–[Ag]–[Ab-POD]–[TMB]–A_{450}

Assay type: EIA (non-competitive)
Detection: Colorimetric A_{450}
Format: Microtitre well, Ab coated
Sample type: Serum, plasma
Sample pre-treatment:
  None
Sample volume: 100 µl
Number of tests: 96
Controls - standards run in assay:
  Standards: I (1), II (2), III (1)
Incubation:
  1 hr (RT) + 1 hr (RT) + 1 hr (RT) + 30 min (RT)
Washes: 3

CONTENTS

Antibodies, antigens, labelled components:
  Anti-human IgM MAb bound to well
  Rubella virus Ag
  Anti-Rubella virus Ab POD conjugated
Substrate: TMB
Controls - standards supplied:
  Standards: 3 (human serum with defined Ab values)
Additional reagents:
  None
Special equipment required:
  None

INTERPRETATION

Comments on interpretation:
  Equivocal: where IgM index is beween 0.8 and 1.2; retest
  a fresh sample in 1–2 weeks or test using an
  alternative method
No. of references: None

NOTES

131732.0

Rubella virus
ANTIBODY DETECTION (IgM)

Manufacturer: Labsystems Oy
Cat. No./Trade name: 61 089 201/Rubella IgM EIA

SUMMARY

[Well-Ag]–**Ab**–[AHIgM-AP]–[PNP]–A_{405}

Assay type: EIA (non-competitive)
Detection: Colorimetric A_{405}
Format: Microtitre well, Ag coated
Sample type: Serum*
Sample pre-treatment:
  None
Sample volume: 100 µl of 1:100 dilution x 2
Number of tests: 96
Controls - standards run in assay:
  Controls: Neg (2), Pos (2)
Incubation:
  1 hr (37°C) + 1 hr (37°C) + 30 min (37°C)
Washes: 2

CONTENTS

Antibodies, antigens, labelled components:
  Rubella virus Ag bound to well
  Anti-human IgM Ab (sh) AP conjugated
Substrate: PNP
Controls - standards supplied:
  Controls: Neg and Pos (human serum)
Additional reagents:
  NaOH
Special equipment required:
  Auto-EIA II analyser (optional)

INTERPRETATION

Comments on interpretation:
  Results are expressed quantitatively in Enzyme
  Immunoassay Units (EIU)
  Retest all positive sera after treatment with Labsystems
  IgG blocking reagent (Cat. no. 6106020) to eliminate
  false positive due to RF and IgG in sample
No. of references: 9

NOTES

131337.0

*Heat treatment may slightly change assay results

Rubella virus
ANTIBODY DETECTION (IgM)

Manufacturer: Medix Biotech Inc
Cat. No./Trade name: KBF 2079-2/Rubella IgM EIA Kit

SUMMARY

[Well-Ag]–**Ab**–[AHIgM-HRP]–[OPD]–A_{490}

Assay type: EIA (non-competitive)
Detection: Colorimetric A_{490}
Format: Microtitre well, Ag coated
Sample type: Serum (do not heat inactivate)
Sample pre-treatment:
Add absorbent solution provided to diluted controls, calibrator and samples, incubate 20 min (RT) to eliminate interference due to RF & IgG
Sample volume: 100 µl of treated sample x 2*
Number of tests: 96
Controls - standards run in assay:
Controls: Neg (2), high Pos (2)*
Calibrators: low Pos (4)
Incubation:
20 min (RT) + 20 min (RT) + 10 min (RT)
Washes: 2

CONTENTS

Antibodies, antigens, labelled components:
Rubella virus Ag bound to well
Control Ag bound to well
Anti-human IgM Ab (g) HRP conjugated
Substrate: OPD
Controls - standards supplied:
Controls: Neg and high Pos (human serum or plasma)
Calibrator: low Pos (human serum or plasma)
Additional reagents:
H_2SO_4
Special equipment required:
None

INTERPRETATION

Comments on interpretation:
Equivocal: where ratio of sample OD:cut-off value is between 0.9-1.10; report as equivocal
No. of references: None

NOTES

131164.0

*Samples and controls are assayed simultaneously in Ag coated and Control Ag coated wells

Rubella virus
ANTIBODY DETECTION (IgM)

Manufacturer: Melotec S.A.
Cat. No./Trade name: /Melotest Rubella IgM

SUMMARY

[Well-Ag]–**Ab**–[AHIgM-HRP]–[TMB]–A_{450}

Assay type: EIA (non-competitive)
Detection: Colorimetric A_{450}
Format: Microtitre well, Ag coated
Sample type: Serum, plasma
Sample pre-treatment:
Add absorbent solution provided to controls and diluted samples, incubate 20 min (RT) to eliminate interference due to RF & IgG in sample
Sample volume: 100 µl of treated sample x 2*
Number of tests: 96
Controls - standards run in assay:
Controls: Neg (1), low Pos (2), high Pos (1)
Incubation:
20 min (RT) + 20 min (RT) + 10 min (RT)
Washes: 2

CONTENTS

Antibodies, antigens, labelled components:
Rubella Ag bound to well
Tissue culture Control Ag bound to well
Anti-human IgM Ab HRP conjugated
Substrate: TMB
Controls - standards supplied:
Controls: Neg, low Pos and high Pos (sera)
Additional reagents:
None
Special equipment required:
None

INTERPRETATION

Comments on interpretation:
Equivocal: ratio of sample OD:cut-off value between 0.9 and 1.1; retest a fresh sample
Repeatably equivocal: retest a fresh sample taken in 2-4 weeks
No. of references: 9

NOTES

131130.0

*Samples and controls are assayed simultaneously in Ag coated and Control Ag coated wells to eliminate any interference due to ANA not neutralized by absorbent solution

© KLUWER ACADEMIC PUBLISHERS 1995, ISSN 1381-5067

Rubella virus
ANTIBODY DETECTION (IgM)

Manufacturer: Menarini Diagnostics
Cat. No./Trade name: M6139/HF Rubella IgM

SUMMARY

[Well-AHIgM]–**Ab**–[Ag]–[MAb-POD]–[TMB]–A$_{450}$

Assay type: EIA (non-competitive)
Detection: Colorimetric A$_{450}$
Format: Microtitre well, Ab coated
Sample type: Serum
Sample pre-treatment:
   None
Sample volume: 100 µl of 1:101 dilution
Number of tests: 96
Controls - standards run in assay:
   Controls: Neg (1), cut-off (1), Pos (1)
Incubation:
   45 min (37°C) + 45 min (37°C) + 15 min (RT)
Washes: 2

CONTENTS

Antibodies, antigens, labelled components:
   Anti-human IgM MAb bound to well
   Rubella virus Ag
   Anti-Rubella virus MAb POD conjugated
Substrate: TMB
Controls - standards supplied:
   Controls: Neg, cut-off and Pos (serum)
Additional reagents:
   None
Special equipment required:
   None

INTERPRETATION

Comments on interpretation:
   No data is provided for interpretation of results
No. of references: None

NOTES

131094.0

Rubella virus
ANTIBODY DETECTION (IgM)

Manufacturer: Radim
Cat. No./Trade name: K2RM/Rubella IgM EIA Well

SUMMARY

[Well-AHIgM]–**Ab**–[Ag]–[MAb-biotin]–[strept-HRP]–
[TMB]–A$_{450}$

Assay type: EIA (non-competitive)
Detection: Colorimetric A$_{450}$
Format: Microtitre well, Ab coated
Sample type: Serum, plasma
Sample pre-treatment:
   None
Sample volume: 100 µl of 1:100 dilution x 2
Number of tests: 96
Controls - standards run in assay:
   Controls: Neg (2), cut-off (2), Pos (2)
Incubation:
   1 hr (37°C) + 60 min (37°C) + 30 min (37°C) + 10 min
      (37°C) or 15 min (RT)
Washes: 3

CONTENTS

Antibodies, antigens, labelled components:
   Anti-human IgM MAb bound to well
   Rubella virus Ag
   Anti-Rubella MAb biotinylated
   Streptavidin HRP conjugated
Substrate: TMB
Controls - standards supplied:
   Controls: Neg, cut-off and Pos (serum)
Additional reagents:
   None
Special equipment required:
   None

INTERPRETATION

Comments on interpretation:
   Grey area: within cut-off ± 10%; retest to confirm
No. of references: 3

NOTES

131274.0

© *KLUWER ACADEMIC PUBLISHERS 1995, ISSN 1381-5067*

Rubella virus
ANTIBODY DETECTION (IgM)

Manufacturer: Roche Diagnostic Systems
Cat. No./Trade name: 07 3491 8/COBAS® CORE
Rubella IgM EIA

SUMMARY

[Bead-AHIgM]–**Ab**–[Ag]–[MAb-HRP]–[TMB]–A_{450}

Assay type: EIA (non-competitive)
Detection: Colorimetric A_{450}
Format: Tube, Ab coated bead
Sample type: Serum, plasma (do not heat inactivate)
Sample pre-treatment:
  None
Sample volume: 10 µl (+500 µl diluent)
Number of tests: 50
Controls - standards run in assay:
  Controls: Neg (1), Pos (3)
Incubation:
  15 min (37°C) + 1 hr (37°C) + 15 min (37°C) (all with
  shaking)
Washes: 2

CONTENTS

Antibodies, antigens, labelled components:
  Anti-human IgM MAb (m) bound to bead
  Rubella virus Ag
  Anti-Rubella virus MAb (m) HRP conjugated
Substrate: TMB (bought separately)
Controls - standards supplied:
  Controls: Neg and Pos (human serum)
Additional reagents:
  Cobas® Core TMB kit
  H_2SO_4 (for manual method only)
Special equipment required:
  Cobas® Core immunoassay analyser (for automated
    method - optional)
  Cobas® Core EIA shaking incubator, tube washer and
    photometer (optional)
  Cobas® Core TORC negative control bead dispenser
  Cobas® Core reaction tubes

INTERPRETATION

Comments on interpretation:
  Equivocal: within cut-off ±10%; retest as indeterminate
    and repeat with a sample taken in a few weeks
  Positive: ≥cut-off +10% and confirmed using
    supplemental tests
No. of references: None

NOTES

131157.0

Rubella virus
ANTIBODY DETECTION (IgM)

Manufacturer: Sigma Diagnostics
Cat. No./Trade name: SIA 414/SIA® Rubella IgM

SUMMARY

[Well-Ag]–**Ab**–[AHIgM-AP]–[PMP]–A_{550}

Assay type: EIA (non-competitive)
Detection: Colorimetric A_{550}
Format: Microtitre well, Ag coated
Sample type: Serum (do not heat inactivate)
Sample pre-treatment:
  Add Pretreatment Solution to sample, incubate 30 min
    (RT) to eliminate interference due to RF and IgG from
    sample
Sample volume: 50 µl (+200 µl diluent)
Number of tests: 96
Controls - standards run in assay:
  Calibrators: 1 (2), 2 (2), 3 (2)
Incubation:
  30 min (RT) + 30 min (RT) + 30 min (RT)
Washes: 2 (+preliminary plate wash)

CONTENTS

Antibodies, antigens, labelled components:
  Rubella virus Ag bound to well
  Anti-human IgM Ab (g) AP conjugated
Substrate: PMP
Controls - standards supplied:
  Calibrators: 1, 2 and 3 (IgM Ab titre in AU/ml – human
    serum)
Additional reagents:
  None
Special equipment required:
  Sigma EIA Multiwell Plate Reader (Cat. no. EQ104 –
    optional)

INTERPRETATION

Comments on interpretation:
  Equivocal: 0.24–0.27 SIA® Rubella IgM Value; probably
    positive
  Equivocal: 0.20–0.23 SIA® Rubella IgM Value; report as
    probably negative; retest a fresh sample 5–7 days
    later or retest using an alternative method
No. of references: 15

NOTES

131435.0

© *KLUWER ACADEMIC PUBLISHERS 1995, ISSN 1381-5067*

Rubella virus
ANTIBODY DETECTION (IgM)

Manufacturer: Sigma Diagnostics
Cat. No./Trade name: SIA 302/SIA® Rubella IgM (Capture)

SUMMARY

[Well-AHIgM]–**Ab**–[Ag:MAb-biotin: strept-HRP]–[TMB]–A_{450}

Assay type: EIA (non-competitive)
Detection: Colorimetric A_{450}
Format: Microtitre well, Ab coated
Sample type: Serum
Sample pre-treatment:
　None
Sample volume: 10 μl (+1 ml diluent)
Number of tests: 96
Controls - standards run in assay:
　Controls: Neg (1), low Pos (3), high Pos (1)
Incubation:
　1 hr (37°C) + 1 hr (37°C) + 30 min (RT)
Washes: 2

CONTENTS

Antibodies, antigens, labelled components:
　Anti-human IgM Ab bound to well
　Rubella virus Ag
　Anti-Rubella virus MAb biotinylated
　Streptavidin HRP conjugated
Substrate: TMB
Controls - standards supplied:
　Controls: Neg, low Pos and high Pos (human serum)
Additional reagents:
　H_2SO_4
Special equipment required:
　Sigma EIA Multiwell Plate Reader (Cat. no. EQ104 – optional)

INTERPRETATION

Comments on interpretation:
　Equivocal: where Index Value is 0.9–1.1; retest to confirm
　Repeatably equivocal: repeat test with a fresh sample
No. of references: 10

NOTES

131433.0

Rubella virus
ANTIBODY DETECTION (IgM)

Manufacturer: Sorin Biomedica
Cat. No./Trade name: P2471/ETI-RUBEK-M reverse

SUMMARY

[Well-AHIgM]–**Ab**–[Ag]–[MAb-HRP]–[TMB]–A_{450}

Assay type: EIA (non-competitive)
Detection: Colorimetric A_{450}
Format: Microtitre well, Ab coated
Sample type: Serum, plasma
Sample pre-treatment:
　None
Sample volume: 100 μl of 1:101 dilution x 2
Number of tests: 96
Controls - standards run in assay:
　Controls: Neg (2), cut-off (4), Pos (2)
Incubation:
　1 hr (37°C) + 1 hr (37°C) + 30 min (RT)
Washes: 2

CONTENTS

Antibodies, antigens, labelled components:
　Anti-human IgM IgG Ab (r) bound to well
　Rubella virus Ag
　Anti-Rubella virus IgG MAb (m) HRP conjugated
Substrate: TMB
Controls - standards supplied:
　Controls: Neg, cut-off and Pos (human serum)
Additional reagents:
　None
Special equipment required:
　ETI-System Reader and ETI-System Washer (optional)

INTERPRETATION

Comments on interpretation:
　Equivocal: within cut-off ±10%; retest to confirm
No. of references: 10

NOTES

131216.0

© KLUWER ACADEMIC PUBLISHERS 1995, ISSN 1381-5067

Rubella virus
ANTIBODY DETECTION (IgM)

Manufacturer: United Biotech Inc
Cat. No./Trade name: 1A-302/MAGIWEL® Rubella IgM

SUMMARY

[Well-Ag]–**Ab**–[AHIgM-HRP]–[TMB]–A_{450}

Assay type: EIA (non-competitive)
Detection: Colorimetric A_{450}
Format: Microtitre well, Ag coated
Sample type: Serum
Sample pre-treatment:
   None
Sample volume: 100 µl of 1:101 dilution
Number of tests: 96
Controls - standards run in assay:
   Reference Standard Set: Neg (1), Pos (1)
   Calibrators: 1 (1)
Incubation:
   30 min (RT) + 30 min (RT) + 15 min (RT)
Washes: 2

CONTENTS

Antibodies, antigens, labelled components:
   Rubella virus Ag bound to well
   Anti-human IgM Ab (g) HRP conjugated
Substrate: TMB
Controls - standards supplied:
   Reference Standard Set: Neg, Pos
   Calibrator: 1 (100 EU/ml)
Additional reagents:
   None
Special equipment required:
   None

INTERPRETATION

Comments on interpretation:
   Equivocal: between 80–100 EU/ml
No. of references: 4

NOTES

131148.0

Rubella virus
ANTIBODY DETECTION (IgG)

Manufacturer: Hemagen Diagnostics Inc
Cat. No./Trade name: 2700/VIRGO® Rubella IgG IFA

SUMMARY

[Slide well-Ag]–**Ab**–[AHIgG-FITC]–fluorescence

Assay type: Immunofluorescence assay (indirect)
Detection: Fluorescence microscopy
Format: Slide well, Ag coated
Sample type: Serum
Sample pre-treatment:
   None
Sample volume: 10–20 µl of 1:4 dilution x 2
Number of tests: 96
Controls - standards run in assay:
   None
Incubation:
   30 min (37°C) + 30 min (37°C)
Washes: 2

CONTENTS

Antibodies, antigens, labelled components:
   Rubella virus infected and uninfected cells bound to slide
   well
   Anti-human IgG Ab (g) FITC conjugated
Substrate:
Controls - standards supplied:
   Controls: Neg, low Pos and high Pos (human serum)
Additional reagents:
   None
Special equipment required:
   None

INTERPRETATION

Comments on interpretation:
   Positive: ≥(1+) fluorescence at 1:4 dilution
No. of references: 31

NOTES

131107.0

© KLUWER ACADEMIC PUBLISHERS 1995, ISSN 1381-5067

Rubella virus
ANTIBODY DETECTION (IgG)

Manufacturer: Stellar Bio Systems Inc
Cat. No./Trade name: Indirect Fluorescent Assay for Rubella Virus IgG Antibody

SUMMARY

[Slide-Ag]–**Ab**–[AHIgG-FITC]–fluorescence

Assay type: Immunofluorescence assay (indirect)
Detection: Fluorescence microscopy
Format: Slide well, Ag coated
Sample type: Serum
Sample pre-treatment:
  None
Sample volume: 20 μl of 1:10 dilution
Number of tests:
Controls - standards run in assay:
  Controls: Neg (1), Pos (1)
Incubation:
  30 min (37°C) + 30 min (37°C)
Washes: 2

CONTENTS

Antibodies, antigens, labelled components:
  Rubella virus infected and uninfected cells (fibroblasts) bound to slide
  Anti-human IgG Ab (g) FITC conjugated
Substrate:
Controls - standards supplied:
  Controls: Neg and Pos (human serum)
Additional reagents:
  None
Special equipment required:
  None

INTERPRETATION

Comments on interpretation:
  Positive: (1 +) – (4 +) bright green fluorescent staining of infected cells
No. of references: 18

NOTES

131635.0

Rubella virus
ANTIBODY DETECTION (IgM)

Manufacturer: Stellar Bio Systems Inc
Cat. No./Trade name: Indirect Fluorescent Assay for Rubella virus IgM Antibody

SUMMARY

[Slide-Ag]–**Ab**–[AHIgM-FITC]–fluorescence

Assay type: Immunofluorescence assay (indirect)
Detection: Fluorescence microscopy
Format: Slide well, Ag coated
Sample type: Serum
Sample pre-treatment:
  Removal of interference due to RF and IgG in sample using commercially available pretreatment devices (not provided)
Sample volume: 20 μl of 1:5 dilution
Number of tests:
Controls - standards run in assay:
  Controls: Neg (1), Pos (1)
Incubation:
  1 hr (37°C) + 30 min (37°C)
Washes: 2

CONTENTS

Antibodies, antigens, labelled components:
  Rubella virus infected and uninfected cells (fibroblasts) bound to slide
  Anti-human IgM Ab (g) FITC conjugated
Substrate:
Controls - standards supplied:
  Controls: Neg and Pos (human serum)
Additional reagents:
  Pretreatment devices for removal of RF and IgG in sample
Special equipment required:
  None

INTERPRETATION

Comments on interpretation:
  Positive: yellow-green fluorescence around entire periphery of organism
No. of references: 18

NOTES

131640.0

Rubella virus
ANTIBODY DETECTION (IgG)

Manufacturer: Diagnostic Products Corporation
Cat. No./Trade name: LKRBZ/Immulite® Rubella IgG

SUMMARY

[Bead-Ag]–**Ab**–[AHIgG-AP]–[substrate]–luminescence

Assay type: Luminometric immunoassay (non-competitive)
Detection: Luminometric
Format: Reaction tube, Ag coated beads
Sample type: Serum
Sample pre-treatment:
  None
Sample volume: 10 µl of 1:21 dilution
Number of tests: 50, 200
Controls - standards run in assay:
  Controls: Neg (1), Pos (1)
Incubation:
  Automated
Washes: Automated

CONTENTS

Antibodies, antigens, labelled components:
  Rubella virus Ag bound to beads
  Anti-human IgG MAb (m) AP conjugated
Substrate: Adamantyl dioxetane (phosphate ester) (bought separately)
Controls - standards supplied:
  Controls: Neg and Pos
Additional reagents:
  Immulite® Chemiluminescence Substrate Module (LSUBX)
  Immulite® Probe Wash Module (LPWS2)
Special equipment required:
  Immulite® Automated Analyzer
  Immulite® Sample cups, holders and caps

INTERPRETATION

Comments on interpretation:
  This is an automated assay. Results designated equivocal; retest a fresh sample in 1 week or use an alternative method
No. of references: 12

NOTES

131738.0

Rubella virus
ANTIBODY DETECTION (IgG)

Manufacturer: Diagnostic Products Corporation
Cat. No./Trade name: LKRBQ1/Immulite® Rubella Quantitative IgG

SUMMARY

[Bead-Ag]–**Ab**–[AHIgG-AP]–[substrate]–luminescence

Assay type: Luminometric immunoassay (non-competitive)
Detection: Luminometric
Format: Reaction tube, Ag coated beads
Sample type: Serum
Sample pre-treatment:
  None
Sample volume: 10 µl of 1:21 dilution
Number of tests: 100, 500
Controls - standards run in assay:
  Controls: Neg (1), Pos (1)
Incubation:
  Automated
Washes: Automated

CONTENTS

Antibodies, antigens, labelled components:
  Rubella virus Ag bound beads
  Anti-human IgG MAb (m) AP conjugated
Substrate: Adamantyl dioxetane (phosphate ester) (bought separately)
Controls - standards supplied:
  Controls: Neg and Pos (human serum)
Additional reagents:
  Immulite® Chemiluminescence Substrate Module (LSUBX)
  Immulite® Probe Wash Module (LPWS2)
Special equipment required:
  Immulite® Automated Analyzer
  Immulite® Sample cups, holders and caps

INTERPRETATION

Comments on interpretation:
  This is an automated assay
  Negative: < 10 IU/ml
  Positive: ≥ 10 IU/ml indicates past infection or protective vaccination level
  A result of < 10 IU/ml may also indicate recent infection
No. of references: 11

NOTES

131798.0

Two adjustors – low and high – (Cat. nos. LBQL and LBQH) are provided with the kit to prepare the kit prior to use

Rubella virus
ANTIBODY DETECTION (IgG)

Manufacturer: Johnson & Johnson Clinical Diagnostics Inc
Cat. No./Trade name: LAN0207/95a/AMERLITE Rubella IgG Assay

SUMMARY

[Well-Ag]–**Ab**–[AHIgG-HRP]–[signal reagent]–
luminescence

Assay type: Luminometric immunoassay (non-competitive)
Detection: Luminometric
Format: Microtitre well, Ag coated
Sample type: Serum, plasma
Sample pre-treatment:
 None
Sample volume: 10 μl (+100 μl diluent)*
Number of tests: 240
Controls - standards run in assay:
 Standards:
 Qualitative: Neg (1), Borderline (3)
 Quantitative: A (2), B (2), C (2), D (2), E (2), F (2)
Incubation:
 1 hr (37°C) + 30 min (37°C) + 2-20 min (RT) (first two incubations with shaking)
Washes: 2

CONTENTS

Antibodies, antigens, labelled components:
 Rubella virus Ag bound to well
 Anti-human IgG Ab (sh) HRP conjugated
Substrate: AMERLITE Signal Reagent
Controls - standards supplied:
 Standards: A-F (0, 10, 50, 100, 200, 500 IU/ml where A is used as Neg control and B as Borderline control for qualitative assay - human plasma + bovine serum)
Additional reagents:
 AMERLITE Signal Reagent, AMERLITE Serology Wash Reagent
Special equipment required:
 AMERLITE Processing Center (optional)
 AMERLITE Immunoassay System

INTERPRETATION

Comments on interpretation:
 This is an automated assay
 Results designated equivocal; retest to confirm
 Repeatably equivocal: repeat test with a fresh sample
No. of references: 4

NOTES

131221.0

*Samples may be assayed singly or in duplicate

Rubella virus
ANTIBODY DETECTION (IgG)

Manufacturer: Sanofi Diagnostics Pasteur
Cat. No./Trade name: 34430/Access® Rubella IgG

SUMMARY

[Particles-Ag]–**Ab**–[AHIgG-AP]–[substrate]–
luminescence

Assay type: Luminometric immunoassay (non-competitive)
Detection: Luminometric
Format: Reaction vessel, Ag coated magnetic particles
Sample type: Serum
Sample pre-treatment:
 None
Sample volume: 20 μl
Number of tests: 50
Controls - standards run in assay:
 Not specified
Incubation:
 Automated (total time 35 min at 37°C)
Washes: Automated

CONTENTS

Antibodies, antigens, labelled components:
 Rubella virus Ag bound to magnetic particles
 Anti-human IgG MAb (m) AP conjugated
Substrate: Lumi-Phos® 500 (bought separately)
Controls - standards supplied:
 None
Additional reagents:
 Access® Rubella IgG Calibrators (Cat. no. 34435)
 Access® Substrate (Cat. no. 81906)
 Access® Wash Buffer (Cat. no. 81907)
 Access® QC Rubella IgG (Cat. no. 34439) or similar products
Special equipment required:
 Access® Immunoassay System

INTERPRETATION

Comments on interpretation:
 Results are calculated automatically
 Equivocal: titre between 10–15 IU/ml; report as indeterminate immune status
No. of references: 12

NOTES

131763.0

© KLUWER ACADEMIC PUBLISHERS 1995, ISSN 1381-5067

Rubella virus
ANTIBODY DETECTION (IgM)

Manufacturer: Johnson & Johnson Clinical Diagnostics Inc
Cat. No./Trade name: LAN1208/95a/AMERLITE Rubella IgM Assay

SUMMARY

[Well-AHIgM]–**Ab**–[Ag]–[MAb-HRP]–[signal reagent]–
luminescence

Assay type: Luminometric immunoassay (non-competitive)
Detection: Luminometric
Format: Microtitre well, Ab coated
Sample type: Serum, plasma
Sample pre-treatment:
  None
Sample volume: 10 μl*
Number of tests: 96
Controls - standards run in assay:
  Controls: Neg (1), Pos (3)
Incubation:
  1 hr (37°C) + 1 hr (37°C) + 5-20 min (RT) (first two incubations with shaking)
Washes: 2

CONTENTS

Antibodies, antigens, labelled components:
  Anti-human IgM MAb (m) bound to well
  Rubella virus Ag
  Anti-Rubella virus MAb (m) HRP conjugated
Substrate: AMERLITE Signal Reagent
Controls - standards supplied:
  Controls: Neg and Pos (human plasma)
Additional reagents:
  AMERLITE Signal Reagent, AMERLITE Serology Wash Reagent
Special equipment required:
  AMERLITE Processing Center (optional)
  AMERLITE Immunoassay System

INTERPRETATION

Comments on interpretation:
  This is an automated assay
  Results designated equivocal; retest to confirm
  Repeatably equivocal: repeat test with a fresh sample
No. of references: 3

NOTES

131224.0

*Samples may be assayed singly or in duplicate

Rubella virus
ANTIBODY DETECTION (IgM)

Manufacturer: Sanofi Diagnostics Pasteur
Cat. No./Trade name: 34440/Access® Rubella IgM

SUMMARY

[Particles-AHIgM]–**Ab**–[Ag:MAb-AP]–[substrate]–
luminescence

Assay type: Luminometric immunoassay (non-competitive)
Detection: Luminometric
Format: Reaction vessel, Ab coated magnetic particles
Sample type: Serum
Sample pre-treatment:
  None
Sample volume: 20 μl
Number of tests: 50
Controls - standards run in assay:
  Not specified
Incubation:
  Automated (total time 35 min at 37°C)
Washes: Automated

CONTENTS

Antibodies, antigens, labelled components:
  Anti-human IgM Ab (g) bound to magnetic particles
  Rubella virus Ag
  Anti-Rubella virus MAb (m) AP conjugated
Substrate: Lumi-Phos® 500 (bought separately)
Controls - standards supplied:
  None
Additional reagents:
  Access® Rubella IgG Controls
  Access® Substrate (Cat. no. 81906)
  Access® Wash Buffer (Cat. no. 81907)
  Access® QC Rubella IgM (Cat. no. 34449) or similar products
Special equipment required:
  Access® Immunoassay System

INTERPRETATION

Comments on interpretation:
  Results are calculated automatically
  Equivocal: titre between 5–15 AU/ml; retest to confirm
No. of references: 14

NOTES

131765.0

Rubella virus
ANTIBODY DETECTION

Manufacturer: Becton Dickinson
Cat. No./Trade name: 8619-26/Rubascan®

SUMMARY

[Latex-Ag]–**Ab**–agglutination

Assay type: Particle agglutination assay
Detection: Visual
Format: Latex particles, Ag coated (test done on card)
Sample type: Serum (do not heat inactivate)
Sample pre-treatment:
 None
Sample volume: 25 µl (undiluted or 1:10 dilution)
Number of tests: 30, 100, 500
Controls - standards run in assay:
 Controls: Neg (1), low Pos (1), high Pos (1)
Incubation:
 Read result after 8 min mechranical rotation (95-110
  rpm)
Washes:

CONTENTS

Antibodies, antigens, labelled components:
 Rubella virus Ag bound to latex particles (Latex Reagent)
Substrate:
Controls - standards supplied:
 Controls: Neg, low Pos and high Pos (human serum)
Additional reagents:
 None
Special equipment required:
 Rubascan® Quantitative Test Cards (for quantitative
  assay)

INTERPRETATION

Comments on interpretation:
 Positive: any agglutination of sample and Latex Reagent
 Serial dilutions of sample (1:5–1:160) may be tested
  using the Rubascan® Quantitative Test Cards
No. of references: 17

NOTES

131020.0

Rubella virus
ANTIBODY DETECTION

Manufacturer: BIOKIT SA
Cat. No./Trade name: Rubagen

SUMMARY

[Latex-Ag]–**Ab**–agglutination

Assay type: Particle agglutination assay
Detection: Visual
Format: Latex particles, Ag coated (test done on slide)
Sample type: Serum
Sample pre-treatment:
 None
Sample volume: 50 µl of 1:5 dilution
Number of tests:
Controls - standards run in assay:
 Controls: Neg (1), Pos (1)
Incubation:
 8 min (RT) on rotary shaker
Washes:

CONTENTS

Antibodies, antigens, labelled components:
 Rubella virus Ag bound to latex particles (Latex Reagent)
Substrate:
Controls - standards supplied:
 Controls: Neg and Pos (rabbit serum, IU/ml with
  reference to WHO reference reagents)
Additional reagents:
 None
Special equipment required:
 None

INTERPRETATION

Comments on interpretation:
 Positive: agglutination of sample with Latex Reagent
No. of references: 7

NOTES

131370.0

© *KLUWER ACADEMIC PUBLISHERS 1995, ISSN 1381-5067*

Rubella virus
ANTIBODY DETECTION

Manufacturer: Carter-Wallace Inc
Cat. No./Trade name: Wampole Impact Rubella™ Slide Test

SUMMARY

[Latex-Ag]–**Ab**–agglutination

Assay type: Particle agglutination assay
Detection: Visual
Format: Latex particles, Ag coated (test done on slide)
Sample type: Serum
Sample pre-treatment:
  None
Sample volume: 10 µl of 1:8 dilution
Number of tests:
Controls - standards run in assay:
  Controls: Neg (1), low Pos (1), high Pos (1)
Incubation:
  10 min (RT) on rotary shaker
Washes:

CONTENTS

Antibodies, antigens, labelled components:
  Rubella virus Ag bound to latex particles (Reagent Latex)
Substrate:
Controls - standards supplied:
  Controls: Neg, low Pos and high Pos (human serum)
Additional reagents:
  None
Special equipment required:
  None

INTERPRETATION

Comments on interpretation:
  Equivocal: possible low titre sample; retest fresh sample using alternative procedures
  Positive: agglutination of sample with Reagent Latex
No. of references: 9

NOTES

131379.0

Rubella virus
ANTIBODY DETECTION

Manufacturer: Orion Diagnostica
Cat. No./Trade name: 67272/Rubalex®

SUMMARY

[Latex-Ag]–**Ab**–agglutination

Assay type: Particle agglutination assay
Detection: Visual
Format: Latex particles, Ag coated (test done on card)
Sample type: Serum
Sample pre-treatment:
  None
Sample volume: 25 µl
Number of tests: 25, 100
Controls - standards run in assay:
  Controls: Neg (1), Pos (1)
Incubation:
  Read result within 3 min of mixing sample and reagents
Washes:

CONTENTS

Antibodies, antigens, labelled components:
  Rubella virus Ag bound to latex particles (Test Latex)
Substrate:
Controls - standards supplied:
  Controls: Neg and Pos
Additional reagents:
  None
Special equipment required:
  None

INTERPRETATION

Comments on interpretation:
  Positive: partial or complete agglutination of sample with Test Latex
No. of references: None

NOTES

131264.0

© KLUWER ACADEMIC PUBLISHERS 1995, ISSN 1381-5067

Rubella virus
ANTIBODY DETECTION

Manufacturer: S.A. Scientific™
Cat. No./Trade name: Rubella Latex Kit

SUMMARY

[Latex-Ag]–**Ab**–agglutination

Assay type: Particle agglutination assay
Detection: Visual
Format: Latex particles, Ag coated (test done on card)
Sample type: Serum
Sample pre-treatment:
  None
Sample volume: 25 µl (undiluted or 1:10 dilution)
Number of tests: 100
Controls - standards run in assay:
  Controls: Neg (1), low Pos (1), high Pos (1)
Incubation:
  8 min (RT) with rotation
Washes:

CONTENTS

Antibodies, antigens, labelled components:
  Rubella virus Ag bound to latex particles (Latex Reagent)
Substrate:
Controls - standards supplied:
  Controls: Neg, low Pos and high Pos (human serum)
Additional reagents:
  None
Special equipment required:
  None

INTERPRETATION

Comments on interpretation:
  Positive: agglutination of sample with Latex Reagent
No. of references: 11

NOTES

131694.0

Rubella virus
ANTIBODY DETECTION

Manufacturer: Seradyn
Cat. No./Trade name: 0369579/Seratest® Rubella

SUMMARY

[Latex-Ag]–**Ab**–agglutination

Assay type: Particle agglutination assay
Detection: Visual
Format: Latex particles, Ag coated (test done on card)
Sample type: Serum
Sample pre-treatment:
  None
Sample volume: 25 µl (undiluted or 1:10 dilution)
Number of tests: 100
Controls - standards run in assay:
  Controls: Neg (1), low Pos (1), high Pos (1)
Incubation:
  8 min (RT) with rotation
Washes:

CONTENTS

Antibodies, antigens, labelled components:
  Rubella virus Ag bound to latex particles (Latex Reagent)
Substrate:
Controls - standards supplied:
  Controls: Neg, low Pos (titre $\geqslant$1:10) and high Pos (titre $\geqslant$1:60) (human serum)
Additional reagents:
  None
Special equipment required:
  None

INTERPRETATION

Comments on interpretation:
  Positive: agglutination of sample with Latex Reagent
No. of references: 13

NOTES

131655.0

Salmonella species

Organisms of the genus *Salmonella* are facultatively anaerobic, motile Gram-negative bacilli, and remain a major cause of enteric disease worldwide. Taxonomically speaking, all salmonellae belong to a single species *S.cholaeraesuis*, and are classified into 7 subgroups on the basis of DNA homology and host range, with virtually all human pathogens belonging to Subgroup 1. Although those organisms previously identified as distinct species are now recognised as different serotypes within those subgroups, for most practical purposes the various serotypes tend to be referred to as if they were different species e.g. *S.typhimurium, S.dublin* etc. More than 2000 distinct serotypes are now recognised on the basis of somatic 'O' and flagellar 'H' antigens.

S.typhi and *S.paratyphi* only infect humans and cause enteric fever, an acute illness with an incubation period of 10–14 days which causes remittent fever, malaise, anorexia, headache, hepatosplenomegaly and myalgia, and in which diarrhoea is not the major clinical feature. More severe complications such as perforation of the bowel and seizures may occur. As with other salmonellae a chronic carrier state may develop, but this is much more common with *S.typhi* and *S.paratyphi*. Infection with these organisms is acquired by ingestion of contaminated water, food or milk. Since man is the only source, typhoid fever may be controlled by eliminating human faecal contamination of water supplies and foods.

Other salmonellae are ubiquitous amongst warm-blooded animals and are transmitted incidentally to man. Poultry and egg products account for about 50% of human cases; meats, dairy products and person-to-person spread account for most of the remainder. Many cases are sporadic, but outbreaks traceable to a common source occur frequently. The large animal reservoir, changes in the production and distribution of food, and altered consumer habits and preferences have made it impossible to prevent a steady increase in incidence during recent years. Asymptomatic infection is extremely common, and is important in food handlers who are at high risk due to occupational exposure and may be responsible for further transmission. The main clinical presentation is acute gastroenteritis with an incubation period of around 1–3 days. Severe bacteraemic illness with metastic infection of bone vascular and other sites may occur, particularly in children and the elderly, and is more common with some serotypes.

Diagnosis

The diagnosis of typhoid fever can be made by isolation of the organism from the blood during the first weeks of the illness. Stool cultures are positive during the subsequent weeks. In infections due to other species of *Salmonella*, the organism may be isolated from the stools, from the blood in patients with bacteraemia, or from sites of metastatic infection. The organisms are non-lactose-fermenters and colonies may be differentiated from other enterobacteriaceae on MacConkey agar and other selective media. A number of specific selective media for the rapid, selective presumptive isolation and identification of salmonellae have been described. These media and selective enrichment broths such as selenite, tetrathionate and brilliant green may be useful, particularly for the examination of suspect food and waters. Presumptive salmonellae are identified biochemically and by slide agglutination with polyvalent antisera. More precise serological identification is usually performed in reference centres.

The Widal agglutination test may be used for the serodiagnosis of enteric fever, and other commercial serodiagnostic systems are available for these and other *Salmonella* infections. A number of commercial systems based on a variety of cultural and non-cultural techniques, particularly ELISA-based systems, are being developed for the rapid detection of salmonellae in foods and waters.

References

Baird-Parker AC. Foodborne salmonellosis. Lancet. 1990;336:1231–35.

Miller SJ, Hohmann EL, Pegues DA. *Salmonella*. In: Mandell GL, Bennett JE, Dolin R, eds. Principles and Practice of Infectious Diseases, 4th edn. New York, Edinburgh, London: Churchill Livingstone; 1995:2013–33.

See also Multipathogen Assays under: Gastrointestinal pathogens

Salmonella species
ANTIGEN DETECTION

Manufacturer: Mast Diagnostics Ltd
Cat. No./Trade name: EIA 701/Mastazyme® Salmonella EIA

SUMMARY

[Well-Ab]–**Ag**–[Ab-HRP]–[TMB]–A_{450}

Assay type: EIA (non-competitive)
Detection: Colorimetric A_{450}
Format: Microtitre well, Ab coated
Sample type: Food products, animal feedstuff
Sample pre-treatment:
  Enrichment: incubate 25 g of sample 16-24 hr (37°C) in Mast Buffered Peptone Water. Further incubate 16-24 hr (42°C) in Mast Rappaport - Vassiliadis Broth, boil 20 min, cool
Sample volume: 100 µl of treated sample
Number of tests: 96
Controls - standards run in assay:
  Controls: Neg (1), Pos (1)
Incubation:
  30 min (37°C) + 30 min (37°C) + 15 min (RT)
Washes: 2

CONTENTS

Antibodies, antigens, labelled components:
  Anti-Salmonella species Ab (r) bound to well
  Anti-Salmonella species Ab (g) HRP conjugated
Substrate: TMB
Controls - standards supplied:
  Controls: Neg and Pos
Additional reagents:
  Mast Rappaport-Vassiliadis Broth (Cat. no. DM269)
  Mast Buffered Peptone Water (Cat. no. DM494)
Special equipment required:
  None

INTERPRETATION

Comments on interpretation:
  Borderline: within designated range; retest to confirm
  Positive: OD of sample > 0.200 and confirmed by bacterial culture
No. of references: 8

NOTES

131268.0

Salmonella species
ANTIGEN DETECTION

Manufacturer: Tecra® Diagnostics
Cat. No./Trade name: SALVIA 48/TECRA® Salmonella Visual Immunoassay

SUMMARY

[Well-Ab]–**Ag**–[Ab-enzyme]–[substrate]–green colour

Assay type: EIA (non-competitive)
Detection: Visual or colorimetric
Format: Microtitre well, Ab coated
Sample type: Food, food-related and environmental samples
Sample pre-treatment:
  Samples require culture in enrichment broth (42–52 hr) prior to assay
Sample volume: 0.2 ml of prepared sample
Number of tests: 48, 96
Controls - standards run in assay:
  Controls: Neg (1), Pos (1)
Incubation:
  30 min (37°C) + 30 m in (37°C)
Washes: 2

CONTENTS

Antibodies, antigens, labelled components:
  Anti-Salmonella species (motile and non-motile) Ab bound to well
  Anti-Salmonella species (motile and non-motile) Ab enzyme conjugated
Substrate: ABTS
Controls - standards supplied:
  Controls: Neg (diluent) and Pos
Additional reagents:
  Pre-enrichment broth, selective enrichment broth, M broth
Special equipment required:
  None

INTERPRETATION

Comments on interpretation:
  Positive: green colouration in well confirmed by standard tests
No. of references: 4

NOTES

131725.0

An independent evaluation report of this kit and a list of international organisations and professional groups who have given an approval to it are available – please contact manufacturer for details

© *KLUWER ACADEMIC PUBLISHERS 1995, ISSN 1381-5067*

Salmonella species
ANTIGEN DETECTION

Manufacturer: Tecra® Diagnostics
Cat. No./Trade name: SALUNQ10 TECRA® UNIQUE®
Salmonella Screening

SUMMARY

[Dipstick-Ab]–**Ag**–[Ab-enzyme]–[substrate]–purple colour

Assay type: EIA (non-competitive)
Detection: Visual
Format: Dipstick, Ab coated
Sample type: Food, food related and environmental
  samples
Sample pre-treatment:
  Samples require culture in enrichment broth (16 hr) prior
    to assay
Sample volume: 4 ml of prepared sample
Number of tests: 10
Controls - standards run in assay:
  Controls: Integral Neg (1), Pos (1)
Incubation:
  20 min (37°C) + 4 hr (37°C) + 30 min (37°C)*
Washes: 2

CONTENTS

Antibodies, antigens, labelled components:
  Anti-Salmonella species (motile and non-motile) Ab
    bound to dipstick
  Anti-Salmonella species (motile and non-motile) Ab
    enzyme conjugated
Substrate: Not specified
Controls - standards supplied:
  Controls: Integral Neg and Pos incorporated in dipstick
Additional reagents:
  None
Special equipment required:
  None

INTERPRETATION

Comments on interpretation:
  Positive: purple colour on dipstick with white negative
    band across top of dipstick
No. of references: 3

NOTES

131727.0

An independent evaluation report of this kit is available –
  please contact manufacturer for details
*The second incubation step in the protocol is a 4 hr
  incubation of the Ab-coated dipstick and sample in M-
  broth for replication

Salmonella species
ANTIGEN DETECTION

Manufacturer: PanBio
Cat. No./Trade name: RSO-25/RAP-SAL Salmonella
(Single Step)

SUMMARY

[Membrane-binding reagent]–**Ag**–[Ab-gold]–red/purple
line

Assay type: Immunochromatographic assay
Detection: Visual
Format: Test device, membrane treated with binding
reagent
Sample type: Faecal specimens, rectal swab
Sample pre-treatment:
  Mix with buffer or sample may be enriched in selinite/
    cystine broth prior to testing
Sample volume: 3 drops of prepared sample or 120 µl of
  enriched sample
Number of tests: 25
Controls - standards run in assay:
  Controls: Neg (1), Pos (1)
Incubation:
  Read result within 30 min of adding sample to device
Washes: 0

CONTENTS

Antibodies, antigens, labelled components:
  Binding reagent (not specified) bound to membrane (Test
    Zone)
  Anti-salmonella Ab gold conjugated (incorporated in
    device)
Substrate:
Controls - standards supplied:
  Integral procedural controls incorporated in Test device
    (Control Zone)
Additional reagents:
  Stock culture of Salmonella for use as Pos control
Special equipment required:
  None

INTERPRETATION

Comments on interpretation:
  Negative: red/purple line in Control Zone and no line in
    Test Zone
  Positive: red/purple line in Control Zone and Test Zone
  Invalid: no line in Control Zone; repeat test with another
    device
No. of references: None

NOTES

131407.0

Salmonella species
ANTIGEN DETECTION

Manufacturer: Boule Diagnostics AB
Cat. No./Trade name: Phadebact® Salmonella Test

SUMMARY

[Staph-Ab]–**Ag**–agglutination

Assay type: Particle agglutination assay
(coagglutination)
Detection: Visual
Format: Staphylococci, Ab coated (test done on slide)
Sample type: Colonies from solid medium culture
Sample pre-treatment:
Prepare heavy suspension of colonies in PBS or saline
Sample volume: 1 drop of prepared sample x 6*
Number of tests: 50
Controls - standards run in assay:
None
Incubation:
Read results within 1 min of mixing sample and reagents
Washes:

CONTENTS

Antibodies, antigens, labelled components:
Test Reagents bound to Staphylococci (6 separate
reagents)
1. Anti-salmonella (group 0:2) Ab (r)
2. Anti-salmonella (group 0:4) Ab (r)
3. Anti-salmonella (group 0:7) Ab (r)
4. Anti-salmonella (group 0:8) Ab (r)
5. Anti-salmonella (group 0:9) Ab (r)
6. Anti-salmonella (group 0:3) Ab (r)
Substrate:
Controls - standards supplied:
None
Additional reagents:
None
Special equipment required:
None

INTERPRETATION

Comments on interpretation:
Positive: coagglutination of sample with the appropriate
reagent and no coagglutination with the other
reagents identifies the Salmonella group present in
the sample
No. of references: None

NOTES

131393.0

*Samples are tested with all six reagents simultaneously

Salmonella species
ANTIGEN DETECTION

Manufacturer: Microgen Bioproducts
Cat. No./Trade name: M42/MicroScreen® Salmonella
Latex Slide Agglutination Test

SUMMARY

[Latex-Ab]–**Ag**–agglutination

Assay type: Particle agglutination assay
Detection: Visual
Format: Latex particles, Ab coated (test done on slide)
Sample type: Broth culture or colonies from solid media
culture (isolated from faecal specimens)
Sample pre-treatment:
None
Sample volume: 1 drop of broth or 2-3 colonies (+ 1 drop
saline)
Number of tests: 50
Controls - standards run in assay:
Controls: Pos (1)
Incubation:
Read result within 2 min of mixing sample and reagents
Washes:

CONTENTS

Antibodies, antigens, labelled components:
Anti-Salmonella Ab bound to latex particles (Test Latex)
Substrate:
Controls - standards supplied:
None
Additional reagents:
Known Salmonella culture for use as Positive Control
Special equipment required:
None

INTERPRETATION

Comments on interpretation:
Positive: agglutination of sample with Test Latex
No. of references: 4

NOTES

131195.0

© KLUWER ACADEMIC PUBLISHERS 1995, ISSN 1381-5067

Salmonella species
ANTIGEN DETECTION

Manufacturer: Unipath Limited
Cat. No./Trade name: Salmonella Latex Test

SUMMARY

[Latex-Ab]–**Ag**–agglutination

Assay type: Particle agglutination assay
Detection: Visual
Format: Latex particles, Ab coated (test done on card)
Sample type: Positive material from Unipath Salmonella Rapid Test*
Sample pre-treatment:
    None
Sample volume: 1 drop of tested material from Rapid Test x 2
Number of tests: 30
Controls - standards run in assay:
    Controls: Pos (1)
Incubation:
    Read result within 2 min of mixing sample and reagent
Washes:

CONTENTS

Antibodies, antigens, labelled components:
    Anti-Salmonella IgG Ab (r) bound to latex particles (Test Latex)
    Non-immune rabbit Ig Ab bound to latex particles (Control Latex)
Substrate:
Controls - standards supplied:
    Controls: Pos
Additional reagents:
    None
Special equipment required:
    None

INTERPRETATION

Comments on interpretation:
    Positive: agglutination of sample with Test Latex and no agglutination with Control Latex
No. of references: 5

NOTES

131612.0

*This kit is for use only with presumptively positive material from food materials, food products and factory environmental samples identified using Unipath Salmonella Rapid Test (Cat. no. FT201)

Salmonella typhi
ANTIBODY DETECTION (IgG AND/OR IgM))

Manufacturer: Immuno Pharmacology Research
Cat. No./Trade name: ELISA TIFO IgG/IgM ELISA

SUMMARY

[Well-Ag]–**Ab**–[AHIgG/IgM-AP]–[PNP]–A_{405}

Assay type: EIA (non-competitive)
Detection: Colorimetric A_{405}
Format: Microtitre well, Ag coated
Sample type: Serum
Sample pre-treatment:
    None
Sample volume: 100 µl of 1:200 dilution x 2
Number of tests: 48, 96
Controls - standards run in assay:
    Controls: Neg (1), Pos (1)
Incubation:
    1 hr (37°C) + 1 hr (37°C) + 30 min (37°C)
Washes: 2

CONTENTS

Antibodies, antigens, labelled components:
    Salmonella typhi Ag bound to well
    Anti-human IgG Ab AP conjugated
    Anti-human IgM Ab AP conjugated
Substrate: PNP
Controls - standards supplied:
    Controls: Neg and Pos
Additional reagents:
    None
Special equipment required:
    None

INTERPRETATION

Comments on interpretation:
    Classification of samples is according to cut-off; no further testing
No. of references: 2

NOTES

131329.0

This kit can detect IgG and IgM antibodies and can differentiate between them

Schistosoma species

The schistosomes or blood flukes are parasitic flatworms of the class Trematoda. The majority of human infections are attributable to 5 species: *Schistosoma mansoni*, *S.japonicum*, *S.haematobium*, *S.mekongi* and *S.intercalatum*. More than 200 million people are infected worldwide and the prevalence is increasing. Man is the principal definitive host for all 5 species of schistosomes. The adult schistosome worms, unlike most other flukes, exist as separate sexes and are 1–2 cm in length. Adult forms of *S.mansoni*, *S.japonicum*, *S.mekongi*, and *S.intercalatum* inhabit the portal and mesenteric veins; *S.haematobium* inhabits the vesical plexus of the bladder. The lateral edges of the male fluke are enfolded to form a groove which is inhabited by the female worm, and several hundred or more eggs may be produced daily. The eggs migrate into the gut or bladder (*S.haematobium*) and are excreted in the faeces or urine. In fresh water the eggs hatch and release motile ciliated miracidia which pentrate the body of a snail intermediate host, which is specific for each schistosome species and whose distribution is important in determining the endemicity of the schistosomes. The miracidia multiply asexually inside the snail and within a few weeks yield large numbers of motile fork-tailed cercariae. When the cercariae encounter human skin they penetrate it and change into schistosomula forms which migrate to the lungs and liver, mature into adult worms which mate and pass through the venous system to their final habitat.

Three major disease syndromes occur in schistosomiasis, and correlate with the stages of development of the parasite. Schistosome dermatitis results from the penetration of cercariae, but is more commonly associated with a reaction to non-human species which die in the dermal layers. Acute schistosomiasis (Katayama fever), with fever, lymphadenopathy, hepatosplenomegaly and eosinophilia, is usually of a few weeks duration and is associated with a serum-sickness like reaction to the large egg burden. Chronic disease is the result of eggs becoming retained in the tissues with subsequent granuloma formation. *S.mansoni*, *S.japonicum* and *S.mekongi* most commonly affect the liver and intestines and may produce pain and dysenteric symptoms. Ultimately hepatic involvement results in portal hypertension, hypersplenism, and liver failure. *S.haematobium* mainly affects the bladder and ureters, resulting in scarring and obstructive lesions which may subsequently produce infection, uraemia and even bladder cancer.

Diagnosis

The definitive diagnosis of schistosomiasis is obtained by the demonstration of the characteristic schistosome eggs in the faeces, in the urine or in a biopsy specimen, usually of the rectal mucosa. The morphology of the eggs is important in speciation. Egg detection in urine and faecal samples may be assisted by a variety of concentration and filtration techniques.

A variety of serodiagnostic tests have been described, but clinical experience with some of the more recently developed systems is limited. ELISA-based assays may be particularly useful in the follow-up of travellers returned from endemic areas, those with light infections not easily diagnosed microscopically, and those with CNS or ectopic infection in whom faeces, urine or rectal infection may not be demonstrable.

Reference

Mahmoud AAF. Trematodes and other flukes. In: Mandell GL, Bennett JE, Dolin R, eds. Principles and Practice of Infectious Diseases, 4th edn. New York, Edinburgh, London: Churchill Livingstone. 1995:2538–44.

Schistosoma species
ANTIBODY DETECTION (IgG)

Manufacturer: Amico Laboratories Inc
Cat. No./Trade name: 5800G/AMIZYME® Schistosoma species IgG Assay

SUMMARY

[Well-Ag]–**Ab**–[AHIgG-HRP]–[ABTS]–A_{405}

Assay type: EIA (non-competitive)
Detection: Colorimetric A_{405}
Format: Microtitre well, Ag coated
Sample type: Serum
Sample pre-treatment:
  None
Sample volume: 5 µl (+245 µl diluent)
Number of tests: 96
Controls - standards run in assay:
  Controls: Neg (1)
  Calibrators: I (1)
Incubation:
  25 min (RT) + 25 min (RT) + 25 min (RT)
Washes: 2

CONTENTS

Antibodies, antigens, labelled components:
  Schistosoma species (cercarial) Ag bound to well
  Anti-human IgG Ab (g) HRP conjugated
Substrate: ABTS
Controls - standards supplied:
  Controls: Neg
  Calibrator: 1
Additional reagents:
  None
Special equipment required:
  None

INTERPRETATION

Comments on interpretation:
  Equivocal: where sample OD is between 0.375–0.399 (95–99 U/ml); retest to confirm
No. of references: 9

NOTES

131788.0

Schistosoma species
ANTIBODY DETECTION (IgM)

Manufacturer: Amico Laboratories Inc
Cat. No./Trade name: 5800M/AMIZYME® Schistosoma species IgM Assay

SUMMARY

[Well-Ag]–**Ab**–[AHIgM-HRP]–[ABTS]–A_{405}

Assay type: EIA (non-competitive)
Detection: Colorimetric A_{405}
Format: Microtitre well, Ag coated
Sample type: Serum
Sample pre-treatment:
  None
Sample volume: 5 µl (+245 µl diluent)*
Number of tests: 96
Controls - standards run in assay:
  Controls: Neg (1)
  Calibrators: I (1)
Incubation:
  25 min (RT) + 25 min (RT) + 25 min (RT)
Washes: 2

CONTENTS

Antibodies, antigens, labelled components:
  Schistosoma species (cercarial) Ag bound to well
  Anti-human IgM Ab (g) HRP conjugated
Substrate: ABTS
Controls - standards supplied:
  Controls: Neg
  Calibrator: 1
Additional reagents:
  None
Special equipment required:
  None

INTERPRETATION

Comments on interpretation:
  Equivocal: where sample OD is between 0.275–0.299 (92–99 U/ml); retest to confirm
No. of references: 9

NOTES

131796.0

*Sample diluted in IgM serum diluent provided to eliminate interference due to RF and IgG

© KLUWER ACADEMIC PUBLISHERS 1995, ISSN 1381-5067

Schistosoma species
ANTIBODY DETECTION (IgM)

Manufacturer: Amico Laboratories Inc
Cat. No./Trade name: 5800MC/AMIZYME®
Schistosoma species IgM Capture Assay

SUMMARY

[Well-Ag]–**Ab**–[AHIgM]–[AMIg-HRP]–[ABTS]–A_{405}

Assay type: EIA (non-competitive)
Detection: Colorimetric A_{405}
Format: Microtitre well, Ag coated
Sample type: Serum
Sample pre-treatment:
   None
Sample volume: 5 µl (+245 µl diluent)
Number of tests: 96
Controls - standards run in assay:
   Controls: Neg (1)
   Calibrators: I (1)
Incubation:
   25 min (RT) + 25 min (RT) + 25 min (RT) + 25 min
      (RT)
Washes: 3

CONTENTS

Antibodies, antigens, labelled components:
   Schistosoma species (cercarial) Ag bound to well
   Anti-human IgM MAb (m)
   Anti-mouse Ig Ab (g) HRP conjugated
Substrate: ABTS
Controls - standards supplied:
   Controls: Neg
   Calibrator: 1
Additional reagents:
   None
Special equipment required:
   None

INTERPRETATION

Comments on interpretation:
   Equivocal: where sample OD is between 0.285–0.299;
      retest to confirm
No. of references: 2

NOTES

131806.0

Shigella species

Organisms of the genus *Shigella* are small, facultatively-anaerobic bacilli of the Enterobacteriaceae family. Shigellae are non-sporing, non-motile and non-capsulate and are the causal organism of shigellosis, more commonly known as bacillary dysentery. Approximately 40 serotypes of the organism are recognised which may be divided into four groups on the basis of serological and biochemical similarity: group A (*S.dysenteriae*), group B (*S.flexneri*), group C (*S.boydii*) and group D (*S.sonnei*). Disease is characterised by severe cramping abdominal pain and diarrhoea with blood and mucous. The organisms invade the intestinal mucosa where they multiply and cause death and sloughing of the mucosal cells. Extraintestinal isolations are rare, and deep invasiveness is not a feature. Enterotoxin production plays a key role in the pathogenesis of disease. Epidemics of bacillary dysentery tend to occur in predictable cycles which are thought to reflect the waxing and waning of herd immunity within the population. The peak incidence of infection is during the summer months. Shigellae are highly host-adapted to man, and the mimimum infectious dose is only about 200 organisms. Person-to-person transmission therefore occurs readily in crowded conditions, for example in military barracks, institutions for the mentally retarded, infant schools and day-care centres. Infection is most common amongst infants and young children, but secondary spread within the household to other children and adults is common. Disease is more prevalent in tropical countries, and poor standards of hygiene and nutrition increase the incidence and severity. Although most spread is person-to-person via contamination of the hands with infected faeces, outbreaks due to contamination of food and water supplies have occurred and there is good evidence that flies can play a role in transmission. *S.dysenteriae* is associated with a more severe clinical presentation, although this serogroup is now uncommon in outbreaks in the industrialised countries where *S.sonnei* is by far the most common type isolated.

Diagnosis

Definitive diagnosis is based upon the isolation of the causative organism from a stool specimen. The organisms are readily isolated on routine enteric laboratory media yielding non- or late-lactose fermenting colonies on MacConkey agar and non-xylose fermenting colonies on xylose-lysine-deoxycholate (XLD) agar. Various enrichment broths have been described to aid diagnosis in late disease when the number of organisms present in the specimen may be somewhat reduced. Suspect colonies are often screened using abbreviated identification systems such as triple sugar iron (TSI) slants from which presumptive shigellae (alkaline slant, acid butt with no H_2S or gas production) may subsequently be confirmed by full biochemical and serological testing. Serological examination does not generally play a role in acute diagnosis, although it may be useful in epidemiological studies.

References

DuPont HL. *Shigella* species. In: Mandell GL, Bennett JE, Dolin R, eds. Principles and Practice of Infectious Diseases, 4th edn. New York, Edinburgh, London: Churchill Livingstone. 1995:2033–9.

Farmer JJ, Kelly MT. Enterobacteriaceae. In: Balows A, Hausler WJ, Herrmann KL, Isenberg HD, Shadomy HJ, eds. Manual of Clinical Microbiology, 5th edn. Washington: ASM. 1991:360–83.

See Multipathogen Assays section under: Gastrointestinal pathogens

Taenia species

Taenia saginata and *Taenia solium* are parasitic cestodes (tapeworms), which may be transmitted to humans. Infection with the beef tapeworm, *T.saginata*, is acquired when inadequately cooked meat of infected cattle is eaten. Man is the definitive host for this organism, which has a worldwide distribution. The encysted larval form (cysticercus) in the meat is released in the intestine and develops into an adult worm. It attaches to the intestinal wall by specialised structures on the head (scolex) and releases eggs into the gut lumen from the ribbon-like chain of proglottid segments, each of which contains both male and female sex organs, connected to the head by a short neck. If ingested by cattle, these eggs hatch to yield oncospheres which penetrate the intestinal wall and the larval forms develop in the muscle tissue. Human tissue does not support the growth of cysticerci, so man is unable to serve as an intermediate host for this organism. Symptoms are generally minimal comprising mild abdominal discomfort, although the passage per rectum of proglottids or larger segments of worm up to several feet in length may be particularly distressing. Rarely a large mass of proglottids causes intestinal obstruction.

In contrast with *T.saginata*, man may be both definitive and intermediate host of the pork tapeworm, *T.solium*. The pathogenesis and clinical presentation of intestinal infection with the adult tapeworm are as described above. If food or water contaminated with infected human faeces is ingested, or if autoinfection occurs in an infected individual, man may become the intermediate host for this organism as outlined for cattle infected with *T.saginata* resulting in the condition known as cysticercosis. The cysts can develop virtually anywhere within the body, and cause local inflammation. Serious symptoms may arise due to CNS lesions causing seizures and symptoms due to compression of local structures, or cardiac involvement.

Diagnosis

Intestinal infection is normally diagnosed by the detection of eggs or proglottids in faeces. The sensitivity of egg detection may be improved by various techniques such as formyl-ethyl acetate concentration. Since the eggs of *T.saginata* and *T.solium* are indistinguishable, proglottids should be sought to permit speciation. As with other parasitic infections patients frequently have an eosinophilic blood picture.

A variety of serological tests, predominantly ELISA-based, are now available for the detection of antibodies to these organisms, and may be particularly useful in the diagnosis of neurocysti- cercosis and may be performed on blood and CSF. There is cross-reactivity in some of these assays with antibodies to other cestode species, and results must be interpreted with care in endemic areas. Tests based upon detection of cestode DNA and parasite antigens in stool and other samples may be important in the future.

References

Richards F, Schantz PM. Laboratory diagnosis of cysticercosis. Clinics in Laboratory Medicine. 1991;11:1011–28.

Cook GC. Protozoan and helminthic infections. In: Lambert HP ed. Infections of the central nervous system. Philadelphia, London: Decker. 1991:264–82.

King CH. Cestodes. In: Mandell GL, Bennett JE, Dolin R, eds. Principles and Practice of Infectious Diseases, 4th edn. New York, Edinburgh, London: Churchill Livingstone. 1995:2544–53.

Taenia solium
ANTIBODY DETECTION (IgG)

Manufacturer: Immunetics
Cat. No./Trade name: Human Cysticercosis Western Blot Kit

SUMMARY

[Strip-Ag]–**Ab**–[AHIgG-AP]–[BCIP/NBT]–bands

Assay type: Immunoblot assay
Detection: Visual
Format: Reaction tray, Ag coated nitrocellulose strip
Sample type: Serum, plasma or CSF
Sample pre-treatment:
   None
Sample volume: 1 ml of 1:100 dilution
Number of tests: 24
Controls - standards run in assay:
   Controls: Neg (1), Pos (1)
Incubation:
   1 hr (RT) + 2 hr (RT)
Washes: 2

CONTENTS

Antibodies, antigens, labelled components:
   T. solium (50, 42-39, 24, 21, 18, 14 and 13kD) Ags bound to nitrocellulose strip as discrete bands
   Anti-human IgG Ab AP conjugated
Substrate: BCIP/NBT
Controls - standards supplied:
   Controls: Neg and Pos (human serum)
Additional reagents:
   None
Special equipment required:
   None

INTERPRETATION

Comments on interpretation:
   Positive: coloured bands on membrane. For interpretation of band pattern see manufacturer's instructions
No. of references: 3

NOTES

131775.0

Kit licensed to Immunetics by CDC under USA Patent

Taenia solium
ANTIBODY DETECTION

Manufacturer: LMD Laboratories Inc
Cat. No./Trade name: TS-8/Cysticercosis Serology Microwell ELISA Kit

SUMMARY

[Well-Ag]–**Ab**–[Protein A-HRP]–[TMB]–A_{450}

Assay type: EIA (non-competitive)
Detection: Colorimetric A_{450}
Format: Microtitre well, Ag coated
Sample type: Serum (do not heat inactivate)
Sample pre-treatment:
   None
Sample volume: 5 µl (+315 µl diluent)
Number of tests: 48, 96
Controls - standards run in assay:
   Controls: Neg (1), weak Pos (1), strong Pos (1)
Incubation:
   10 min (RT) + 5 min (RT) + 5 min (RT)
Washes: 2

CONTENTS

Antibodies, antigens, labelled components:
   T. solium (vesicular cyst) Ag bound to well
   Protein A HRP conjugated
Substrate: TMB
Controls - standards supplied:
   Controls: Neg (human serum), weak Pos and strong Pos (rabbit serum)
Additional reagents:
   None
Special equipment required:
   None

INTERPRETATION

Comments on interpretation:
   Confirm all positive results by a more specific test
No. of references: 4

NOTES

131641.0

Cross-reactivity with Echinococcus species may occur

© KLUWER ACADEMIC PUBLISHERS 1995, ISSN 1381-5067

Taenia solium
ANTIBODY DETECTION (IgG)

Manufacturer: Melotec S.A.
Cat. No./Trade name: Melotest Cysticercosis

SUMMARY

[Well-Ag]–**Ab**–[AHIgG-HRP]–[TMB]–A_{450}

Assay type: EIA (non-competitive)
Detection: Colorimetric A_{450}
Format: Microtitre well, Ag coated
Sample type: Serum, plasma
Sample pre-treatment:
    None
Sample volume: 100 µl of 1:64 dilution x 2
Number of tests: 96
Controls - standards run in assay:
    Controls: Neg (1), low Pos (2), high Pos (1)
Incubation:
    10 min (RT) + 5 min (RT) + 10 min (RT)
Washes: 2

CONTENTS

Antibodies, antigens, labelled components:
    T. solium Ag bound to well
    Anti-human IgG Ab HRP conjugated
Substrate: TMB
Controls - standards supplied:
    Controls: Neg, low Pos and high Pos (sera)
Additional reagents:
    None
Special equipment required:
    None

INTERPRETATION

Comments on interpretation:
    Equivocal: ratio of sample OD:cut-off value between 0.9
        and 1.1; retest a fresh sample
    Repeatably equivocal: retest a fresh sample taken in 2-4
        weeks
No. of references: 4

NOTES

131133.0

Significant cross-reactivity has been reported between
    Echinococcus and Taenia solium

© *KLUWER ACADEMIC PUBLISHERS 1995, ISSN 1381-5067*

Toxocara canis

Toxocariasis in man is the result of infection with the nematode (roundworm) parasite of dogs, *Toxocara canis*. The infection has a worldwide distribution, and between 2 and 90% of dogs may be affected. The life cycle of *Toxocara* species in dogs following ingestion of eggs is similar to that of *Ascaris lumbricoides* in man with larval migration via the lungs, trachea, posterior pharynx, and thence to the bowel where the adults mature, with the unique variation that the bowel of fetal puppies may be infected by transplacental larval spread with resultant shedding of numerous eggs from birth. If human infection occurs, usually as a result of the ingestion of contaminated soil by young children, development of the parasite is arrested and second stage larvae may migrate into various tissues and evoke a granulomatous response and become encysted. The process of migration in humans is called visceral larva migrans. In most cases infection is asymptomatic or produces only mild non-specific features, however more specific symptoms may occur due to granulomatous change within the liver, lungs, kidneys, heart, striated muscle, brain and eye and fulminant presentations can occur; there is usually a marked accompanying eosinophilia. Ocular involvement may occur in the absence of any other features of the disease, presenting with a posterior or peripheral inflammatory mass and associated loss of vision. As with visceral larva migrans, children are most commonly affected.

Diagnosis

The clinical diagnosis suggested by the presence of marked eosinophilia, hepatomegaly and signs/symptoms of other organ involvement, or the detection of the characteristic ocular lesions, may be definitively confirmed by the detection of larvae in affected tissues on histologic examination of the characteristic granulomatous lesions. However larvae are not always detectable, and the use of ELISA-based serodiagnostic tests provides a useful adjunct in the diagnosis of this condition. The test appears relatively specific, but the sensitivity may vary according to the background level of antibodies in the test population. In ocular larva migrans some patients have low or negative antibody titres, but aqueous or vitreous fluid titres may be of use in this situation.

Reference

Nash TE. Visceral larva migrans and other unusual helminth infections. In: Mandell GL, Bennett JE, Dolin R, eds. Principles and Practice of Infectious Diseases, 4th edn. New York, Edinburgh, London: Churchill Livingstone. 1995:2553–7.

Toxocara canis
ANTIBODY DETECTION

Manufacturer: Cellabs Pty Ltd
Cat. No./Trade name: KT2/TOXOCARA Ab CELISA®

SUMMARY

[Well-Ag]–**Ab**–[AHIg-HRP]–[TMB]–A_{450}

Assay type: EIA (non-competitive)
Detection: Colorimetric A_{450}
Format: Microtitre well, Ag coated
Sample type: Serum, plasma
Sample pre-treatment:
　None
Sample volume: 100 μl
Number of tests: 192
Controls - standards run in assay:
　Controls: Neg (1), high Pos (1)
Incubation:
　1 hr (RT) + 1 hr (RT) + 15 min (RT)
Washes: 2

CONTENTS

Antibodies, antigens, labelled components:
　T. canis (E/S) Ag bound to well
　Anti-human Ig Ab HRP conjugated
Substrate: TMB
Controls - standards supplied:
　Controls: Neg and high Pos
Additional reagents:
　None
Special equipment required:
　None

INTERPRETATION

Comments on interpretation:
　Equivocal: Samples that give values just above the cut-off point should be considered as being associated with a past infection, or a current light infection, repeat test in one month
　Positive: Samples that give values well above the cut-off point are associated with recent infection
No. of references: 3

NOTES

131779.0

This test system is not designed as a diagnostic method for active clinical toxocariasis and must not be used for this purpose

Toxocara canis
ANTIBODY DETECTION

Manufacturer: LMD Laboratories Inc
Cat. No./Trade name: TC-1/Toxocara Serology Microwell ELISA Kit

SUMMARY

[Well-Ag]–**Ab**–[Protein A-HRP]–[TMB]–A_{450}

Assay type: EIA (non-competitive)
Detection: Colorimetric A_{450}
Format: Microtitre well, Ag coated
Sample type: Serum (do not heat inactivate)
Sample pre-treatment:
　None
Sample volume: 3 μl (+300 μl diluent)
Number of tests: 48, 96
Controls - standards run in assay:
　Controls: Neg (1), weak Pos (1), strong Pos (1)
Incubation:
　10 min (RT) + 5 min (RT) + 5 min (RT)
Washes: 2

CONTENTS

Antibodies, antigens, labelled components:
　Toxocara canis (excretory) Ag bound to well
　Protein A HRP conjugated
Substrate: TMB
Controls - standards supplied:
　Controls: Neg (human serum), weak Pos and strong Pos (rabbit serum)
Additional reagents:
　None
Special equipment required:
　None

INTERPRETATION

Comments on interpretation:
　Classification of samples is according to cut-off; no further testing
No. of references: 7

NOTES

131644.0

Toxoplasma gondii

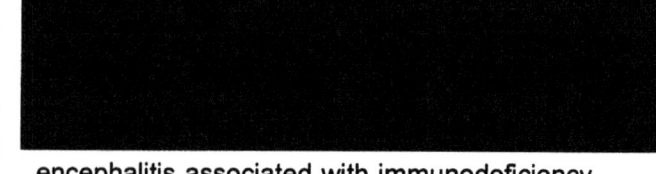

Natural history

Toxoplasma gondii is an uniquitous obligate intracellular protozoan parasite of crescentic shape, 6 μm long and containing a single nucleus. The natural hosts are members of the cat family. Infection occurs by direct or indirect contact with contaminated cat faeces, generally by ingestion of cysts from contaminated vegetables or under-cooked meat. The parasite replicates both sexually and asexually to form oocysts excreted in faeces. The oocysts sporulate and can remain viable in soil for over a year. Ingestion of the sporocyst leads to release of tachyzoites which invade and replicate within nucleated cells of all vertebrate species. Tissue cysts develop and these persist for the life of the host, particularly in the brain, heart and skeletal muscle. In immunocompromised patients reactivation of the latent organism can occur.

Most cases of toxoplasmosis are asymptomatic or show a mild disease with lymphadenopathy and malaise. Occasional cases may have more pro-longed disease with symptoms reminiscent of glandular fever (1% of cases). The prevalence of toxoplasmosis in the UK is low with a seroconver-sion rate < 1%, whereas in some countries such as France it is much higher and is particularly associated with different dietary habits. The main risk associated with *Toxoplasma gondii* is asso-ciated with infection during pregnancy and trans-placental spread rising from 25% in the first trimester to 65% in the third trimester. Severe fetal infection however is mainly associated with first trimester infection and abnormalities can include hydrocephalus, retinochoroiditis, cerebral calcifica-tion as well as hepatitis, pneumonia and myocardi-tis. Ocular infection is particularly common and toxoplasmosis is an important cause of chorio-retinitis. In immunocompromised patients, severe and often fatal toxoplasmosis can occur either by primary infection of by reactivation of latent infection, with most symptoms being referable to the central nervous system.

Diagnosis

In countries with a high prevalence of infection, antenatal screening for determination of past infection is carried out usually by ELISA or latex agglutination, both of which lend themselves to automation in the screening situation. The gold standard for acute infection is the Sabin and Feldman dye test which relies on the killing of live tachyzoites by neutralising antibody followed by the uptake of dye into dead parasites. Recently, ELISA tests which measure specific IgM have become available. Molecular methods for detecting reacti-vated *Toxoplasma gondii*, for example in cases of encephalitis associated with immunodeficiency, may be available in specialist centres.

Comment

Toxoplasma infection can be treated in pregnancy and the neonatal period with alternating courses of spiromycin and sulphadiazine with pyrimethamine without significant toxicity. Congenital toxoplasmo-sis must be differentiated from the other agents associated with TORCH, usually by serological means. Women who are shown to be seronegative in pregnancy should be warned of the dangers of handling cat litter, of gardening without the protection of gloves, of eating unwashed vegetables and fruit and of not cooking meat thoroughly. Attempts should be made to control vectors implicated in spread and it is likely that an effective vaccine particularly for the cat hosts will be developed.

References

Holliman RE. *Toxoplasma gondii*. In: Greenough A, Osborne J, Sutherland S, eds. Congenital, Perinatal and Neonatal Infections. Churchill Livingstone. 1992:209–21.

Beaman M, McCabe RE, Wong S, Remington RS. *Toxoplasma gondii*. In: Mandell GL, Bennett JE, Dolin R, eds. Principles and Practice of Infectious Diseases, 5th edn. Churchill Livingstone. 1995:2455–75.

*See also Multipathogen Assays section under:
Infectious Mononucleosis Screening Test
TECH Screening Panel
TORCH Screening Panel*

© *KLUWER ACADEMIC PUBLISHERS* 1995, ISSN 1381-5067

Toxoplasma gondii
ANTIGEN DETECTION

Manufacturer: Cellabs Pty Ltd
Cat. No./Trade name: KT1/TOXO-CEL® IF Test

SUMMARY

[Slide]–**Ag**–[MAb-FITC]–fluorescence

Assay type: Immunofluorescence assay (direct)
Detection: Fluorescence microscopy
Format: Slide, specimen coated
Sample type: Body fluids or tissues (human or veterinary specimens)
Sample pre-treatment:
Body fluids: centrifuge, resuspend in PBS, apply to slide, dry, fix
Tissue and biopsy specimens: prepare impression smear, dry, fix
Sample volume: 20 μl of prepared specimen
Number of tests: 50
Controls - standards run in assay:
Controls: Pos (1)
Incubation:
30 min (RT)
Washes: 1

CONTENTS

Antibodies, antigens, labelled components:
Anti-*T. gondii* (tachyzoites) MAb FITC conjugated
Substrate:
Controls - standards supplied:
Controls: Pos (slide)
Additional reagents:
None
Special equipment required:
None

INTERPRETATION

Comments on interpretation:
Positive: bright green fluorescent oval bodies or crescent shaped tachyzoites
No. of references: 12

NOTES

131408.0

Toxoplasma gondii
ANTIBODY DETECTION

Manufacturer: Diesse
Cat. No./Trade name: ENZYWELL Toxoplasma Screen

SUMMARY

$$[\text{Well-Ag}] \left\langle \begin{matrix} \textbf{Ab} \\ [\text{MAb-POD}] \end{matrix} \right. -[\text{TMB}]-A_{450}$$

Assay type: EIA (competitive)
Detection: Colorimetric A_{450}
Format: Microtitre well, Ag coated
Sample type: Serum (do not heat inactivate)
Sample pre-treatment:
None
Sample volume: 30 μl x 2
Number of tests: 96
Controls - standards run in assay:
Calibrator: 1 (2)
Incubation:
1 hr (37°C) + 15 min (RT)
Washes: 1

CONTENTS

Antibodies, antigens, labelled components:
T. gondii Ag bound to well
Anti-*T. gondii* MAb POD conjugated
Substrate: TMB
Controls - standards supplied:
Calibrator: 1 (bovine serum)
Additional reagents:
None
Special equipment required:
None

INTERPRETATION

Comments on interpretation:
Classification of samples is according to cut-off; no further testing
No. of references: 3

NOTES

131447.0

© KLUWER ACADEMIC PUBLISHERS 1995, ISSN 1381-5067

Toxoplasma gondii
ANTIBODY DETECTION

Manufacturer: LMD Laboratories Inc
Cat. No./Trade name: TOX-8/Toxoplasma IgG Serology Microwell ELISA Kit

SUMMARY

[Well-Ag]–**Ab**–[Protein A-HRP]–[TMB]–A_{450}

Assay type: EIA (non-competitive)
Detection: Colorimetric A_{450}
Format: Microtitre well, Ag coated
Sample type: Serum (do not heat inactivate)
Sample pre-treatment:
    None
Sample volume: 10 µl (+310 µl diluent)
Number of tests: 48, 96
Controls - standards run in assay:
    Controls: Neg (1), weak Pos (1), strong Pos (1)
Incubation:
    10 min (RT) + 5 min (RT) + 5 min (RT)
Washes: 2

CONTENTS

Antibodies, antigens, labelled components:
    T. gondii (RH strain) Ag bound to well
    Protein A HRP conjugated
Substrate: TMB
Controls - standards supplied:
    Controls: Neg (human serum), weak Pos and strong Pos (rabbit serum)
Additional reagents:
    None
Special equipment required:
    None

INTERPRETATION

Comments on interpretation:
    Classification of samples is according to cut-off; no further testing
No. of references: 4

NOTES

131642.0

Toxoplasma gondii
ANTIBODY DETECTION (IgA)

Manufacturer: Abbott Laboratories
Cat. No./Trade name: 9A46-20/IMx® System Toxo IgA

SUMMARY

[Particle-Ag]–**Ab**–[AHIgA-AP]–[4-MP]–fluorescence

Assay type: EIA (non-competitive)
Detection: Fluorometric
Format: Microparticles, Ag coated
Sample type: Serum, plasma
Sample pre-treatment:
    None
Sample volume: 150 µl
Number of tests: 100
Controls - standards run in assay:
    Controls: Neg (1), Pos (1)
    Calibrators: A (2)
Incubation:
    Automated
Washes: Automated

CONTENTS

Antibodies, antigens, labelled components:
    T. gondii Ag bound to microparticles
    Anti-human IgA Ab (g) AP conjugated
Substrate: 4-MP
Controls - standards supplied:
    Calibrator: A (human serum)
Additional reagents:
    IMx® Toxo IgA Controls (Cat. no. 9A46-10)
    IMx® Probe Cleaning Solution (Cat. no. 1A71-02)
Special equipment required:
    IMx® System

INTERPRETATION

Comments on interpretation:
    This is an automated assay
No. of references: 19

NOTES

131691.0

© KLUWER ACADEMIC PUBLISHERS 1995, ISSN 1381-5067

Toxoplasma gondii
ANTIBODY DETECTION (IgA)

Manufacturer: Bouty SpA
Cat. No./Trade name: 21875/BEIA TOXO-IgA Capture

SUMMARY

[Well-AHIgA]–**Ab**–[Ag-biotin]–[strept–HRP]–[TMB]–A_{450}

Assay type: EIA (non-competitive)
Detection: Colorimetric A_{450}
Format: Microtitre well, Ab coated
Sample type: Serum
Sample pre-treatment:
　None
Sample volume: 10 μl (+800 μl diluent)*
Number of tests: 96
Controls - standards run in assay:
　Controls: Neg (2), cut-off (2), Pos (2)
Incubation:
　1 hr (RT) + 1 hr (RT) + 30 min (RT)
Washes: 2

CONTENTS

Antibodies, antigens, labelled components:
　Anti-human IgA IgG Ab bound to well
　T. gondii (purified P30) Ag biotinylated
　Streptavidin HRP conjugated
Substrate: TMB
Controls - standards supplied:
　Controls: Neg, cut-off and Pos (human serum)
Additional reagents:
　None
Special equipment required:
　None

INTERPRETATION

Comments on interpretation:
　Equivocal: within cut-off ±5%; retest to confirm
　Repeatably equivocal: repeat test with a fresh sample
No. of references: None

NOTES

131243.0

*Samples can be assayed singly or in duplicate

Toxoplasma gondii
ANTIBODY DETECTION (IgA)

Manufacturer: Chimica Diagnostica
Cat. No./Trade name: 41360/Toxoplasma Gondii IgA

SUMMARY

[Well-Ag]–**Ab**–[AHIgA-POD]–[TMB]–A_{450}

Assay type: EIA (non-competitive)
Detection: Colorimetric A_{450}
Format: Microtitre well, Ag coated
Sample type: Serum
Sample pre-treatment:
　None
Sample volume: 100 μl of 1:101 dilution x 2*
Number of tests: 96
Controls - standards run in assay:
　Controls: Neg (2), cut-off (2), Pos (2)
Incubation:
　1 hr (RT) + 30 min (RT) + 15 min (RT)
Washes: 2

CONTENTS

Antibodies, antigens, labelled components:
　T. gondii Ag bound to well
　Anti-human IgA Ab (r) POD conjugated
Substrate: TMB
Controls - standards supplied:
　Controls: Neg, cut-off and Pos
Additional reagents:
　None
Special equipment required:
　None

INTERPRETATION

Comments on interpretation:
　Equivocal: where Index (ratio of OD sample:OD cut-off
　control) is between >0.9 and <1.1; report as
　equivocal
No. of references: 2

NOTES

131368.0

*Diluent contains AHIgG to eliminate interference due to RF
　and IgG in sample

© KLUWER ACADEMIC PUBLISHERS 1995, ISSN 1381-5067

Toxoplasma gondii
ANTIBODY DETECTION (IgA)

Manufacturer: Diesse
Cat. No./Trade name: ENZYWELL Toxoplasma IgA

SUMMARY

[Well-Ag]–**Ab**–[AHIgA-POD]–[TMB]–A$_{450}$

Assay type: EIA (non-competitive)
Detection: Colorimetric A$_{450}$
Format: Microtitre well, Ag coated
Sample type: Serum (do not heat inactivate)
Sample pre-treatment:
   None
Sample volume: 100 µl of 1:101 dilution x 2
Number of tests: 96
Controls - standards run in assay:
   Controls: Neg (2), cut-off (3), Pos (2)
Incubation:
   45 min (37°C) + 45 min (37°C) + 15 min (RT)
Washes: 2

CONTENTS

Antibodies, antigens, labelled components:
   T. gondii Ag bound to well
   Anti-human IgA MAb POD conjugated
Substrate: TMB
Controls - standards supplied:
   Controls: Neg, cut-off and Pos (human serum)
Additional reagents:
   None
Special equipment required:
   None

INTERPRETATION

Comments on interpretation:
   Equivocal: within cut-off ±10%; retest to confirm
   Repeatably equivocal: repeat with a fresh sample
No. of references: 5

NOTES

131445.0

Toxoplasma gondii
ANTIBODY DETECTION (IgA)

Manufacturer: SFRI Laboratoire
Cat. No./Trade name: IIE 020/Toxo IgA

SUMMARY

[Well-AHIgA]–**Ab**–[Ag:MAb-POD]–[OPD]–A$_{492}$

Assay type: EIA (non-competitive)
Detection: Colorimetric A$_{492}$
Format: Microtitre well, Ab coated
Sample type: Serum, plasma
Sample pre-treatment:
   None
Sample volume: 10 µl (+1 ml diluent)
Number of tests: 96
Controls - standards run in assay:
   Controls: Neg (1), Pos (1)
Incubation:
   30 min (RT) + 2 hr (RT) + 15 min (RT)
Washes: 2 (+ preliminary plate wash)

CONTENTS

Antibodies, antigens, labelled components:
   Anti-human IgA Ab bound to well
   T. gondii Ag
   Anti-*T. gondii* MAb POD conjugated
Substrate: OPD
Controls - standards supplied:
   Controls: Neg and Pos (serum)
Additional reagents:
   None
Special equipment required:
   None

INTERPRETATION

Comments on interpretation:
   Equivocal: where cut-off is between 0.5 and 0.7; retest to
   confirm
   Repeatably equivocal: repeat test at a later date
No. of references: 9

NOTES

131650.0

Toxoplasma gondii
ANTIBODY DETECTION (IgG)

Manufacturer: Abbott Laboratories
Cat. No./Trade name: 6317-24/ABBOTT TOXO-G EIA Kit

SUMMARY

[Bead-Ag]–**Ab**–[AHIgG-HRP]–[OPD]–A$_{492}$

Assay type: EIA (non-competitive)
Detection: Colorimetric A$_{492}$
Format: Reaction wells, Ag coated beads
Sample type: Serum, plasma
Sample pre-treatment:
 None
Sample volume: 20 μl (+200 μl diluent)
Number of tests: 100, 500
Controls - standards run in assay:
 Qualitative: Controls: Neg (1), low Pos (1), high Pos (1)
 Quantitative: Standard (1) in addition to Controls
Incubation:
 1 hr (37°C) + 1 hr (37°C) + 30 min (RT)
Washes: 2

CONTENTS

Antibodies, antigens, labelled components:
 T. gondii Ag bound to beads
 Anti-human IgG Ab (g) HRP conjugated
Substrate: OPD
Controls - standards supplied:
 Controls: Neg, low Pos and high Pos (human serum)
 Standards: 1 (12 IU/ml, WHO standard - human serum)
Additional reagents:
 H$_2$SO$_4$
Special equipment required:
 Pentawash® II and Gast® Vacuum Pump (or similar)
 Quantum Analyser (optional)

INTERPRETATION

Comments on interpretation:
 Assessment of immune status by calculation of TOXO-G Index
 A quantitative procedure is available using Controls and Standard provided
No. of references: 6

NOTES

131352.0

Toxoplasma gondii
ANTIBODY DETECTION (IgG)

Manufacturer: Abbott Laboratories
Cat. No./Trade name: 2254-66/IMx® System Toxo IgG

SUMMARY

[Particle-Ag]–**Ab**–[AHIgG-AP]–[4-MP]–fluorescence

Assay type: EIA (non-competitive)
Detection: Fluorometric
Format: Microparticles, Ag coated
Sample type: Serum, plasma
Sample pre-treatment:
 None
Sample volume: 150 μl
Number of tests: 100
Controls - standards run in assay:
 Controls: Neg (1), Pos (1)
 Calibrators: 1 (1), 2 (1), 3 (1), 4 (1), 5 (1), 6 (1)
Incubation:
 Automated
Washes: Automated

CONTENTS

Antibodies, antigens, labelled components:
 T. gondii Ag bound to microparticles
 Anti-human IgG Ab AP conjugated
Substrate: 4-MP
Controls - standards supplied:
 Calibrator: Mode 1 (human serum)
Additional reagents:
 IMx® Toxo IgG Calibrators (Cat. no. 2254-01 – Ab titre 0, 10–300 IU/ml with ref to WHO Standard)
 IMx® Toxo IgG Controls (Cat. no. 2254-10)
 IMx® Probe Cleaning Solution (Cat. no. 1A71-02)
Special equipment required:
 IMx® System

INTERPRETATION

Comments on interpretation:
 This is an automated assay
No. of references: 15

NOTES

131692.0

© KLUWER ACADEMIC PUBLISHERS 1995, ISSN 1381-5067

Toxoplasma gondii
ANTIBODY DETECTION (IgG)

Manufacturer: Alfa Biotech
Cat. No./Trade name: 05772910/Toxoplasma IgG Elisa System

SUMMARY

[Well-Ag]–**Ab**–[AHIgG-HRP]–[TMB]–A_{450}

Assay type: EIA (non-competitive)
Detection: Colorimetric A_{450}
Format: Microtitre well, Ag coated
Sample type: Serum
Sample pre-treatment:
   None
Sample volume: 100 µl of 1:100 dilution x 2
Number of tests: 96
Controls - standards run in assay:
   Calibrators: 1 (2), 2 (2), 3 (2), 4 (2), 5 (2)
Incubation:
   20 min (37°C) + 20 min (37°C) + 20 min (37°C)
Washes: 2

CONTENTS

Antibodies, antigens, labelled components:
   T. gondii Ag bound to well
   Anti-human IgG Ab (g) HRP conjugated
Substrate: TMB
Controls - standards supplied:
   Calibrators: 5 (human serum; range 0–100 IU/ml)
Additional reagents:
   None
Special equipment required:
   None

INTERPRETATION

Comments on interpretation:
   Equivocal: where Ab concentration is between 10 and 12.5 IU/ml; retest to confirm
   Where value is > 100 IU/ml; report as high Pos or retest at higher serum dilution
No. of references: 10

NOTES

131101.0

Toxoplasma gondii
ANTIBODY DETECTION (IgG)

Manufacturer: Amico Laboratories Inc
Cat. No./Trade name: 5600G/AMIZYME® Toxoplasma IgG Assay

SUMMARY

[Well-Ag]–**Ab**–[AHIgG-HRP]–[ABTS]–A_{405}

Assay type: EIA (non-competitive)
Detection: Colorimetric A_{405}
Format: Microtitre well, Ag coated
Sample type: Serum
Sample pre-treatment:
   None
Sample volume: 10 µl (+490 µl diluent)
Number of tests: 96
Controls - standards run in assay:
   Controls: Neg (1)
   Calibrator: 1 (1)
Incubation:
   25 min (RT) + 25 min (RT) + 25 min (RT)
Washes: 2

CONTENTS

Antibodies, antigens, labelled components:
   T. gondii Ag bound to well
   Anti-human IgG Ab HRP conjugated
Substrate: ABTS
Controls - standards supplied:
   Controls: Neg
   Calibrator: 1
Additional reagents:
   None
Special equipment required:
   None

INTERPRETATION

Comments on interpretation:
   Equivocal: where sample OD is between 0.385–0.399; retest to confirm
No. of references: 3

NOTES

131015.0

Toxoplasma gondii
ANTIBODY DETECTION (IgG)

Manufacturer: Amico Laboratories Inc
Cat. No./Trade name: 5600G/AMIZYME® Toxoplasma IgG Assay

SUMMARY

[Well-Ag]–**Ab**–[AHIgG-HRP]–[ABTS]–A_{405}

Assay type: EIA (non-competitive)
Detection: Colorimetric A_{405}
Format: Microtitre well, Ag coated
Sample type: Serum
Sample pre-treatment:
　None
Sample volume: 5 μl (+245 μl diluent)
Number of tests: 96
Controls - standards run in assay:
　Controls: Neg (1)
　Calibrators: I (1)
Incubation:
　25 min (RT) + 25 min (RT) + 25 min (RT)
Washes: 2

CONTENTS

Antibodies, antigens, labelled components:
　T. gondii Ag bound to well
　Anti-human IgG Ab (g) HRP conjugated
Substrate: ABTS
Controls - standards supplied:
　Controls: Neg
　Calibrator: 1
Additional reagents:
　None
Special equipment required:
　None

INTERPRETATION

Comments on interpretation:
　Equivocal: where sample OD is between 0.375–0.399
　　(95–99 U/ml); retest to confirm
No. of references: 3

NOTES

131789.0

Toxoplasma gondii
ANTIBODY DETECTION (IgG)

Manufacturer: Behringwerke AG
Cat. No./Trade name: /Enzygnost® Toxoplasmosis/IgG

SUMMARY

[Well-Ag]–**Ab**–[AHIgG-POD]–[TMB]–A_{450}

Assay type: EIA (non-competitive)
Detection: Colorimetric A_{450}
Format: Microtitre well, Ag coated
Sample type: Serum, plasma
Sample pre-treatment:
　None
Sample volume: 20 μl of 1:21 dilution*
Number of tests: 96
Controls - standards run in assay:
　Controls: Neg (1), Pos (2)
Incubation:
　1 hr (37°C) + 1 hr (37°C) + 30 min (RT)
Washes: 2

CONTENTS

Antibodies, antigens, labelled components:
　T. gondii Ag bound to well
　Anti-human IgG Ab (r) POD conjugated
Substrate: TMB (bought separately)
Controls - standards supplied:
　Controls: Neg and Pos (human serum – assigned values)
Additional reagents:
　Supplementary Reagents for Enzygnost®/TMB (Code
　　No. OUVP)
Special equipment required:
　Behring ELISA Processor II (optional)
　Behring ELISA Processor III (optional)

INTERPRETATION

Comments on interpretation:
　Equivocal: retest to confirm
　Repeatably equivocal: report as equivocal and retest a
　　fresh specimen at least 8 days later
No. of references: 2

NOTES

131025.0

*If high Ab levels are expected a further serum dilution of
　1:10 is required

© *KLUWER ACADEMIC PUBLISHERS 1995, ISSN 1381-5067*

Toxoplasma gondii
ANTIBODY DETECTION (IgG)

Manufacturer: BIOKIT SA
Cat. No./Trade name: 3000-1214/bioelisa TOXO IgG

SUMMARY

[Well-Ag]–**Ab**–[AHIgG-HRP]–[TMB]–A_{450}

Assay type: EIA (non-competitive)
Detection: Colorimetric A_{450}
Format: Microtitre well, Ag coated
Sample type: Serum (do not heat inactivate)
Sample pre-treatment:
 None
Sample volume: 100 µl of 1:100 dilution
Number of tests: 96
Controls - standards run in assay:
 Calibrators: Neg (2), low Pos (2), high Pos (2)
Incubation:
 1 hr (37°C) + 30 min (37°C) + 30 min (RT)
Washes: 2

CONTENTS

Antibodies, antigens, labelled components:
 T. gondii Ag bound to well
 Anti-human IgG Ab (r) HRP conjugated
Substrate: TMB
Controls - standards supplied:
 Calibrators: Neg, low Pos (10 IU/ml) and high Pos (200 IU/ml) – human serum
Additional reagents:
 None
Special equipment required:
 None

INTERPRETATION

Comments on interpretation:
 Classification of samples is according to cut-off; no further testing. Positive samples are considered to have IgG Ab levels of ≥10 IU/ml
No. of references: 11

NOTES

131372.0

Toxoplasma gondii
ANTIBODY DETECTION (IgG)

Manufacturer: bioMerieux
Cat. No./Trade name: 30 209/VIDAS TOXO IgG

SUMMARY

[Solid phase-Ag]–**Ab**–[AHIgG-AP]–[4MP]–fluorescence

Assay type: EIA (non-competitive)
Detection: Fluorometric
Format: Solid phase receptacle (SPR), Ag coated
Sample type: Serum (do not heat inactivate)
Sample pre-treatment:
 None
Sample volume: 100 µl
Number of tests: 60
Controls - standards run in assay:
 Controls: Neg (1), Pos (1)
 Calibrators: Pos (1)
Incubation:
 Automated - total time 40 min
Washes: Automated

CONTENTS

Antibodies, antigens, labelled components:
 T. gondii (membrane and cytoplasmic) Ags bound to solid phase receptacle
 Anti-human IgG Ab (m) AP conjugated
Substrate: 4-MP
Controls - standards supplied:
 Controls: Neg and Pos (human plasma)
 Calibrators: Pos (titre in IU/ml - human serum)
Additional reagents:
 None
Special equipment required:
 Vitek immunodiagnostic assay system (VIDAS)

INTERPRETATION

Comments on interpretation:
 This is an automated assay
 Results designated equivocal; retest to confirm
No. of references: None

NOTES

131309.0

© KLUWER ACADEMIC PUBLISHERS 1995, ISSN 1381-5067

Toxoplasma gondii
ANTIBODY DETECTION (IgG)

Manufacturer: Biotecx Laboratories Inc
Cat. No./Trade name: OptiCoat® Toxo (IgG) Elisa Kit

SUMMARY

[Well-Ag]–**Ab**–[AHIgG-AP]–[PNP]–A_{405}

Assay type: EIA (non-competitive)
Detection: Colorimetric A_{405}
Format: Microtitre well, Ag coated
Sample type: Serum
Sample pre-treatment:
   None
Sample volume: 100 µl of 1:41 dilution x 2
Number of tests: 96
Controls - standards run in assay:
   Controls: Neg (2), Pos (2)
   Calibrators: 1 (2)
Incubation:
   30 min (RT) + 30 min (RT) + 30 min (RT)
Washes: 2

CONTENTS

Antibodies, antigens, labelled components:
   T. gondii Ag bound to well
   Anti-human IgG Ab AP conjugated
Substrate: PNP
Controls - standards supplied:
   Controls: Neg and Pos (human serum)
   Calibrators: 1
Additional reagents:
   None
Special equipment required:
   None

INTERPRETATION

Comments on interpretation:
   Classification of samples is according to cut-off; no
      further testing
No. of references: 3

NOTES

131759.0

Toxoplasma gondii
ANTIBODY DETECTION (IgG)

Manufacturer: BioWhittaker Inc
Cat. No./Trade name: 30-628U/Toxoelisa II

SUMMARY

[Well-Ag]–**Ab**–[AHIgG-AP]–[PMP]–A_{550}

Assay type: EIA (non-competitive)
Detection: Colorimetric A_{550}
Format: Microtitre well, Ag coated
Sample type: Serum (do not heat inactivate)
Sample pre-treatment:
   None
Sample volume: 10 µl (+ 200 µl diluent)
Number of tests: 192
Controls - standards run in assay:
   Controls: A (1), B (1)
   Calibrators: Neg (1), Pos 1 (1), Pos 2 (1)
Incubation:
   45 min (RT) + 45 min (RT) + 45 min (RT)
Washes: 2 (+ preliminary plate wash)

CONTENTS

Antibodies, antigens, labelled components:
   T. gondii Ag bound to well
   Anti-human IgG Ab (g) AP conjugated
Substrate: PMP
Controls - standards supplied:
   Controls: A and B
   Calibrators: Neg, Pos 1, Pos 2 (all human serum)
Additional reagents:
   None
Special equipment required:
   BioWhittaker automated system and software (optional)

INTERPRETATION

Comments on interpretation:
   Equivocal: Elisa value between 0.15–0.16 or Index value
      between 0.88–0.89; retest to confirm
   Repeatably equivocal: retest a fresh sample or use an
      alternative method
No. of references: 3

NOTES

131742.0

Toxoplasma gondii
ANTIBODY DETECTION (IgG)

Manufacturer: BioWhittaker Inc
Cat. No./Trade name: 30-338U/Toxostat

SUMMARY

[Well-Ag]–**Ab**–[AHIgG-AP]–[PMP]–A_{550}

Assay type: EIA (non-competitive)
Detection: Colorimetric A_{550}
Format: Microtitre well, Ag coated
Sample type: Serum (do not heat inactivate)
Sample pre-treatment:
　None
Sample volume: 10 μl (+200 μl diluent)
Number of tests: 192
Controls - standards run in assay:
　Controls: Neg (1), Pos (1)
　Standards: Neg (1), low Pos (1), high Pos (1)
Incubation:
　15 min (RT) + 15 min (RT) + 15 min (RT)
Washes: 2 (+preliminary plate wash)

CONTENTS

Antibodies, antigens, labelled components:
　T. gondii Ag bound to well
　Anti-human IgG Ab (g) AP conjugated
Substrate: PMP
Controls - standards supplied:
　Controls: Neg and Pos
　Standards: Neg, low Pos and high Pos (all human serum)
Additional reagents:
　None
Special equipment required:
　BioWhittaker automated system and software (optional)

INTERPRETATION

Comments on interpretation:
　Equivocal: Predicted Index value (PIV) between 0.80–
　　0.90; retest to confirm
　Repeatably equivocal: retest a fresh sample or use
　　alternative method
　Low Positive: PIV between 1.00–1.74
　Mid Positive: PIV between 1.75–3.89
　High Positive: PIV ⩾3.90
No. of references: 16

NOTES

131749.0

Toxoplasma gondii
ANTIBODY DETECTION (IgG)

Manufacturer: Bouty SpA
Cat. No./Trade name: 21879/BEIA TOXO IgG Quant

SUMMARY

[Well-Ag]–**Ab**–[AHIgG-HRP]–[TMB]–A_{450}

Assay type: EIA (non-competitive)
Detection: Colorimetric A_{450}
Format: Microtitre well, Ag coated
Sample type: Serum
Sample pre-treatment:
　None
Sample volume: 10 μl (+800 μl diluent)*
Number of tests: 96, 192
Controls - standards run in assay:
　Controls: Neg (2)
　Calibrators: 1 (2), 2 (2), 3 (2), 4 (2)
Incubation:
　30 min (RT) + 30 min (RT) + 15 min (RT)
Washes: 2

CONTENTS

Antibodies, antigens, labelled components:
　T. gondii Ag bound to well
　Anti-human IgG Ab HRP conjugated
Substrate: TMB
Controls - standards supplied:
　Controls: Neg (human serum)
　Calibrators: 4 (15, 30, 100 and 250 IU/ml - human serum)
Additional reagents:
　None
Special equipment required:
　None

INTERPRETATION

Comments on interpretation:
　Negative: < 15 IU/ml
　Low Positive: between 15 and 30 IU/ml
　Positive: ⩾30 IU/ml
No. of references: 7

NOTES

131241.0

*Samples can be assayed singly or in duplicate

© KLUWER ACADEMIC PUBLISHERS 1995, ISSN 1381-5067

Toxoplasma gondii
ANTIBODY DETECTION (IgG)

Manufacturer: Centocor Inc
Cat. No./Trade name: M405/800-905/CAPTIA® Toxo-G

SUMMARY

[Well-Ag]–**Ab**–[AHIgG-HRP]–[TMB]–A$_{450}$

Assay type: EIA (non-competitive)
Detection: Colorimetric A$_{450}$
Format: Microtitre well, Ag coated
Sample type: Serum (do not heat inactivate)
Sample pre-treatment:
  None
Sample volume: 100 µl of 1:101 dilution
Number of tests: 96
Controls - standards run in assay:
  Controls: Neg (1), low Pos (2), high Pos (1)*
Incubation:
  1 hr (37°C) + 1 hr (37°C) + 30 min (RT)
Washes: 2

CONTENTS

Antibodies, antigens, labelled components:
  T. gondii (trophozoites from mouse infections) Ag bound
    to well
  Anti-human IgG Ab (r) HRP conjugated
Substrate: TMB
Controls - standards supplied:
  Controls: Neg, low Pos and high Pos (human serum - IU/
    ml - WHO reference standard)
Additional reagents:
  H$_2$SO$_4$
Special equipment required:
  None

INTERPRETATION

Comments on interpretation:
  Equivocal: where Antibody Index is between 0.9 and 1.1;
    retest to confirm
  Repeatably equivocal: repeat test using fresh sample
No. of references: 6

NOTES

131619.0

*If results are to be expressed in IU/ml, use high Pos control
  in duplicate

Toxoplasma gondii
ANTIBODY DETECTION (IgG)

Manufacturer: Chimica Diagnostica
Cat. No./Trade name: 41300/Toxoplasma Gondii IgG

SUMMARY

[Well-Ag]–**Ab**–[AHIgG-POD]–[TMB]–A$_{450}$

Assay type: EIA (non-competitive)
Detection: Colorimetric A$_{450}$
Format: Microtitre well, Ag coated
Sample type: Serum
Sample pre-treatment:
  None
Sample volume: 100 µl of 1:101 dilution x 2
Number of tests: 96
Controls - standards run in assay:
  Standards: 1 (2), 2 (2), 3 (2), 4 (2), 5 (2), 6 (2)
Incubation:
  1 hr (37°C) + 1 hr (37°C) + 20 min (RT)
Washes: 2

CONTENTS

Antibodies, antigens, labelled components:
  T. gondii Ag bound to well
  Anti-human IgG Ab (r) POD conjugated
Substrate: TMB
Controls - standards supplied:
  Standards: 1–6 (titres 0, 10, 20, 50, 100 and 250 IU/ml)
Additional reagents:
  None
Special equipment required:
  None

INTERPRETATION

Comments on interpretation:
  Results are expressed quantitatively in IU/ml with
    reference to the Standards provided
No. of references: 2

NOTES

131361.0

Toxoplasma gondii
ANTIBODY DETECTION (IgG)

Manufacturer: Clark Laboratories
Cat. No./Trade name: Toxoplasma gondii IgG Elisa Test

SUMMARY

[Well-Ag]–**Ab**–[AHIgG-HRP]–[OPD]–A_{490}

Assay type: EIA (non-competitive)
Detection: Colorimetric A_{490}
Format: Microtitre well, Ag coated
Sample type: Serum (do not heat inactivate)
Sample pre-treatment:
   None
Sample volume: 10 µl (+200 µl diluent)
Number of tests: 96
Controls - standards run in assay:
   Controls: Neg (1), Pos (1)
   Calibrators: 1 (2)
Incubation:
   20 min (RT) + 20 min (RT) + 10 min (RT)
Washes: 2

CONTENTS

Antibodies, antigens, labelled components:
   T. gondii Ag bound to well
   Anti-human IgG Ab (g) HRP conjugated
Substrate: TMB*
Controls - standards supplied:
   Controls: Neg and Pos (human serum or plasma)
   Calibrator: 1 (human serum or plasma)
Additional reagents:
   H_2SO_4
Special equipment required:
   None

INTERPRETATION

Comments on interpretation:
   Classification of samples is according to cut-off; no
   further testing
No. of references: None

NOTES

131662.0

*OPD can be supplied as an alternative

Toxoplasma gondii
ANTIBODY DETECTION (IgG)

Manufacturer: Dade International Inc
(Bartels Division)
Cat. No./Trade name: B1029-310/Bartels PRIMA
System® Toxoplasma IgG EIA

SUMMARY

[Well-Ag]–**Ab**–[AHIgG-enzyme]–[PNP]–A_{405}

Assay type: EIA (non-competitive)
Detection: Colorimetric A_{405}
Format: Microtitre well, Ag coated
Sample type: Serum
Sample pre-treatment:
   None
Sample volume: 100 µl of 1:21 dilution
Number of tests: 96
Controls - standards run in assay:
   Controls: Neg (1), Pos (1)
   Reference serum: (1)
Incubation:
   30 min (37°C) + 30 min (37°C) + 30 min (37°C)
Washes: 2

CONTENTS

Antibodies, antigens, labelled components:
   T. gondii (RH strain) Ag bound to well
   Anti-human IgG Ab enzyme conjugated
Substrate: PNP
Controls - standards supplied:
   Controls: Neg and Pos
   Reference serum: 1
Additional reagents:
   None
Special equipment required:
   None

INTERPRETATION

Comments on interpretation:
   Classification of samples is according to cut-off; no
   further testing
No. of references: 7

NOTES

131048.0

© *KLUWER ACADEMIC PUBLISHERS 1995, ISSN 1381-5067*

Toxoplasma gondii
ANTIBODY DETECTION (IgG)

Manufacturer: Denka Seiken Co. Ltd
Cat. No./Trade name: Toxoplasma IgG - EIA "SEIKEN"

SUMMARY

[Well-Ag]–**Ab**–[AHIgG-POD]–[TMB]–A_{450}

Assay type: EIA (non-competitive)
Detection: Colorimetric A_{450}
Format: Microtitre well, Ag coated
Sample type: Serum
Sample pre-treatment:
  None
Sample volume: 10 µl (+2 ml diluent)
Number of tests: 96
Controls - standards run in assay:
  Standards: I (1), II (2), III (1)
Incubation:
  1 hr (RT) + 1 hr (RT) + 30 min (RT)
Washes: 2

CONTENTS

Antibodies, antigens, labelled components:
  T. gondii Ag bound to well
  Anti-human IgG Ab (g) POD conjugated
Substrate: TMB
Controls - standards supplied:
  Standards: I (Neg), II (low titre) and III (high titre) – all human serum
Additional reagents:
  None
Special equipment required:
  None

INTERPRETATION

Comments on interpretation:
  Equivocal: where the calculated IgG antibody titre is between 0.5 and 1.0; repeat test with a fresh specimen in 1–2 weeks
No. of references: 10

NOTES

131800.0

An independent evaluation of this kit is available – please contact manufacturer for details

Toxoplasma gondii
ANTIBODY DETECTION (IgG)

Manufacturer: Diesse
Cat. No./Trade name: 91040/ENZYWELL Toxoplasma IgG

SUMMARY

[Well-Ag]–**Ab**–[AHIgG-POD]–[TMB]–A_{450}

Assay type: EIA (non-competitive)
Detection: Colorimetric A_{450}
Format: Microtitre well, Ag coated
Sample type: Serum (do not heat inactivate)
Sample pre-treatment:
  None
Sample volume: 100 µl of 1:101 dilution x 2
Number of tests: 96
Controls - standards run in assay:
  Controls: Neg (2), cut-off (3), Pos (2)
Incubation:
  45 min (37°C) + 45 min (37°C) + 15 min (RT)
Washes: 2

CONTENTS

Antibodies, antigens, labelled components:
  T. gondii Ag bound to well
  Anti-human IgG MAb POD conjugated
Substrate: TMB
Controls - standards supplied:
  Controls: Neg, cut-off and Pos (human serum)
Additional reagents:
  Toxoplasma gondii IgG Calibration Set (Cat. no. 91940 - for optional quantitative results)
Special equipment required:
  None

INTERPRETATION

Comments on interpretation:
  Equivocal: within cut-off ±10%; retest to confirm
  Repeatably equivocal: repeat with a fresh sample
  Results may be expressed quantitatively in IU/ml using 6 calibrators bought separately
No. of references: 3

NOTES

131440.0

© KLUWER ACADEMIC PUBLISHERS 1995, ISSN 1381-5067

Toxoplasma gondii
ANTIBODY DETECTION (IgG)

Manufacturer: E. Merck
Cat. No./Trade name: 14069/Toxoplasma gondii IgG MAGIA®

SUMMARY

[Particles-Ag]–**Ab**–[AHIgG-AP]–[substrate]–A_{405}

Assay type: EIA (non-competitive)
Detection: Colorimetric A_{405}
Format: Cuvette, Ag coated particles (magnetisable)
Sample type: Serum
Sample pre-treatment:
　None
Sample volume: 10 µl
Number of tests: 100
Controls - standards run in assay:
　Calibrators: zero, sero G (automated)
Incubation:
　Automated
Washes: Automated

CONTENTS

Antibodies, antigens, labelled components:
　T. gondii Ag bound to particles (magnetisable)
　Anti-human IgG Ab AP conjugated
Substrate: Bought separately
Controls - standards supplied:
　Bought separately
Additional reagents:
　Particle wash and system solutions (Cat. nos. 14097, 14096)
　Substrate (Cat. no. 14095), Calibrators (Cat. nos. 14090, 14092)
　Anti-foam (Cat. no. 14098)
Special equipment required:
　MAGIA® 7000 Immunoanalyzer (Cat. no. 14011) or MAGIA® 8000 (Cat. no. 114526)
　Reaction cuvettes (Cat. no. 14085)

INTERPRETATION

Comments on interpretation:
　This is an automated assay
No. of references: 2

NOTES

131063.0

Toxoplasma gondii
ANTIBODY DETECTION (IgG)

Manufacturer: Gull Laboratories Inc
Cat. No./Trade name: Toxo IgG (Quantitative) ELISA Test

SUMMARY

[Well-Ag]–**Ab**–[AHIgG-AP]–[PNP]–A_{405}

Assay type: EIA (non-competitive)
Detection: Colorimetric A_{405}
Format: Microtitre well, Ag coated
Sample type: Serum
Sample pre-treatment:
　None
Sample volume: 100 µl of 1:21 dilution
Number of tests: 96
Controls - standards run in assay:
　Controls: Neg (1), Pos (1)
　Calibrators: 1 (1), 2 (1), 3 (1)*
　Reference serum: 1 (3)
Incubation:
　30 min (37°C) + 30 min (37°C) + 30 min (37°C)
Washes: 2

CONTENTS

Antibodies, antigens, labelled components:
　T. gondii Ag bound to well
　Anti-human IgG Ab (caprine) AP conjugated
Substrate: PNP
Controls - standards supplied:
　Controls: Neg and Pos
　Calibrators: 3
　Reference serum: 1 (all human serum)
Additional reagents:
　None
Special equipment required:
　None

INTERPRETATION

Comments on interpretation:
　Equivocal: where sample OD is between mean OD of Reference Serum x 0.91 and mean OD of Reference Serum x 0.99; retest to confirm
　Repeatably equivocal: retest a fresh specimen or use an alternative method
No. of references: 8

NOTES

131698.0

*Omit calibrators for qualitative assay

© KLUWER ACADEMIC PUBLISHERS 1995, ISSN 1381-5067

Toxoplasma gondii
ANTIBODY DETECTION (IgG)

Manufacturer: Human GmbH
Cat. No./Trade name: 51209/Toxoplasma gondii - IgG - ELISA

SUMMARY

[well-Ag]–**Ab**–[AHIgG-HRP]–[TMB]–A$_{450}$

Assay type: EIA (non-competitive)
Detection: Colorimetric A$_{450}$
Format: Microtitre well, Ag coated
Sample type: Serum
Sample pre-treatment:
   None
Sample volume: 100 μl of 1:100 dilution
Number of tests: 96
Controls - standards run in assay:
   Controls: Neg (2), cut-off (2), low Pos (2), Pos (2), high
      Pos (2)
Incubation:
   30 min (RT) + 30 min (RT) + 15 min (RT)
Washes: 2

CONTENTS

Antibodies, antigens, labelled components:
   T. gondii Ag bound to well
   Anti-human IgG Ab (r) HRP conjugated
Substrate: TMB
Controls - standards supplied:
   Controls: Neg, cut-off, low Pos, high Pos and Pos
      (human)
Additional reagents:
   None
Special equipment required:
   None

INTERPRETATION

Comments on interpretation:
   Equivocal: within cut-off ±15%; retest in parallel with
      fresh sample taken 7–10 days later
No. of references: 5

NOTES

131341.0

Toxoplasma gondii
ANTIBODY DETECTION (IgG)

Manufacturer: IFCI Clone Systems
Cat. No./Trade name: 08.1000/EIAGEN Toxoplasmosis IgG

SUMMARY

[Well-Ag]–**Ab**–[AHIgG-POD]–[TMB]–A$_{450}$

Assay type: EIA (non-competitive)
Detection: Colorimetric A$_{450}$
Format: Microtitre well, Ag coated
Sample type: Serum
Sample pre-treatment:
   None
Sample volume: 10 μl (+ 1 ml diluent)*
Number of tests: 96
Controls - standards run in assay:
   Controls: Neg (2), cut-off (3), Pos (2)
Incubation:
   45 min (37°C) + 45 min (37°C) + 15 min (RT)
Washes: 2

CONTENTS

Antibodies, antigens, labelled components:
   T. gondii Ag bound to well
   Anti-human IgG MAb POD conjugated
Substrate: TMB
Controls - standards supplied:
   Controls: Neg, cut-off and Pos (human serum)
Additional reagents:
   T. gondii Standard Set (6 calibrators) is available for a
      performance of a quantitative assay (optional)
Special equipment required:
   None

INTERPRETATION

Comments on interpretation:
   Equivocal: within cut-off ±10%; retest to confirm
   Repeatably equivocal: retest a fresh sample
No. of references: 3

NOTES

131674.0

*Samples are assayed in duplicate

© KLUWER ACADEMIC PUBLISHERS 1995, ISSN 1381-5067

Toxoplasma gondii
ANTIBODY DETECTION (IgG)

Manufacturer: Immuno Pharmacology Research
Cat. No./Trade name: TOXOPLASMA IgG ELISA

SUMMARY

[Well-Ag]–**Ab**–[AHIgG-AP]–[PNP]–A_{405}

Assay type: EIA (non-competitive)
Detection: Colorimetric A_{405}
Format: Microtitre well, Ag coated
Sample type: Serum
Sample pre-treatment:
  None
Sample volume: 200 µl of 1:100 dilution x 2
Number of tests: 48, 96
Controls - standards run in assay:
  Controls: Neg (1), Pos (1)
Incubation:
  1 hr (37°C) + 1 hr (37°C) + 30 min (37°C)
Washes: 2

CONTENTS

Antibodies, antigens, labelled components:
  T. gondii Ag bound to well
  Anti-human IgG Ab AP conjugated
Substrate: PNP
Controls - standards supplied:
  Controls: Neg and Pos
Additional reagents:
  None
Special equipment required:
  None

INTERPRETATION

Comments on interpretation:
  Equivocal: where sample OD is between 0.150 and
    0.300; retest to confirm
  Repeatably equivocal: retest a fresh specimen in 15 days
No. of references: None

NOTES

131319.0

Toxoplasma gondii
ANTIBODY DETECTION (IgG)

Manufacturer: Immunobiological Laboratories
Cat. No./Trade name: VE 57191/Toxoplasma gondii IgG

SUMMARY

[Well-Ag]–**Ab**–[AHIgG-HRP]–[TMB]–A_{450}

Assay type: EIA (non-competitive)
Detection: Colorimetric A_{450}
Format: Microtitre well, Ag coated
Sample type: Serum, plasma
Sample pre-treatment:
  None
Sample volume: 100 µl of 1:100 dilution x 2
Number of tests: 96
Controls - standards run in assay:
  Controls: 1 (2)
  Standards: 1 (2), 2 (2), 3 (2), 4 (2), 5 (2)
Incubation:
  1 hr (RT) + 30 min (RT) + 15 min (RT)
Washes: 2

CONTENTS

Antibodies, antigens, labelled components:
  T. gondii Ag bound to well
  Anti-human IgG Ab HRP conjugated
Substrate: TMB
Controls - standards supplied:
  Controls: 1
  Standards: 5 (50, 150, 500, 1000 and 2000 IU/ml)
Additional reagents:
  None
Special equipment required:
  None

INTERPRETATION

Comments on interpretation:
  Negative: 0–15 IU/ml
  Weak Positive: 15–50 IU/ml
  Positive: 50–160 IU/ml
  Strong Positive: > 160 IU/ml
No. of references: 9

NOTES

131252.0

Toxoplasma gondii
ANTIBODY DETECTION (IgG)

Manufacturer: Incstar
Cat. No./Trade name: 4560/Toxoplasma IgG Clin-ELISA

SUMMARY

[Well-Ag]–**Ab**–[AHIgG-AP]–[PNP]–A_{450}

Assay type: EIA (non-competitive)
Detection: Colorimetric A_{450}
Format: Microtitre well, Ag coated
Sample type: Serum
Sample pre-treatment:
　None
Sample volume: 10 µl (+ 500 µl diluent)
Number of tests: 96
Controls - standards run in assay:
　Controls: low Pos (1), high Pos (1)
　Calibrators: I (1), II (1), III (1)
Incubation:
　30 min (RT) + 30 min (RT) + 45 min (RT)
Washes: 2

CONTENTS

Antibodies, antigens, labelled components:
　T. gondii (RH strain) Ag bound to well
　Anti-human IgG Ab AP conjugated
Substrate: PNP
Controls - standards supplied:
　Controls: low Pos, high Pos (human serum)
　Calibrators: I-Neg, II-low Pos, III-high Pos
Additional reagents:
　IgG Diluent Colorizer (Cat. no. 4506 – optional)
Special equipment required:
　None

INTERPRETATION

Comments on interpretation:
　Equivocal: values between 101 and 110; retest to confirm
　　and/or retest a fresh sample in 2–3 weeks
　Repeatably equivocal: report as negative
No. of references: 18

NOTES

131718.0

This product is not available in the UK but is sold in the rest
　of Europe

Toxoplasma gondii
ANTIBODY DETECTION (IgG)

Manufacturer: Kreatech Diagnostics
Cat. No./Trade name: EL-2001-gTOX/Toxoplasma IgG
EIA

SUMMARY

[Well-Ag]–**Ab**–[AHIgG–HRP]–[TMB]–A_{450}

Assay type: EIA (non-competitive)
Detection: Colorimetric A_{450}
Format: Microtitre well, Ag coated
Sample type: Serum (do not heat inactivate)
Sample pre-treatment:
　None
Sample volume: 10 µl (+ 1 ml diluent)*
Number of tests: 96
Controls - standards run in assay:
　Controls: Neg (2), cut-off (4), Pos (2)
Incubation:
　1 hr (37°C) + 1 hr (37°C) + 30 min (RT)
Washes: 2

CONTENTS

Antibodies, antigens, labelled components:
　T. gondii Ag from infected Hep-2 cells bound to well
　Anti-human IgG Ab HRP conjugated
Substrate: TMB
Controls - standards supplied:
　Controls: Neg, cut-off and Pos (human serum)
Additional reagents:
　None
Special equipment required:
　None

INTERPRETATION

Comments on interpretation:
　Classification of samples is according to cut-off; no futher
　　testing
No. of references: None

NOTES

131085.0

*Samples are assayed in duplicate

© KLUWER ACADEMIC PUBLISHERS 1995, ISSN 1381-5067

Toxoplasma gondii
ANTIBODY DETECTION (IgG)

Manufacturer: Labsystems Oy
Cat. No./Trade name: 61 00 201S/Toxoplasma gondii IgG EIA

SUMMARY

[Well-Ag]–**Ab**–[AHIgG-AP]–[PNP]–A_{405}

Assay type: EIA (non-competitive)
Detection: Colorimetric A_{405}
Format: Microtitre well, Ag coated
Sample type: Serum*
Sample pre-treatment:
   None
Sample volume: 100 μl of 1:100 dilution x 2
Number of tests: 96
Controls - standards run in assay:
   Controls: Pos (2)
   Calibrators: 1 (2), 2 (2), 3 (2), 4 (2), 5 (2), 6 (2)
Incubation:
   1 hr (37°C) + 1 hr (37°C) + 30 min (37°C)
Washes: 2

CONTENTS

Antibodies, antigens, labelled components:
   T. gondii Ag bound to well
   Anti-human IgG Ab (sh) AP conjugated
Substrate: PNP
Controls - standards supplied:
   Controls: Pos (titre in IU/ml - human serum)
   Calibrators: 1–6 (titres in IU/ml - human serum)
Additional reagents:
   NaOH
Special equipment required:
   Auto-EIA II analyser (optional)

INTERPRETATION

Comments on interpretation:
   Results are expressed quantitatively in Enzyme Immunoassay Units (EIU) or International Units per ml (IU/ml)
No. of references: 10

NOTES

131335.0

*Heat treatment may slightly change assay results

Toxoplasma gondii
ANTIBODY DETECTION (IgG)*

Manufacturer: Labsystems Oy
Cat. No./Trade name: 61 00 202/Toxoplasma gondii IgG Avidity

SUMMARY

[Well-Ag]–**Ab**–[AHIgG-AP]–[PNP]–A_{405}

Assay type: EIA (non-competitive)
Detection: Colorimetric A_{405}
Format: Microtitre well, Ag coated
Sample type: Serum**
Sample pre-treatment:
   None
Sample volume: 100 μl of serial dilutions (four-fold: 1:50–1:12 000)
Number of tests: 96
Controls - standards run in assay:
   Controls: low avidity (serial dilutions), high avidity (serial dilutions)
Incubation:
   1 hr (37°C) + 1 hr (37°C) + 30 min (37°C)
Washes: 2*

CONTENTS

Antibodies, antigens, labelled components:
   T. gondii Ag bound to well
   Anti-human IgG Ab (sh) AP conjugated
Substrate: PNP
Controls - standards supplied:
   Controls: low avidity and high avidity (human serum)
Additional reagents:
   NaOH
Special equipment required:
   Auto-EIA II analyser (optional)

INTERPRETATION

Comments on interpretation:
   Results are expressed as a ratio of high affinity Ig Ab titre
   Low avidity: < 15%; acute primary infection within last 3 months
   Borderline avidity: 15–30%; retest and report average
   High avidity: > 30%; no primary infection within the last 3 months
No. of references: 12

NOTES

131340.0

*This assay distinguishes low affinity IgG Abs (recent primary infection) from high affinity IgG Ab (pre-existing immunity) by means of washing steps for paired samples with and without protein denaturant
*Heat treatment may slightly change assay results

© KLUWER ACADEMIC PUBLISHERS 1995, ISSN 1381-5067

Toxoplasma gondii
ANTIBODY DETECTION (IgG)

Manufacturer: Medix Biotech Inc
Cat. No./Trade name: KBF 2096/Toxoplasma gondii IgG
EIA Kit

SUMMARY

[Well-Ag]–**Ab**–[AHIgG-HRP]–[OPD]–A_{490}

Assay type: EIA (non-competitive)
Detection: Colorimetric A_{490}
Format: Microtitre well, Ag coated
Sample type: Serum (do not heat inactivate)
Sample pre-treatment:
　None
Sample volume: 10 μl (+ 200 μl diluent)
Number of tests: 96
Controls - standards run in assay:
　Controls: Neg (1), low Pos (2), high Pos (1)
Incubation:
　20 min (RT) + 20 min (RT) + 10 min (RT)
Washes: 2

CONTENTS

Antibodies, antigens, labelled components:
　T. gondii Ag bound to well
　Anti-human IgG Ab (g) HRP conjugated
Substrate: OPD
Controls - standards supplied:
　Controls: Neg and high Pos (human serum or plasma)
　Calibrator: low Pos (human serum or plasma)
Additional reagents:
　H_2SO_4
Special equipment required:
　None

INTERPRETATION

Comments on interpretation:
　Equivocal: where ratio of sample OD:cut-off value is
　　between 0.09-1.10; report as equivocal
No. of references: None

NOTES

131161.0

Toxoplasma gondii
ANTIBODY DETECTION (IgG)

Manufacturer: Melotec S.A.
Cat. No./Trade name: /Melotest Toxo IgG

SUMMARY

[Well-Ag]–**Ab**–[AHIgG-HRP]–[TMB]–A_{450}

Assay type: EIA (non-competitive)
Detection: Colorimetric A_{450}
Format: Microtitre well, Ag coated
Sample type: Serum, plasma
Sample pre-treatment:
　None
Sample volume: 100 μl of 1:20 dilution x 2
Number of tests: 96
Controls - standards run in assay:
　Controls: Neg (1), low Pos (2), high Pos (1)
Incubation:
　20 min (RT) + 20 min (RT) + 10 min (RT)
Washes: 2

CONTENTS

Antibodies, antigens, labelled components:
　T. gondii Ag bound to well
　Anti-human IgG Ab HRP conjugated
Substrate: TMB
Controls - standards supplied:
　Controls: Neg, low Pos and high Pos (sera)
Additional reagents:
　None
Special equipment required:
　None

INTERPRETATION

Comments on interpretation:
　Equivocal: ratio of sample OD:cut-off value between 0.9
　　and 1.1; retest a fresh sample
　Repeatably equivocal: retest a fresh sample taken in 2-4
　　weeks
No. of references: 8

NOTES

131128.0

© KLUWER ACADEMIC PUBLISHERS 1995, ISSN 1381-5067

Toxoplasma gondii
ANTIBODY DETECTION (IgG)

Manufacturer: Menarini Diagnostics
Cat. No./Trade name: M6118/HF Toxoplasma IgG

SUMMARY

[Well-Ag]–**Ab**–[AHIgG-POD]–[TMB]–A_{450}

Assay type: EIA (non-competitive)
Detection: Colorimetric A_{450}
Format: Microtitre well, Ag coated
Sample type: Serum
Sample pre-treatment:
  None
Sample volume: 100 µl of 1:101 dilution
Number of tests: 96
Controls - standards run in assay:
  Controls: Neg (1), cut-off (1), Pos (1)
Incubation:
  45 min (37°C) + 45 min (37°C) + 15 min (RT)
Washes: 2

CONTENTS

Antibodies, antigens, labelled components:
  T. gondii Ag bound to well
  Anti-human IgG MAb POD conjugated
Substrate: TMB
Controls - standards supplied:
  Controls: Neg, cut-off and Pos (serum)
Additional reagents:
  None
Special equipment required:
  None

INTERPRETATION

Comments on interpretation:
  No data is provided for interpretation of results
No. of references: None

NOTES

131091.0

Toxoplasma gondii
ANTIBODY DETECTION (IgG)

Manufacturer: Organon Teknika NV
Cat. No./Trade name: Toxonostika® IgG II

SUMMARY

[Well-Ag]–**Ab**–[AHIgG-HRP]–[TMB]–A_{450}

Assay type: EIA (non-competitive)
Detection: Colorimetric A_{450}
Format: Microtitre well, Ag coated
Sample type: Serum, plasma (do not heat inactivate)
Sample pre-treatment:
  None
Sample volume: 10 µl of 1:101 dilution (adult); 1:20
  (newborn)
Number of tests: 96, 192
Controls - standards run in assay:
  Controls:
  Neg (1), Pos (1), strong Pos (1) (for one strip)
  Neg (2), Pos (2), strong Pos (2), (for two or more strips)
Incubation:
  1 hr (37°C) + 1 hr (37°C) + 30 min (RT)
Washes: 2

CONTENTS

Antibodies, antigens, labelled components:
  T. gondii Ag bound to well
  Anti-human IgG Ab (sh) HRP conjugated
Substrate: TMB
Controls - standards supplied:
  Controls: Neg, Pos and strong Pos (human serum)
Additional reagents:
  H_2SO_4
Special equipment required:
  None

INTERPRETATION

Comments on interpretation:
  Classification of samples is according to cut-off; no
  further testing
No. of references: 3

NOTES

131200.0

© KLUWER ACADEMIC PUBLISHERS 1995, ISSN 1381-5067

Toxoplasma gondii
ANTIBODY DETECTION (IgG)

Manufacturer: Radim
Cat. No./Trade name: K1TG/Toxoplasma IgG EIA Well

SUMMARY

[Well-Ag]–**Ab**–[AHIgG-HRP]–[TMB]–A_{450}

Assay type: EIA (non-competitive)
Detection: Colorimetric A_{450}
Format: Microtitre well, Ag coated
Sample type: Serum, plasma
Sample pre-treatment:
    None
Sample volume: 100 μl of 1:300 dilution
Number of tests: 96, 192
Controls - standards run in assay:
    Qualitative: Control: Neg (1): Standards: 1 (1)
    Quantitative: Control: Neg (1): Standards: 1 (1), 2 (1), 3
        (1), 4 (1), 5 (1)
Incubation:
    1 hr (37°C) + 30 min (37°C) + 10 min (37°C) or 15 min
        (RT)
Washes: 2

CONTENTS

Antibodies, antigens, labelled components:
    T. gondii Ag bound to well
    Anti-human IgG Ab (g) HRP conjugated
Substrate: TMB
Controls - standards supplied:
    Controls: Neg (0, IU/ml - serum)
    Standards: 5 (15, 30, 60, 120, 240 IU/ml - serum)
Additional reagents:
    None
Special equipment required:
    None

INTERPRETATION

Comments on interpretation:
    Qualitative:
    Equivocal: within cut-off ± 10%; retest to confirm
    Quantitative:
    Negative: < 15 IU/ml
    Weak Positive: between 15 and 30 IU/ml
    Positive: > 30 IU/ml
No. of references: 5

NOTES

131271.0

Toxoplasma gondii
ANTIBODY DETECTION (IgG)

Manufacturer: Roche Diagnostic Systems
Cat. No./Trade name: 07 5558 3/COBAS® CORE Toxo IgG EIA II

SUMMARY

[Bead-Ag]–**Ab**–[AHIgG-HRP]–[TMB]–A_{450}

Assay type: EIA (non-competitive)
Detection: Colorimetric A_{450}
Format: Tube, Ag coated bead
Sample type: Serum, plasma
Sample pre-treatment:
    None
Sample volume: 3 μl
Number of tests: 100
Controls - standards run in assay:
    Controls: low Pos 1 (1), high Pos 2 (1)
    Calibrators: a (1), b (1), c (1), d (1)
Incubation:
    15 min (37°C) + 30 min (37°C) + 15 min (37°C) (all with
        shaking)
Washes: 2

CONTENTS

Antibodies, antigens, labelled components:
    T. gondii Ag bound to bead
    Anti-human IgG MAb (m) HRP conjugated
Substrate: TMB (bought separately)
Controls - standards supplied:
    Controls: 1 (low Pos) and 2 (high Pos) (human serum)
    Calibrators: a, b, c, d (0-300 IU/ml, human serum)
Additional reagents:
    Cobas® Core TMB Kit
    H_2SO_4 (for manual assay only)
Special equipment required:
    Cobas® Core immunoassay analyser (for automated
        method - optional)
    Cobas® Core EIA shaking incubator, tube washer and
        photometer (optional)
    Cobas® Core TORC negative control bead dispenser
    Cobas® Core reaction tubes

INTERPRETATION

Comments on interpretation:
    Equivocal: between 2 and 6 IU/ml; retest a fresh sample
        taken in 2-3 weeks time in parallel with initial sample
No. of references: None

NOTES

131155.0

© KLUWER ACADEMIC PUBLISHERS 1995, ISSN 1381-5067

Toxoplasma gondii
ANTIBODY DETECTION (IgG)

Manufacturer: Sanofi Diagnostics Pasteur
Cat. No./Trade name: /Platelia® Toxo IgG

SUMMARY

[Well-Ag]–**Ab**–[AHIgG-HRP]–[OPD]–A_{492}

Assay type: EIA (non-competitive)
Detection: Colorimetric A_{492}
Format: Microtitre well, Ag coated
Sample type: Serum
Sample pre-treatment:
    None
Sample volume: 10 µl (+1 ml diluent)
Number of tests: 96
Controls - standards run in assay:
    Standards: Neg (1), Pos I (2), Pos II (1), Pos III (1)
Incubation:
    1 hr (37–40°C) + 1 hr (37–40°C) + 30 min (RT)
Washes: 2 (+ preliminary plate wash)

CONTENTS

Antibodies, antigens, labelled components:
    T. gondii (RH strain) Ag bound to well
    Anti-human IgG MAb (m) HRP conjugated
Substrate: OPD
Controls - standards supplied:
    Standards: Neg, Pos I, Pos II and Pos III (human serum –
        calibrated against WHO Tox S 60 reference serum)
Additional reagents:
    None
Special equipment required:
    None

INTERPRETATION

Comments on interpretation:
    Equivocal: within designated range; retest a fresh sample
        in 3 weeks in parallel with initial specimen
No. of references: 7

NOTES

131761.0

Toxoplasma gondii
ANTIBODY DETECTION (IgG)

Manufacturer: SFRI Laboratoire
Cat. No./Trade name: IIE 018/Toxo IgG

SUMMARY

[Well-Ag]–**Ab**–[AHIgG-POD]–[TMB]–A_{450}

Assay type: EIA (non-competitive)
Detection: Colorimetric A_{450}
Format: Microtitre well, Ag coated
Sample type: Serum, plasma
Sample pre-treatment:
    None
Sample volume: 10 µl (+1 ml diluent)
Number of tests: 96
Controls - standards run in assay:
    Standards: 1 (1), 2 (1), 3 (1), 4 (1)
Incubation:
    30 min (37°C) + 15 min (37°C) + 15 min (37°C)
Washes: 2

CONTENTS

Antibodies, antigens, labelled components:
    T. gondii Ag bound to well
    Anti-human IgG Ab POD conjugated
Substrate: TMB
Controls - standards supplied:
    Standards: 4 (0, 10, 60 and 240 IU/ml)
Additional reagents:
    None
Special equipment required:
    None

INTERPRETATION

Comments on interpretation:
    Equivocal: where Ab titre is between 7–10 IU/ml; retest a
        fresh sample in 3 weeks
No. of references: 11

NOTES

131648.0

Toxoplasma gondii
ANTIBODY DETECTION (IgG)

Manufacturer: Sigma Diagnostics
Cat. No./Trade name: SIA 404/SIA® Toxoplasma IgG

SUMMARY

$[\text{Well-Ag}]–\textbf{Ab}–[\text{AHIgG-AP}]–[\text{PMP}]–A_{550}$

Assay type: EIA (non-competitive)
Detection: Colorimetric A_{550}
Format: Microtitre well, Ag coated
Sample type: Serum (do not heat inactivate)
Sample pre-treatment:
 None
Sample volume: 10 µl (+200 µl diluent)
Number of tests: 96
Controls - standards run in assay:
 Controls: A (1), B (1)
 Calibrators: 1 (2), 2 (2), 3 (2)
Incubation:
 45 min (RT) + 45 min (RT) + 45 min (RT)
Washes: 2 (+ preliminary plate wash)

CONTENTS

Antibodies, antigens, labelled components:
 T. gondii Ag bound to well
 Anti-human IgG Ab (g) AP conjugated
Substrate: PMP
Controls - standards supplied:
 Controls: A & B (human serum)
 Calibrators: 1, 2 and 3 (IgG Ab titre in AU/ml – human
 serum)
Additional reagents:
 None
Special equipment required:
 Sigma EIA Multiwell Plate Reader (Cat. no. EQ104 –
 optional)

INTERPRETATION

Comments on interpretation:
 Equivocal: 0.19–0.20 SIA® Toxoplasma IgG Values, or
 0.90–0.99 Index Value; retest to confirm
 Repeatably equivocal: report as such and retest using
 alternative method or fresh sample
 For information on reporting in IU/ml, contact
 manufacturers
No. of references: 4

NOTES

131438.0

Toxoplasma gondii
ANTIBODY DETECTION (IgG)

Manufacturer: Sorin Biomedica
Cat. No./Trade name: P2858/ETI-TOXOK-G

SUMMARY

$[\text{Well-Ag}]–\textbf{Ab}–[\text{AHIgG-HRP}]–[\text{TMB}]–A_{450}$

Assay type: EIA (non-competitive)
Detection: Colorimetric A_{450}
Format: Microtitre well, Ag coated
Sample type: Serum, plasma
Sample pre-treatment:
 None
Sample volume: 100 µl of 1:505 dilution x 2
Number of tests:
Controls - standards run in assay:
 Controls: Neg (2)
 Calibrators: 1 (2), 2 (2), 3 (2), 4 (2)
Incubation:
 1 hr (37°C) + 1 hr (37°C) + 30 min (RT)
Washes: 2

CONTENTS

Antibodies, antigens, labelled components:
 T. gondii (RH strain) Ag bound to well
 Anti-human IgG IgG Ab (g) HRP conjugated
Substrate: TMB
Controls - standards supplied:
 Controls: Neg (human serum)
 Calibrators: 4 (15, 30, 75 and 150 IU/ml)
Additional reagents:
 None
Special equipment required:
 ETI-System Reader and ETI-System Washer (optional)

INTERPRETATION

Comments on interpretation:
 Equivocal: within cut-off ± 10%; retest to confirm
No. of references: 14

NOTES

131213.0

© KLUWER ACADEMIC PUBLISHERS 1995, ISSN 1381-5067

Toxoplasma gondii
ANTIBODY DETECTION (IgG)

Manufacturer: United Biotech Inc
Cat. No./Trade name: 1A-101/MAGIWELL® Toxo IgG

SUMMARY

[Well-Ag]–**Ab**–[AHIgG-HRP]–[TMB]–A_{450}

Assay type: EIA (non-competitive)
Detection: Colorimetric A_{450}
Format: Microtitre well, Ag coated
Sample type: Serum
Sample pre-treatment:
    None
Sample volume: 100 µl of 1:101 dilution
Number of tests: 96
Controls - standards run in assay:
    Reference Standard Set: Neg (1), Pos (1)
    Calibrators: 1 (1)
Incubation:
    30 min (RT) + 30 min (RT) + 15 min (RT)
Washes: 2

CONTENTS

Antibodies, antigens, labelled components:
    T. gondii Ag (m) bound to well
    Anti-human IgG Ab (g) HRP conjugated
Substrate: TMB
Controls - standards supplied:
    Reference Standard Set: Neg, Pos
    Calibrator: 1 (100 EU/ml)
Additional reagents:
    H_2SO_4
Special equipment required:
    None

INTERPRETATION

Comments on interpretation:
    Weak Positive: between 10–20 EU/ml
    Positive: > 20 EU/ml
No. of references: 7

NOTES

131145.0

Toxoplasma gondii
ANTIBODY DETECTION (IgM)

Manufacturer: Abbott Laboratories
Cat. No./Trade name: 6084-22/ABBOTT TOXO-M EIA Kit

SUMMARY

[Bead-AHIgM]–**Ab**–[Ag:Ab-HRP]–[OPD]–A_{492}

Assay type: EIA (non-competitive)
Detection: Colorimetric A_{492}
Format: Reaction wells, Ab coated beads
Sample type: Serum
Sample pre-treatment:
    None
Sample volume: 20 µl (+ 200 µl diluent)
Number of tests: 50, 100
Controls - standards run in assay:
    Controls: Neg (1), low Pos (3), high Pos (1)
Incubation:
    2 hr (37°C) + 1 hr (37°C) + 30 min (RT)
Washes: 2

CONTENTS

Antibodies, antigens, labelled components:
    Anti-human IgM Ab (g) bound to bead
    T. gondii Ag
    Anti-*T. gondii* IgG Ab (r) HRP conjugated
Substrate: OPD
Controls - standards supplied:
    Controls: Neg, low Pos and high Pos (human plasma)
Additional reagents:
    H_2SO_4
Special equipment required:
    Pentawash® II and Gast® Vacuum Pump (or similar)
    Quantum Analyser (optional)

INTERPRETATION

Comments on interpretation:
    Equivocal: where TOXO-M Index is between 0.400 and 0.499; retest to confirm
No. of references: 5

NOTES

131354.0

© KLUWER ACADEMIC PUBLISHERS 1995, ISSN 1381-5067

Toxoplasma gondii
ANTIBODY DETECTION (IgM)

Manufacturer: Abbott Laboratories
Cat. No./Trade name: 7A82-66/IMx® System Toxo IgM

SUMMARY

[Particle-Ag]–**Ab**–[AHIgM-AP]–[4-MP]–fluorescence

Assay type: EIA (non-competitive)
Detection: Fluorometric
Format: Microparticles, Ag coated
Sample type: Serum, plasma
Sample pre-treatment:
   None
Sample volume: 150 µl
Number of tests: 100
Controls - standards run in assay:
   Controls: Neg (1), Pos (1)
   Calibrators: 1 (2)
Incubation:
   Automated
Washes: Automated

CONTENTS

Antibodies, antigens, labelled components:
   T. gondii Ag bound to microparticles
   Anti-human IgM Ab AP conjugated
Substrate: 4-MP
Controls - standards supplied:
   Calibrator: 1 (human serum)
Additional reagents:
   IMx® Toxo IgM Controls (Cat. no. 7A82-10)
   IMx® Rheumatoid Factor Neutralization Reagent (Cat. no. 1A14-22)
   IMx® Probe Cleaning Solution (Cat. no. 1A71-02)
Special equipment required:
   IMx® System

INTERPRETATION

Comments on interpretation:
   Sample reactivity is calculated automatically
   All sera giving Toxo IgM indexes $\geqslant 0.500$ must be treated with Rheumatoid Factor Neutralizing Reagent and rerun using alternative protocol (see Manufacturer's instructions)
No. of references: 28

NOTES

131693.0

Toxoplasma gondii
ANTIBODY DETECTION (IgM)

Manufacturer: Alfa Biotech
Cat. No./Trade name: 05772955/Toxoplasma IgM Elisa System Capture Method

SUMMARY

[Well-AHIgM]–**Ab**–[Ag:MAb]–[AMIg-HRP]–[TMB]–A$_{450}$

Assay type: EIA (non-competitive)
Detection: Colorimetric A$_{450}$
Format: Microtitre well, Ab coated
Sample type: Serum
Sample pre-treatment:
   None
Sample volume: 100 µl of 1:100 dilution x2
Number of tests: 96
Controls - standards run in assay:
   Controls: Neg (2), cut-off (4), Pos (2)
Incubation:
   20 min (37°C) + 20 min (37°C) + 20 min (37°C)
Washes: 2

CONTENTS

Antibodies, antigens, labelled components:
   Anti-human IgM Ab bound to well
   T. gondii Ag
   Anti-*T. gondii* MAb
   Anti-mouse Ig Ab (g) HRP conjugated
Substrate: TMB
Controls - standards supplied:
   Controls: Neg, cut-off and Pos (human serum)
Additional reagents:
   None
Special equipment required:
   None

INTERPRETATION

Comments on interpretation:
   Equivocal: within cut-off ± 15%; retest to confirm
   Repeatably equivocal: retest a fresh sample in 2 weeks
No. of references: 10

NOTES

131102.0

© KLUWER ACADEMIC PUBLISHERS 1995, ISSN 1381-5067

Toxoplasma gondii
ANTIBODY DETECTION (IgM)

Manufacturer: Amico Laboratories Inc
Cat. No./Trade name: 5600M/AMIZYME® Toxoplasma gondii IgM Assay

SUMMARY

[Well-Ag]–**Ab**–[AHIgM-HRP]–[ABTS]–A$_{405}$

Assay type: EIA (non-competitive)
Detection: Colorimetric A$_{405}$
Format: Microtitre well, Ag coated
Sample type: Serum
Sample pre-treatment:
    None
Sample volume: 5 µl (+245 µl diluent)*
Number of tests: 96
Controls - standards run in assay:
    Controls: Neg (1)
    Calibrators: I (1)
Incubation:
    25 min (RT) + 25 min (RT) + 25 min (RT)
Washes: 2

CONTENTS

Antibodies, antigens, labelled components:
    T. gondii Ag bound to well
    Anti-human IgM Ab (g) HRP conjugated
Substrate: ABTS
Controls - standards supplied:
    Controls: Neg
    Calibrator: 1
Additional reagents:
    None
Special equipment required:
    None

INTERPRETATION

Comments on interpretation:
    Equivocal: where sample OD is between 0.275–0.299
    (92–99 U/ml); retest to confirm
No. of references: 12

NOTES

131797.0

*Sample diluted in IgM serum diluent provided to eliminate interference due to RF and IgG

Toxoplasma gondii
ANTIBODY DETECTION (IgM)

Manufacturer: Amico Laboratories Inc
Cat. No./Trade name: 5600MC/AMIZYME® Toxoplasma gondii IgM Capture Assay

SUMMARY

[Well-Ag]–**Ab**–[AHIgM]–[AMIg-HRP]–[ABTS]–A$_{405}$

Assay type: EIA (non-competitive)
Detection: Colorimetric A$_{405}$
Format: Microtitre well, Ag coated
Sample type: Serum
Sample pre-treatment:
    None
Sample volume: 5 µl (+245 µl diluent)
Number of tests: 96
Controls - standards run in assay:
    Controls: Neg (1)
    Calibrators: I (1)
Incubation:
    25 min (RT) + 25 min (RT) + 25 min (RT) + 25 min (RT)
Washes: 3

CONTENTS

Antibodies, antigens, labelled components:
    T. gondii Ag bound to well
    Anti-human IgM MAb (m)
    Anti-mouse Ig Ab (g) HRP conjugated
Substrate: ABTS
Controls - standards supplied:
    Controls: Neg
    Calibrator: 1
Additional reagents:
    None
Special equipment required:
    None

INTERPRETATION

Comments on interpretation:
    Equivocal: where sample OD is between 0.285–0.299;
    retest to confirm
No. of references: 2

NOTES

131807.0

© KLUWER ACADEMIC PUBLISHERS 1995, ISSN 1381-5067

Toxoplasma gondii
ANTIBODY DETECTION (IgM)

Manufacturer: Behringwerke AG
Cat. No./Trade name: Enzygnost® Toxoplasmosis/IgM

SUMMARY

[Well-AHIgM]–**Ab**–[Ag-POD]–[TMB]–A_{450}

Assay type: EIA (non-competitive)
Detection: Colorimetric A_{450}
Format: Microtitre well, Ab coated
Sample type: Serum, plasma
Sample pre-treatment:
  Add dilution buffer (contains RF absorbent) to diluted
    sample, incubate 15 min (RT) or overnight (2–8°C) to
    reduce interference due to RF and IgG
Sample volume: 20 µl of treated sample
Number of tests: 96
Controls - standards run in assay:
  Controls: Neg (2), Pos (2)
Incubation:
  1 hr (37°C) + 30 min (37°C)
Washes: 1

CONTENTS

Antibodies, antigens, labelled components:
  Anti-human IgM Ab (g) bound to well
  T. gondii Ag POD conjugated
Substrate: TMB (bought separately)
Controls - standards supplied:
  Controls: Neg and Pos (human serum – assigned values)
Additional reagents:
  Supplementary reagents for Enzygnost/TMB (Cat. no.
    OUVP)
Special equipment required:
  Behring ELISA Processor II (optional)
  Behring ELISA Processor III (optional)

INTERPRETATION

Comments on interpretation:
  Equivocal: retest to confirm
  Repeatably equivocal: report as equivocal and retest a
    fresh specimen at a later date
No. of references: 5

NOTES

131029.0

Toxoplasma gondii
ANTIBODY DETECTION (IgM)

Manufacturer: BIOKIT SA
Cat. No./Trade name: 3000-1210/bioelisa TOXO IgM
(Immunocapture)

SUMMARY

[Well-AHIgM]–**Ab**–[Ag-HRP]–[TMB]–A_{450}

Assay type: EIA (non-competitive)
Detection: Colorimetric A_{450}
Format: Microtitre well, Ab coated
Sample type: Serum (do not heat inactivate)
Sample pre-treatment:
  None
Sample volume: 100 µl of 1:101 dilution
Number of tests: 96
Controls - standards run in assay:
  Controls: Neg (2), cut-off (4), Pos (2)
Incubation:
  1 hr (37°C) + 1 hr (37°C) + 30 min (RT)
Washes: 2

CONTENTS

Antibodies, antigens, labelled components:
  Anti-human IgM Ab (r) bound to well
  T. gondii Ag (from HEP-2 cell line) HRP conjugated*
Substrate: TMB
Controls - standards supplied:
  Neg, cut-off and Pos (human serum)
Additional reagents:
  None
Special equipment required:
  None

INTERPRETATION

Comments on interpretation:
  Equivocal: within cut-off ±10%; retest to confirm
No. of references: 11

NOTES

131374.0

*Control Ag consisting of uninfected cellular components is
added to conjugate to reduce non-specific activity

© *KLUWER ACADEMIC PUBLISHERS 1995, ISSN 1381-5067*

Toxoplasma gondii
ANTIBODY DETECTION (IgM)

Manufacturer: bioMerieux
Cat. No./Trade name: 30 202/VIDAS TOXO IgM

SUMMARY

[Solid phase-AHIgM]–**Ab**–[Ag:MAb-AP]–[4MP]–
fluorescence

Assay type: EIA (non-competitive)
Detection: Fluorometric
Format: Solid phase receptacle (SPR), Ab coated
Sample type: Serum
Sample pre-treatment:
None
Sample volume: 100 µl
Number of tests: 60
Controls - standards run in assay:
Controls: Neg (1), Pos (1)
Standards: Pos (1)
Incubation:
Automated - total time 40 min
Washes: Automated

CONTENTS

Antibodies, antigens, labelled components:
Anti-human IgM Ab (g) bound to solid phase receptacle
T. gondii
Anti-*T. gondii* (P30) MAb AP conjugated
Substrate:
Controls - standards supplied:
Controls: Neg and Pos (human serum)
Standards: Pos (human serum)
Additional reagents:
None
Special equipment required:
Vitek immunodiagnostic assay system (VIDAS)

INTERPRETATION

Comments on interpretation:
This is an automated assay
Samples designated equivocal; retest to confirm
Repeatably equivocal: retest using a fresh sample in 2-3
weeks
No. of references: None

NOTES

131311.0

Toxoplasma gondii
ANTIBODY DETECTION (IgM)

Manufacturer: BioWhittaker Inc
Cat. No./Trade name: 30-332U/Toxo Stat M

SUMMARY

[Well-Ag]–**Ab**–[AHIgM-AP]–[PMP]–A_{550}

Assay type: EIA (non-competitive)
Detection: Colorimetric A_{550}
Format: Microtitre well, Ag coated
Sample type: Serum (do not heat inactivate)
Sample pre-treatment:
Mix sample with Pretreatment Serum, incubate 30 min
(RT) and centrifuge to eliminate interference due to
IgG and RF
Sample volume: 25 µl of treated sample (+ 100 µl diluent)
Number of tests: 96
Controls - standards run in assay:
Calibrators: Neg (1), Pos 1 (1), Pos 2 (1)
Incubation:
45 min (RT) + 30 min (RT) + 30 min (RT) (all with
shaking)
Washes: 2 (+preliminary plate wash)

CONTENTS

Antibodies, antigens, labelled components:
T. gondii Ag bound to well
Anti-human IgM Ab (g) AP conjugated
Substrate: PMP
Controls - standards supplied:
Calibrators: Neg, Pos 1, Pos 2 (all human serum)
Rheumatoid Factor Control (human serum)
Additional reagents:
None
Special equipment required:
BioWhittaker automated system and software (optional)

INTERPRETATION

Comments on interpretation:
Equivocal: test value between 0.10–0.12; indicates
probable positive, retest in duplicate to confirm
No. of references: 11

NOTES

131756.0

Toxoplasma gondii
ANTIBODY DETECTION (IgM)

Manufacturer: Bouty SpA
Cat. No./Trade name: 21876/BEIA TOXO-IgM Capture

SUMMARY

[Well-AHIgM]–**Ab**–[Ag-biotin]–[strept–HRP]–[TMB]–A$_{450}$

Assay type: EIA (non-competitive)
Detection: Colorimetric A$_{450}$
Format: Microtitre well, Ab coated
Sample type: Serum
Sample pre-treatment:
    None
Sample volume: 10 µl (+800 µl diluent)*
Number of tests: 96
Controls - standards run in assay:
    Controls: Neg (2), cut-off (2), Pos (2)
Incubation:
    1 hr (RT) + 1 hr (RT) + 30 min (RT)
Washes: 2

CONTENTS

Antibodies, antigens, labelled components:
    Anti-human IgM IgG Ab bound to well
    T. gondii (purified P30) Ag biotinylated
    Streptavidin HRP conjugated
Substrate: TMB
Controls - standards supplied:
    Controls: Neg, cut-off and Pos (human serum)
Additional reagents:
    None
Special equipment required:
    None

INTERPRETATION

Comments on interpretation:
    Equivocal: within cut-off ±5%; retest to confirm
    Repeatably equivocal: repeat test with a fresh sample
No. of references: None

NOTES

131242.0

*Samples can be assayed singly or in duplicate

Toxoplasma gondii
ANTIBODY DETECTION (IgM)

Manufacturer: Centocor Inc
Cat. No./Trade name: M400/800-910/CAPTIA™ Toxo-M

SUMMARY

[Well:AHIgM]–**Ab**–[Ag]–[MAb-biotin]–[strept-HRP]–[TMB]–A$_{450}$

Assay type: EIA (non-competitive)
Detection: Colorimetric A$_{450}$
Format: Microtitre well, Ab coated
Sample type: Serum (do not heat inactivate)
Sample pre-treatment:
    None
Sample volume: 100 µl of 1:101 dilution
Number of tests: 96
Controls - standards run in assay:
    Controls: Neg (1), low Pos (2), high Pos (1)
Incubation:
    1 hr (37°C) + 1 hr (37°C) + 30 min (RT)
Washes: 2

CONTENTS

Antibodies, antigens, labelled components:
    Anti-human IgM Ab (r) bound to well
    T. gondii Ag
    Anti-*T. gondii* MAb biotinylated
    Streptavidin HRP conjugated
Substrate: TMB
Controls - standards supplied:
    Controls: Neg, low Pos and high Pos (human serum)
Additional reagents:
    H$_2$SO$_4$
Special equipment required:
    None

INTERPRETATION

Comments on interpretation:
    Equivocal: where Antibody Index is between 0.9 and 1.1;
    retest to confirm
No. of references: 8

NOTES

131622.0

© *KLUWER ACADEMIC PUBLISHERS 1995, ISSN 1381-5067*

Toxoplasma gondii
ANTIBODY DETECTION (IgM)

Manufacturer: Chimica Diagnostica
Cat. No./Trade name: 41350/Toxoplasma Gondii IgM

SUMMARY

[Well-AHIgM]–**Ab**–[Ag]–[MAb-POD]–[TMB]–A$_{450}$

Assay type: EIA (non-competitive)
Detection: Colorimetric A$_{450}$
Format: Microtitre well, Ab coated
Sample type: Serum
Sample pre-treatment:
　None
Sample volume: 100 µl of 1:50 dilution x 2*
Number of tests: 96
Controls - standards run in assay:
　Standards: 1 (2), 2 (2), 3 (2), 4 (2), 5 (2), 6 (2)
Incubation:
　1 hr (37°C) + 1 hr (37°C) + 20 min (RT)
Washes: 2

CONTENTS

Antibodies, antigens, labelled components:
　Anti-human IgM MAb bound to well
　T. gondii Ag
　Anti-*T. gondii* IgG MAb POD conjugated
Substrate: TMB
Controls - standards supplied:
　Standards: 1–6 (titres 0, 5, 20, 50, 100 and 500 IU/ml)
Additional reagents:
　None
Special equipment required:
　None

INTERPRETATION

Comments on interpretation:
　Results are expressed quantitatively in IU/ml with
　reference to the standards provided
No. of references: 2

NOTES

131366.0

*Diluent contains AHIgG to eliminate interference due to RF
　and IgG in sample

Toxoplasma gondii
ANTIBODY DETECTION (IgM)

Manufacturer: Clark Laboratories
Cat. No./Trade name: Toxoplasma gondii IgM Elisa Test

SUMMARY

[Well-Ag]–**Ab**–[AHIgM-HRP]–[OPD]–A$_{490}$

Assay type: EIA (non-competitive)
Detection: Colorimetric A$_{490}$
Format: Microtitre well, Ag coated
Sample type: Serum (do not heat inactivate)
Sample pre-treatment:
　Add Absorbent Solution provided to diluted serum and
　　incubate 20 min (RT) to eliminate interference due to
　　RF and IgG
Sample volume: 100 µl of treated sample
Number of tests: 96
Controls - standards run in assay:
　Controls: Neg (1), Pos (1)
　Calibrators: 1 (2)
Incubation:
　20 min (RT) + 20 min (RT) + 10 min (RT)
Washes: 2

CONTENTS

Antibodies, antigens, labelled components:
　T. gondii Ag bound to well
　Anti-human IgM Ab (g) HRP conjugated
Substrate: TMB*
Controls - standards supplied:
　Controls: Neg and Pos (human serum or plasma)
　Calibrator: 1 (human serum or plasma)
Additional reagents:
　H$_2$SO$_4$
Special equipment required:
　None

INTERPRETATION

Comments on interpretation:
　Classification of samples is according to cut-off; no
　further testing
No. of references: None

NOTES

131668.0

*OPD can be supplied as an alternative

Toxoplasma gondii
ANTIBODY DETECTION (IgM)

Manufacturer: Dade International Inc (Bartels Division)
Cat. No./Trade name: B1029-315/Bartels PRIMA System® Toxoplasma IgM EIA

SUMMARY

[Well-Ag]–**Ab**–[AHIgM-enzyme]–[PNP]–A_{405}

Assay type: EIA (non-competitive)
Detection: Colorimetric A_{405}
Format: Microtitre well, Ag coated
Sample type: Serum
Sample pre-treatment:
  None
Sample volume: 100 µl of 1:11 dilution*
Number of tests: 96
Controls - standards run in assay:
  Controls: Neg (1), Pos (1)
  Reference serum: (1)
Incubation:
  30 min (37°C) + 30 min (37°C) + 30 min (37°C)
Washes: 2

CONTENTS

Antibodies, antigens, labelled components:
  T. gondii (RH strain) Ag bound to well
  Anti-human IgM Ab enzyme conjugated
Substrate: PNP
Controls - standards supplied:
  Controls: Neg and Pos
  Reference serum: 1
Additional reagents:
  None
Special equipment required:
  None

INTERPRETATION

Comments on interpretation:
  Classification of samples is according to cut-off; no further testing
No. of references: 8

NOTES

131045.0

*Specimen diluent contains an absorbent to eliminate interference due to RF and IgG

Toxoplasma gondii
ANTIBODY DETECTION (IgM)

Manufacturer: Denka Seiken Co. Ltd
Cat. No./Trade name: Toxoplasma IgM – EIA "SEIKEN"

SUMMARY

[Well-AHIgM]–**Ab**–[Ag]–[MAb-POD]–[TMB]–A_{450}

Assay type: EIA (non-competitive)
Detection: Colorimetric A_{450}
Format: Microtitre well, Ab coated
Sample type: Serum, plasma
Sample pre-treatment:
  None
Sample volume: 10 µl (+2 ml diluent)
Number of tests: 96
Controls - standards run in assay:
  Standards: I (1), II (2), III (1)
Incubation:
  1 hr (RT) + 1 hr (RT) + 1 hr (RT) + 30 min (RT)
Washes: 3

CONTENTS

Antibodies, antigens, labelled components:
  Anti-human IgM MAb (m) bound to well
  T. gondii Ag
  Anti-*T. gondii* MAb (m) POD conjugated
Substrate: TMB
Controls - standards supplied:
  Standards: I (Neg), II (low titre) and III (high titre) (all human serum)
Additional reagents:
  None
Special equipment required:
  None

INTERPRETATION

Comments on interpretation:
  Equivocal: where the calculated IgM Antibody Index is between 0.8 and 1.2; repeat test with a fresh specimen in 1–2 weeks or use an alternative method
No. of references: None

NOTES

131690.0

An independent evaluation of this kit is available – please contact manufacturer for details

© KLUWER ACADEMIC PUBLISHERS 1995, ISSN 1381-5067

Toxoplasma gondii
ANTIBODY DETECTION (IgM)

Manufacturer: Diesse
Cat. No./Trade name: 91041/ENZYWELL Toxoplasma IgM*

SUMMARY

[Well-AHIgM]–**Ab**–[Ag]–[MAb-POD]–[TMB]–A$_{450}$

Assay type: EIA (non-competitive)
Detection: Colorimetric A$_{450}$
Format: Microtitre well, Ab coated
Sample type: Serum
Sample pre-treatment:
 None
Sample volume: 100 µl of 1:101 dilution x 2
Number of tests: 96
Controls - standards run in assay:
 Controls: Neg (2), cut-off (3), Pos (2)
Incubation:
 45 min (37°C) + 45 min (37°C) + 15 min (RT)
Washes: 2

CONTENTS

Antibodies, antigens, labelled components:
 Anti-human IgM MAb bound to well
 T. gondii Ag
 Anti-*T. gondii* MAb POD conjugated
Substrate: TMB
Controls - standards supplied:
 Controls: Neg, cut-off and Pos (human serum)
Additional reagents:
 None
Special equipment required:
 None

INTERPRETATION

Comments on interpretation:
 Equivocal: within cut-off ± 10%; retest to confirm
 Repeatably equivocal: repeat with a fresh sample
No. of references: 4

NOTES

131448.0

*In the ENZYWELL Toxoplasma IgM Plus Kit (Cat. no. 91042) the contents and method are the same but the Immunocomplex remains stable for 2 weeks after reconstitution

Toxoplasma gondii
ANTIBODY DETECTION (IgM)

Manufacturer: E. Merck
Cat. No./Trade name: 14081/Toxoplasma gondii IgM MAGIA®

SUMMARY

[Particles-AHIgM]–**Ab**–[Ag-AP]–[substrate]–A$_{405}$

Assay type: EIA (non-competitive)
Detection: Colorimetric A$_{405}$
Format: Cuvette, Ab coated particles (magnetisable)
Sample type: Serum
Sample pre-treatment:
 None
Sample volume: 10 µl
Number of tests: 100
Controls - standards run in assay:
 Calibrators: zero, sero G (automated)
Incubation:
 Automated
Washes: Automated

CONTENTS

Antibodies, antigens, labelled components:
 Anti-human IgM Ab bound to particles (magnetisable)
 T. gondii Ag AP conjugated
Substrate: Bought separately
Controls - standards supplied:
 Bought separately
Additional reagents:
 Particle wash and system solutions (Cat. nos. 14097, 14096
 Substrate (Cat. no. 14095), calibrators (Cat. nos. 14090, 14092)
 Anti-foam (Cat. no. 14098)
Special equipment required:
 MAGIA® 7000 Immunoanalyzer (Cat. no. 14011) or MAGIA® 8000 (Cat. no. 114526)
 Reaction cuvettes (Cat. no. 14085)

INTERPRETATION

Comments on interpretation:
 This is an automated assay
No. of references: 2

NOTES

131064.0

Toxoplasma gondii
ANTIBODY DETECTION (IgM)

Manufacturer: Gull Laboratories Inc
Cat. No./Trade name: TXE150/Toxo IgM ELISA Test

SUMMARY

[Well-Ag]–**Ab**–[AHIgM-AP]–[PNP]–A_{405}

Assay type: EIA (non-competitive)
Detection: Colorimetric A_{405}
Format: Microtitre well, Ag coated
Sample type: Serum
Sample pre-treatment:
   None
Sample volume: 100 µl of 1:11 dilution*
Number of tests: 96
Controls - standards run in assay:
   Controls: Neg (1), Pos (1)
   Reference Serum: 1 (3)
Incubation:
   30 min (37°C) + 30 min (37°C) + 30 min (37°C)
Washes: 2

CONTENTS

Antibodies, antigens, labelled components:
   T. gondii (strain RH) Ag bound to well
   Anti-human IgM Ab (caprine) AP conjugated
Substrate: PNP
Controls - standards supplied:
   Controls: Neg and Pos
   Reference serum: 1 (all human serum)
Additional reagents:
   None
Special equipment required:
   None

INTERPRETATION°

Comments on interpretation:
   Equivocal: where sample OD is between mean OD of
      Reference Serum and mean OD of Reference Serum
      x 0.9; retest to confirm
   Repeatably equivocal: retest using an alternative method
No. of references: 12

NOTES

131702.0

*Diluent contains an absorbent to eliminate interference due
   to RF and IgG in serum

Toxoplasma gondii
ANTIBODY DETECTION (IgM)

Manufacturer: Human GmbH
Cat. No./Trade name: 51109/Toxoplasma gondii - IgM -
Elisa

SUMMARY

[Well-Ag]–**Ab**–[AHIgM-HRP]–[TMB]–A_{450}

Assay type: EIA (non-competitive)
Detection: Colorimetric A_{450}
Format: Microtitre well, Ag coated
Sample type: Serum
Sample pre-treatment:
   Dilute sample with Dilution Buffer IgM and incubate 5 min
      (RT) to eliminate interference due to RF and IgG
Sample volume: 100 µl of prepared sample
Number of tests: 96
Controls - standards run in assay:
   Controls: Neg (2), Pos (2)
Incubation:
   30 min (RT) + 30 min (RT) + 15 min (RT)
Washes: 2

CONTENTS

Antibodies, antigens, labelled components:
   T. gondii Ag bound to well
   Anti-human IgM Ab (r) HRP conjugated
Substrate: TMB
Controls - standards supplied:
   Controls: Neg and Pos (human)
Additional reagents:
   None
Special equipment required:
   None

INTERPRETATION

Comments on interpretation:
   Equivocal: within cut-off ± 15%; retest in parallel with a
      fresh sample taken 7–14 days later
No. of references: 5

NOTES

131345.0

© KLUWER ACADEMIC PUBLISHERS 1995, ISSN 1381-5067

IgM Elisa

IgM

SUMMARY

[Well-AHIgM]–**Ab**–[Ag-HRP]–[TMB]–A_{450}

Assay type: EIA (non-competitive)
Detection: Colorimetric A_{450}
Format: Microtitre well, Ab coated
Sample type: Serum
Sample pre-treatment:
 None
Sample volume: 10 µl
Number of tests: 96
Controls - standards run in assay:
 Controls: Neg (1), cut-off (2), Pos (1)
Incubation:
 30 min (37°C) + 30 min (37°C) + 15 min (RT)
Washes: 2

CONTENTS

Antibodies, antigens, labelled components:
 Anti-human IgM Ab (g) bound to well
 T. gondii Ag HRP conjugated
Substrate: TMB
Controls - standards supplied:
 Controls: Neg, cut-off, and Pos (human serum)
Additional reagents:
 None
Special equipment required:
 None

INTERPRETATION

Comments on interpretation:
 Equivocal: where ratio of mean OD of cut-off control:OD
 of sample (patient index) is between 0.9–1.1; retest in
 parallel with a fresh sample taken 7–14 days later
No. of references: 6

NOTES

131349.0

SUMMARY

[Well-AHIgM]–**Ab**–[Ag:MAb-POD]–[TMB]–A_{450}

Assay type: EIA (non-competitive)
Detection: Colorimetric A_{450}
Format: Microtitre well, Ab coated
Sample type: Serum
Sample pre-treatment:
 None
Sample volume: 10 µl (+ 1 ml diluent)*
Number of tests: 96
Controls - standards run in assay:
 Controls: Neg (2), cut-off (3), Pos (2)
Incubation:
 45 min (37°C) + 45 min (37°C) + 15 min (RT)
Washes: 2

CONTENTS

Antibodies, antigens, labelled components:
 Anti-human IgM MAb bound to well
 T. gondii (tachyzoites) Ag
 Anti-*T. gondii* MAb POD conjugated
Substrate: TMB
Controls - standards supplied:
 Controls: Neg, cut-off and Pos (human serum)
Additional reagents:
 None
Special equipment required:
 None

INTERPRETATION

Comments on interpretation:
 Equivocal: within cut-off ±10%; retest to confirm
 Repeatably equivocal: retest a fresh sample
No. of references: 4

NOTES

131677.0

*Samples are assayed in duplicate

© KLUWER ACADEMIC PUBLISHERS 1995, ISSN 1381-5067

Toxoplasma gondii
ANTIBODY DETECTION (IgM)

Manufacturer: Immuno Pharmacology Research
Cat. No./Trade name: TOXOPLASMA IgM ELISA

SUMMARY

[Well-Ag]–**Ab**–[AHIgM-AP]–[PNP]–A_{405}

Assay type: EIA (non-competitive)
Detection: Colorimetric A_{450}
Format: Microtitre well, Ag coated
Sample type: Serum
Sample pre-treatment:
   None
Sample volume: 200 µl of 1:100 dilution x 2
Number of tests: 48
Controls - standards run in assay:
   Controls: Neg (1), Pos (1)
Incubation:
   1 hr (37°C) + 1 hr (37°C) + 30 min (37°C)
Washes: 2

CONTENTS

Antibodies, antigens, labelled components:
   T. gondii Ag bound to well
   Anti-human IgM Ab AP conjugated
Substrate: PNP
Controls - standards supplied:
   Controls: Neg and Pos
Additional reagents:
   None
Special equipment required:
   None

INTERPRETATION

Comments on interpretation:
   Equivocal: where sample OD is between 0.150 and
      0.300; retest to confirm
   Repeatably equivocal: retest a fresh specimen in 15 days
No. of references: None

NOTES

131324.0

Toxoplasma gondii
ANTIBODY DETECTION (IgM)

Manufacturer: Incstar
Cat. No./Trade name: 5560/Toxoplasma IgM Clin-ELISA

SUMMARY

[Well-Ag]–**Ab**–[AHIgM-AP]–[PNP]–A_{450}

Assay type: EIA (non-competitive)
Detection: Colorimetric A_{450}
Format: Microtitre well, Ag coated
Sample type: Serum
Sample pre-treatment:
   Add Serum Pretreatment Reagent to sample and
      incubate 30 min (RT) to reduce interference due to
      RF and IgG
Sample volume: 200 µl of treated sample
Number of tests: 96
Controls - standards run in assay:
   Controls: Neg (1), Pos (1)
   Calibrators: I (1), II (1), III (1)
Incubation:
   30 min (RT) + 30 min (RT) + 45 min (RT)
Washes: 2

CONTENTS

Antibodies, antigens, labelled components:
   T. gondii (RH strain) Ag bound to well
   Anti-human IgM Ab AP conjugated
Substrate: PNP
Controls - standards supplied:
   Controls: Neg, Pos (human serum)
   Calibrators: I-Neg, II-low Pos, III-high Pos
Additional reagents:
   IgG Diluent Colorizer (Cat. no. 4506 – optional)
Special equipment required:
   None

INTERPRETATION

Comments on interpretation:
   Equivocal: values between 101 and 119; retest to confirm
      and/or retest a fresh sample in 5–7 days
   Repeatably equivocal: report as negative
No. of references: 33

NOTES

131719.0

This product is not available in the UK but is sold in the rest
   of Europe

© *KLUWER ACADEMIC PUBLISHERS 1995, ISSN 1381-5067*

Toxoplasma gondii
ANTIBODY DETECTION (IgM)

Manufacturer: Kreatech Diagnostics
Cat. No./Trade name: EL-2002-mTOX/Toxoplasma IgM EIA

SUMMARY

[Well-AHIgM]–**Ab**–[Ag–HRP]–[TMB]–A_{450}

Assay type: EIA (non-competitive)
Detection: Colorimetric A_{450}
Format: Microtitre well, Ab coated
Sample type: Serum (do not heat inactivate)
Sample pre-treatment:
　None
Sample volume: 10 μl (+1 ml diluent)*
Number of tests: 96
Controls - standards run in assay:
　Controls: Neg (2), cut-off (4), Pos (2)
Incubation:
　1 hr (37°C) + 1 hr (37°C) + 30 min (RT)
Washes: 2

CONTENTS

Antibodies, antigens, labelled components:
　Anti-human IgM Ab bound to well
　T. gondii Ag from infected Hep-2 cells HRP conjugated**
Substrate: TMB
Controls - standards supplied:
　Controls: Neg, cut-off and Pos (human serum)
Additional reagents:
　None
Special equipment required:
　None

INTERPRETATION

Comments on interpretation:
　Classification of samples is according to cut-off; no
　　further testing
No. of references: None

NOTES

131087.0

*Samples are assayed in duplicate
**Control antigen from uninfected Hep-2 cells is added to
　conjugate to reduce non-specific activity

Toxoplasma gondii
ANTIBODY DETECTION (IgM)

Manufacturer: Labsystems Oy
Cat. No./Trade name: 61 01 201/Toxoplasma gondii IgM EIA

SUMMARY

[Well-Ag]–**Ab**–[AHIgM-AP]–[PNP]–A_{405}

Assay type: EIA (non-competitive)
Detection: Colorimetric A_{405}
Format: Microtitre well, Ag coated
Sample type: Serum*
Sample pre-treatment:
　None
Sample volume: 100 μl of 1:100 dilution x 2
Number of tests: 96
Controls - standards run in assay:
　Controls: Neg (2), Pos (2)
Incubation:
　1 hr (37°C) + 1 hr (37°C) + 30 min (37°C)
Washes: 2

CONTENTS

Antibodies, antigens, labelled components:
　T. gondii Ag bound to well
　Anti-human IgM Ab (r) AP conjugated
Substrate: PNP
Controls - standards supplied:
　Controls: Neg and Pos (human serum)
Additional reagents:
　NaOH
Special equipment required:
　Auto-EIA II analyser (optional)

INTERPRETATION

Comments on interpretation:
　Results are expressed quantitatively in Enzyme
　　Immunoassay Units (EIU)
　Retest all positive sera after treatment with Labsystems
　　IgG blocking reagent (Cat. no. 6106020) to eliminate
　　false positive due to RF and IgG in sample
No. of references: 10

NOTES

131338.0

*Heat treatment may slightly change assay results

Toxoplasma gondii
ANTIBODY DETECTION (IgM)

Manufacturer: Medix Biotech Inc
Cat. No./Trade name: KBF 2097-2/Toxoplasma gondii IgM EIA Kit

SUMMARY

[Well-Ag]–**Ab**–[AHIgM-HRP]–[OPD]–A_{490}

Assay type: EIA (non-competitive)
Detection: Colorimetric A_{490}
Format: Microtitre well, Ag coated
Sample type: Serum (do not heat inactivate)
Sample pre-treatment:
  Add absorbent solution provided to diluted controls, calibrator and samples, incubate 20 min (RT) to eliminate interference due to RF & IgG
Sample volume: 100 µl of treated sample x 2*
Number of tests: 96
Controls - standards run in assay:
  Controls: Neg (2), high Pos (2)*
  Calibrator: low Pos (4)
Incubation:
  20 min (RT) + 20 min (RT) + 10 min (RT)
Washes: 2

CONTENTS

Antibodies, antigens, labelled components:
  T. gondii Ag bound to well
  Control Ag bound to well
  Anti-human IgM Ab (g) HRP conjugated
Substrate: OPD
Controls - standards supplied:
  Controls: Neg and high Pos (human serum or plasma)
  Calibrator: low Pos (human serum or plasma)
Additional reagents:
  H_2SO_4
Special equipment required:
  None

INTERPRETATION

Comments on interpretation:
  Equivocal: where ratio of sample OD:cut-off value is between 0.9-1.10; report as equivocal
No. of references: None

NOTES

131162.0

*Samples and controls are assayed simultaneously in Ag coated and Control Ag coated wells

Toxoplasma gondii
ANTIBODY DETECTION (IgM)

Manufacturer: Melotec S.A.
Cat. No./Trade name: /Melotest Toxo IgM

SUMMARY

[Well-Ag]–**Ab**–[AHIgM-HRP]–[TMB]–A_{450}

Assay type: EIA (non-competitive)
Detection: Colorimetric A_{450}
Format: Microtitre well, Ag coated
Sample type: Serum, plasma
Sample pre-treatment:
  Add absorbent solution provided to controls and diluted samples, incubate 20 min (RT) to eliminate interference due to RF & IgG in sample
Sample volume: 100 µl of treated sample x 2*
Number of tests: 96
Controls - standards run in assay:
  Controls: Neg (1), low Pos (2), high Pos (1)
Incubation:
  20 min (RT) + 20 min (RT) + 10 min (RT)
Washes: 2

CONTENTS

Antibodies, antigens, labelled components:
  T. gondii Ag bound to well
  Tissue culture Control Ag bound to well
  Anti-human IgM Ab HRP conjugated
Substrate: TMB
Controls - standards supplied:
  Controls: Neg, low Pos and high Pos (sera)
Additional reagents:
  None
Special equipment required:
  None

INTERPRETATION

Comments on interpretation:
  Equivocal: ratio of sample OD:cut-off value between 0.9 and 1.1; retest a fresh sample
  Repeatably equivocal: retest fresh sample in 2–4 weeks
No. of references: 8

NOTES

131129.0

*Samples and controls are assayed simultaneously in Ag coated and Control Ag coated wells to eliminate any interference due to ANA not neutralized by absorbent solution

© *KLUWER ACADEMIC PUBLISHERS 1995, ISSN 1381-5067*

Toxoplasma gondii
ANTIBODY DETECTION (IgM)

Manufacturer: Menarini Diagnostics
Cat. No./Trade name: M6119/HF Toxoplasma IgM

SUMMARY

[Well-AHIgM]–**Ab**–[Ag]–[MAb-POD]–[TMB]–A_{450}

Assay type: EIA (non-competitive)
Detection: Colorimetric A_{450}
Format: Microtitre well, Ab coated
Sample type: Serum
Sample pre-treatment:
  None
Sample volume: 100 µl of 1:101 dilution
Number of tests: 96
Controls - standards run in assay:
  Controls: Neg (1), cut-off (1), Pos (1)
Incubation:
  45 min (37°C) + 45 min (37°C) + 15 min (RT)
Washes: 2

CONTENTS

Antibodies, antigens, labelled components:
  Anti-human IgM MAb bound to well
  T. gondii Ag
  Anti-T. gondii MAb POD conjugated
Substrate: TMB
Controls - standards supplied:
  Controls: Neg, cut-off and Pos (serum)
Additional reagents:
  None
Special equipment required:
  None

INTERPRETATION

Comments on interpretation:
  No data is provided for interpretation of results
No. of references: None

NOTES

131095.0

Toxoplasma gondii
ANTIBODY DETECTION (IgM)

Manufacturer: Organon Teknika NV
Cat. No./Trade name: Toxonostika® IgM II

SUMMARY

[Well-AHIgM]–**Ab**–[Ag]–[Ab-HRP]–[TMB]–A_{450}

Assay type: EIA (non-competitive)
Detection: Colorimetric A_{450}
Format: Microtitre well, Ab coated
Sample type: Serum, plasma (do not heat inactivate)
Sample pre-treatment:
  None
Sample volume: 10 µl of 1:101 dilution (adult); 1:20
  (newborn)
Number of tests: 96, 192
Controls - standards run in assay:
  Controls:
  Neg (1), Pos (1), strong Pos (1) (for one strip)
  Neg (2), Pos (2), strong Pos (2), (for two or more strips)
Incubation:
  1 hr (37°C) + 1 hr (37°C) + 30 min (RT)
Washes: 2

CONTENTS

Antibodies, antigens, labelled components:
  Anti-human IgM Ab (sh) bound to well
  T. gondii Ag
  Anti-T. gondii Ab (sh) HRP conjugated
Substrate: TMB
Controls - standards supplied:
  Controls: Neg, Pos and strong Pos (human serum)
Additional reagents:
  H_2SO_4
Special equipment required:
  None

INTERPRETATION

Comments on interpretation:
  Classification of samples is according to cut-off
No. of references: 3

NOTES

131201.0

© KLUWER ACADEMIC PUBLISHERS 1995, ISSN 1381-5067

Toxoplasma gondii
ANTIBODY DETECTION (IgM)

Manufacturer: Radim
Cat. No./Trade name: K1TM/Toxoplasma IgMEIA Well

SUMMARY

[Well-AHIgM]–**Ab**–[Ag]–[MAb-biotin]–[strept-HRP]–
[TMB]–A_{450}

Assay type: EIA (non-competitive)
Detection: Colorimetric A_{450}
Format: Microtitre well, Ab coated
Sample type: Serum, plasma
Sample pre-treatment:
  None
Sample volume: 100 µl of 1:100 dilution x 2
Number of tests: 96
Controls - standards run in assay:
  Controls: Neg (2), cut-off (2), Pos (2)
Incubation:
  1 hr (37°C) + 60 min (37°C) + 30 min (37°C) + 10 min
    (37°C) or 15 min (RT)
Washes: 3

CONTENTS

Antibodies, antigens, labelled components:
  Anti-human IgM MAb bound to well
  T. gondii Ag
  Anti-T. gondii MAb biotinylated
  Streptavidin HRP conjugated
Substrate: TMB
Controls - standards supplied:
  Controls: Neg, cut-off and Pos (serum)
Additional reagents:
  None
Special equipment required:
  None

INTERPRETATION

Comments on interpretation:
  Grey area: within cut-off ±10%; retest to confirm
No. of references: 5

NOTES

131275.0

Toxoplasma gondii
ANTIBODY DETECTION (IgM)

Manufacturer: Roche Diagnostic Systems
Cat. No./Trade name: 07 3493 4/COBAS® CORE Toxo
IgM EIA

SUMMARY

[Bead-AHIgM]–**Ab**–[Ag:MAb-HRP]–[TMB]–A_{450}

Assay type: EIA (non-competitive)
Detection: Colorimetric A_{450}
Format: Tube, Ab coated bead
Sample type: Serum, plasma
Sample pre-treatment:
  None
Sample volume: 10 µl (+500 µl diluent)
Number of tests: 50
Controls - standards run in assay:
  Controls: Neg (1), Pos (3)
Incubation:
  15 min (37°C) + 1 hr (37°C) + 15 min (37°C) (all with
    shaking)
Washes: 2

CONTENTS

Antibodies, antigens, labelled components:
  Anti-human IgM MAb (m) bound to bead
  T. gondii and anti-T. gondii MAb (m) HRP conjugated
Substrate: TMB (bought separately)
Controls - standards supplied:
  Controls: Neg and Pos (human serum)
Additional reagents:
  Cobas® Core TMB kit
  H_2SO_4 (for manual assay only)
Special equipment required:
  Cobas® Core immunoassay analyser (for automated
    method - optional)
  Cobas® Core EIA shaking incubator, tube washer and
    photometer (optional)
  Cobas® Core TORC negative control bead dispenser
  Cobas® Core reaction tubes

INTERPRETATION

Comments on interpretation:
  Equivocal: within cut-off ±10%; report as indeterminate
    and repeat with a sample taken in a few weeks
  Positive: ≥cut-off +10% and confirmed using
    supplemental tests
No. of references: None

NOTES

131158.0

© KLUWER ACADEMIC PUBLISHERS 1995, ISSN 1381-5067

ANTIBODY DETECTION (IgM)

958

Toxoplasma gondii
ANTIBODY DETECTION (IgM)

Manufacturer: Sanofi Diagnostics Pasteur
Cat. No./Trade name: 72750/Platelia® Toxo IgM

SUMMARY

[Well-AHIgM]–**Ab**–[Ag: MAb-HRP]–[OPD]–A$_{492}$

Assay type: EIA (non-competitive)
Detection: Colorimetric A$_{492}$
Format: Microtitre well, Ab coated
Sample type: Serum
Sample pre-treatment:
 None
Sample volume: 10 μl (+1 ml diluent)
Number of tests: 92
Controls - standards run in assay:
 Controls: Neg (1), cut-off (2), Pos (1)
Incubation:
 1 hr (37°C) + 1 hr (37°C) + 30 min (RT)
Washes: 2 (+ preliminary plate wash)

CONTENTS

Antibodies, antigens, labelled components:
 Anti-human IgM MAb (m) bound to well
 T. gondii (RH strain) Ag
 Anti-T. gondii (P30) MAb (m) HRP conjugated
Substrate: OPD
Controls - standards supplied:
 Controls: Neg, cut-off and Pos (human serum)
Additional reagents:
 None
Special equipment required:
 None

INTERPRETATION

Comments on interpretation:
 Equivocal: between cut-off and cut-off–20%; retest a
 fresh sample in 1 week
No. of references: 11

NOTES

131762.0

Toxoplasma gondii
ANTIBODY DETECTION (IgM)

Manufacturer: SFRI Laboratoire
Cat. No./Trade name: IIE 019/Toxo IgM

SUMMARY

[Well-AHIgM]–**Ab**–[Ag:MAb-POD]–[TMB]–A$_{450}$

Assay type: EIA (non-competitive)
Detection: Colorimetric A$_{450}$
Format: Microtitre well, Ab coated
Sample type: Serum, plasma
Sample pre-treatment:
 None
Sample volume: 20 μl (+400 μl diluent)
Number of tests: 96
Controls - standards run in assay:
 Controls: Neg (1), Pos 1 (1), Pos 2 (1)
Incubation:
 15 min (37°C) + 15 min (37°C) + 15 min (37°C)
Washes: 2

CONTENTS

Antibodies, antigens, labelled components:
 Anti-human IgM Ab bound to well
 T. gondii Ag
 Anti-T. gondii MAb POD conjugated
Substrate: TMB
Controls - standards supplied:
 Controls: Neg, Pos 1 and Pos 2 (human serum)
Additional reagents:
 None
Special equipment required:
 None

INTERPRETATION

Comments on interpretation:
 Equivocal: where cut-off is between 1 and 1.2; retest to
 confirm
 Repeatably equivocal: repeat test at a later date
No. of references: 15

NOTES

131649.0

© KLUWER ACADEMIC PUBLISHERS 1995, ISSN 1381-5067

Toxoplasma gondii
ANTIBODY DETECTION (IgM)

Manufacturer: Sigma Diagnostics
Cat. No./Trade name: SIA 408/SIA® Toxoplasma IgM

SUMMARY

[Well-Ag]–**Ab**–[AHIgM-AP]–[PMP]–A$_{550}$

Assay type: EIA (non-competitive)
Detection: Colorimetric A$_{550}$
Format: Microtitre well, Ag coated
Sample type: Serum (do not heat inactivate)
Sample pre-treatment:
　Add Pretreatment Solution to sample, incubate 30 min
　　(RT) to eliminate interference due to RF and IgG
Sample volume: 50 µl (+200 µl diluent)
Number of tests: 96
Controls - standards run in assay:
　Calibrators: 1 (2), 2 (2), 3 (2)
Incubation:
　30 min (RT) + 30 min (RT) + 30 min (RT)
Washes: 2 (+preliminary plate wash)

CONTENTS

Antibodies, antigens, labelled components:
　T. gondii Ag bound to well
　Anti-human IgM Ab (g) AP conjugated
Substrate: PMP
Controls - standards supplied:
　Calibrators: 1, 2 and 3 (IgM Ab titre in AU/ml – human
　　serum)
Additional reagents:
　None
Special equipment required:
　Sigma EIA Multiwell Plate Reader (Cat. no. EQ104 –
　　optional)

INTERPRETATION

Comments on interpretation:
　Equivocal: 0.25–0.35 SIA® Toxoplasma IgM Value;
　　report as probably positive and retest in parallel with a
　　fresh specimen 5–7 days later
No. of references: 10

NOTES

131437.0

Toxoplasma gondii
ANTIBODY DETECTION (IgM)

Manufacturer: Sigma Diagnostics
Cat. No./Trade name: SIA 303/SIA® Toxoplasma IgM
(Capture)

SUMMARY

[Well-AHIgM]–**Ab**–[Ag:MAb-biotin: strept-HRP]–[TMB]–
　　　　　　　　　　　　　　　　　　　A$_{450}$

Assay type: EIA (non-competitive)
Detection: Colorimetric A$_{450}$
Format: Microtitre well, Ab coated
Sample type: Serum
Sample pre-treatment:
　None
Sample volume: 10 µl (+1 ml diluent)
Number of tests: 96
Controls - standards run in assay:
　Controls: Neg (1), low Pos (3), high Pos (1)
Incubation:
　1 hr (37°C) + 1 hr (37°C) + 30 min (RT)
Washes: 2

CONTENTS

Antibodies, antigens, labelled components:
　Anti-human IgM Ab bound to well
　T. gondii Ag
　Anti-*T. gondii* MAb biotinylated
　Streptavidin HRP conjugated
Substrate: TMB
Controls - standards supplied:
　Controls: Neg, low Pos and high Pos (human serum)
Additional reagents:
　H$_2$SO$_4$
Special equipment required:
　Sigma EIA Multiwell Plate Reader (Cat. no. EQ104 –
　　optional)

INTERPRETATION

Comments on interpretation:
　Equivocal: where Index Value is 0.9–1.1; retest to
　　confirm
　Repeatably equivocal: repeat test with a fresh sample
No. of references: 8

NOTES

131436.0

© KLUWER ACADEMIC PUBLISHERS 1995, ISSN 1381-5067

Toxoplasma gondii
ANTIBODY DETECTION (IgM)

Manufacturer: Sorin Biomedica
Cat. No./Trade name: P2708/ETI-TOXOK-M reverse

SUMMARY

[Well-AHIgM]–**Ab**–[Ag]–[MAb-HRP]–[TMB]–A_{450}

Assay type: EIA (non-competitive)
Detection: Colorimetric A_{450}
Format: Microtitre well, Ab coated
Sample type: Serum, plasma
Sample pre-treatment:
   None
Sample volume: 100 µl of 1:101 dilution x 2
Number of tests: 96
Controls - standards run in assay:
   Controls: Neg (2), cut-off (4), Pos (2)
Incubation:
   1 hr (37°C) + 1 hr (37°C) + 30 min (RT)
Washes: 2

CONTENTS

Antibodies, antigens, labelled components:
   Anti-human IgM IgG Ab (r) bound to well
   T. gondii (RH strain) Ag
   Anti-*T. gondii* IgG MAb (m) HRP conjugated
Substrate: TMB
Controls - standards supplied:
   Controls: Neg, cut-off and Pos (human serum)
Additional reagents:
   None
Special equipment required:
   ETI-System Reader and ETI-System Washer (optional)

INTERPRETATION

Comments on interpretation:
   Equivocal: within cut-off ± 10%; retest to confirm
No. of references: 14

NOTES

131215.0

Toxoplasma gondii
ANTIBODY DETECTION (TOTAL Ab AND/OR IgG AND/OR IgM)

Manufacturer: bioMerieux
Cat. No./Trade name: 75471/Toxo-Spot IF

SUMMARY

[Slide-Ag]–**Ab**–[AHIg/IgG/IgM-FITC]–fluorescence

Assay type: Immunofluorescence assay (indirect)
Detection: Fluorescence microscopy
Format: Slide, Ag coated
Sample type: Serum
Sample pre-treatment:
   None
Sample volume: 20 µl of diluted serum x 2*
Number of tests: 100
Controls - standards run in assay:
   Controls: Neg (1), Pos (1) - for each dilution
Incubation:
   30 min (37°C) + 30 min (37°C)
Washes: 2

CONTENTS

Antibodies, antigens, labelled components:
   T. gondii Ag (from mouse ascitic fluid) bound to slide
   Anti-human Ig Ab (g) FITC conjugated
   Anti-human IgG Ab (g) FITC conjugated
   Anti-human IgM Ab (g) FITC conjugated
Substrate:
Controls - standards supplied:
   Controls: Neg (PBS) and Pos - serial dilutions (titre for IgG Ab in U/ml - human serum)
Additional reagents:
   None
Special equipment required:
   None

INTERPRETATION

Comments on interpretation:
   Positive: green fluorescence of whole or periphery of organism with the appropriate conjugate identifies the antibody class present in sample
No. of references: 3

NOTES

131282.0

This kit can detect total antibodies and IgG and IgM antibodies and can differentiate between them
*Dilutions should be chosen to provide serum titres of ⩾ 10 IU/ml and ⩾ 300 IU/ml with reference to Positive control

© KLUWER ACADEMIC PUBLISHERS 1995, ISSN 1381-5067

Toxoplasma gondii
ANTIBODY DETECTION

Manufacturer: Zeus Scientific Inc
Cat. No./Trade name: 8000/Toxoplasma IFA Test System

SUMMARY

[Slide-Ag]–**Ab**–[AHIg-FITC]–fluorescence

Assay type: Immunofluorescence assay (indirect)
Detection: Fluorescence microscopy
Format: Slide, Ag coated
Sample type: Serum
Sample pre-treatment:
   None
Sample volume: 20 µl of 1:16 and 1:64 dilution
Number of tests: 120
Controls - standards run in assay:
   Controls: Neg (1), low Pos (1), high Pos (1)
Incubation:
   30 min (RT) + 30 min (RT)
Washes: 2

CONTENTS

Antibodies, antigens, labelled components:
   T. gondii Ag (RH strain from mouse peritoneal fluid)
      bound to slide
   Anti-human Ig Ab (g) FITC conjugated
Substrate:
Controls - standards supplied:
   Controls: Neg, low Pos (titre 1:32-1:128) and high Pos
      (titre 1:512-1:2048) all human serum
Additional reagents:
   None
Special equipment required:
   None

INTERPRETATION

Comments on interpretation:
   Positive: peripheral or total yellow green fluorescence of
      organism
No. of references: 12

NOTES

131295.0

Toxoplasma gondii
ANTIBODY DETECTION (IgG)

Manufacturer: GenBio
Cat. No./Trade name: 1200/ImmunoFA® Toxo IgG Test

SUMMARY

[Slide-Ag]–**Ab**–[AHIgG-FITC]–fluorescence

Assay type: Immunofluorescence assay (indirect)
Detection: Fluorescence microscopy
Format: Slide well, Ag coated
Sample type: Serum
Sample pre-treatment:
   None
Sample volume: 30 µl of 1:16 and 1:256 dilution
Number of tests: 100
Controls - standards run in assay:
   Controls: Neg (1), low Pos (1), high Pos (1)
Incubation:
   30 m in (37°C) + 30 min (37°C)
Washes: 2

CONTENTS

Antibodies, antigens, labelled components:
   T. gondii (RH strain from mouse peritoneal fluid) bound to
      slide well
   Anti-human IgG Ab (g) FITC conjugated
Substrate:
Controls - standards supplied:
   Controls: Neg, low Pos and high Pos (serum)
Additional reagents:
   None
Special equipment required:
   None

INTERPRETATION

Comments on interpretation:
   Positive: (1+)–(4+) yellow green fluorescence around
      entire periphery of organism
No. of references: 15

NOTES

131419.0

Toxoplasma gondii
ANTIBODY DETECTION (IgG)

Manufacturer: Gull Laboratories Inc
Cat. No./Trade name: TX100/Toxo IgG Test

SUMMARY

[Slide-Ag]–**Ab**–[AHIgG-FITC]–fluorescence

Assay type: Immunofluorescence assay (indirect)
Detection: Fluorescence microscopy
Format: Slide well, Ag coated
Sample type: Serum
Sample pre-treatment:
 None
Sample volume: 15 µl of 1:10 dilution
Number of tests: 100
Controls - standards run in assay:
 Controls: Neg (1), low Pos (1), high Pos (1)
Incubation:
 30 min (RT) + 30 min (RT)
Washes: 2

CONTENTS

Antibodies, antigens, labelled components:
 T. gondii (tachyzoites) from mouse peritoneum or human
  fibroblasts bound to slide well
 Anti-human IgG Ab (caprine) FITC conjugated
Substrate:
Controls - standards supplied:
 Controls: Neg, low Pos and high Pos (human serum)
Additional reagents:
 None
Special equipment required:
 None

INTERPRETATION

Comments on interpretation:
 Positive: yellow-green fluorescence of entire cell
  periphery in the majority of organisms per field at
  ≥1:10 dilution
No. of references: 13

NOTES

131706.0

Toxoplasma gondii
ANTIBODY DETECTION (IgG)

Manufacturer: Hemagen Diagnostics Inc
Cat. No./Trade name: 2010/VIRGO® TOXO IgG IFA

SUMMARY

[Slide well-Ag]–**Ab**–[AHIgG-FITC]–fluorescence

Assay type: Immunofluorescence assay (indirect)
Detection: Fluorescence microscopy
Format: Slide well, Ag coated
Sample type: Serum
Sample pre-treatment:
 None
Sample volume: 10–20 µl of 1:16 and 1:64 dilutions x2
Number of tests: 96
Controls - standards run in assay:
 Controls: Neg (1), low Pos (1), high Pos (1)
Incubation:
 30 min (RT) + 30 min (RT)
Washes: 2

CONTENTS

Antibodies, antigens, labelled components:
 T. gondii infected and uninfected cells bound to slide well
 Anti-human IgG Ab (g) FITC conjugated*
Substrate:
Controls - standards supplied:
 Controls: Neg, low Pos and high Pos (human serum)*
Additional reagents:
 None
Special equipment required:
 None

INTERPRETATION

Comments on interpretation:
 Low Positive: ≥(1+) fluorescence at 1:16–1:64 dilution
 Medium Positive: ≥(1+) fluorescence at 1:256 dilution
 High Positive: ≥(1+) fluorescence at 1:1024 dilution
No. of references: 14

NOTES

131106.0

*For IgM antibody assay, anti-human IgM-FITC conjugate
 and Neg and Pos controls are available

© *KLUWER ACADEMIC PUBLISHERS* 1995, ISSN 1381-5067

Toxoplasma gondii
ANTIBODY DETECTION (IgG)

Manufacturer: LD, Labor Diagnostika GmbH
Cat. No./Trade name: 63333/LD Toxo IFA Kit

SUMMARY

[Slide-Ag]–**Ab**–[AHIgG-FITC]–fluorescence

Assay type: Immunofluorescence assay (indirect)
Detection: Fluorescence microscopy
Format: Slide, Ag coated
Sample type: Serum
Sample pre-treatment:
   None
Sample volume: 20 µl of 1:16 dilution)
Number of tests: 122
Controls - standards run in assay:
   Controls: Neg (1), Pos (1)
Incubation:
   30 min (RT) + 30 min (RT)
Washes: 2

CONTENTS

Antibodies, antigens, labelled components:
   T. gondii (RH Strain) Ag bound to slide
   Anti-human IgG Ab FITC conjugated
Substrate:
Controls - standards supplied:
   Controls: Neg and Pos
Additional reagents:
   None
Special equipment required:
   None

INTERPRETATION

Comments on interpretation:
   Positive: brilliant green fluorescence of entire organism
      or entire periphery of organism
No. of references: None

NOTES

131777.0

Toxoplasma gondii
ANTIBODY DETECTION (IgG)

Manufacturer: Stellar Bio Systems Inc
Cat. No./Trade name: Indirect Fluorescent Assay for
Toxoplasma gondii IgG Antibody

SUMMARY

[Slide-Ag]–**Ab**–[AHIgG-FITC]–fluorescence

Assay type: Immunofluorescence assay (indirect)
Detection: Fluorescence microscopy
Format: Slide well, Ag coated
Sample type: Serum
Sample pre-treatment:
   None
Sample volume: 20 µl of 1:16 dilution
Number of tests:
Controls - standards run in assay:
   Controls: Neg (1), Pos (1)
Incubation:
   30 min (37°C) + 30 min (37°C)
Washes: 2

CONTENTS

Antibodies, antigens, labelled components:
   T. gondii organisms bound to slide
   Anti-human IgG Ab (g) FITC conjugated
Substrate:
Controls - standards supplied:
   Controls: Neg and Pos (human serum)
Additional reagents:
   None
Special equipment required:
   None

INTERPRETATION

Comments on interpretation:
   Positive: yellow green fluorescence around entire
      periphery of organisms
No. of references: 6

NOTES

131634.0

© *KLUWER ACADEMIC PUBLISHERS 1995, ISSN 1381-5067*

Toxoplasma gondii
ANTIBODY DETECTION (IgM)

Manufacturer: GenBio
Cat. No./Trade name: 1300/ImmunoFA® Toxo IgM Test

SUMMARY

[Slide-Ag]–**Ab**–[AHIgM-FITC]–fluorescence

Assay type: Immunofluorescence assay (indirect)
Detection: Fluorescence microscopy
Format: Slide well, Ag coated
Sample type: Serum
Sample pre-treatment:
  None
Sample volume: 30 μl of serial dilutions (initial dilution: 1:8 for adults; 1:2 for neonates)
Number of tests: 100
Controls - standards run in assay:
  Controls: Neg (1), Pos (1)
Incubation:
  30 m in (37°C) + 30 min (37°C)
Washes: 2

CONTENTS

Antibodies, antigens, labelled components:
  T. gondii (RH strain from mouse peritoneal fluid) bound to slide well
  Anti-human IgM Ab (g) FITC conjugated
Substrate:
Controls - standards supplied:
  Controls: Neg and Pos (human serum)
Additional reagents:
  None
Special equipment required:
  None

INTERPRETATION

Comments on interpretation:
  Positive: (1 +)–(4 +) yellow green fluorescence around entire periphery of organism
No. of references: 14

NOTES

131420.0

False positive reactions may occur in sera containing RF and IgG
Where interpretation of results is difficult, repeat test after fractionation of serum IgG & IgM or selective removal of IgG Ab

Toxoplasma gondii
ANTIBODY DETECTION (IgM)

Manufacturer: Gull Laboratories Inc
Cat. No./Trade name: TX150/Toxo IgM Test

SUMMARY

[Slide-Ag]–**Ab**–[AHIgM-FITC]–fluorescence

Assay type: Immunofluorescence assay (indirect)
Detection: Fluorescence microscopy
Format: Slide well, Ag coated
Sample type: Serum
Sample pre-treatment:
  None
Sampie volume: 15 μl of 1:10 dilution
Number of tests: 100
Controls - standards run in assay:
  Controls: Neg (1), Pos (1)
Incubation:
  30 min (37°C) + 30 min (37°C)
Washes: 2

CONTENTS

Antibodies, antigens, labelled components:
  T. gondii (RH strain – tachyzoites) from mouse peritoneum or human fibroblasts bound to slide well
  Anti-human IgM Ab (caprine) FITC conjugated
Substrate:
Controls - standards supplied:
  Controls: Neg and Pos (human serum)
Additional reagents:
  None
Special equipment required:
  None

INTERPRETATION

Comments on interpretation:
  Positive: yellow-green fluorescence of entire cell periphery in the majority of organisms per field at ≥1:10 dilution
No. of references: 13

NOTES

131708.0

Toxoplasma gondii
ANTIBODY DETECTION (IgM)

Manufacturer: Stellar Bio Systems Inc
Cat. No./Trade name: Indirect Fluorescent Assay for Toxoplasma gondii IgM Antibody

SUMMARY

[Slide-Ag]–**Ab**–[AHIgM-FITC]–fluorescence

Assay type: Immunofluorescence assay (indirect)
Detection: Fluorescence microscopy
Format: Slide well, Ag coated
Sample type: Serum
Sample pre-treatment:
    Removal of interference due to RF and IgG in sample using commercially available pretreatment devices (not provided)
Sample volume: 20 µl of 1:8 dilution
Number of tests:
Controls - standards run in assay:
    Controls: Neg (1), Pos (1)
Incubation:
    1 hr (37°C) + 30 min (37°C)
Washes: 2

CONTENTS

Antibodies, antigens, labelled components:
    T. gondii organisms bound to slide
    Anti-human IgM Ab (g) FITC conjugated
Substrate:
Controls - standards supplied:
    Controls: Neg and Pos (human serum)
Additional reagents:
    Pretreatment device for removal of RF and IgG in sample
Special equipment required:
    None

INTERPRETATION

Comments on interpretation:
    Positive: yellow-green fluorescence around entire periphery of organism
No. of references: 6

NOTES

131639.0

Toxoplasma gondii
ANTIBODY DETECTION (IgG)

Manufacturer: Diagnostic Products Corporation
Cat. No./Trade name: LKTX2/Immulite® Toxoplasma IgG

SUMMARY

[Bead-Ag]–**Ab**–[AHIgG-AP]–[substrate]–luminescence

Assay type: Luminometric immunoassay (non-competitive)
Detection: Luminometric
Format: Reaction tube, Ag coated beads
Sample type: Serum
Sample pre-treatment:
    None
Sample volume: 10 µl of 1:21 dilution
Number of tests: 50, 200
Controls - standards run in assay:
    Controls: Neg (1), Pos (1)
Incubation:
    Automated
Washes: Automated

CONTENTS

Antibodies, antigens, labelled components:
    T. gondii Ag bound to beads
    Anti-human IgG MAb (m) AP conjugated
Substrate: Adamantyl dioxetane (phosphate ester) (bought separately)
Controls - standards supplied:
    Controls: Neg and Pos
Additional reagents:
    Immulite® Chemiluminescence Substrate Module (LSUBX)
    Immulite® Probe Wash Module (LPWS2)
Special equipment required:
    Immulite® Automated Analyzer
    Immulite® Sample cups, holders and caps

INTERPRETATION

Comments on interpretation:
    This is an automated assay. Results designated equivocal; retest a fresh sample in 1 week or use an alternative method
No. of references: 10

NOTES

131739.0

Toxoplasma gondii
ANTIBODY DETECTION (IgG)

Manufacturer: Diagnostic Products Corporation
Cat. No./Trade name: LKTXQ1/Immulite® Toxoplasma Quantitative IgG

SUMMARY

[Bead-Ag]–**Ab**–[AHIgG-AP]–[substrate]–luminescence

Assay type: Luminometric immunoassay (non-competitive)
Detection: Luminometric
Format: Reaction tube, Ag coated beads
Sample type: Serum
Sample pre-treatment:
None
Sample volume: 10 µl of 1:21 dilution
Number of tests: 100, 500
Controls - standards run in assay:
Controls: Neg (1), Pos (1)
Incubation:
Automated
Washes: Automated

CONTENTS

Antibodies, antigens, labelled components:
T. gondii (strain RH, tachyzoites from mouse peritoneum) Ag bound to beads
Anti-human IgG MAb (m) AP conjugated
Substrate: Adamantyl dioxetane (phosphate ester) (bought separately)
Controls - standards supplied:
Controls: Neg and Pos (human serum)
Additional reagents:
Immulite® Chemiluminescence Substrate Module (LSUBX)
Special equipment required:
Immulite® Automated Analyzer
Immulite® Sample cups, holders and caps

INTERPRETATION

Comments on interpretation:
This is an automated assay
Negative: < 8 IU/ml
Positive: ≥ 8 IU/ml indicates past infection
No. of references: 10

NOTES

131799.0

Two adjustors – low and high – (Cat. nos. LBQL and LBQH) are provided with the kit to prepare the kit prior to use

Toxoplasma gondii
ANTIBODY DETECTION (IgG)

Manufacturer: Johnson & Johnson Clinical Diagnostics Inc
Cat. No./Trade name: LAN2301/AMERLITE Toxo IgG Assay

SUMMARY

[Well-Ag]–**Ab**–[AHIgG-HRP]–[signal reagent]–luminescence

Assay type: Luminometric immunoassay (non-competitive)
Detection: Luminometric
Format: Microtitre well, Ag coated
Sample type: Serum, plasma
Sample pre-treatment:
None
Sample volume: 10 µl*
Number of tests: 144
Controls - standards run in assay:
Controls: 1 (1), 2 (3), 3 (1)
Incubation:
1 hr (37°C) + 30 min (37°C) + 2-20 min (RT) (first two incubations with shaking)
Washes: 2

CONTENTS

Antibodies, antigens, labelled components:
T. gondii (tachyzoites) Ag bound to well
Anti-human IgG Ab (r) HRP conjugated
Substrate: AMERLITE Signal Reagent
Controls - standards supplied:
Controls: 1 (0 IU/ml), 2 (5-20 IU/ml), 3 (50-250 IU/ml) (human plasma)
Additional reagents:
AMERLITE Signal Reagent, AMERLITE Serology Wash Reagent
Special equipment required:
AMERLITE Processing Center (optional)
AMERLITE Immunoassay System

INTERPRETATION

Comments on interpretation:
This is an automated assay
Results designated equivocal; retest to confirm
Repeatably equivocal: repeat test with a fresh sample
No. of references: 4

NOTES

131222.0

*Samples may be assayed singly or in duplicate

Toxoplasma gondii
ANTIBODY DETECTION (IgG)

Manufacturer: Sanofi Diagnostics Pasteur
Cat. No./Trade name: 34450/Access® Toxo IgG

SUMMARY

[Particles-Ag]–**Ab**–[AHIgG-AP]–[substrate]–
luminescence

Assay type: Luminometric immunoassay (non-competitive)
Detection: Luminometric
Format: Reaction vessel, Ag coated magnetic particles
Sample type: Serum
Sample pre-treatment:
None
Sample volume: 10 µl
Number of tests: 50
Controls - standards run in assay:
Not specified
Incubation:
Automated (total time 35 min at 37°C)
Washes: Automated

CONTENTS

Antibodies, antigens, labelled components:
T. gondii Ag bound to magnetic particles
Anti-human IgG Ab AP conjugated
Substrate: Lumi-Phos® 500 (bought separately)
Controls - standards supplied:
None
Additional reagents:
Access® Toxo IgG Calibrators (Cat. no. 34455)
Access® Substrate (Cat. no. 81906)
Access® Wash Buffer (Cat. no. 81907)
Access® QC Toxo IgG (Cat. no. 34459) or similar products
Special equipment required:
Access® Immunoassay System

INTERPRETATION

Comments on interpretation:
Results are calculated automatically
No. of references: 15

NOTES

131764.0

Toxoplasma gondii
ANTIBODY DETECTION (IgM)

Manufacturer: Johnson & Johnson Clinical Diagnostics Inc
Cat. No./Trade name: LAN2300T/95a/AMERLITE Toxo IgM Assay

SUMMARY

[Well-AHIgM]–**Ab**–[Ag]–[MAb-HRP]–[signal reagent]–
luminescence

Assay type: Luminometric immunoassay (non-competitive)
Detection: Luminometric
Format: Microtitre well, Ab coated
Sample type: Serum, plasma
Sample pre-treatment:
None
Sample volume: 10 µl*
Number of tests: 144
Controls - standards run in assay:
Controls: Neg (1), Pos (3)
Incubation:
30 min (37°C) + 30 min (37°C) + 5-20 min (RT) (first two incubations with shaking)
Washes: 2

CONTENTS

Antibodies, antigens, labelled components:
Anti-human IgM MAb (m) bound to well
T. gondii Ag
Anti-*T. gondii* MAb (m) HRP conjugated
Substrate: AMERLITE Signal Reagent
Controls - standards supplied:
Controls: Neg (0 IU/ml), Pos (100-130 IU/ml) (human plasma)
Additional reagents:
AMERLITE Signal Reagent, AMERLITE Serology Wash Reagent, AMERLITE TOXO IgG Diluent and Quantitation Panel (for quantitative assay)
Special equipment required:
AMERLITE Processing Center (optional)
AMERLITE Immunoassay System

INTERPRETATION

Comments on interpretation:
This is an automated assay
Results designated equivocal: retest to confirm
Repeatably equivocal: repeat test with a fresh sample
No. of references: 7

NOTES

131223.0

*Samples may be assayed singly or in duplicate

© *KLUWER ACADEMIC PUBLISHERS 1995, ISSN 1381-5067*

Toxoplasma gondii
ANTIBODY DETECTION (IgM)

Manufacturer: Sanofi Diagnostics Pasteur
Cat. No./Trade name: 34460/Access® Toxo IgM

SUMMARY

[Particles-AHIgM]–**Ab**–[Ag:MAb-AP]–[substrate]–
luminescence

Assay type: Luminometric immunoassay (non-competitive)
Detection: Luminometric
Format: Reaction vessel, Ab coated magnetic particles
Sample type: Serum
Sample pre-treatment:
    None
Sample volume: 10 µl
Number of tests: 50
Controls - standards run in assay:
    Not specified
Incubation:
    Automated (total time 35 min at 37°C)
Washes: Automated

CONTENTS

Antibodies, antigens, labelled components:
    Anti-human IgM Ab (g) bound to magnetic particles
    T. gondii Ag
    Anti-*T. gondii* MAb (m) AP conjugated
Substrate: Lumi-Phos® 500 (bought separately)
Controls - standards supplied:
    None
Additional reagents:
    Access® Toxo IgG Controls (Cat. no. 34465)
    Access® Substrate (Cat. no. 81906)
    Access® Wash Buffer (Cat. no. 81907)
    Access® QC Toxo IgM (Cat. no. 34469) or similar
        products
Special equipment required:
    Access® Immunoassay System

INTERPRETATION

Comments on interpretation:
    Results are calculated automatically
    Equivocal: titre between 100–140 AU/ml; retest fresh
        sample in 10–20 days
No. of references: 15

NOTES

131766.0

Toxoplasma gondii
ANTIBODY DETECTION

Manufacturer: BIOKIT SA
Cat. No./Trade name: toxocell latex

SUMMARY

[Latex-Ag]–**Ab**–agglutination

Assay type: Particle agglutination assay
Detection: Visual
Format: Latex particles, Ag coated (test done on slide)
Sample type: Serum
Sample pre-treatment:
    None
Sample volume: 50 µl
Number of tests:
Controls - standards run in assay:
    Controls: Neg (1), Pos (1)
Incubation:
    5 min (RT) on rotary shaker
Washes:

CONTENTS

Antibodies, antigens, labelled components:
    T. gondii Ag bound to latex particles (Latex Reagent)
Substrate:
Controls - standards supplied:
    Controls: Neg and Pos (rabbit serum)
Additional reagents:
    None
Special equipment required:
    None

INTERPRETATION

Comments on interpretation:
    Positive: agglutination of sample with Latex Reagent
    Where clinical findings suggest toxoplasmosis then
        negative result may be due to prozone effect with
        strong Pos sera; therefore retest using 1:5 dilution of
        sample
No. of references: 8

NOTES

131371.0

Toxoplasma gondii
ANTIBODY DETECTION

Manufacturer: Eiken Chemical Co. Ltd
Cat. No./Trade name: V-ST00/Toxoreagent 'Eiken'

SUMMARY

[Latex-Ag]–**Ab**–agglutination

Assay type: Particle agglutination assay
Detection: Visual
Format: Latex particles, Ag coated (test done in microtitre wells)
Sample type: Serum (do not heat inactivate)*
Sample pre-treatment:
    None
Sample volume: 25 μl of serial dilutions (1:8 initial dilution)
Number of tests: 50
Controls - standards run in assay:
    Controls: Neg (1), Pos (1)
Incubation:
    At least 12 hr (RT)
Washes:

CONTENTS

Antibodies, antigens, labelled components:
    T. gondii Ag bound to latex particles (Latex Suspension)
Substrate:
Controls - standards supplied:
    Controls: Pos (serum)
Additional reagents:
    None
Special equipment required:
    None

INTERPRETATION

Comments on interpretation:
    Agglutination of sample and latex suspension at the
        following titres:
    Human: Negative: Ab titre < 1:16
     Weak Positive: Ab titre 1:16
    Positive: Ab titre > 1:32
    Swine and cat: Negative: Ab titre < 1:32
    Weak Positive: Ab titre 1:32
    Positive: Ab titre > 1:64
No. of references: 17

NOTES
131269.0

*This assay is suitable for swine and cat serum in addition to
    human serum

Toxoplasma gondii
ANTIBODY DETECTION

Manufacturer: Laboratoire Fumouze
Cat. No./Trade name: Fumouze Toxolatex

SUMMARY

[Latex-Ag]–**Ab**–agglutination

Assay type: Particle agglutination assay
Detection: Visual
Format: Latex particles, Ag coated
Sample type: Serum
Sample pre-treatment:
    None
Sample volume: 25 μl
Number of tests: 120
Controls - standards run in assay:
    Controls: Neg (1), Pos (1)
Incubation:
    Read result within 6 min of mixing sample and reagent
Washes:

CONTENTS

Antibodies, antigens, labelled components:
    T. gondii Ag bound to latex particles (Latex reagent)
Substrate:
Controls - standards supplied:
    Controls: Neg and Pos (serum)
Additional reagents:
    None
Special equipment required:
    None

INTERPRETATION

Comments on interpretation:
    Positive: agglutination of sample with Latex reagent*
No. of references: **

NOTES
131235.0

*If a prozone effect is suspected, repeat test with 1.3 or 1.5
    dilution of serum
**Available upon request

© KLUWER ACADEMIC PUBLISHERS 1995, ISSN 1381-5067

Toxoplasma gondii
ANTIBODY DETECTION (IgA AND/OR IgM)

Manufacturer: bioMerieux
Cat. No./Trade name: 75362/Toxo ISAGA Plus IgA/IgM

SUMMARY

[Well-AHIgA/IgM]–**Ab**–[Toxoplasma]–agglutination

Assay type: Particle agglutination assay
Detection: Visual
Format: Microtitre well, Ab coated
Sample type: Serum
Sample pre-treatment:
  None
Sample volume: 100 µl of 1:100 dilution (adults), 1:20 (neonates) and 1:10 (known Positive sera)*
Number of tests: 48
Controls - standards run in assay:
  Controls: Neg (2), Pos (2)
Incubation:
  2 hr (37°C) + 12-24 hr (37°C)
Washes: 1

CONTENTS

Antibodies, antigens, labelled components:
  Anti-human IgA MAb (m) bound to well
  Anti-human IgM MAb (m) bound to well
  T. gondii Ag (suspension of organisms from mouse ascitic fluid - separate suspensions for IgA and IgM assays)
Substrate:
Controls - standards supplied:
  Controls: Neg (PBS) and Pos (ISAGA Index = 11–12, human serum)
Additional reagents: None
Special equipment required: None

INTERPRETATION

Comments on interpretation:
  Negative: sedimentation button formed in well base (ISAGA Index 0– < 6)
  Positive: agglutination mat formed across cell base (ISAGA Index ⩾ 6), and confirmed by retesting with increased Toxoplasma Ag concentrations
  Borderline: ISAGA Index 6–9; retest fresh sample in 3 weeks
  Specific interpretation for newborns – see manufacturer's notes
No. of references: 10

NOTES

131288.0

This kit can detect IgA and IgM antibodies and can differentiate between them
*Duplicate samples are tested with two concentrations of both Toxoplasma Ag suspensions for both IgA and IgM assays

Toxoplasma gondii
ANTIBODY DETECTION (IgM)

Manufacturer: bioMerieux
Cat. No./Trade name: 75361/Toxo-ISAGA

SUMMARY

[Well-AHIgM]–**Ab**–[Toxoplasma]–agglutination

Assay type: Particle agglutination assay
Detection: Visual
Format: Microtitre well, Ab coated
Sample type: Serum
Sample pre-treatment:
  None
Sample volume: 100 µl of 1:100 dilution (adults), 1:20 (neonates) and 1:10 (known Positive sera)*
Number of tests: 192
Controls - standards run in assay:
  Controls: Neg (2), Pos (2)
Incubation:
  2 hr (37°C) + 12-24 hr (37°C)
Washes: 1

CONTENTS

Antibodies, antigens, labelled components:
  Anti-human IgM MAb (m) bound to well
  T. gondii Ag (suspension of organisms from mouse ascitic fluid)
Substrate:
Controls - standards supplied:
  Controls: Neg (PBS) and Pos (ISAGA Index = 12, human serum)
Additional reagents:
  None
Special equipment required:
  None

INTERPRETATION

Comments on interpretation:
  Negative: sedimentation button in well base (ISAGA Index 0– < 6)
  Positive: agglutination mat across cell base (ISAGA Index ⩾ 6) and confirmed by retesting with increased Toxoplasma Ag concentrations
  Borderline: ISAGA Index 6–9; retest fresh sample in 3 weeks
No. of references: 5

NOTES

131287.0

*Duplicate samples are tested with two concentrations of Toxoplasma Ag suspension

© KLUWER ACADEMIC PUBLISHERS 1995, ISSN 1381-5067

Trichinella spiralis

Trichinella spiralis is a nematode (roundworm) parasite and is the causal organism of trichinosis, a disease of worldwide distribution affecting both man and other animals. Infection is acquired by consumption of inadequately cooked meat, usually pork, containing encysted larvae. Following ingestion, the organisms excyst in the stomach by acid/pepsin digestion. The larvae pass into the small intestine, where they attach to the mucosa at the bases of the villi and mature into adult worms. Fertilized females release several hundred larvae over a two week period prior to excretion in the faeces. Newborn larvae make their way via the intestinal lymphatics and the thoracic duct to the general circulation, by which means they are distributed throughout the body. Encystment occurs within skeletal muscle, most commonly of the diaphragm, chest wall, biceps and gastrocnemius, and the organisms may remain viable and infectious for several years. The cyst wall may become calcified with the passage of time. The majority of infections are sub-clinical but there may be diarrhoea and abdominal pain associated with a heavy worm burden, fever, muscle pain, and periorbital oedema due to larvae in the muscles. In very severe infection a fatal myocarditis or encephalitis may occur. The systemic symptoms are usually maximal after 2–3 weeks and then subside, although in some cases more persistent myalgia and cardiac abnormalities may occur.

Diagnosis

Non-specific findings which may suggest the diagnosis in the appropriate clinical settings are marked eosinophilia and elevated muscle creatine phosphokinase and lactate dehydrogenase. The diagnosis may be confirmed by the observation of larvae in muscle biopsy specimens. Antibodies to the organism are detectable approximately 3 weeks following infection. Classically these are detected by means of the bentonite flocculation test, but more recently immunofluorescence and ELISA-based tests have become more commonly used.

Reference

Grove DI. Tissue nematodes. In: Mandell GL, Bennett JE, Dolin R, eds. Principles and Practice of Infectious Diseases, 4th edn. New York, Edinburgh, London: Churchill Livingstone. 1995:2531–7.

© KLUWER ACADEMIC PUBLISHERS 1995, ISSN 1381-5067

Trichinella spiralis
ANTIBODY DETECTION

Manufacturer: LMD Laboratories Inc
Cat. No./Trade name: TN-1/Trichinella Serology
Microtitre ELISA Kit

SUMMARY

[Well-Ag]–**Ab**–[Protein A-HRP]–[TMB]–A_{450}

Assay type: EIA (non-competitive)
Detection: Colorimetric A_{450}
Format: Microtitre well, Ag coated
Sample type: Serum (do not heat inactivate)
Sample pre-treatment:
　None
Sample volume: 5 µl (+315 µl diluent)
Number of tests: 48, 96
Controls - standards run in assay:
　Controls: Neg (1), weak Pos (1), strong Pos (1)
Incubation:
　10 min (RT) + 5 min (RT) + 5 min (RT)
Washes: 2

CONTENTS

Antibodies, antigens, labelled components:
　Trichinella spiralis (excretory-secretory) Ag bound to well
　Protein A HRP conjugated
Substrate: TMB
Controls - standards supplied:
　Controls: Neg (human serum), weak Pos and strong Pos
　(rabbit serum)
Additional reagents:
　None
Special equipment required:
　None

INTERPRETATION

Comments on interpretation:
　Classification of samples is according to cut-off; no
　further testing
No. of references: 6

NOTES

131643.0

Trichinella spiralis
ANTIBODY DETECTION

Manufacturer: Melotec S.A.
Cat. No./Trade name: Melotest Trichinosis

SUMMARY

[Well-Ag]–**Ab**–[AHIgG&IgM-HRP]–[TMB]–A_{450}

Assay type: EIA (non-competitive)
Detection: Colorimetric A_{450}
Format: Microtitre well, Ag coated
Sample type: Serum, plasma
Sample pre-treatment:
　None
Sample volume: 100 µl of 1:32 dilution x 2
Number of tests: 96
Controls - standards run in assay:
　Controls: Neg (1), low Pos (2), high Pos (1)
Incubation:
　10 min (RT) + 5 min (RT) + 10 min (RT)
Washes: 2

CONTENTS

Antibodies, antigens, labelled components:
　Trichinella spiralis (excretory-secretory, ES) Ag bound to
　well
　Anti-human IgG & IgM Abs HRP conjugated
Substrate: TMB
Controls - standards supplied:
　Controls: Neg, low Pos and high Pos (sera)
Additional reagents:
　None
Special equipment required:
　None

INTERPRETATION

Comments on interpretation:
　Equivocal: ratio of sample OD:cut-off value between 0.9
　and 1.1; retest a fresh sample
　Repeatably equivocal: retest fresh sample in 2–4 weeks
No. of references: 6

NOTES

131136.0

© KLUWER ACADEMIC PUBLISHERS 1995, ISSN 1381-5067

Trypanosoma species

Trypanosomiasis describes a group of conditions caused by flagellated protozoan parasites of the genus *Trypanosoma*. African trypanosomiasis is caused by two subspecies of the *T.brucei* complex, *T.brucei gambiense* (West African), and *T.brucei rhodesiense* (East African). Organisms are transmitted from a range of wild and domestic animals and other human cases by flies of the genus *Glossina* (tsetse), whose geographical distribution determines the epidemiology of the infection. Ingested 'short form' trypomastigotes undergo change to procyclic forms which undergo division and migrate to the salivary glands and become epimastigotes. Ultimately after further cycles of division and multiplication, these yield infective non-dividing metacyclic forms. Following inoculation by fly bite, local multiplication occurs. Multiplying trypomastigotes invade the bloodstream, often weeks after initial infection, and produce fever, general malaise, myalgias, headache, generalised lymphadenopathy, splenomegaly and hepatomegaly. Fever is episodic in nature at this time, with the level of parasitaemia fluctuating in response to variation in parasite surface antigens circumventing the humoral immune response. Ultimately meningoencephalitis supervenes with persistent headache, disturbance of sleep pattern, ataxia, abnormal behaviour and decreased consciousness which progresses to coma and death.

South American trypanosomiasis (Chaga's disease) is caused by *Trypanosoma cruzi*. Rodents and armadillos form the animal reservoir and the organism is usually transmitted from animals to man in the faeces of triatomid bugs when the host scratches the site of the insect bite, although transmission by infected blood transfusion is well described in endemic areas. The insect vector is only found in the Southern Americas, and the disease is associated with poor social and hygienic conditions where humans, reservoir animals and triatomid bugs are all found in close proximity. Insects are infected by ingestion of trypomastigotes whilst feeding on parasitised animals and humans. Flagellated epimastigotes multiply in the midgut and form infective metacyclic trypomastigotes in the hind gut. Following inoculation into the host, the parasite invades a variety of host cells which become filled with multiplying amastigote forms, prior to differentiation into trypomastigotes which are released when the cell ruptures. These may invade locally or spread haematogenously to distant sites. During the acute phase the trypanosomes parasitise muscles, including the myocardium and there is fever and lymphadenopathy; in chronic disease the heart is predominantly affected, with atrial and ventricular dilatation, conduction defects and aneurysm formation. The gastrointestinal tract may also be affected with the development of dilatation, obstruction and megadisease of the oesophagus and colon. Death is usually due to cardiac or gastrointestinal complications.

Diagnosis

Diagnosis of African trypanosomiasis is usually made by demonstration of trypanosomes in peripheral Giemsa-stained blood smears during the febrile stage. Organisms may also be found in aspirates of lymph nodes, bone marrow or skin lesions at the inoculation site, and examination of CSF is important in the early diagnosis of CNS involvement. A variety of serodiagnostic methods have been described, including direct agglutination, immunofluorescence and ELISA methods and are relatively specific, although some tests are difficult to interpret in endemic areas.

In acute Chaga's disease definitive diagnosis is made by the detection of trypanosomes in Giemsa-stained blood films, CSF and tissue aspirates, either directly or following concentration. Xenodiagnosis, where a triatomid bug is fed on the patient's blood and the faeces are examined for the organism, may also be performed and liquid media are available to attempt isolation. Laboratory-acquired infection is well documented. Serological tests play a key role in the diagnosis of chronic disease. Complement fixation, immunofluorescence, haemagglutination and ELISA-based systems have all been described and numerous commercial kits are available. Problems with cross-reactivity often necessitate the use of more than one assay which can cause logistic problems, particularly in the screening of donated blood. DNA-based diagnostic methods show considerable promise in this area.

References

Kirchhoff LV. Agents of african trypanosomiasis. In: Mandell GL, Bennett JE, Dolin R, eds. Principles and Practice of Infectious Diseases, 4th edn. New York, Edinburgh, London: Churchill Livingstone. 1995:2450–5.

Trypanosoma cruzi
ANTIBODY DETECTION (IgG)

Manufacturer: Cellabs Pty Ltd
Cat. No./Trade name: KT3/T CRUZI CELISA™

SUMMARY

[Well-Ag]–**Ab**–[AHIgG-POD]–[TMB]–A_{450}

Assay type: EIA (non-competitive)
Detection: Colorimetric A_{450}
Format: Microtitre well, Ag coated
Sample type: Serum, plasma
Sample pre-treatment:
  None
Sample volume: 100 μl of 1:100 dilution
Number of tests: 192
Controls - standards run in assay:
  Controls: Neg (1), Pos (1)
Incubation:
  1 hr (RT) + 1 hr (RT) + 15 min (RT)
Washes: 2

CONTENTS

Antibodies, antigens, labelled components:
  T. cruzi Ag bound to well
  Anti-human IgG Ab POD conjugated
Substrate: TMB
Controls - standards supplied:
  Controls: Neg and Pos (human serum)
Additional reagents:
  None
Special equipment required:
  None

INTERPRETATION

Comments on interpretation:
  A positive result usually indicates that the donor is or has been infected with *T. cruzi*. It does not necessarily mean that the donor has clinical Chagas Disease
No. of references: None

NOTES

131415.0

For Research use only
Cross-reactivity may occur with *T. rangeli* infections or leishmaniasis

Vibrio species

Vibrio species are comma-shaped Gram-negative bacilli, closely related to the other members of the Enterobacteriaceae. They exhibit rapid motility by means of a single polar flagellum, and are widely distributed in surface waters throughout the world. *V.cholerae* is the causal organism of cholera, a diarrhoeal disease characterised by profuse watery diarrhoea, with rapid depletion of fluid and electrolytes. If there is inadequate rehydration, hypovolaemia with shock, confusion, uraemia, hypokalaemia and acidosis develop, and may be rapidly fatal. The severe fluid depletion observed in cholera is the result of a potent protein enterotoxin comprising 5 B (binding) subunits, which facilitate specific attachment to the GM_1 ganglioside of the enterocyte, and a catalytic A subunit which increases the activity of intracellular adenylate cyclase. The resultant increase in intracellular cAMP stimulates active chloride secretion with concomitant fluid loss. Man is the only known natural host for *V.cholerae*. The disease is most prevalent in southern Asia but small endemic foci exist in other areas. Since the early 19th century a number of pandemics of cholera have been described. The disease is characteristically spread via contamination of the water supply, but faecal–oral transmission within households and contamination of food may also be important. Classical cholera vibrios (01 serotype) agglutinate with Ogawa or Inaba antisera, but similar diarrhoeal disease may be due to non-01 (non-agglutinating) organisms, although these are associated with sporadic rather than epidemic disease. The relationship between the various epidemic and non-epidemic strains, and their role in the emergence of cholera pandemics is poorly understood.

Vibrio parahaemolyticus is a halophilic vibrio which occurs in coastal waters worldwide and is a common cause of diarrhoea wherever seafood which has either been inadequately cooked, or recontaminated with sea water, is consumed. This organism is an important cause of diarrhoeal disease in Japan and has recently been implicated in outbreaks of disease along the Atlantic and Gulf coasts of the USA and on Caribbean cruise ships. The organism produces an enterotoxin and also invades the colonic mucosa, stimulating a brisk inflammatory response. The incubation period is usually less than 24 hours, and the onset is abrupt with explosive watery diarrhoea and cramping abdominal pain. Fever, chills and headache may also be present. A variety of other halophilic non-cholera vibrios have been associated with human disease. *V.vulnificus* and *V.alginolyticus* are also part of the normal marine flora, however they are primarily associated with soft tissue infection and

septicaemia following contamination of superficial wounds with sea water, rather than diarrhoeal disease. *V.fluvialis*, *V.hollisae*, and *V.damsela* are less clearly associated with human disease, the former two with clinical syndromes similar to the non-01 *V.cholerae* and the latter with disease analogous to *V.vulnificus*.

Diagnosis

Definitive diagnosis of *Vibrio* infection may be made by isolation of the causative organism. *Vibrio* species grow well on common laboratory media, including blood, chocolate and MacConkey agars. However differentiation of the organisms, particularly in faecal specimens, may be difficult. Thiosulphate-citrate-bile-sucrose (TCBS) medium is an alkaline, selective and differential medium which is useful in the isolation of pathogenic vibrios. *V.cholerae*, *V.fluvialis* and *V.alginolyticus* yield yellow colonies on this medium, whilst *V.parahaemolyticus*, *V.hollisae* and *V.damsela* produce green colonies. The organisms are characteristically oxidase-positive and can be fully identified by biochemical testing, although some commercial systems may require the use of NaCl-supplemented solutions for the preparation of organism suspensions.

Serological screening with polyvalent antisera for serogroup 01 organisms is important in the rapid, presumptive identification of *V.cholerae* isolates but the possibility of infection with novel strains, such as 0139 Bengal, which may not react with conventional antisera should be considered. Confirmation of identity and biotyping are usually performed at a reference facility.

References

Greenough WB. *Vibrio cholerae* and cholera. In: Mandell GL, Bennett JE, Dolin R, eds. Principles and Practice of Infectious Diseases, 4th edn. New York, Edinburgh, London: Churchill Livingstone. 1995:1934–45.

Carpenter CCJ. Other pathogenic vibrios. In: Mandell GL, Bennett JE, Dolin R, eds. Principles and Practice of Infectious Diseases, 4th edn. New York, Edinburgh, London: Churchill Livingstone. 1995:1945–8.

Kelly MT, Hickman-Brenner FW, Farmer JJ. *Vibrio*. In: Balows A, Hausler WJ, Herrmann KL, Isenberg HD, Shadomy HJ, eds. Manual of Clinical Microbiology, 5th edn. Washington: ASM. 1991:384–95.

See also Multipathogen Assays section under: Gastrointestinal pathogens

Vibrio parahaemolyticus (haemolytic toxin)
ANTIGEN DETECTION

Manufacturer: Denka Seiken Co. Ltd
Cat. No./Trade name: 341001/KAP-RPLA "SEIKEN"

SUMMARY

[Latex-Ab]–**Ag**–agglutination

Assay type: Particle agglutination assay
Detection: Visual
Format: Latex particles, Ab coated (test done in well)
Sample type: Culture fluid
Sample pre-treatment:
  Sample is prepared by culture in Mannit-Peptone*
  solution
Sample volume: 25 μl of supernatant (serial dilutions)
Number of tests: 20
Controls - standards run in assay:
  Controls: Pos (1)
Incubation:
  18–20 hr (RT)
Washes:

CONTENTS

Antibodies, antigens, labelled components:
  Anti-*V. paralyticus* heat stable haemolytic toxin IgG Ab (r)
  bound to latex particle (Latex Reagent)
  Non-immune rabbit IgG Ab bound to latex particle
  (Control Latex)
Substrate:
Controls - standards supplied:
  Controls: Pos
Additional reagents:
  Polypeptone, D-Mannit, NaCl and distilled water for
  preparation of Mannit-Peptone solution*
Special equipment required:
  None

INTERPRETATION

Comments on interpretation:
  Positive: agglutination of sample with Latex Reagent and
  no agglutination with Control Latex
No. of references: 2

NOTES

131816.0

*For instructions for preparation of solution see
manufacturer's notes

© KLUWER ACADEMIC PUBLISHERS 1995, ISSN 1381-5067

Yersinia enterocolitica

Yersinia enterocolitica is a Gram-negative, non-lactose fermenting, urease-positive bacillus which is the causal organism of diarrhoeal illness and an appendicitis-like syndrome in man. The organism is commonly found in the environment and the gastrointestinal tracts of a wide variety of animals, but with the exceptions of porcine isolates, the majority of types are infrequently associated with human infection. Although of worldwide distribution, infection has been more frequently reported from the cooler regions of Canada and Northern Europe, where rates of infection may exceed those of *Shigella*, and are close to those of *Salmonella* or *Campylobacter*. Human infection is thought to result from the ingestion of contaminated food (especially pork), water or milk, with person-to-person spread infrequent although nosocomial infection has been documented. The most common presentation in children under 5 years is enterocolitis with fever, diarrhoea and abdominal pain following an incubation period of between one and 10 days, and may rarely be complicated by more severe inflammation and even perforation. In older children the disease more commonly presents as mesenteric adenitis/terminal ileitis. Some patients may develop a reactive arthritis, persistence of which is often associated with the presence of the HLA-B27 antigen. Erythema nodosum is a relatively common finding and pharyngitis, pneumonia, septicaemia and a variety of other manifestations have all been described, and are more common in adults.

The organism grows on standard laboratory media, including blood and MacConkey agars, although selective CIN agar is preferred by some laboratories. Although *Y. enterocolitica* grows well at 37°C it is a psychotroph and will grow in buffered saline at 4°C. This property is of use for selective enrichment and, as with *Listeria* infections, may be of importance in the epidemiology of the infection. Some workers recommend isolation at 25°C, and the organism demonstrates motility in broth cultures at this temperature which is not observed at higher temperatures. Colonies appear translucent, are smaller than those of other Enterobacteriaceae, and may be further identified by biochemical testing.

Diagnosis

Definitive diagnosis is based on isolation of the organism from faecal and other clinical samples, as outlined above. Serology is useful for epidemiological purposes and certain serotypes, particularly 0:3, 0:8 and 0:9, more commonly result in invasive infection. Virulence appears to be associated with plasmids which encode specific outer membrane proteins.

Yersinia infection is often considered in the differential diagnosis of a variety of clinical conditions. In the absence of a positive culture, serological investigation may be of value. Agglutinating antibodies are detectable within 1 week of illness, and reach a peak in the second week. A variety of techniques including microagglutination, complement fixation, and ELISA-based assays have been described for the detection of antibodies, and a variety of commercial systems are available. Cross-reactions may occur between antigens of *Y. enterocolitica* and those of *Brucella abortus*, rickettsiae, salmonellae and thyroid tissue.

References

Doyle MP. Pathogenic *Escherichia coli, Yersinia enterocolitica* and *Vibrio parahaemolyticus*. Lancet. 1990;336:1111–5.

Farmer JJ, Kelly MT. Enterobacteriaceae. In: Balows A, Hausler WJ, Herrmann KL, Isenberg HD, Shadomy HJ, eds. Manual of Clinical Microbiology, 5th edn. Washington: ASM. 1991:360–83.

Butler T. *Yersini* species. In: Mandell GL, Bennett JE, Dolin R, eds. Principles and Practice of Infectious Diseases, 4th edn. New York, Edinburgh, London: Churchill Livingstone. 1995:2070–8.

Yersinia enterocolitica
ANTIBODY DETECTION (IgA AND/OR IgG)

Manufacturer: Dako A/S
Cat. No./Trade name: K463/DAKO Yersinia
enterocolitica 0:3 ELISA

SUMMARY

[Well-Ag]–**Ab**–[AHIgA/IgG-HRP]–[OPD]–A_{490}

Assay type: EIA (non-competitive)
Detection: Colorimetric A_{490}
Format: Microtitre well, Ag coated
Sample type: Serum
Sample pre-treatment:
    None
Sample volume: IgA: 100 µl of 1:600 dilution x 2;
    IgG: 100 µl of 1:6000 dilution x 2
Number of tests: 192
Controls - standards run in assay:
    Controls: cut-off IgA (2), cut-off IgG (2), Pos IgA (2), Pos
        IgG (2)
Incubation:
    2 hr (RT) + 1 hr (RT) + 15 min (RT)
Washes: 2

CONTENTS

Antibodies, antigens, labelled components:
    Y. enterocolitica (0:3 LPS) Ag bound to well
    Anti-human IgA Ab (r) HRP conjugated
    Anti-human IgG Ab (r) HRP conjugated
Substrate: OPD
Controls - standards supplied:
    Controls: cut-off IgA, cut-off IgG, Pos IgA and Pos IgG
        (human serum)
Additional reagents:
    None
Special equipment required:
    None

INTERPRETATION

Comments on interpretation:
    Equivocal: within cut-off ±10%; retest to confirm
    Repeatably equivocal: retest with a fresh sample at a
        later date
No. of references: 8

NOTES

131039.0

This kit can detect IgA and IgM antibodies and can
    differentiate between them

Multipathogen assays

The following products are supplied with reagents that permit testing for a range of organisms.

Gastrointestinal pathogens: Adenovirus and Rotavirus
ANTIGEN DETECTION

Manufacturer: Orion Diagnostica
Cat. No./Trade name: 67491/Diarlex™ Rota-Adeno

SUMMARY

[Latex spot-Ab]–**Ag**–agglutination

Assay type: Particle agglutination assay
Detection: Visual
Format: Dry latex spot on card, Ab coated
Sample type: Faecal specimens
Sample pre-treatment:
   Filter, using Faecal Specimen Filtration Vial, or centrifuge diluted samples
Sample volume: 50 µl of prepared sample x 3
Number of tests: 20
Controls - standards run in assay:
   Controls: Pos Adenovirus (1), Pos Rotavirus (1)
Incubation:
   Read result within 2 min of mixing sample and reagents
Washes:

CONTENTS

Antibodies, antigens, labelled components:
   Anti-Rotavirus Ab bound to dry latex spot on test card (Rota Latex)
   Anti-Adenovirus Ab bound to dry latex spot on test card (Adeno Latex)
   Non-immune Ig Ab bound to dry latex spot on test card (Control Latex)
Substrate:
Controls - standards supplied:
   Controls: Pos Adenovirus and Pos Rotavirus
Additional reagents:
   None
Special equipment required:
   Faecal Specimen Filtration Vials Kit (Cat. no 68312 – optional)

INTERPRETATION

Comments on interpretation:
   Positive: agglutination of sample in appropriate dry latex test spot and no agglutination in dry latex Control spot identifies the organism present in sample
   Non-specific: agglutination of sample with Control Latex only; retest a fresh sample
No. of references: None

NOTES

131267.0

This kit can detect Rotavirus and Adenovirus and can differentiate between them

Gastrointestinal pathogens: Cryptosporidium species and Giardia species
ANTIGEN DETECTION

Manufacturer: Cellabs Pty Ltd
Cat. No./Trade name: KR2/Crypto/Giardia CEL® IF Test

SUMMARY

[Slide]–**Ag**–[MAbs-FITC]–fluorescence

Assay type: Immunofluorescence assay (direct)
Detection: Fluorescence microscopy
Format: Slide, specimen coated
Sample type: Faecal specimens
Sample pre-treatment:
   Specimens may be concentrated prior to use by filtration or with Faecal Parasite Concentration Kit*. Otherwise dilute samples with 10% formalin or PBS, stand before use, apply to slide, dry, fix
Sample volume: 20 µl of prepared specimen
Number of tests: 50
Controls - standards run in assay:
   Controls: Pos (1)
Incubation:
   30 min (37°C) + 10 min (37°C)
Washes: 1

CONTENTS

Antibodies, antigens, labelled components:
   Anti-Cryptosporidium (oocysts) MAb and anti-Giardia (cysts) MAb FITC conjugated
Substrate:
Controls - standards supplied:
   Controls: Pos (slide of oocysts and cysts)
Additional reagents:
   None
Special equipment required:
   None

INTERPRETATION

Comments on interpretation:
   Positive: Cryptosporidium - green fluorescence of ≥one oocyst
   Positive: Giardia - green fluorescence of ≥one cyst
No. of references: 21

NOTES

131409.0

This kit can detect Cryptosporidium oocysts and Giardia cysts and can differentiate between them in rat, dog, cattle and sheep in addition to human specimens
*Kit purchased from Evergreen Scientific, CA 90058, USA (Cat. no. 240-3074-030)

Gastrointestinal pathogens: Cryptosporidium species and Giardia species
ANTIGEN DETECTION

Manufacturer: Meridian Diagnostics Inc
Cat. No./Trade name: 250050/Merifluor®
Cryptosporidium/Giardia

SUMMARY

[Slide]–**Ag**–[MAbs-FITC]–fluorescence

Assay type: Immunofluorescence assay (direct)
Detection: Fluorescence microscopy
Format: Slide, specimen coated
Sample type: Faecal specimens
Sample pre-treatment:
  Collect specimen in Para-Pak® container* with formalin 10% or SAF. Stand 30 min prior to use, apply to slide, dry, fix
Sample volume: 1 drop of sample in preservative
Number of tests: 30
Controls - standards run in assay:
  Controls: Neg (1), Pos (1)
Incubation:
  30 min (RT)
Washes: 1

CONTENTS

Antibodies, antigens, labelled components:
  Anti-Cryptosporidium (oocysts) MAb and anti-Giardia (cysts) MAb FITC conjugated
Substrate:
Controls - standards supplied:
  Controls: Neg and Pos
Additional reagents:
  None
Special equipment required:
  None

INTERPRETATION

Comments on interpretation:
  Positive: Cryptosporidium – green fluorescence of ⩾one oocyst
  Positive: Giardia – green fluorescence of ⩾one cyst
No. of references: None

NOTES

131391.0

This kit can detect Cryptosporidium oocysts and Giardia cysts and can differentiate between them
*Para-Pak® available from Meridian Technical Services Dept.

Gastrointestinal pathogens: *E. coli* (enterotoxin, LT) and *V. cholerae* (enterotoxin, CT)
ANTIGEN DETECTION

Manufacturer: Denka Seiken Co. Ltd
Cat. No./Trade name: TD920/VET-RPLA

SUMMARY

[Latex-Ab]–**Ag**–agglutination

Assay type: Particle agglutination assay
Detection: Visual
Format: Latex particles, Ab coated (test done in V-well microtitre plate)
Sample type: Broth culture*
Sample pre-treatment:
  Centrifuge or filter broth prior to use
Sample volume: 25 µl of prepared sample (serial dilutions)
Number of tests:
Controls - standards run in assay:
  Controls: Pos (1)
Incubation:
  20–24 hr (RT)
Washes:

CONTENTS

Antibodies, antigens, labelled components:
  Anti-*V. cholerae* enterotoxin (CT) IgG Ab (r) bound to latex particles (Test Latex)
  Non-immune rabbit Ig Ab bound to latex particles (Control Latex)
Substrate:
Controls - standards supplied:
  Controls: Pos (*V. cholerae* enterotoxin)
Additional reagents:
  Peptone water (Cat. no. OXOID CM9) and ingredients for preparation of Mundells Medium*
Special equipment required:
  Microtitre plate (V-well) and lid

INTERPRETATION

Comments on interpretation:
  Positive: agglutination of sample with Test Latex at a higher dilution than with Control Latex
No. of references: 14

NOTES

131772.0

This kit detects *Vibrio cholerae* enterotoxin (CT) and *E. coli* enterotoxin (LT) but does not differentiate between them as they are antigenically similar
*Use Peptone water for *Vibrio cholerae* culture and Mundell's Medium (see manufacturer's instructions for formulation) for *E. coli* culture

Gastrointestinal pathogens: Salmonella spp and Shigella spp
ANTIGEN DETECTION

Manufacturer: Carter-Wallace Inc
Cat. No./Trade name: Bactigen® Salmonella- Shigella

SUMMARY

[Latex-Ab]–**Ag**–agglutination

Assay type: Particle agglutination assay
Detection: Visual
Format: Latex particles, Ab coated (test done on slide)
Sample type: Enrichment broth culture or colonies from solid media culture (isolated from faecal specimens)
Sample pre-treatment:
  Broth: agitate to mix broth and organisms
  Solid culture: mix few colonies with buffer
Sample volume: 1 drop of prepared specimen x 3
Number of tests:
Controls - standards run in assay:
  Controls: Pos A (3), Pos B (3)
Incubation:
  10 min (RT) on rotary shaker
Washes:

CONTENTS

Antibodies, antigens, labelled components:
  Latex Reagents bound to latex particles (3 separate reagents)
  1. Anti-Salmonella (> 80 common serotypes) Ab (g)
  2. Anti-Shigella group B (serotypes 1–6) Ab (r) and group D (forms I & II) Ab (r)
  3. Anti-Shigella group A (serotypes 1–7) Ab (r) and group C (serotypes 1–7) Ab (r)
Substrate:
Controls - standards supplied:
  Controls: Pos A (Salmonella enteritidis) and Pos B (Shigella groups C and D)
Additional reagents:
  Sterile enrichment broth or buffer for use as Neg control
Special equipment required: None

INTERPRETATION

Comments on interpretation:
  Positive: agglutination of sample with the appropriate Latex Reagent and not with the other two Latex Reagents identifies the organism present in sample
  Unsatisfactory: agglutination of sample with two or more Latex Reagents; repeat test
No. of references: 15

NOTES

131378.0

This kit can detect Salmonella, Shigella (groups B & D) and Shigella (groups A & C) and can differentiate between them

Infectious Mononucleosis Syndrome Test: CMV, EBV, heterophile Ag and *T. gondii*
ANTIBODY DETECTION

Manufacturer: GenBio
Cat. No./Trade name: 2325/ImmunoDot® Infectious Mononucleosis Screening Panel

SUMMARY

[Dipstick-Ag]–**Ab**–[AHIgA,IgG&IgM-AP]–[BCIP]–dot

Assay type: EIA (non-competitive)
Detection: Visual
Format: Reaction vessel, Ag coated dipstick
Sample type: Serum, plasma, heparinized whole blood
Sample pre-treatment:
  None
Sample volume: 10 µl serum or 20 µl whole blood (+2 ml diluent)
Number of tests: 25, 50, 100
Controls - standards run in assay:
  Controls: Pos (1)
  Integral procedural: Neg (1), Pos (1)
Incubation:
  5 min (44–48°C) + 5 min (44–48°C) + 15 min (44–48°C) + 5 min (44–48°C)
Washes: 4

CONTENTS

Antibodies, antigens, labelled components:
  Heterophile Ag, EBV (HR1), CMV (AD 169) and *T. gondii* (RH) Ags bound to dipstick as 4 discrete dots
  Anti-human IgA, IgG & IgM Abs (g) AP conjugated
Substrate: BCIP
Controls - standards supplied:
  Integral Procedural Controls: Neg and Pos (bound to dipstick as separate dots)
Additional reagents:
  Positive Control Reagent (Cat. no. 2217)
Special equipment required:
  GenBio Workstation (Cat. no. 4001 or 4990)

INTERPRETATION

Comments on interpretation:
  Positive: reactivity in the appropriate Ag dot identifies the antibodies present in the sample
No. of references: 8

NOTES

131423.0

This kit can detect antibodies to heterophile antigen, EBV, CMV and *T. gondii* and can differentiate between them

Meningitis pathogens: *E. coli, N. meningitis*
ANTIGEN DETECTION

Manufacturer: Becton Dickinson
Cat. No./Trade name: Directigen® Neisseria meningitidis Group B and E. coli K1

SUMMARY

[Latex-Ab]–**Ag**–agglutination

Assay type: Particle agglutination assay
Detection: Visual
Format: Latex particles, Ab coated (test done on slide)
Sample type: CSF, serum, urine, blood culture
Sample pre-treatment:
   CSF, serum, urine, blood culture: dilute, heat (100°C) and centrifuge
Sample volume: 50 µl of treated sample x 2
Number of tests: 90
Controls - standards run in assay:
   Controls: Neg (2), Pos (2)
Incubation:
   Read result 10 min after mixing sample and reagents
Washes:

CONTENTS

Antibodies, antigens, labelled components:
   Anti-N. meningitidis (group B) and *E. coli* (K1) MAbs (m) bound to latex particles (Test Latex)
   Non-immune rabbit Ig Ab bound to latex particles (Control Latex)
Substrate:
Controls - standards supplied:
   Controls: Neg and Pos (polyvalent)
Additional reagents:
   Directigen® Specimen Buffer (Cat. no. 8563-91)
Special equipment required:
   Directigen® Test Slides (Cat. no. 8507-79)

INTERPRETATION

Comments on interpretation:
   Positive: agglutination with the Test Latex and not with the Control Latex
No. of references: 15

NOTES

131207.0

This kit can detect *N. meningitidis* (group B) and *E. coli* (K1) but does not differentiate between them

Meningitis pathogens: *E. coli, H. influenzae, N. meningitidis, S. pneumoniae* and Strep group B
ANTIGEN DETECTION

Manufacturer: Becton Dickinson
Cat. No./Trade name: Directigen® Meningitis Combo Kit

SUMMARY

[Latex-Ab]–**Ag**–agglutination

Assay type: Particle agglutination assay
Detection: Visual
Format: Latex particles, Ab coated (test done on slide)
Sample type: CSF, serum, urine, blood culture
Sample pre-treatment:
   Serum, urine, blood culture: dilute, heat (100°C) and centrifuge
   CSF: heat (100°C) unless *N. meningitidis* (group B) or *E. coli* (K1) are suspected
Sample volume: 50 µl of treated sample x 6
Number of tests: 90
Controls - standards run in assay:
   Controls: Neg (6), Pos (6)
Incubation:
   Read result 10 min after mixing sample and reagents
Washes:

CONTENTS

Antibodies, antigens, labelled components:
   Test Latex Reagents bound to latex particles (5 separate reagents)
   1. Anti-*H. influenzae* (type b) Ab (r)
   2. Anti-*S. pneumoniae* Ab (r)
   3. Anti-*N. meningitidis* (groups C, WI35) Ab (r)
   4. Anti-*N. meningitidis* (groups A,Y) MAb (m)
   5. Anti-*N.meningitidis* (group B) and *E.coli* (K1) MAbs (m)
   Control Latex Reagents bound to latex particles (2 reagents)
   Non-immune rabbit Ig Ab; Non-immune mouse Ig Ab
Substrate:
Controls - standards supplied:
   Controls: Neg and Pos (polyvalent)
Additional reagents:
   Directigen® Specimen Buffer (Cat. no. 8563-91)
Special equipment required:
   Directigen® Combo Glass Slides (Cat. no. 8524-60)

INTERPRETATION

Comments on interpretation:
   Positive: agglutination of sample with appropriate Test Reagent and not appropriate Control Reagent identifies the organism present in the sample
No. of references: 15

NOTES

131202.0

This kit can detect 5 major meningitis pathogens and can differentiate between them

© KLUWER ACADEMIC PUBLISHERS 1995, ISSN 1381-5067

Meningitis pathogens: *H. influenzae, N. meningitidis, S. agalactiae, S. pneumoniae*
ANTIGEN DETECTION

Manufacturer: Boule Diagnostics AB
Cat. No./Trade name: Phadebact® CSF Test

SUMMARY

[Staph-Ab]–**Ag**–coagglutination

Assay type: Particle agglutination assay (coagglutination)
Detection: Visual
Format: Staphylococci; Ab coated (test done on slide)
Sample type: CSF, urine, serum
Sample pre-treatment:
　CSF: heat in waterbath 5 min (80°C) and remove RBC by centrifugation
Sample volume: 1 drop of prepared sample x 4*
Number of tests: 20
Controls - standards run in assay:
　None
Incubation:
　Read result within 1 min of mixing sample and reagent
Washes:

CONTENTS

Antibodies, antigens, labelled components:
　Anti-*S. pneumoniae* Ab (r) bound to Staphylococci
　Anti-*H. influenzae* type b Ab (r) bound to Staphylococci
　Anti-*Strep. agalactiae* Ab (r) bound to Staphylococci
　Anti-*N. meningitidis* groups A,B,C,Y,W135 Ab (r) bound to Staphylococci
Substrate:
Controls - standards supplied:
　None
Additional reagents:
　Phadebact® CSF Pos Controls
Special equipment required:
　None

INTERPRETATION

Comments on interpretation:
　Positive: coagglutination of sample with the appropriate reagent and not with the other reagents identifies the organism present in the sample
No. of references: None

NOTES

131416.0

This kit can detect *H. influenzae* type b, *N. meningitidis* groups A,B,C,Y, and W135, and *Strep. agalactiae* and *S. pneumoniae* and can differentiate between them
*Samples are tested with all 4 reagents simultaneously

TECH Screening Panel: *T. gondii,* EBV, CMV and HSV
ANTIBODY DETECTION

Manufacturer: GenBio
Cat. No./Trade name: 2225/ImmunoDot® TECH Screening Panel

SUMMARY

[Dipstick-Ag]–**Ab**–[AHIg-AP]–[BCIP]–dot

Assay type: EIA (non-competitive)
Detection: Visual
Format: Reaction vessel, Ag coated dipstick
Sample type: Serum, heparinized whole blood
Sample pre-treatment:
　None
Sample volume: 10 µl serum or 20 µl whole blood (+2 ml diluent)
Number of tests: 25, 50, 100
Controls - standards run in assay:
　Controls: Pos (1)
　Integral procedural: Neg (1), Pos (1)
Incubation:
　5 min (44–48°C) + 5 min (44–48°C) + 15 min (44–48°C) + 5 min (44–48°C)
Washes: 4

CONTENTS

Antibodies, antigens, labelled components:
　T. gondii (RH), EBV (HR 1), CMV (AD-169) and HSV (MacIntyre strain) Ags bound to dipstick as 4 discrete dots
　Anti-human Ig Ab (g) AP conjugated
Substrate: BCIP
Controls - standards supplied:
　Integral procedural controls: Neg and Pos (bound to dipstick as separate dots)
Additional reagents:
　Positive Control Reagent (Cat. no. 2215)
Special equipment required:
　GenBio Workstation (Cat. no. 4001 or 4990)

INTERPRETATION

Comments on interpretation:
　Positive: reactivity in the appropriate Ag dot identifies the antibodies present in the sample
No. of references: 17

NOTES

131628.0

This kit can detect antibodies to *T. gondii*, EBV, CMV and HSV and can differentiate between them

TORCH Screening Panel: *T. gondii*, Rubella virus, CMV and HSV
ANTIBODY DETECTION

Manufacturer: GenBio
Cat. No./Trade name: 4125/ImmunoDot® Preconception Screening (TORCH) Panel

SUMMARY

[Dipstick-Ag]–**Ab**–[AHIg-AP]–[BCIP]–dot

Assay type: EIA (non-competitive)
Detection: Visual
Format: Reaction vessel, Ag coated dipstick
Sample type: Serum, heparinized whole blood
Sample pre-treatment:
    None
Sample volume: 10 µl serum or 20 µl whole blood (+2 ml diluent)
Number of tests: 25, 50, 100
Controls - standards run in assay:
    Controls: Pos (1)
    Integral procedural: Neg (1), Pos (1)
Incubation:
    5 min (44–48°C) + 5 min (44–48°C) + 15 min (44–48°C) + 5 min (44–48°C)
Washes: 4

CONTENTS

Antibodies, antigens, labelled components:
    T. gondii (RH), Rubella virus (HPV-77), CMV (AD-169) and HSV (MacIntyre strain) Ags bound to dipstick as 4 discrete dots
    Anti-human Ig Ab (g) AP conjugated
Substrate: BCIP
Controls - standards supplied:
    Integral procedural controls: Neg and Pos (bound to dipstick as separate dots)
Additional reagents:
    Positive Control Reagent (Cat. no. 2215)
Special equipment required:
    GenBio Workstation (Cat. no. 4001 or 4990)

INTERPRETATION

Comments on interpretation:
    Positive: reactivity in the appropriate Ag dot identifies the antibodies present in the sample
No. of references: 13

NOTES

131629.0

This kit can detect antibodies to *T. gondii*, Rubella virus, CMV and HSV and can differentiate between them

© *KLUWER ACADEMIC PUBLISHERS 1995, ISSN 1381-5067*

LIST OF MANUFACTURERS AND DISTRIBUTORS

Abbott Laboratories

Corporate
Abbott Laboratories
Diagnostics Division
Abbott Park
North Chicago, IL 60064
USA
Tel: +1-312-937-6100

By region
AUSTRALIA
Abbott Australasia Pty Ltd
Diagnostics Division
PO Box 394
North Ryde, NSW 2113
Australia
Tel: +61-2-8881111
Fax: +61-2-8873948

AUSTRIA
Abbott GmbH
Diagnostics Division
Diefenbachgasse 35
1150 Vienna
Austria
Tel: +43-1-89124
Fax: +43-1-8941747

BELGIUM
Abbott SA
Diagnostics Division
Rue du Bosquet 2
1348 Ottognies-
Louvain-la-Neuve
Belgium
Tel: +32-10-475311
Fax: +32-10-475334

CANADA
Abbott Laboratories Ltd
Diagnostics Division
7115 Milcreek Drive
Mississauga
Ontario, L5N 3R3
Canada
Tel: +1-416-858-2450
Fax: +1-416-858-2462

DENMARK
Abbott A/S
Diagnostics Division
Bygstubben 16
Troeroed
2950 Vedbaek
Denmark
Tel: +45-42-894452
Fax: +45-42-890132

FRANCE
Laboratoires Abbott
Division Diagnostic
12 rue de la Couture
Silic 203
94518 Rungis Cedex
France
Tel: +33-1-45602500
Fax: +33-1-45600498

GERMANY
Abbott Diagnostic Products
GmbH
Max-Planck-Ring 2
65205 Wiesbaden-
Delkenheim

Germany
Tel: +49-61-2250101
Fax: +49-61-22501244
Telex: 4182 598

GREECE
Abbott Laboratories (Hellas)
SA
Diagnostics Division
194 Syngrou Ave
Athens
Greece
Tel: +30-1-9519019
Fax: +30-1-9592790

HONG KONG
Abbott Laboratories Ltd
Unit No. 3005, 30/F
West Tower, Shun Tak Centre
200 Connaught Road
Hong Kong
Tel: +852-5490019
Fax: +852-8580498

IRELAND
Abbott Laboratories Ireland
Ltd
70 Broomhill Road
Tallaght, Dublin 24
Ireland
Tel: +353-16-517388
Fax: +353-16-517765

ITALY
Abbott SpA
Divisione Diagnostici
Via Mar della Ciria 262
00144 Rome
Italy
Tel: +39-6-529911
Fax: +39-6-52991436

JAPAN
Dainabot Co Ltd
33 Mori Building 6th floor
8-21 Toranomon 3-chome
Minato-ku
Tokyo 105
Japan
Tel: +81-3-34379441
Fax: +81-3-34379367

MEXICO
Abbott Laboratories de
Mexico SA
Diagnostics Division
Apartado Postal No 44-983
Mexico 03100, D.F.
Mexico
Tel: +52-5-7264600
Fax: +52-5-7264644

NETHERLANDS
Abbott NV
Maalderij 21
1185 ZB Amstelveen
Netherlands
Tel: +31-20-5454540
Fax: +31-20-6402231

NEW ZEALAND
Abbott Laboratories (NZ) Ltd
Diagnostics Division
146 Harris Road
PO Box 58-611
Greenmount, Auckland
New Zealand
Tel: +64-9-2749886

Fax: +64-9-2746633

PHILIPPINES
Abbott Laboratories (Phils)
Inc
POB 29
Commercial Center
Makati
Metro Manila
Philippines
Tel: +63-2-6318471
Fax: +63-2-6318488

PORTUGAL
ABBOTT Laboratorios Ltd
Apartado 20/Alfragide
2700 Amadora
Portugal
Tel: +351-1-4712575
Fax: +351-1-4712725

PUERTO RICO
Abbott Laboratories
Diagnostics Division
PO Box 4706
Carolina 00628
Puerto Rico
Tel: +1-809-7623366
Fax: +1-809-7505454

SINGAPORE
Abbott Laboratories Private
Ltd
GPO Box 1016
Singapore 9020
Singapore
Tel: +65-2788343
Fax: +65-2708873

SOUTH AFRICA
Abbott LABS South Africa
(Pty) Ltd
Diagnostics Division
PO Box 1616
Johannesburg 2000
South Africa
Tel: +27-11-4941957
Fax: +27-11-4943041

SPAIN
Abbott Cientifica SA
Costa Brava, 13
28034 Madrid
Spain
Tel: +34-1-3373400
Fax: +34-1-7349664

SWEDEN
Abbott AB
Diagnostics Division
PO Box 1074
164 21 Kista-Stockholm
Sweden
Tel: +46-8-7036700
Fax: +46-8-7527123

SWITZERLAND
Abbott AG
Diagnostics Division
Gewerbestrasee 5
6330 Cham/Zug
Switzerland
Tel: +41-42-444400
Fax: +41-42-415140

UNITED KINGDOM
Abbott Diagnostics
Abbott House

Norden Road
Maidenhead
Berkshire, SL6 4XF
United Kingdom
Tel: +44-01628-784041
Fax: +44-01628-644205

USA
Abbott Laboratories
Diagnostics Division
1921 Hurd
PO Box 152020
Irving, TX 75015-2020
USA
Tel: +1-214-518-6000

Abbott Laboratories
Diagnostics Division
Abbott Park
North Chicago, IL 60064
USA
Tel: +1-312-937-6100

Abbott Laboratories
Diagnostics Division
820 Mission St
South Pasadena, CA 91030
USA
Tel: +1-818-440-0700

Alexon

Corporate
Alexon
1190 Borregas Avenue
Sunnyvale, CA 94089-1302
USA
Tel: +1-408-747-7000
Fax: +1-408-747-7011

Alfa Biotech

Corporate
Alfa Biotech
Viale Sarca 223
20126 Milan
Italy
Tel: +39-2-661381
Fax: +39-2-66138443

Amico Laboratories Inc

Corporate
Amico Laboratories Inc
PO Box 90203
5012 Illinois Avenue
Nashville, TN 37209
USA
Tel: +1-615-385-3114
Fax: +1-615-385-3114

AMRAD Biotech

Corporate
AMRAD Biotech
34 Wadhurst Drive
Boronia, Victoria 3155
Australia
Tel: +61-3-9887-3909
Fax: +61-3-9887-3007

© *KLUWER ACADEMIC PUBLISHERS* 1995, ISSN 1381-5067

By region
AUSTRALIA
AMRAD Biotech
34 Wadhurst Drive
Boronia, Victoria 3155
Australia
Tel: +61-3-9887-3909
Fax: +61-3-9887-3007

Trace Scientific
1860 Princes Highway
Clayton, Victoria 3168
Australia
Tel: +61-3-9543-1255
Fax: +61-3-9543-6719

JAPAN
Sumitomo Metal Industries
Ltd
16-1 Sunayama
Hasaki-Cho Kashima-Gun
Ibaraki 314-02
Japan
Tel: +81-479-465136
Fax: +71-479-465146

MALAYSIA
Kyowa Hakko (Kalupa)
20 Jalan SS 19/5
47500 Subang Jaya
Selangor Darul Ehsan
Malaysia
Tel: +60-3-7340669
Fax: +60-3-7340990

SINGAPORE
Kyowa Hakko Industry (S) Pte
Ltd
300 Orchard Road #06-02
The Promenade
Singapore 0923
Singapore
Tel: +65-7334948
Fax: +65-7330819

TAIWAN
Target Science Inc
2F No. 142-1, Sec 1
Shing Shang S Rd
Taipei
Taiwan
Tel: +886-2-3417866
Fax: +886-2-3416192

Argene-Biosoft

Corporate
Argene-Biosoft
Parc Technologique
Delta Sud BP 24
Varilhes 09120
France
Tel: +33-61696100
Fax: +33-61696101

Bard Diagnostic Sciences Inc

Corporate
Bard Diagnostic Sciences Inc
12277 134th CT NE
Redmond, WA 98052
USA
Tel: +1-206-814-1604
Fax: +1-206-814-1607

By region
USA
Bard Diagnostic Sciences Inc
12277 134th CT NE
Redmond, WA 98052
USA
Tel: +1-206-814-1604
Fax: +1-206-814-1607

Seradyn Inc
PO Box 1210
Indianapolis, IN 46206
USA
Tel: +1-317-266-2000
Fax: +1-317-266-2991

Becton Dickinson Immuno-diagnostics

Corporate
Becton Dickinson
Immunodiagnostics
Mountain View Avenue
Orangeburg, NY 10962-1294
USA
Tel: +1-914-359-2700
Fax: +1-415-966-8914

By region
AUSTRIA
Laevosan GmbH
Estermannstrasse 17
4020 Linz
Austria
Tel: +43-732-7651
Fax: +43-732-782833

BELGIUM
Becton Dickinson Benelux NV
Denderstraat 24
POB 13
9320 Erembodegem-Aalst
Belgium
Tel: +32-53-720211
Fax: +32-53-720200

Beckton Dickinson
Middle East and African
Operations
Denderstrat 24
Postbus 13
9320 Erembodegem
Belgium
Tel: +32-61-3225717
Fax: +32-61-3225830

EASTERN EUROPE
Becton Dickinson GmbH
Postfach 10 16 29
69006 Heidelberg
Germany
Tel: +49-6221-3050
Fax: +49-6221-305254

FINLAND
Becton Dickinson OY
Takomotie 8A
00380 Helsinki
Finland
Tel: +358-90-5062100
Fax: +358-90-5062088

FRANCE
Becton Dickinson France SA
Division Diagnostic

11 rue Aristide Berges
38800 Pont de Claix
France
Tel: +33-76683636
Fax: +33-76683604

Becton Dickinson Europe
Diagnostics Division
5 chemin des Sources
BP 37
38240 Meylan Cedex
France
Tel: +33-76416464
Fax: +33-76418560

GERMANY
Becton Dickinson GmBH
Postfach 10 16 29
69006 Heidelberg
Germany
Tel: +49-6221-3050
Fax: +49-6221-305216

GREECE
Becton Dickinson Hellas SA
36A Ionias Street
17456 Alimos
Athens 138
Greece
Tel: +30-1-9941558
Fax: +30-1-9941672

ITALY
Becton Dickinson Italia SpA
Via Caldera 21
20153 Milan
Italy
Tel: +39-2-482401
Fax: +39-2-48203520

NETHERLANDS
Becton Dickinson BV
Mon Plaisir 89B
Postbus 514
4870 AM Etten-Leur
Netherlands
Tel: +31-1608-37720
Fax: +31-1608-14133

SPAIN
Becton Dickinson SA
Camino de Valdeoliva s/n
San Agustin de Guadalix
28750 Madrid
Spain
Tel: +34-1-8418311
Fax: +34-1-8418113

SWEDEN
Becton Dickinson AB
Box 32045
126 11 Hagersten
Sweden
Tel: +46-8-180030
Fax: +46-8-6450808

SWITZERLAND
Becton Dickinson AG
PO Box
4002 Basel
Switzerland
Tel: +41-61-3225717
Fax: +41-61-3225830

UNITED ARAB EMIRATES
Becton Dickinson Diagnostics
Division
PO Box 52279

Dubai
United Arab Emirates
Tel: +971-4-379525
Fax: +971-4-379551

UNITED KINGDOM
Becton Dickinson UK Ltd
Between Towns Road
Cowley
Oxford, OX4 3LY
United Kingdom
Tel: +44-01865-748844
Fax: +44-01865-787313

USA
Becton Dickinson
Immunodiagnostics
Mountain View Avenue
Orangeburg, NY 10962-1294
USA
Tel: +1-914-359-2700
Fax: +1-415-966-8914

Behringwerke AG

Corporate
Behringwerke AG
Postfach 1140
35001 Marburg
Germany
Tel: +49-6421-390
Fax: +49-6421-66064
Telex: 48232001

By region
ANTILLES
Curacao Pharmacal Company
NV
Ontarioweg 8
Salinja Curacao, NA
PO Box 252
Willemstad, Curacao
Antilles
Fax: +599-9-614819
Telex: 1256

ARGENTINA
Merck Quimica Argentina
SAIC
Behring Diagnosticos
Artilleros 2436
1428 Buenos Aires
Argentina
Tel: +54-1-7886558
Fax: +54-1-7883365

AUSTRALIA
Behring Diagnostics Australia
Pty Ltd
103 Vanessa St. Kingsgrove
2208
Locked Mail Bag 200
Kingsgrove, NSW 2208
Australia
Tel: +61-2-5541600
Fax: +61-2-5024256

AUSTRIA
Behring Institut GmbH
Altmannsdorfer Str 104
1127 Vienna
Austria
Tel: +43-1-80101
Telex: 133701

BAHAMAS
Lowe's Pharmacy Ltd
PO Box N-7504
Nassau
Bahamas
Tel: +1-809-3228006
Telex: 20477

BAHRAIN
Bahrain Pharmacy & General
  Store
PO Box 403
Bahrain
Tel: +973-53491/255506
Fax: +973-275195
Telex: 8979

BANGLADESH
Hoechst Bangladesh Co. Ltd
PO Box 534
133 Kakrail
New Bailey Road
Dhaka 1000
Bangladesh
Tel: +880-2834928
Fax: +880-2-834902
Telex: 632253

BARBADOS, W INDIES
Collins Ltd
PO Box 203
28 Broad Street
Bridgetown
Barbados, W Indies
Tel: +1-809-4249182
Fax: +1-809-4264246
Telex: 2595

BELGIUM
Hoechst Belgium SA
Chaussee de Charleroi 111-
  113
1060 Brussels
Belgium
Tel: +32-2-5364111
Fax: +32-2-5334265
Telex: 21396

BOLIVIA
Corimex Ltda
Calle Goita 135
La Paz
Bolivia
Tel: +591-2-359337
Fax: +591-2-391230
Telex: 3508

BRAZIL
Hoechst do Brasil
Quimica e Farmaceutica SA
Setor Diagnosticos Behring
Av. das Nacoes Unidas
  18.001
8 Andar - Sala 801
CEP 04795 Sao Paulo
Brazil
Tel: +55-11-5257893
Fax: +55-11-5257895
Telex: 1170501

BULGARIA
Hoechst Handelsvertreting
Pionerski Pat No. 2
1618 Sofia
Bulgaria
Tel: +359-2-550119
Fax: +359-2-550116
Telex: 22060

CANADA
Hoechst Roussel Canada Inc
4045 Cote Vertu
Montreal
Quebec H4R ZE8
Canada
Tel: +1-514-333-3605
Fax: +1-514-333-3769

CHILE
Merck Quimica Chilens Ltd
Casilla 4232
Santiago de Chile
Chile
Tel: +56-2-2381160
Fax: +56-2-2383527
Telex: 440197

COLOMBIA
Merck Colombia SA
Apartado Aereo 9896
Bogota, D.E.
Colombia
Tel: +57-1-2907855
Fax: +57-1-2628881
Telex: 41405

COSTA RICA
Hoechst Costa Rica SA
Diagnosticos Behring
Apartado 10158
1000 San Jose
Costa Rica
Tel: +506-334722
Fax: +506-228221
Telex: 2111

CROATIA
Behring Institut GmbH
Altmannsdorfer Str 104
1121 Vienna
Austria
Tel: +43-1-80101
Telex: 133701

CZECH REPUBLIC
Behring Institut GmbH
Altmannsdorfer Str 104
1121 Vienna
Austria
Tel: +43-1-80101
Telex: 133701

DENMARK
Hoechst Danmark A/S
Islevdalveij 110
2610 Rodovre
Denmark
Tel: +45-44888227
Fax: +45-44888352
Telex: 35366

DOMINICAN REPUBLIC
Hoechst Dominicana SA
Apartado Postal 830
Zona Postal 1
Santo Domingo
Dominican Republic
Tel: +1809-5302988
Fax: +1809-5302976
Telex: 3460305

ECUADOR
Merck Ecuador CA
Avda America 1735
Casilla 17-01-2574
Quito
Ecuador

Tel: +593-2-562703
Fax: +593-2-651150
Telex: 2252

EGYPT
Hoechst Orient SAA
PO Box 1486
3 Sharia El Massaneh
  (Zeitoun)
Cairo
Egypt
Tel: +20-2-2575840
Fax: +20-2-694755
Telex: 02331

EL SALVADOR
Quimica Hoechst de El
  Salvador SA
Km 9,5 carre. al Puerto de la
  Libertad
Nuevo San Salvador
El Salvador
Tel: +503-780533
Fax: +503-780909
Telex: 20143

FINLAND
Oy Hoechst Fennica Ab
Postbox 237
00101 Helsinki 10
Finland
Tel: +35-80-8705721
Fax: +35-80-8709812
Telex: 12-4801

FRANCE
Tour Roussel Hoechst
1 Terrasse Bellini
92800 Puteaux
France

Behring Diagnostics
Division de la S.a.p.b.,
  Hoechst-Behring
260, Avenue Napoleon
  Bonaparte
92504 Rueil Malmaison
France
Tel: +33-1-47082661
Fax: +33-1-47082661
Telex: 631183

GERMANY
Behringwerke AG
PO Box 1212
65835 Liederbach
Germany
Tel: +49-69-34020
Fax: +49-69-303834
Telex: 416853

Behringwerke AG
Postfach 1140
35001 Marburg
Germany
Tel: +49-6421-390
Fax: +49-6421-66064
Telex: 48232001

GREECE
Hoechst Hellas AG
Pharma Division
10240 Athens
Greece
Tel: +30-1-809395
Fax: +30-1-8071577
Telex: 0215890

GUATEMALA
Quimica Hoechst de
  Guatemala SA
Diagnosticos Behring
Apartado 155
km 15.5 Carretera Roosevelt
Guatemala City
Guatemala
Tel: +502-2-910011/15
Fax: +502-2-954016
Telex: 5946

GUYANA
Guyana Pharmaceutical Corp
  Ltd
PO Box 160
La Penitence, Demerara
Guyana
Tel: +592-632815
Telex: 2203

HONG KONG
Hoechst China Ltd
23/F Shell Tower
Times Square, 1 Matheson
  St.
Causeway Bay
Hong Kong
Tel: +852-5068333
Fax: +852-5062537

HUNGARY
Hoechst AG
Magyarorszagi kepviselet
Bajcsy-Zs ut. 12
1051 Budapest
Hungary
Tel: +36-1-179011
Fax: +36-1-182822
Telex: 225975

INDIA
E Merck (India) Ltd
Atus House
87 Dr Annie Besant Road,
  Worli
Bombay 400018
India
Tel: +91-22-4922855
Fax: +91-22-4950354

INDONESIA
PT Mensa Binasukses
Pulo Bueran Kav. DD 1A
Jakarta Indust. Est.
  Pulgadung
Jakarta 13939
Indonesia
Tel: +62-21-4601950
Fax: +62-21-4601953

IRELAND
Hoechst Ireland Ltd
Cookstown
Tallaght, Dublin 24
Ireland
Tel: +353-1-511544
Fax: +353-1-596068
Telex: 25158

ISRAEL
Chemipharm Ltd
43 Brodelzky Street
PO Box 17025
Tel Aviv 61170
Israel
Tel: +972-3-6422031
Fax: +972-3-6428762

Telex: 342204

ITALY
Istituto Behring SpA
Sistemi Diagnostici
Sede di Milano
Piazzale Stefano Turr 5
20149 Milan
Italy
Tel: +39-2-31071
Fax: +39-2-31072828
Telex: 320165

JAMAICA, W INDIES
HD Hopwood & Company Ltd
PO Box 165
3 Cari ftd. Ave
Kingston 11
Jamaica, W Indies
Tel: +1-809-92384816
Fax: +1-809-9236351
Telex: 2122

JAPAN
Hoechst Japan Ltd
10-16 Akasaka 8-chome
Pharma Division
Toyko 107
Japan
Tel: +81-3-4795111
Fax: +81-3-4791524
Telex: 242-2342

JORDAN
Salim Sabbagh Sons
Company
PO Box 346
Amman
Jordan
Tel: +962-6-6623158
Fax: +962-6-627376
Telex: 21529

KENYA
Hoechst East Africa Ltd
PO Box 30467
Nairobi
Kenya
Tel: +254-2-557744
Fax: +254-2-545837
Telex: 24031

KOREA
Han-Dok Remedia Ind. C., Ltd
Kangnam PO Box 1560
Seoul 135-615
Korea
Tel: +82-2-5275114
Fax: +82-2-5275002

KUWAIT
Yusuf Ibrahim Alghanim & Co.
W.L.L.
PO Box 435
13005 Safat
Kuwait
Tel: +965-5812148
Fax: +965-5844954
Telex: 23607

LEBANON
Union Pharmacy
PO Box 113-5461
Beyrouth
Lebanon
Telex: 21665/20680

MALAYSIA
Hoechst Malaysia SDN, BHD
12th Floor, Wisma,
Damansara
Jalan Semantan
PO Box 10540
Kuala Lumpur
Malaysia
Tel: +60-3-2549100
Fax: +60-3-2559420
Telex: 30370

MEXICO
Quimica Hoechst de Mexico
SA de CV
Direccion Farma Agro-Vet
Div. Behring Diagnostika
Tecoyotitla 412
Deleg. Alvaro Obregon
01050 Mexico DF
Mexico
Tel: +52-5-6612987
Fax: +52-5-6611612

NETHERLANDS
Hoechst Holland NV
Postbus 12987
1100 AZ Amsterdam
Netherlands
Tel: +31-20-5908514
Fax: +31-20-6910412
Telex: 12051

NEW ZEALAND
Hoechst New Zealand Ltd
CPO Box 67
21-39 Jellicoe Road
Panmure, Auckland
New Zealand
Tel: +64-9-5700700
Fax: +64-9-5709511
Telex: 2338

NIGERIA
Nigerian Hoechst plc
Pharma Division
Plot 144, Oba Akran Avenue
PO Box 261
Ikeja, Lagos
Nigeria
Tel: +234-1-900130/9
Fax: +234-1-964474
Telex: 26381

NORWAY
Norske Hoechst A/S
Okernveien 145
PO Box 177 Okern
0509 Oslo 5
Norway
Tel: +47-2-22727513
Fax: +47-2-22631828
Telex: 71744

OMAN
Muscat Pharmacy
PO Box 438
Muscat
Oman
Tel: +968-722297/702542
Fax: +968-795202
Telex: 8361

PAKISTAN
Progressive Medicals Ltd
415 Muhammadi House
1.1 Chundrigar Road
Karachi

Pakistan
Tel: +92-21-2402601
Fax: +92-21-2414323
Telex: 2603

PARAGUAY
C.E.I.S.A.
Casilla Correo 427
Asuncion
Paraguay
Tel: +595-21-490-25113

PERU
Merck Peruana SA
Avda. Monterrico Chico
Cuadra 2,
Urb. Santa Teresa
Lima 33
Peru
Tel: +51-14-499523
Fax: +51-14-499480

PHILIPPINES
Zuellig Pharma Corporation
PO Box 604
Manila
Philippines
Tel: +63-2-8191561
Fax: +63-2-7610138

POLAND
Hoechst Polska Sp. z.o.o.
ul. Stawki 2, 25 pietro
PO Box 15
00 950 Warsaw
Poland
Tel: +48-2-6353332
Fax: +48-2-6351546
Telex: 812298

PORTUGAL
Hoechst Portuguesa SA
Pharma Division
Apartado 6
2726 Mem Martins Codex
Portugal
Tel: +351-1-9212160
Fax: +351-1-3556762
Telex: 16380

QATAR
Doha Drug Store
PO Box 3331
Doha
Qatar
Tel: +974-426201
Fax: +974-423484
Telex: 4234

ROMANIA
Hoechst AG
Reprezentanta Tehnica
Hotel Bucuresti
Str. Luterana 2-4, Scara D2
Etaj 7, Ap. 19/20
Bukarest
Romania
Tel: +40-1-3125340
Fax: +40-1-3120075
Telex: 11646

RUSSIA
Hoechst
Pharma Moscow
Trjochprudnyj per 11/13
Moscow
Russia
Tel: +7-095-9236001

Fax: +7-095-2002206
Telex: 413138

SAUDI ARABIA
Fuad Al Fadhli
Trading Establishment
PO Box 3506
Riyadh
Saudi Arabia
Tel: +966-1-4550807
Fax: +966-1-4543356
Telex: 400071

SINGAPORE
Hoechst Singapore Pte Ltd
Tanjong Pagar
PO Box 102
Singapore
Tel: +65-2257227
Fax: +65-2257228
Telex: 21662

SLOVAKIA
Behring Institut GmbH
Altmannsdorfer Str 104
1127 Vienna
Austria
Tel: +43-1-80101
Telex: 133701

SLOVENIA
Behring Institut GmbH
Altmannsdorfer Str 104
1127 Vienna
Austria
Tel: +43-1-80101
Telex: 133701

SOUTH AFRICA
Novistan Ltd Behring
Diagnostics
326 Mark Street
Waltloo, Pretoria 0001
South Africa
Tel: +27-12-835372
Fax: +27-12-834121
Telex: 4-25879

SPAIN
Hoechst Iberica SA
Airbau 197-199
08021 Barcelona
Spain
Tel: +34-3-34198438
Fax: +34-3-34198256
Telex: 53160/52496

SRI LANKA
Hoechst (Ceylon) Company
Ltd
PO Box 1127
114 Ward Place
Colombo 8
Sri Lanka
Tel: +94-1-547620
Fax: +94-1-699006
Telex: 21648

SWEDEN
Svenska Hoechst AB
Box 42026
126 12 Stockholm 42
Sweden
Tel: +46-8-7757104
Fax: +46-8-198014
Telex: 10456

SWITZERLAND
Hoechst Pharma AG
Behring Abteilung
Postbox 8048
8048 Zurich
Switzerland
Tel: +41-1-4342627
Fax: +41-1-4318463
Telex: 822189

SYRIA
Tabasch Commerce
PO Box 1323
Aleppo
Syria
Fax: +963-21-235821

TAIWAN
Cheng Hsin Medical Supply
Co. Ltd
3th Fl, 50 Chang Av E. Rd
Section 1
Taipei
Taiwan
Tel: +886-2-5118022
Fax: +886-2-5118026

THAILAND
Hoechst Thai Limited
PO Box 1495
Bangkok
Thailand
Tel: +66-2-23329818
Fax: +66-2-2368155
Telex: 82312

TRINIDAD, W INDIES
Alstons Marketing Company
Ltd
PO Box 1256, Port of Spain
Uriah Butler Highway
Chaguanas
Trinidad, W Indies
Tel: +1809-6712713/20
Fax: +1809-6712857
Telex: 22284

TURKEY
Turk Hoechst Sanayi ve
Ticaret AS
Davutpasa Cad. No. 145
Topkapi 34020
Istanbul
Turkey
Tel: +90-1-5679500
Fax: +90-1-5679532
Telex: 30315/30316

UNITED ARAB EMIRATE
Al Ittihad Pharmacy Co.
PO Box 602
Abu Dhabi
United Arab Emirates
Tel: +971-2-211600
Fax: +971-2-211535
Telex: 22386

UNITED ARAB EMIRATES
Al Ittihad Drug Store
PO Box 5374
Dubai
United Arab Emirates
Tel: +971-6-594585
Fax: +971-6-596490
Telex: 68292

Hoechst Consulting
Arabian Gulf Office

PO Box 2326
Dubai
United Arab Emirates
Tel: +971-4-221153/54
Telex: 47155

UNITED KINGDOM
Behring, A Division of
Hoechst UK Ltd
Hoechst House
50 Salisbury Road
PO Box 18
Hounslow
Middlesex, TW4 6JH
United Kingdom
Tel: +44-0181-5707712
Fax: +44-0181-5694834
Telex: 23284

URUGUAY
EMOL SA
Colon 1563, Of. 001
C.C. 6632
Montevideo
Uruguay
Tel: +598-2-962444
Fax: +598-2-963611
Telex: 23933

USA
Behring Diagnostics Inc
141 University Avenue
Westwood, MA 02090
USA
Tel: +1-617-320-3000
Fax: +1-617-320-3193

VENEZUELA
Laboratorios Geminis SRL
Apartado Postal 61290
Caracas 1060-A
Venezuela
Tel: +58-2-2621897
Fax: +58-2-335011
Telex: 27839

YUGOSLAVIA
Jugohemija
Farmacija
General Zdanova 31
11000 Belgrade
Yugoslavia
Tel: +38-11-341140
Fax: +38-11-331674
Telex: 11390

ZIMBABWE
Hoechst Zimbabwe (Pvt) Ltd
PO Box 3229
Harare
Zimbabwe
Tel: +263-4-700063
Fax: +263-4-704987
Telex: 26076

Biokit

Corporate
Biokit
Llissa d'Amunt
08186 Barcelona
Spain
Tel: +34-3-8414250

BAG - Biologische Analysensystem GmbH

Corporate
BAG - Biologische
Analysensystem GmbH
Amstgerichtstrasse 1-5
35423 Lich
Germany
Tel: +49-6404-925-0
Fax: +49-6404-62554

Biomerica Inc

Corporate
Biomerica Inc
1533 Monrovia Avenue
Newport Beach, CA 92663
USA
Tel: +1-714-645-2111
Fax: +1-714-722-6674

By region
USA
Allergy Immuno Technologies
Inc
1527 Monrovia Avenue
Newport Beach, CA 92663
USA
Tel: +1-714-645-3703

Biomerica Inc
1533 Monrovia Avenue
Newport Beach, CA 92663
USA
Tel: +1-714-645-2111
Fax: +1-714-722-6674

bioMerieux sa

Corporate
bioMerieux sa
Marcy-L'Etoile 69280
France
Tel: +33-78872000
Fax: +33-78872090
Telex: 330 967

By region
AUSTRALIA
bioMerieux Vitek-Australia Pty
Ltd
Unit 1/6 Gladstone Road
Castle Hill, NSW 2154
Australia
Tel: +61-2-8994600
Fax: +61-2-8991232

AUSTRIA
bioMerieux Austria GmbH
Lamezanstrasse 17
1232 Vienna
Austria
Tel: +43-1-6155519
Fax: +43-1-6155519-33

BELGIUM
bioMerieux Benelux sa
Victor-Hugostraat 215
1040 Brussels
Belgium
Tel: +32-2-7365949

Fax: +32-2-7335597
Telex: 61909

BRAZIL
Biolab Merieux S/A
Estrada Mapua
491 Jacarepagua
22713 Rio de Janerio
Brazil
Tel: +55-21-4455454
Fax: +55-21-3426099
Telex: 2131211

CANADA
bioMerieux Canada Inc
8114-B Route
Transcanadienne
Saint-Laurent Quebec, H4S
1M5
Canada
Tel: +1-514-3367321
Fax: +1-514-3366450

CHILE
bioMerieux Chile SA
Av. Ricardo Lyon 1899
Providencia/Casilla 2291
Santiago de Chile
Chile
Tel: +56-2-2514047
Fax: +56-2-2091604

FRANCE
bioMerieux sa
Marcy-L'Etoile 69280
France
Tel: +33-78872000
Fax: +33-78872090
Telex: 330 967

GERMANY
bioMerieux Deutschland
GmbH
Weberstrasse 8
72622 Nurtingen
Germany
Tel: +49-702-230070
Fax: +49-702-236110
Telex: 7267414

HONG KONG
bioMerieux Vitek - Hong Kong
Ltd
8/F Parkview Commercial
Building
9-11 Shelter St., Causeway
Bay
Hong Kong
Tel: +852-8906067
Fax: +852-8955174

ITALY
bioMerieux Italia srl
Via G Moscati 9
00168 Rome
Italy
Tel: +39-6-3014454
Fax: +39-6-3050079
Telex: 625458

JAPAN
bioMerieux Vitek - Japan Ltd
Yanagisawa bldg
14-12 Hirakawacho 2-chome
Chiyoda-ku
Tokyo 102
Japan
Tel: +81-3-5210-3221

Fax: +81-3-5210-3150

KOREA
bioMerieux Korea
Suit 1012 Samhung Bldg.,
705-9
Yuksam-Dong, Kangnam-ku
Seoul
Korea
Tel: +822-5628260
Fax: +822-5658262

MEXICO
bioMerieux Mexico SA de CV
Av. Cuauhtemoc n.1338 of
404A
Delegacion Benito Juarez
C.P. 03310 Mexico, DF
Mexico
Tel: +52-5-6040243
Fax: +52-5-6040243

NETHERLANDS
bioMerieux Benelux BV
Bruistensingel 620
5232 AJ's Hertogenbosch
Pays-Bas
Netherlands
Tel: +31-73-441818
Fax: +31-73-442211

POLAND
bioMerieux Polska Sp z.o.o
ul. Zelazna 68
00866 Warsaw
Poland
Tel: +48-22-204081
Fax: +48-22-242434

PORTUGAL
bioMerieux Portuguesa Lda
Calcada de Santa Catarina
9-C r/c Dto
Cruz Quebrada - 1495 Lisbon
Portugal
Tel: +351-1-4150278
Fax: +351-1-4150118

SPAIN
bioMerieux Espana sa
Manuel Tovar 36
Madrid 28034
Spain
Tel: +34-1-3581142
Fax: +34-1-3580629
Telex: 46620

SWITZERLAND
bioMerieux Suise sa
51 avenue Blanc
1202 Geneva
Switzerland
Tel: +41-22-7325760
Fax: +41-22-7388842

UNITED KINGDOM
bioMerieux UK Ltd
Grafton Way
Basingstoke
Hampshire, RG22 6HY
United Kingdom
Tel: +44-01256-461881
Fax: +44-01256-816863

USA
bioMerieux Vitek Inc
595 Anglum Drive
Hazelwood

Missouri 63042-2395
USA
Tel: +1-314-731-8500
Fax: +1-314-731-8800

Bion Enterprises Ltd

Corporate
Bion Enterprises Ltd
674 Busse Highway
Park Ridge, IL 60068
USA
Tel: +1-708-692-9333
Fax: +1-708-692-4747

Bio-Rad Laboratories Inc

Corporate
Bio-Rad Laboratories Inc
1000 Alfred Nobel Drive
Hercules, CA 94547
USA
Tel: +1-415-724-7000
Fax: +1-415-724-3167

By region
ARGENTINA
Alfatron SRL
San Martin 683
1004 Buenos Aires
Argentina
Tel: +54-1-3131730
Fax: +54-1-7869391
Telex: 24042

AUSTRALIA
Bio-Rad Laboratories Pty Ltd
Unit 11, 112-118 Talavera Rd
PO Box 371
North Ryde, NSW 2113
Australia
Tel: +61-2-8055000
Fax: +61-2-8051920

AUSTRIA
Bio-Rad Laboratories GmbH
Auhofstrasse 78D
1130 Vienna
Austria
Tel: +43-222-8778901
Fax: +43-222-87

BELGIUM
Bio-Rad Laboratories SA - NV
Begonia Street
9810 Nazareth Eke
Belgium
Tel: +32-91-855511
Fax: +32-91-856554

BRAZIL
Erviegas Instrumental
Cirurgico Ltda
Rua Do Seminario
199-13 Andar Cj. 131
CEP 01034 Sao Paulo
Brazil
Tel: +55-11-2274822
Fax: +55-11-5312566
Telex: 30531

BioAgency Ltda

Rua Estela, 121/51
04011 Sao Paulo
Brazil
Tel: +55-11-2852838
Fax: +55-11-2757596

CANADA
Bio-Rad Laboratories Ltd
5149 Bradco Blvd
Mississauga
Ontario, L4W 2A6
Canada
Tel: +1-416-624-0713
Fax: +1-416-624-3019

CZECHOSLOVAKIA
Bio-Rad Technical Office
Bozejovika 145
14200 Praha
Czechoslovakia
Tel: +42-2-495377
Fax: +42-2-4729792

DENMARK
Bie & Berntsen
Sandbaekvej 7
2610 Roedovre
Denmark
Tel: +45-44948822
Fax: +45-44942709
Telex: 35288

FINLAND
Instrumentarium Corp.
Instrulab
Vitikka 1/PB 63
02631 Espoo
Finland
Tel: +358-0-5281
Fax: +358-0-524986
Telex: 121083

FRANCE
Bio-Rad SA
94/96 rue Victor Hugo
BP 220
94203 Ivry Sur Seine Cedex
Paris
France
Tel: +33-1-49606834
Fax: +33-1-46712467

GERMANY
Bio-Rad Laboratories GmbH
Heidemannstrasse 164
80939 Munchen
Germany
Tel: +49-89-318840
Fax: +49-89-31884100
Telex: 529514

HONG KONG
Bio-Rad Pacific
Unit 1111, 11/F, New
Kowloon Plaza
38 Tai Kok Tsui Road
Tak Kok Tsui
Kowloon
Hong Kong
Tel: +852-7893300
Fax: +852-7891257
Telex: 42814

Line Analytics Ltd
1407B Sea View Estate
2-8 Watson Road
North Point
Hong Kong

Tel: +852-5785839
Fax: +852-8072674
Telex: 61677

Wearmax Ltd
Room 18, 2/F, New Port
Centre
116-118 Ma Tau Kok Road
Kowloon
Hong Kong
Tel: +852-3348791
Fax: +852-7650529

HUNGARY
Bio-Rad Technical Office
Hegyalja ut. 170
1113 Budapest
Hungary
Tel: +36-1-1858446
Fax: +36-1-1858446

INDIA
Chem-Tek India
C-88 Kirtinager
New Delhi 110015
India
Tel: +91-11-5413401
Fax: +91-11-3327526

Everest Enterprise
57 Sarang Street
4th Floor - 36
Bombay 400003
India
Tel: +91-22-345022

Special Instruments
Consortium (SPINCO)
Block G-8, Second Floor
81 Luz Church Road
Alwarpet Circle
Madras 600004
India
Tel: +91-44-76714

Versha Scientific Instruments
148 Gulmohar Enclave
New Delhi 110049
India
Tel: +91-11-6894702
Telex: 73273

INDONESIA
PT Diastika Biotekindo
Jalan Garuda 71-E
Jakarta 10610
Indonesia
Tel: +62-21-414404
Fax: +62-21-4204964

IRELAND
Alpha Analytical
Poulreva House
Grove Road
Dublin 6
Ireland
Tel: +353-1-964219
Fax: +353-1-963751

ISRAEL
Eisenberg Bros. Ltd
Asia House 4
Weizman Street
PO Box 33638
Tel Aviv 61336
Israel
Tel: +972-3-430875
Fax: +972-3-262627

© *KLUWER ACADEMIC PUBLISHERS 1995, ISSN 1381-5067*

Telex: 33511

ITALY
Bio-Rad Laboratories SRL
Via Cellini 18A
20090 Segrate - Milan
Italy
Tel: +39-2-216091
Fax: +39-2-21609399

Bio-Rad Laboratories
(Commercial Office)
Via Nuovo Luce 50A - 1 TRAV
95126 Catania
Italy
Tel: +39-95-7123776
Fax: +39-95-7123849

Bio-Rad Laboratories
(Commercial Office)
Via G. Pallavicini 4
00144 Rome
Italy
Tel: +39-6-5295892
Fax: +39-6-5294240

JAPAN
Nippon Bio-Rad Laboratories
KK
Sumitomo Seimei Kachidoki
Bldg
5-3-6 kachidoki
Chuo Ku
Tokyo 104
Japan
Tel: +81-3-35347240
Fax: +81-3-35348037

Nippon Bio-Rad Laboratories
Tsukuba Shiou-Kaikan
1-15-1 Azuma
Tsukuba-City
Ibaraki 305
Japan
Tel: +81-298-520835
Fax: +81-298-520829

Nippon Bio-Rad Laboratories
1487-2 Tsuruma
Machida-City
Tokyo 194
Japan
Tel: +81-427-995685
Fax: +81-427-995684

Nippon Bio-Rad Laboratories
Osaka Shin-Kitano Dai Ichi
Seimei Bldg
1-14-11 Shin Kitano
Yodogawa-ku
Osaka 532
Japan
Tel: +81-6-3086568
Fax: +81-6-3083064

Nippon Bio-Rad Laboratories
KK
Midori Bldg
3-121-1 Issha
Meitoh-ku, Nagoya City
Aichi 465
Japan
Tel: +81-52-7022358
Fax: +81-52-7022812

Nippon Bio-Rad Laboratories
KK
Morimen Bldg

2-17-5 Hakataeki-Higashi
Hakata-ku, Fukuoka City
Fukuoka 812
Japan
Tel: +81-92-4754856
Fax: +81-92-4754858

KOREA
Sunil Developed Engineering
Co. Ltd
3/F, Korea Bible Society Bldg
1365-16, Seocho-Dong
Seocho-ku
Seoul
Korea
Tel: +82-2-5535721
Fax: +82-2-5568439
Telex: 26603

MALAYSIA
Microstate Separations Sdn
Bhd
41B Lorong Rahim Kajai 13
Taman Tun Dr Ismail
6000 Kuala Lumpur
Malaysia
Tel: +60-7-185530
Fax: +60-7-175953
Telex: 37804

MEXICO
Instrumental Teknion SA de
CV
(Inteksa)
Amores 320
Col. Del Valle
Mexico DF 03100
Mexico
Tel: +52-5-433714
Fax: +52-5-239195

NETHERLANDS
Bio-Rad Laboratories BV
Fokkerstraat 10
3905 KV Veenendaal
Netherlands
Tel: +31-83-8540666
Fax: +31-83-8542216

NEW ZEALAND
Bio-Rad Laboratories Pty Ltd
Unit 15 Poland Court
21 Poland Road
PO Box 100-051
North Shore Mail Ctr
Glenfield, Auckland
New Zealand
Tel: +64-9-4433099
Fax: +64-9-4433097

NORTHERN IRELAND
Alpha Analytical
The Technology Centre
2/3 Curran Point
Larne, Co. Antrim
BT40 1AU
Northern Ireland
Tel: +44-574-60116
Fax: +44-574-60548

NORWAY
Bio-Test A/S
Idrettsveien 2
Postboks 66
1580 Rygge
Norway
Tel: +47-9-261777
Fax: +47-9-261760

Telex: 72591

PAKISTAN
Chemial House
2 Albert Street
GP Box 1138
Lahore 4
Pakistan
Tel: +92-42-212123
Fax: +92-42-213298

PHILIPPINES
Philab Industries Inc.
1158 Chino Roces Avenue
1200 Makati
Metro Manila
Philippines
Tel: +63-2-876658
Fax: +63-2-8163946

REPUBLIC OF CHINA
Bio-Rad Pacific (Beijing
Office)
Yanshan Hotel Office Tower
#1317
138A Haidian Road
Beijing
Republic of China
Tel: +86-1-2564308
Fax: +86-1-2568640

Wearmax Ltd
Guangzhou Office
Rm 702-704 Da Xi Nan Hotel
111 Liuhua Road
Guangzhou
Republic of China
Tel: +86-20-680712
Fax: +86-20-680896

Wearmax Ltd
Shanghai Office
208 Hubei Road
Shanghai
Republic of China
Tel: +86-21-3207141
Fax: +86-21-3207141

Wearmax Ltd (Beijing Office)
Room 1719-1721 Ramada Inn
Asia Hotel
8 Gong Ti Bei Lu
Xin Chong Xi Jie
Beijing
Republic of China
Tel: +86-1-5007788
Fax: +86-1-5008091

RUSSIA
Bio-Rad Technical Office
BioChemMack
Leninskie Gory, GSP 3
Moscow 119899
Russia
Tel: +7-095-939-2421
Fax: +7-095-939-0997
Telex: 411483

SINGAPORE
Diagnostic Biotechnology
(PTE) Ltd
65 Science Park Drive
Singapore Science Park
Singapore 0511
Singapore
Tel: +65-7782855
Fax: +65-7777741

Microstate Separations
187 Goldhill Centre
51 Thomson Road
Singapore 1130
Singapore
Tel: +65-2519561
Fax: +65-2554955

SPAIN
Bio-Rad Laboratories SA
Avda. Valdelaparra 3
Pol Ind Alcobendas
28100 Alcobendas
Madrid
Spain
Tel: +34-1-6617085
Fax: +34-1-6619698

SWEDEN
Kemila Preparat AB
Tingsvagen 19/Box 609
191 26 Sollentuna
Sweden
Tel: +46-8-359040
Fax: +46-8-350545
Telex: 15396

SWITZERLAND
Bio-Rad Laboratories AG
Kanalstrasse 17
8152 Glattbrugg
Switzerland
Tel: +41-1-8101677
Fax: +41-1-8101933

TAIWAN
Long Chain Int'l Corp
3.F, No. 64 T'A Cheng Street
Taipei
Taiwan
Tel: +886-2-5522605
Fax: +886-2-5525461

Pantech Instruments Co. Ltd
4/Fl., No. 3 Lane 338
Sec. 3 Nanking East Road
Taipei
Taiwan
Tel: +886-2-7319211
Fax: +886-2-7319590
Telex: 13819

THAILAND
Diagnostic Biotechnology Co.,
Ltd
526 Ratchada Complex, 5th
Floor
Ratchadapisek Road
Huanykwang
Bangkok 10310
Thailand
Tel: +66-2-5136218
Fax: +66-2-5135818

UNITED KINGDOM
Bio-Rad Laboratories Ltd
Bio-Rad House
Maylands Avenue
Hemel Hempstead
Herts, HP2 7TD
United Kingdom
Tel: +44-01442-232552
Fax: +44-01442-59118
Telex: 837770

USA
Bio-Rad Laboratories Inc
1000 Alfred Nobel Drive

Hercules, CA 94547
USA
Tel: +1-415-724-7000
Fax: +1-415-724-3167

Biotecx Laboratories Inc

Corporate
Biotecx Laboratories Inc
6023 South Loop East
Houston, TX 77033
USA
Tel: +1-713-643-0606
Fax: +1-713-643-3143
Telex: 3791311

Biotest AG

Corporate
Biotest AG
Landsteinerstrasse 5
63303 Dreieich
Germany
Tel: +49-6103-8010
Fax: +49-6103-801130
Telex: 4 185 429

By region
ARGENTINA/PARAGUAY
Boehringer Mannheim
   Argentina SACI e I
Viamonte 2213-15
1056 Buenos Aires
Argentina
Tel: +54-1-9510023
Fax: +54-1-112542

AUSTRALIA
Bio-Mediq DPC Pty (for
   serological products)
PO Box 106
Doncaster, Victoria 3108
Australia
Tel: +61-3-8401800
Fax: +61-3-8402767

Gelman Sciences Pty (for
   microbiological products)
27 Sirius Road
Lane Cove 2066
PO Box 456
Sydney
Australia
Tel: +61-2-4282333
Fax: +61-2-4285610

AUSTRIA
Biotest Pharmazeutika GmbH
Einsiedlergasse 58
1053 Vienna
Austria
Tel: +43-1-5451561
Fax: +43-1-545156137
Telex: 134 644

BAHRAIN
Jaffar Pharmacy
PO Box 122
Manama
Bahrain
Tel: +973-731316
Fax: +973-732533

BANGLADESH
Business Point Ltd
Al-haj Mansion, 3rd floor
82, Motijheel Commercial
   Area
Dhaka 2
Bangladesh
Tel: +880-2-246669
Fax: +880-2-863773
Telex: 642442

BENELUX
NV Biotest Seralc SA
Weiveldlaan 41, bus 31
1930 Zaventem
Belgium
Tel: +32-2-7257560
Fax: +32-2-7257598

BOLIVIA
Abendroth International
   Drogueria Internacional Srl
Calle Yanacocha 344/360
Apartado 3275
La Paz
Bolivia
Tel: +591-2-4571
Telex: 2356

BRAZIL
Marcos Pedrilson Produtos
   Hospitalares Ltda
Estrada Velha da Pavuna, 4-10
Del Castilho
21051-070 Rio de Janeiro, RY
Brazil
Tel: +55-21-5909222
Fax: +55-21-2902596

CAMEROON
Africapharm
PO Box 4741
Yaounde
Cameroon
Tel: +237-220670
Fax: +237-427703
Telex: 8512

CANADA
Gelman Sciences Inc
2535 De Miniac Street
Montreal, Quebec H4S 1E5
Canada
Tel: +1-514-337-2744
Fax: +1-514-337-7114

CHILE
Laboratorio LINSAN SA
Av. Pedro de Valdivia 3078
Santiago de Chile
Chile
Tel: +56-2-2239462
Fax: +56-2-2238329

COLUMBIA
Quimica Internacional Ltda
Transversal 15 No. 119-46
Apartado Aereo 48400
Bogota
Columbia
Tel: +57-1-2137708
Fax: +57-1-2136836

CYPRUS
Cyprus Pharmaceutical Org.
   Ltd
Papaellina House

35 King Paul 1 Street
PO Box 1005
Nicosia 136
Cyprus
Tel: +357-2-443140
Fax: +357-2-365136
Telex: 2 417

CZECHOSLOVAKIA
Merck Spol.
Slavetinska 26
19000 Prague 9 - Klanovice
Czechoslovakia
Tel: +42-2-7882063
Fax: +42-2-7881203

DENMARK
Meda S/A
Dynamovej 11
2730 Herlev
Denmark
Tel: +45-42844211
Fax: +45-845199
Telex: 35 289

Struers A/S (microbiological
   products only)
Valhojs Alle 176
2610 Redovre/Copenhagen
Denmark
Tel: +45-708090
Fax: +45-721320
Telex: 19 625

ECUADOR
AABCOM Techica Cia Ltda
Reina Victoria y Colon
Edif. Banco de Guayaquil
Piso 15, Of. 1501-B
Quito
Ecuador
Tel: +593-2-565772
Fax: +593-2-239728

EGYPT
Gamma Trade Co.
14, El Fath St., Shebab St.
El Mohandsin
PO Box 61
Cairo
Egypt
Tel: +20-2-3480997
Fax: +20-2-3492687
Telex: 20983

ETHIOPIA
Pharma Share Company
PO Box 1122
Addis Ababa
Ethiopia
Tel: +251-1-550207
Fax: +251-1-552799
Telex: 21312

FINLAND
Oy Plastic Trade AB
Erottajankatu 4c
PL 414
00120 Helsinki
Finland
Tel: +358-0-60951
Fax: +358-0-6095410

FRANCE
Biotest S.a.r.l.
Zone Industrielle Centre
80 Rue Helene Boucher
78530 Buc

France
Tel: +33-1-39202080
Fax: +33-1-39202081
Telex: 696229

GERMANY
Biotest AG
Landsteinerstrasse 5
63303 Dreieich
Germany
Tel: +49-6103-8010
Fax: +49-6103-801130
Telex: 4 185 429

GHANA
KY Merchants Ltd
Pasico Elders Bldg.
Guggisberg Road
Korle Lagoon/Accra
Ghana
Tel: +233-21-668881
Telex: 2647

GREECE
Costas A Papaellinas
   (serological products only)
Hellas SA
PO Box 182
26th KM Paenias-
   Markopoulou
19400 Koropi-Attiki
Greece
Tel: +30-6624888
Fax: +30-6626210

Biomedical Hellas
   (microbiological products
   only)
31 Patriarchon Jeremion
   Street
11475 Athens
Greece
Tel: +30-1-6463402
Fax: +30-1-6462492
Telex: 225264

GUATEMALA
Merck Centroamericana SA
Zona 11, Edifcio Merck
Apartado Postal 1651
Guatemala CA
Guatemala
Tel: +502-2-922111
Fax: +502-2-942954

HONDURAS
Representaciones Caceres
Casa No. 1201 Col. Alameda
Contiguo INA
Tegucigalpa DC
Honduras
Tel: +504-321749
Fax: +504-321749

HONG KONG
Jebsen & Co. Ltd, Chemical
   Division
GPO Box 97
28/F, Caroline Centre
2-38 Yun Ping Road
Causeway Bay
Hong Kong
Tel: +922-622222

HUNGARY
Biotest Ungarn
Dobsinai Utca 6/B
1124 Budapest

Hungary
Tel: +36-1-1561697
Fax: +36-1-1561697

ICELAND
Stefan Thorarensen Ltd
Sidumili 32
108 Reykjavic
Iceland
Tel: +354-1-68044
Fax: +354-1-680016
Telex: 2245

INDIA
Lupin Laboratories Ltd
159 CST Road Kalina
Santacruz (East)
Bombay 400 098
India
Tel: +91-22-6124050
Fax: +91-22-6114008

INDONESIA
PT Wigindo Ausadha Graha
Tifa Bldg, 2nd Floor
Jakarta
Indonesia
Tel: +62-21-5780494
Fax: +62-21-5200021

IRAN
Ramisak
Towhid Square Golbar Ave.,
44
PO Box 14 455/141
Tehran
Iran
Tel: +98-21-923019
Fax: +98-21-6424793

IRAQ
Ismailiya Trading Agencies &
Exporting Co.
Hay Al-Mustansiriya - Sec 502
- Lane 14
House 102
PO Box 38092 Palastine Post
Baghdad
Iraq
Tel: +964-1-7615058
Fax: +964-1-4162550

IRELAND
Novapath Supplies Ltd
159 Lr. Rathmines Road
Dublin 6
Ireland
Tel: +353-1-974634
Fax: +353-1-962208

ITALY
Biotest S.r.l.
Via Leonardo da Vinci, 43
20090 Trezzano sul Naviglio
Italy
Tel: +39-2-48401818
Fax: +39-2-48402068
Telex: 315 453

JAPAN
Biotest Japan Liason Office
Cuatro Mita Building
Mita 4-1-4 Minato-ku
Tokyo 108
Japan
Tel: +81-3-52321654
Fax: +81-3-52321655

Dainippon Pharmaceutical
Co. (for serological
products)
6-8 Doshomachi 2-chome
Chuo-ku
Osaka 541
Japan
Tel: +81-6-2035307
Fax: +81-6-2036581

Gunze Sangyo Inc (for
microbiological products)
SID Dept
Aoba Dalichi Bldg 8F
Kudan Minami 2-3-1
Chiyoda-ku
Tokyo 102
Japan
Tel: +81-3-52111807
Fax: +81-3-52111903

JORDAN
Nabih Nabulsi Drugstores
PO Box 1066
Amman
Jordan
Tel: +962-6-678332
Fax: +962-6-678331

KENYA
Sahaj Laboratory Supplies
PO Box 39130
Nairobi
Kenya
Tel: +254-2-744538
Fax: +254-2-750112
Telex: 25112

KOREA
Han Sei International Ltd (for
serological products)
Suite 501, Sammi B/D 386-3
Mangwon-Dong, Mapo-ku
Seoul 121 230
Korea
Tel: +82-2-3234921
Fax: +82-2-3234923

Jin Yang Corporation (for
microbiological products)
Room 506, 2-Kun Shin Bldg,
204-9
Dowha-Dong, Napo-ku
Seoul
Korea
Tel: +82-2-7037567
Fax: +82-2-7031647

KUWAIT
Majlan Trading Company
WLL, PO Box 5722
13058 Safat Kuwait
Kuwait
Tel: +965-2434903
Fax: +965-2446726

LEBANON
Befreco
BP 165959
Beyrouth
Lebanon
Tel: +961-1-200130
Fax: +961-1-422458

MALAWI
Scientific & Technical
Services
PO Box 51385

Limbe
Malawi
Tel: +265-671867
Fax: +265-670368
Telex: 44516

MALAYSIA
Bio-Diagnostics Sdn Bhd
19 Jalan SS 5A/11
47301 Petaling Jaya
Selangor
Malaysia
Tel: +60-3-7754063
Fax: +60-3-7758637

MALTA
Rodel Ltd
6 Lourdes Lane
St Julians
PO Box 308
Valetta
Malta
Tel: +356-313967
Fax: +356-344696
Telex: 1569

MAURITIUS
Chem-Tech
PO Box 742
BV Port Louis
Mauritius
Tel: +230-2-2125235
Fax: +230-2-2085330
Telex: 4610

MEXICO
Productos Biologicas Hyla SA
de CV
Pinon, 158-A Esquina
Begonias
Col. NVA, Sta. Maria
02800 Mexico DF
Mexico
Tel: +52-5-3556987
Fax: +52-5-3558354

MOROCCO
Echosynthese Maghreb
7, Rue Asphodeles Angle
Rue Kadi Bakkar
Casablanca
Morocco
Tel: +212-2-235047
Fax: +212-2-237451
Telex: 21862

NEW ZEALAND
Scianz Corporation Ltd
PO Box 6848
Auckland
New Zealand
Tel: +64-9-4807060
Fax: +64-9-4807090

NIGERIA
ISN Products (Nig) Ltd
Marina GPO
PO Box 4719
Lagos
Nigeria
Tel: +234-1-960984
Fax: +234-1-521767

NORWAY
Kebo A/S
Postboks 195
Leirdal
Jerikoveien 28

1011 Oslo
Norway
Tel: +47-22301718
Fax: +47-22301390
Telex: 76989

OMAN
Muttrah Pharmacy
PO Box 111
Muscat
Oman
Tel: +968-714100
Fax: +968-711573

PAKISTAN
The Eastern Trade &
Distribution Co (for
serological products)
507-508, 5th Floor Commerce
Centre
Hasrat Mohani Road
Karachi
Pakistan
Tel: +92-212903
Fax: +92-218200

Pharma Link International (for
microbiological products)
Suite 303, 3rd Floor Al-Almin
Tower
University Road
Gulshan-e-iqbal, Block 10
PO Box 17641
Karachi 75300
Pakistan
Tel: +92-21-470033
Fax: +92-21-469081

PERU
HW Kessel SA
Avda Corpac 312
San Isidro, Lima
Peru
Tel: +51-14-413485
Fax: +51-14-20286

PHILIPPINES
Dispo Phillipines Inc
Dispophil Bldg, Amber
Avenue
Ortigas Complex, 1601 Pasig
Metro Manila
Philippines
Tel: +63-2-6315283
Fax: +63-2-6315293

POLAND
NOBIPHARM
ul. Rydygiera 8
01793 Warsaw
Poland
Tel: +48-2-6339802
Fax: +48-2-6338295

PORTUGAL
Izasa Portugal
Distribuidores Tecnicas Lda
Av. Ventura Terra No.15
1600 Lisbon
Portugal
Tel: +351-1-7587728
Fax: +351-1-7599529

QATAR
EBN Sina Medical
Establishment
PO Box 337
Doha

© KLUWER ACADEMIC PUBLISHERS 1995, ISSN 1381-5067

Qatar
Tel: +974-417952
Fax: +974-439405

SAUDI ARABIA
Abdulrehman Algosaibi
General Trading Bureau
PO Box 215
Riyadh 1 14 11
Saudi Arabia
Tel: +966-1-4793000
Fax: +966-1-4771374

SINGAPORE
Fisons Instruments
72 Hillview Avenue
Tacam House 04-00
Singapore 0400
Singapore
Tel: +65-7608288
Fax: +65-7628288

SLOVAKIA
Merck spol sro
Hagarova 9/A
PO Box 3
83004 Bratislava
Slovakia
Tel: +42-7-288910
Fax: +42-7-283125

SOUTH AFRICA
Iepsa Edms Pty Ltd
PO Box 3 03 81
Sunnyside 0132
Pretoria
South Africa
Tel: +27-12-3226634
Fax: +27-12-3229193

SPAIN
Izasa SL
Calle Aragon 90
Barcelona 08015
Spain
Tel: +34-3-2548100
Fax: +34-3-510874
Telex: 51027

SUDAN
Atlas Trading Company
PO Box 1024
Khartoum
Sudan
Tel: +249-73050
Telex: 22289

SWEDEN
Labdesign AB
Ritaslingan 16
18366 Taby
Sweden
Tel: +46-6308500
Fax: +46-6300905

SWITZERLAND
Biotest (Schweiz) AG
Bahnhofstr. 18
5504 Othmarsingen
Switzerland
Tel: +41-64-562770
Fax: +41-64-562750

SYRIA
Harastani Brothers - Import/
Export Wholesale
PO Box 175
Damascus

Syria
Tel: +963-11-215553
Fax: +963-11-714154

TAIWAN
Black General Co. (for
serological products)
1F, No.6, Lane 289
Chuang Ching Road
PO Box 11 003
Taipei 105
Taiwan
Tel: +886-2-7233391
Fax: +886-2-7220780

Kelu Trading Co. (for
microbiological products)
50-2 Hsin Sheng, S. Road
Se. 1, 4th Floor
PO Box 783
Taipei 10 047
Taiwan
Tel: +886-2-3964569
Fax: +886-2-3921048

THAILAND
Biosystems Co.
627 Soi Sathupradit 49
Sathupradit Road
Yannawa, Bangkok 10120
Thailand
Tel: +66-2-2952064
Fax: +66-2-2941059

TUNISIA
Central Chirurgical
11 Rue d'Athenes
Tunis
Tunisia
Tel: +216-1-334887
Fax: +216-1-353941

TURKEY
RAK
Med Services Ind &
Investment Ltd
Saglik Hizmetleri Sanayii
ve Yatririm
Tunali Hilmi Ca 89 Kat 3
06700 Kavaklidere
Ankara
Turkey
Tel: +90-312-4673215
Fax: +90-312-4681588

UNITED ARAB EMIRATES
Alphamed
PO Box 11245
Dubai
United Arab Emirates
Tel: +971-4-662042
Fax: +971-4-669388

UNITED KINGDOM
Biotest (UK) Ltd
Unit 21A
Monkspath Business Park
Highlands Road
Shirley, Solihull
West Midlands
B90 4NZ
United Kingdom
Tel: +44-0121-7333393
Fax: +44-0121-7333066
Telex: 335 745

URUGUAY
Boehringer Mannheim de

Uruguay SA
Blvr Gral Artigas 1436
Montevideo
Uruguay
Tel: +598-2-781951
Fax: +598-2-7998094

USA
Biotest Diagnostics Corp
66 Ford Road, Suite 131
Denville, NJ 07834
USA
Tel: +1-201-625-1300
Fax: +1-201-625-9454

VENEZUELA
ABE Quimica de Venezuela
Calle El Recreo
Edif 9, Piso 4, Oficina 943
Bello Monte, Caracas
Venezuela
Tel: +58-2-718790
Fax: +58-2-718790

ZIMBABWE
Grayco (PVT) Ltd
44 The Chase
Mount Pleasant
Harare
Zimbabwe
Tel: +263-4-304053
Fax: +263-4-304053

Biotrin International

Corporate
Biotrin International
93 The Rise
Mount Merrion
County Dublin
Ireland
Tel: +353-1-2831166
Fax: +353-1-2831232

By region
FRANCE
Biotrin International SARL
14 rue Gorge de Loup
69009 Lyon
France
Tel: +33-72530461
Fax: +33-72530476

GERMANY
Biotrin Trignost GmbH
16 Wisenstrasse
74889 Sinsheim
Germany
Tel: +49-7261-62122
Fax: +49-7261-62624

IRELAND
Biotrin International
93 The Rise
Mount Merrion
County Dublin
Ireland
Tel: +353-1-2831166
Fax: +353-1-2831232

USA
Biotrin International (USA)
2 Ridgedale Avenue
Suite 375
Cedarknolls, NJ 07927

USA
Tel: +1-800-866-1398
Fax: +1-210-267-6618

BioWhittaker Inc

Corporate
BioWhittaker Inc
8830 Biggs Ford Road
PO Box 127
Walkersville
MD 21793-0127
USA
Tel: +1-301-8987025
Fax: +1-301-8458291

Boule Diagnostics AB

Corporate
Boule Diagnostics AB
Lunastigen 3
14144 Huddinge
Sweden
Tel: +46-8-7460035
Fax: +46-8-7468496

Bouty SpA

Corporate
Bouty SpA
Viale Casiraghi 471
20099 Seston San Giovanni
Milan
Italy
Tel: +39-2-262891
Fax: +39-2-26221305

Cambridge Biotech Corporation

Corporate
Cambridge Biotech
Corporation
365 Plantation Street
Worcester, MA 01605
USA
Tel: +1-508-797-5777
Fax: +1-508-791-0224

By region
IRELAND
Cambridge Biotech Ltd
Mervue Industrial Estate
Galway
Ireland
Tel: +353-91-757534
Fax: +353-91-752551

USA
Cambridge Biotech
Corporation
365 Plantation Street
Worcester, MA 01605
USA
Tel: +1-508-797-5777
Fax: +1-508-791-0224

Cambridge Diagnostics Ireland Ltd

Corporate
Cambridge Diagnostics
    Ireland Ltd
Mervue Industrial Estate
Galway
Ireland
Tel: +353-91-757534
Fax: +353-91-752551

Carter-Wallace Inc

Corporate
Carter-Wallace Inc
2 Research Way
Princeton, NJ 08540-6628
USA
Tel: +1-609-520-3100
Fax: +1-609-520-3114

By region
FRANCE
Fumouze Laboratoires
26 rue de Freres Chausson
92600 Asnieres
France
Tel: +33-49684100
Fax: +33-49684142

ITALY
Bouty Laboratories
471 Viale Casiraghi
20099 Sesto San Giovanni
Milan
Italy
Tel: +39-2-262891
Fax: +39-2-26221305

USA
Carter-Wallace Inc
2 Research Way
Princeton, NJ 08540-6628
USA
Tel: +1-609-520-3100
Fax: +1-609-520-3114

Cellabs Pty Ltd

Corporate
Cellabs Pty Ltd
Unit 7, 27 Dale Street
PO Box 421
Brookvale, NSW 2100
Australia
Tel: +61-2-9905-0133
Fax: +61-2-9905-6426

Centocor

Corporate
Centocor
200 Great Valley Parkway
Malvern, PA 19355-1307
USA
Tel: +1-215-889-4666
Fax: +1-215-889-4666

By region
AUSTRALIA
Immuno Diagnostics
c/o Palk Freight
12 Ewan Street
Mascot, NSW 2020
Australia

FRANCE
CIS Bioindustries
CIS Bio International
BP No. 32
91192 Gif-Sur-Yvette Cedex
France

GERMANY
BYK Sangtec
c/o Manfred Borchardt
    Luftfract
Frankfurt Int. Airport
Frankfurt
Germany

Buecker
Hermann Hesse Strasse 56
55127 Mainz
Germany

Boehringer Mannheim GmbH
c/o Rhenus Customs Brokers
Frankfurt Int. Airport
Frankfurt
Germany

ITALY
Sorin Biomedica SpA
c/o Mr Baldazzi, Broker
Turin Airport
Turin
Italy

Ares Serono Diag.
Rinaldo Rinaldi/Fiumicino Arpt
For Ares Serono Diag.
    Biodata
via Tiburtina Valeria KM
    19600
Guidonia Monticelio
Rome
Italy

Medical Systems SpA
c/o G Damiano, Agent
Cristoforo Columbia Airport
Genoa
Italy

JAPAN
Toray Fuji Bionics, Inc.
11-12 Kitimachi 1 Chrome
Nerima-ku
Tokyo 176
Japan

SPAIN
Izasa SA
Consigned to Munoz Y
    Cabrero
C Balmes 49
08001 Barcelona
Spain

SWITZERLAND
Hoffman La-Roche
Diagnostic Division
Department DIA/T, Building
    203

4002 Basle
Switzerland

UNITED KINGDOM
Centocor UK
Broadford Park
Shalford, Guildford
Surrey, GU4 8EW
United Kingdom
Tel: +44-01483-448405
Fax: +44-01483-448444

USA
Centocor
200 Great Valley Parkway
Malvern, PA 19355-1307
USA
Tel: +1-215-889-4666
Fax: +1-215-889-4666

Hybritech Inc.
7330 Carroll Road
San Diego, CA 92121
USA

Abbott Diagnostics
Skokie Warehouse Building,
    K-2
Route 41 & 22nd Street
North Chicago, IL 60064
USA

Amersham Corporation
2636 South Clearbrook Drive
Arlington Heights, IL 60005
USA

Chimica Diagnostica

Corporate
Chimica Diagnostica
Via L. Varanini 29/b
20127
Italy
Tel: +39-2-2896742
Fax: +39-2-2896779

By region
CROATIA
Chrono Poduzece SPO
Tijardoviceva 14
58000 Split
Croatia
Tel: +38-58-364664
Fax: +38-58-364901

FRANCE
Labinter SA
15 rue du Quattre Septembre
13100 Aix-en-Provence
Cedex 01
France
Tel: +33-42384858
Fax: +33-42276801

ITALY
Chimica Diagnostica
via L. Varanini 29/b
20127 Milan
Italy
Tel: +39-2-2896742
Fax: +39-2-2896779

JORDAN
Jordan Lab. Equipment Est.

Al-Sayegh Center, 1st Floor
Al-Abdali
Amman
Jordan
Tel: +962-6-615470
Fax: +962-6-601846

PORTUGAL
Rosalina Carvalho
Importacao e Distribuicao de
Equipaamentos e Reagentes
    Lab.
Av. dos Missionarios 62
2735 Cacem
Portugal
Tel: +351-1-9143318
Fax: +351-1-9143318

SAUDI ARABIA
Mada Al-Saudi
Olyah Prince Sultan Bin
    Abdulaziz St.
Behind Arab National Bank
PO Box 9671
Riyadh 11433
Saudi Arabia
Tel: +966-1-4640652
Fax: +966-1-4647379

SPAIN
Coulter Cientifica SA
Poligono Industrial 2
La Fuensanta, Parcela 11
28936 Mostoles (Madrid)
Spain
Tel: +34-91-6453011
Fax: +34-91-6455690

SRI LANKA
Asiri Hospitals Ltd
181 Kirula Road
Colombo 5
Sri Lanka
Tel: +94-1-500608
Fax: +94-1-698315

TURKEY
AB Medikal
Sanayi ve Ticaret Ltd
Mecidiye Koy Cad 24/2
Istanbul
Turkey
Tel: +90-1-2244470
Fax: +90-1-2249767

Clark Laboratories

Corporate
Clark Laboratories
PO Box 1059
Jamestown, NY 14702-1059
USA
Tel: +1-716-483-3851
Fax: +1-716-488-1990

Cortecs Ltd

Corporate
Cortecs Ltd
The Old Blue School
Lower Square
Isleworth, Middlesex
TW7 6RL
United Kingdom
Tel: +44-0181-568-7071

© KLUWER ACADEMIC PUBLISHERS 1995, ISSN 1381-5067

Fax: +44-0181-847-2373

Dade International Inc

Corporate
Dade International Inc
Stratus Immunochemistry
10 555 West Flagler
Miami, FL 33174
USA
Tel: +1-305-222-6686
Fax: +1-305-222-6200

By region
AUSTRALIA
Baxter Healthcare Pty Ltd
PO Box 88
Toongabbie, NSW 2146
Australia
Tel: +61-2-6889111
Fax: +61-2-6889123

BELGIUM
Baxter World Trade SA
Baxter SA
Rue Colonel Bourg, 105B
1140 Brussels
Belgium
Tel: +32-2-7411711
Fax: +32-2-7322381

EUROPE
Biotrin International Ltd
93 The Rise
Mount Merrion
Dublin
Ireland
Tel: +353-1-2831166
Fax: +353-1-2831232

FRANCE
Baxter SA
Avenue Louis-Pasteur
Boite Postale 56
78311 Maurepas Cedex
France
Tel: +33-1-34615050
Fax: +33-1-34615008

GERMANY
Baxter Deutschland GmbH
Edisonstrasse 3
85716 Unterschleissheim
Germany
Tel: +49-89-317010
Fax: +49-89-31701-177

ITALY
Baxter SpA
Via Lampedusa 11/A
20141 Milan
Italy
Tel: +39-2-89501732
Fax: +39-2-89527222

JAPAN
Baxter Limited, Sales Office
Sankaido Building
9-13 Alasaka 1-chome
Minato-ku
Tokyo 107
Japan
Tel: +81-3-32376627
Fax: +81-3-35057811

NETHERLANDS
Baxter BV
PO Box 1536
3600 BM Maarssem
Netherlands
Tel: +31-30-468911
Fax: +31-30-411755

NORWAY
Baxter A/S
Gjerdrumsvei 10B
0486 Oslo 4
Norway
Tel: +47-2-184101
Fax: +47-2-184109

SPAIN
Baxter SA
Calle Solsones, No. 2, Bajo
Local 7
08820 El Prat de Llobregat
Barcelona
Spain
Tel: +34-3-4787162
Fax: +34-3-4787109

SWITZERLAND
Baxter AG
BonnStrasse
CH-3186 Dudingen
Switzerland
Tel: +41-37-438111
Fax: +41-37-438960

UNITED KINGDOM
B.M. Browne (UK) Ltd
9 Commerce Park
Brunel Road
Theale, Reading
RG7 4AB
United Kingdom
Tel: +44-01734-305333
Fax: +44-01734-305111

USA
Dade International Inc
Stratus Immunochemistry
10 555, West Flagler
Miami, FL 33174
USA
Tel: +1-305-222-6200
Fax: +1-305-222-6686

DAKO A/S

Corporate
DAKO A/S
Produktionsvej 42
Postbox 1359
2600 Glostrup
Denmark
Tel: +45-44920044
Fax: +45-42841822
Telex: 35128

By region
ARGENTINA
Chemetron Latinoamericana
 SA
Junin 262
PB 1 Capital Federal
1026 Buenos Aires
Argentina
Tel: +54-1-9538918
Fax: +54-1-9538918
Telex: 9900

AUSTRALIA
Bio Scientific Pty Ltd
PO Box 78
28 Monroe Avenue
Kirrawee
Gymea, NSW 2227
Australia
Tel: +61-2-5212177
Fax: +61-2-5423037

AUSTRIA
Bender and Co GmbH
Dr. Boehringergasse 5-11
PO Box 103
1121 Vienna
Austria
Tel: +43-1-801050
Fax: +43-1-80105488
Telex: 132430

BELGIUM
Prosan bvba sprl
Maurits Sabbestraat 67
9050 Gentbrugge
Belgium
Tel: +32-91-313704
Fax: +32-91-319898

BRAZIL
Embrabio-Empressa
 Brasileira de Biotechnologia
 Ltda
Rua Apinages 1.081
05017 Sao Paulo/Sp
Brazil
Tel: +55-11-2625511
Fax: +55-11-2630272
Telex: 035970

CANADA
Dimension Laboratories Inc.
12 Falconer Drive, Unit 4
Mississauga
Ontario L5N 3L9
Canada
Tel: +1-416-858-8510
Fax: +1-416-858-8801

CHILE
PROLAB
Vergara 24
Oficina 908
Casilla 3645
Santiago
Chile
Tel: +56-2-6987215
Fax: +56-2-6989617

COSTA RICA
Laboratorios Zeiedon SA
Apartado Postal 5236-1000
San Jose
Costa Rica
Tel: +506-353959
Fax: +506-351275

CYPRUS
Medisell Co. Ltd
55A Limassol Avenue
PO Box 8318
Nicosia
Cyprus
Tel: +357-2-311362
Fax: +357-2-494300

DENMARK
DAKO A/S
Produktionsvej 42

Postbox 1359
2600 Glostrup
Denmark
Tel: +45-44920044
Fax: +45-42841822
Telex: 35128

ECUADOR
Proveedores Para
 Laboratorios C, Ltda
Edificio-Pro-Lab.
Luis Urdaneta Y Ave
del Ejercito
Guayaquil
Ecuador
Tel: +593-4-281943
Fax: +593-4-285953
Telex: 42985

EGYPT
LAB TECHNOLOGY
4 Leith Ben Saad Str.
PO Box 5959
Heliopolis West 113351
Cairo
Egypt
Tel: +20-2-2351785
Fax: +20-2-2428366
Telex: 22127

FINLAND
COFACTOR
Suomen biotekniikka Oy
Kavallinmaki 13
02700 Kauniainen
Finland
Tel: +358-0-594822
Fax: +358-0-594864

Soumen biotekniikka Oy
 Cofactor
Kavallinmaki 13
02700 Kauniainen
Finland

FRANCE
DAKO SA
2 rue Albert Einstein
BP 149
78196 Trappes Cedex
France
Tel: +33-30500050
Fax: +33-30500011
Telex: 695029

GERMANY
DAKO Diagnostika GmbH
Am Stadtrand
22047 Hamburg
Germany
Tel: +49-40-6937026
Fax: +49-40-6952741

GREECE
Ange M. Calliphronas
4 Evripidou Street
10559 Athens
Greece
Tel: +30-1-3213272
Fax: +30-1-3218871

HONG KONG
China South Technology Ltd
Rm 1303-4, 13/F, Remex
Centre
42 Wong Chuh Hang Road
Hong Kong
Tel: +852-5528339

Fax: +852-5526883

INDIA
J Mitra & Bros. Pvt. Ltd
A-180, Okhla
Industrial Area, Phase 1
Okhla
New Delhi 110020
India
Tel: +91-11-6818971
Fax: +91-11-6818970
Telex: 031-75314

INDONESIA
PT DWI Marga Sakti
J1 Kemuning No. 17
Tomang Raya
PO Box 4129 Jkt
Jakarta Barat
Indonesia
Tel: +62-21-597102
Fax: +62-21-591286

IRAN
Eskan Teb Tech. Co.
6 Business Section, ESKAN
Buildings
Mirdamad Avenue
PO Box 19395
1836 Teheran
Iran
Tel: +98-21-8087602
Fax: +98-21-4279491
Telex: 214158

ISRAEL
Tzamal Ltd
21 Gonen Street
PO Box 3064
Kiryat Matalon
Petac Tiqva
Israel
Tel: +972-3-9240288
Fax: +972-3-9240259
Telex: 381542

ITALY
DAKO SpA
Via G. Fantoli 21/17
20138 Milan
Italy
Tel: +39-2-58011221
Fax: +39-2-504778

JAPAN
DAKO Japan Co. Ltd
Hiraoka Building
Nishinotouin-higashiiru
Shijo-dori, Shimogyo-ku
Kyoto 600
Japan
Tel: +81-75-2113655
Fax: +81-75-2111755

KOREA
Fine Chemical Co. Ltd
Garden Tower Building, Rm
No. 1103
98-78 Wun Ni-Dong, Jong Ro-
Go
KPO Box 1260
Seoul
Korea
Tel: +82-2-7447859
Fax: +82-2-7445281

KUWAIT
WARBA Medical Supplies Co.

Nakib Building, 4th Floor
Abu Bakir Street, Al Jiblah
PO Box 26267
13123 Safat
Kuwait
Tel: +965-2426939
Fax: +965-2429482
Telex: 44470

MALAYSIA
General Scientific Co. Sdn.
Bhd
No. 7 Jalan 222
Section 51 A
46100 Petaling Jaya
Selangor Darul Ehsan
Malaysia
Tel: +60-3-7575433
Fax: +60-3-7571768
Telex: 374431

MALTA
Michele Perssso Ltd
Catalunya Building, Psaila
Street
B'Kara
Malta
Tel: +356-492191
Fax: +356-482593

MEXICO
Dostym SA de CV
J.GPE. Montenegro No. 2325
Col. Arcos Sur.
CP 44150
Guadalajara
Jalisco
Mexico
Tel: +52-36-6153385
Fax: +52-36-5236153513
Telex: 683226

NETHERLANDS
ITK diagnostics bv
Johan Enschedeweg 13
1422 DR Uithoorn
Netherlands
Tel: +31-2975-68893
Fax: +31-2975-63458

NEW ZEALAND
Med-Bio Enterprises Ltd
PO Box 33-135
Barrington, Christchurch
New Zealand
Tel: +64-3-3381020
Fax: +64-3-3380028

NORWAY
Bio-Test A/S
Idrettsveien 2
PO Box 66
1580 Rygge
Norway
Tel: +47-69261777
Fax: +47-69261760

OMAN
Mustafa & Tawad Trading Co.
LLC
S & I Dept.
PO Box 4918
Ruwi
Oman
Tel: +968-709955
Fax: +968-54005

PANAMA
Importador DMD SA
Apartado 8556 Calle 31
Esto No. 1-95
Panama
Tel: +507-270537/251247
Fax: +507-271246

PARAGUAY
Dr Ruben A Sosky
Calidad, Mexico 923
Ascuncion
Paraguay
Tel: +595-21-447680
Fax: +595-21-447392

PERU
Representaciones Atlanta SA
Av. Republica de Panama
NRO 465
Callao 1
Peru
Tel: +51-14-652421
Fax: +51-14-654833

PHILIPPINES
Levin's International Corp.
3rd Floor R. Syjuco Building
993 E. Delos Santos Ave
cor. Bansalangin St, Diliman
Qeuzon City
Philippines
Tel: +63-2-974475/2
Fax: +63-2-984841

POLAND
ALAB sp. 20.0
Ul. Pasteura 3
02-093 Warsaw
Poland
Tel: +48-2-6598571
Fax: +48-2-6582059

PORTUGAL
Labometer LDA
Rua Duque de Palmela No.
30, 1-G
1200 Lisbon
Portugal
Tel: +351-1-537284
Fax: +351-1-3525066
Telex: 18489

SAUDI ARABIA
Medical Business Centre,
MBC
PO Box 189
Jeddah 21411
Saudi Arabia
Tel: +966-2-6421437
Fax: +966-2-6442502

SINGAPORE
Lab Essentials (S) Pte Ltd
108 Pasir Panjang Road
#02-02, Amcol Warehouse
Singapore 0511
Singapore
Tel: +65-4794009
Fax: +65-4790013

SOUTH AFRICA
Southern Cross
Biotechnology (Pty) Ltd
PO Box 23681
Claremont 7735
South Africa
Tel: +27-21-615166

Fax: +27-21-617734

SPAIN
Atom SA
Passeig d'Amunt 18
08024 Barcelona
Spain
Tel: +34-3-2847904
Fax: +34-3-2108255
Telex: 51227

SWEDEN
DAKOPATTS AB
Box 13
72521 Alosjo
Sweden
Tel: +46-8-996000
Fax: +46-8-996065

SWITZERLAND
IG-Instrumenten Gesellschaft
AG
Raffelstrasse 32
8045 Zurich
Switzerland
Tel: +41-1-4613311
Fax: +41-1-4613001
Telex: 813432

Lablink
PO Box 178
8059 Zurich
Switzerland

THAILAND
Science Tech. Co. Ltd
321/43 Nanglinchee Road
Chongnondsee
Yannawa
Bangkok 10120
Thailand
Tel: +66-2-2854101-3
Fax: +66-2-2854856
Telex: 82731

TURKEY
Hayat Corporation Inc.
Millet Cad. Renk Apt. No 75
Kat:4 D-8
34280 Findilezade
Istanbul
Turkey
Tel: +90-1-5879887
Fax: +90-1-5879402

UNITED ARAB EMIRATES
Al-Zahrawi Medical
PO Box No. 5973
Dubai
United Arab Emirates
Tel: +971-4-622728
Fax: +971-4-625506

UNITED KINGDOM
DAKO Ltd
16 Manor Courtyard
Hughenden Avenue
High Wycombe
Bucks, HP13 5RE
United Kingdom
Tel: +44-01494-452016
Fax: +44-01494-441553

DAKO Diagnostics Ltd
Denmark House
Angel Drove
Ely, Cambridgeshire
CA7 4ET

United Kingdom
Tel: +44-01353-669911
Fax: +44-01353-668989

URUGUAY
Poliuruguay srl
Avda Uruguay 1771
Montevideo
Uruguay
Tel: +598-2-402365
Fax: +598-2-409017

USA
DAKO Corporation
6392 Via Real
Carpinteria, CA 93013
USA
Tel: +1-805-566-6655
Fax: +1-805-566-6688
Telex: 658481

ZIMBABWE
National Diagnostics Ltd
PO Box 3535
Harare
Zimbabwe
Tel: +263-4-791615
Fax: +263-4-728055

Denka Seiken Co. Ltd

Corporate
Denka Seiken Co. Ltd
12-1 Nihonbashikabuto-cho
Chuo-ku
Tokyo 103
Japan
Tel: +81-3-3669-9421
Fax: +81-3-3664-1005

By region
FRANCE
Eurobio Laboratoires
ZA Courtaboeuf
7 Avenue de Scandinavie
91953 Les Ulix Cedex B
France
Tel: +33-1-69079477
Fax: +33-1-69079544

GERMANY
LA Labor Diagnostika GmbH
Industriestrasse 12
46359 Heiden
Germany
Tel: +49-2867-990727
Fax: +49-2867-990729

JAPAN
Denka Seiken Co. Ltd
12-1 Nihonbashikabuto-cho
Chuo-ku
Tokyo 103
Japan
Tel: +81-3-3669-9421
Fax: +81-3-3664-1005

UNITED KINGDOM
Mast International Ltd
Mast House, Derby Road
Bootle, Merseyside
L20 1EA
United Kingdom
Tel: +44-0151-933-7277
Fax: +44-0151-944-1332

Diagast Laboratories

Corporate
Diagast Laboratories
59 Rue de Trevise
BP 2034
59014 Lille Cedex
France
Tel: +33-20526800
Fax: +33-20520610

Diagnostic Products Corporation

Corporate
Diagnostic Products
Corporation
5700 West 96th Street
Los Angeles, CA 90045
USA
Tel: +1-213-776-0180
Fax: +1-213-642-0192

By region
IRELAND
Novapath Supplies Ltd
159 Lower Rathmines Rd
Dublin 6
Ireland
Tel: +353-1-965666
Fax: +353-1-962208

NORTHERN IRELAND
Vector Scientific Ltd
30 Island Street
Belfast, BT4 1DH
Northern Ireland
Tel: +44-232-739248
Fax: +44-232-739249

UNITED KINGDOM
DPL Division, Euro/DPC Ltd
Glyn Rhonwy
Llanberis, Caernarfon
Gwynedd, LL55 4EL
United Kingdom
Tel: +44-01286-871872
Fax: +44-01286-871802

USA
Diagnostic Products
Corporation
5700 West 96th Street
Los Angeles, CA 90045
USA
Tel: +1-213-776-0180
Fax: +1-213-642-0192

Diamedix Corporation

Corporate
Diamedix Corporation
2140 North Miami Avenue
Miami, FL 33127
USA
Tel: +1-305-324-2300
Fax: +1-305-324-2395

Diesse Srl

Corporate
Diesse Srl
50020 Sanbuca
Firenze
Italy
Tel: +39-55-8071371
Fax: +39-55-8071373

Eiken Chemical Co. Ltd. *see* MAST Diagnostics

E. Merck

Corporate
E. Merck
Diagnostics Division
Frankfurter Strasse 250
64293 Darmstadt
Germany
Tel: +49-6151-720
Fax: +49-6151-722000

By region
ARGENTINA
Merck Quimica Argentina
SAIC
Casilla correo 1442
1000 Buenos Aires
Argentina
Tel: +54-1-5510027
Fax: +54-1-112135
Telex: 21733

AUSTRALIA
Merck Pty
207 Colchester Road
Kilsyth, Victoria 3137
Australia
Tel: +61-2-7285855
Fax: +61-2-7281351
Telex: 33596

AUSTRIA
Merck Gesellschaft m.b.H.
Zimbagasse 5
Postfach 700
1147 Vienna
Austria
Tel: +43-1-9716110
Fax: +43-1-975560
Telex: 133188

BANGLADESH
GA Traders Ltd
PO Box 3430
16, Motijheel Commercial
Area
Dhaka 2100
Bangladesh
Tel: +880-227299/236187
Fax: +880-2863472

BELGIUM
Merck Belgolabo NV-SA
Brusselsesteenweg 288
3090 Overijse
Belgium
Tel: +32-6-890711
Fax: +32-6-879120

Telex: 23807

BOLIVIA
Corimex Ltda
Calle Goitia 135
Cajon Postal 4788
La Paz
Bolivia
Tel: +591-320216
Fax: +591-391230
Telex: 3508

BRAZIL
Quimitra
Comercio e Industria SA
Caixa Postal 70556
22741 Rio de Janeiro
Brazil
Tel: +55-21-3424646
Fax: +55-21-3421263
Telex: 2123792

BURMA
Inspection & Agency Corp.
Agency Division
PO Box 404
No. 383 Mahanbandoola
Street
Rangoon
Burma
Tel: +95-1-76045/73169
Telex: 21215

CANADA
BDH Inc.
350 Evans Avenue
Toronto
Ontario M8Z 1K5
Canada
Tel: +1-416-255-8521
Fax: +1-416-255-7453
Telex: 6967678

CHILE
Merck Quimica Chilena Soc.
Lda.
Francisco de Paulo Taforo
1981
Casilla 4232
Santiago de Chile
Chile
Tel: +56-2-381160
Fax: +56-2-383527
Telex: 440197

CHINA
Jebsen & Co. Ltd
China Trade Division, EMO-
Dept.
Prince's Building, 23rd Floor
GPO Box 97
Hongkong
Tel: +852-8437978
Fax: +852-681742
Telex: 73221

COLOMBIA
Merck Colombia SA
Apartado Aereo 9896
Bogota 6 DE
Colombia
Tel: +57-1-2907855
Fax: +57-1-2628881
Telex: 41405

COSTA RICA
Cooperation Cefa SA
Apartado 10300

San Jose de Costa Rica
Costa Rica
Tel: +506-322122/203040
Fax: +506-327125
Telex: 2557

CYPRUS
Markides & Vouros Ltd
PO Box 2002
5A + B Pargas Street
Nicosia
Cyprus
Tel: +357-2-475121/464193
Fax: +357-2-457158
Telex: 2939

CZECHOSLOVAKIA
Merck Spol sro
Slawetinska 26
1900 Prague-Klanovice
Czechoslovakia
Tel: +42-8-7881203

DENMARK
Merck Denmark AS
Skodsborgvej 48
2830 Virum
Denmark
Tel: +45-42856622
Fax: +45-45831224

ECUADOR
Merck Ecuador CA
Casilla 1701 2574
Quito
Ecuador
Tel: +593-2-561150
Fax: +593-2-562703
Telex: 22252

EGYPT
Akhnaton Trading &
   Representation
12 Muntazeh Street
Zamalek, PO Box 1002
Cairo
Egypt
Tel: +20-2-3404945
Fax: +20-2-3412677
Telex: 93882

EL SALVADOR
Merck El Salvador SA
Apartado Postal 2039
11 Avenida Norte Bis 513
San Salvador
El Salvador
Tel: +503-228381
Fax: +503-211277

FRANCE
Laboratoires Merck Clevenot
   SA
Division Laboratoire
5/9 Rue Anquetil
94736 Nogent sur Marne
France
Tel: +33-43945400
Fax: +33-48765815
Telex: 261374

Merck Clevenot SA
Division Laboratoire
5/9 rue Anquetil
94731 Nogent-sur-Marne
France
Tel: +33-43945400
Fax: +33-48765786

Telex: 261374

GERMANY
E. Merck
Diagnostics Division
Frankfurter Strasse 250
64293 Darmstadt
Germany
Tel: +49-6151-720
Fax: +49-6151-722000

GREECE
Merck-Hellas GmbH
PO Box 72545
16410 Argyroupoli
Greece
Tel: +30-9929944
Fax: +30-9953214
Telex: 216101

GUATEMALA
Merck Centroamericana SA
Apartado postal 1651
Ciudad de Guatemala
Guatemala
Tel: +502-2-922111
Fax: +502-2-941543
Telex: 5110

HONDURAS
Drogueria Paysen
SA de CV
Apartado 252
Tegucigalpa
Honduras
Tel: +504-325010/325251
Fax: +504-311110
Telex: 1111

INDIA
E. Merck India Ltd
87 Dr Annie Besant Road
PO Box 16554
Bombay 400018
India
Tel: +91-22-4922855
Fax: +91-22-4922354

INDONESIA
PT Multiredjeki Kita
Jalan Raya Gedong Desa No.
   8
Pasar Rebo
Jakarta Timur 13760
PO Box 4387
Jakarta 10001
Indonesia
Tel: +62-21-8402091/94
Fax: +62-21-8402095
Telex: 4-8472

IRAN
Merck Trading AG
PO Box 15745/653
Tehran
Iran
Tel: +98-21-890116
Fax: +98-21-890116
Telex: 212282

IRELAND
Norman Lauder Ltd
Dunfirth
Butterfield Avenue
Rathfarnham
Dublin 14
Ireland
Tel: +353-1-932643

Fax: +353-1-931606
Telex: 93392

ISRAEL
Mercury Chemical Agencies
   Ltd
7 Hasharoshet St.
PO Box 6
Ramat Hasharon 47100
Israel
Tel: +972-5-401215
Fax: +972-5-408403
Telex: 341302

ITALY
Bracco Industria Chimica SpA
Via E. Folli 50
20134 Milan
Italy
Tel: +39-2-21771
Fax: +39-2-26410678
Telex: 311185

IVORY COAST
Societe Polychimie
01-BP 3907
Abidjan 01
Republic de Cote d'Ivoire
Ivory Coast
Tel: +225-355667/357688
Fax: +225-351385
Telex: 439298

JAPAN
Kanto Chemical Co. Inc
2-8 Nihombashi-Honcho-3-
   chome
Chuo-ku
Tokyo 103
Japan
Tel: +81-3-2706500
Fax: +81-3-6690823
Telex: 2223446

Merck Japan Ltd
ARCO Tower, 5F
8-1 Shimomeguro 1-chome
Meguro-ku
Tokyo 153
Japan
Tel: +81-3-54344722
Fax: +81-3-54344706
Telex: 2226868

JORDAN
Henry Marroum & Sons
Prince Mohammed Street
PO Box 589
Amman
Jordan
Tel: +962-6-623523
Fax: +962-6-638088

KOREA
Merck Korea Ltd
Yeong-Dong
PO Box 312
Seoul
Korea
Tel: +82-2-5555964
Fax: +82-2-5571440
Telex: 26803

KUWAIT
Bader Sultan & Bros. Co. Ltd
6th Ring Road
PO Box 867
13009 Safat

Kuwait
Tel: +965-4332566
Fax: +965-4334217

MALAYSIA
Merck (Malaysia) SDN BHD
No. 25 Jalan 2/71
Taman Tun Dr., Ismail
60000 Kuala Lumpur
Malaysia
Tel: +60-7-178922
Fax: +60-7-178925

MEXICO
Merck-Mexico SA
Apartado 8619
Mexico DF 06000
Mexico
Tel: +52-5-7269015
Fax: +52-5-3590759
Telex: 1761818

NETHERLANDS
E. Merck Netherland BV
Basisweg 34
Postbus 8198
1005 AD Amsterdam
Netherlands
Tel: +31-20-5811511
Fax: +31-20-6149694
Telex: 14382

NEW ZEALAND
BDH Chemicals Ltd
PO Box 1246
Palmerston North
New Zealand
Tel: +64-6-3582038
Fax: +64-6-3567311

NICARAGUA
Comercial Genie Penalba SA
Apartado 694
Managua DN
Nicaragua
Tel: +505-2-26120/22808
Fax: +505-2-25473
Telex: 1330

NORWAY
Merck AS
Postboks 51 Ellingsrudaasen
1006 Oslo 10
Norway
Tel: +47-2-321150
Fax: +47-2-305244
Telex: 77413

PAKISTAN
AD Marker (Pvt) Ltd
D-7 Shaheed-e-Millat Road
PO Box 2027
Karachi 8
Pakistan
Tel: +92-21-449031
Fax: +92-21-4559221
Telex: 25280 or 24166

PARAGUAY
Vincente Scavone & CIA
   CEISA
Casilla de Correo 427
Asuncion
Paraguay
Tel: +595-21-490251
Fax: +595-21-494704
Telex: 44165

PERU
Merck Peruana SA
Casilla 4331
Lima 1
Peru
Tel: +51-14-620142
Fax: +51-14-620298

PHILIPPINES
Merck Inc.
PO Box 1799
Makati 1299
Metro Manila
Philippines
Tel: +63-2-8185860
Fax: +63-2-8103024
Telex: 66371

POLAND
E. Merck
Oddzial w Warszawie
Ul. Prozna 12 A
00950 Warsaw
Poland
Tel: +48-22-205923
Fax: +48-22-205923

PORTUGAL
Merck Portuguesa Lda
Apartado 3185
1304 Lisbon Codex
Portugal
Tel: +351-1-3621434
Fax: +351-1-3621445
Telex: 63916

SAUDI ARABIA
Al-Jeel Medical & Trading Co.
PO Box 5012
Riyadh 11422
Saudi Arabia
Tel: +966-1-4041717
Fax: +966-1-4059052
Telex: 401723

Almura's
PO Box 10735
Riyadh 11443
Saudi Arabia
Tel: +966-1-4652916
Fax: +966-1-4644981
Telex: 5405406

SENEGAL
Societe Polychimie SA
Boite Postale 284
Dakar
Republic du Senegal
Tel: +221-323348
Fax: +221-320125
Telex: 550

SINGAPORE
All Eight Marketing Services
Pte Ltd
BK-212 Hongkong Street 21
03-343 Singapore 1953
Singapore
Tel: +65-2886388
Fax: +65-2849805
Telex: 55461

FE Zuellig (Trading) Pty Ltd
421 Tagore Avenue
PO Box 725
9014 Singapore
Singapore
Tel: +65-4598832

Fax: +65-4582706
Telex: 21750

SOUTH AFRICA
Merck Laboratory Supplies
PO Box 1998
Midrand 1685
South Africa
Tel: +27-11-3151100
Fax: +27-11-3151353
Telex: 424534

SPAIN
Igoda SA
Depart. Diagnosticos
Apartados 47
08100 Mollet del Valles
Barcelona
Spain
Tel: +34-3-5705750
Fax: +34-3-5701656
Telex: 94180

SWEDEN
E. Merck AB
Kungsgatan 65
111 22 Stockholm
Sweden
Tel: +46-8-202114
Fax: +46-8-245594
Telex: 10677

SWITZERLAND
E. Merck (Schweiz) AG
Witikoner Str. 15
Postfach 213
8029 Zurich
Switzerland
Tel: +41-1-559233
Fax: +41-1-559958
Telex: 816135

SYRIA
Droguerie Syrie
Antranik Kaprielian
Boite Postale 5441
Aleppo
Syria
Tel: +963-21-220277
Telex: 331078

TAIWAN
Merck Taiwan Ltd
3. FL, No. 34 Min Chuen West
Road
PO Box 681058
Taipei
Taiwan
Tel: +886-2-5219331
Fax: +886-2-5367734
Telex: 21787

THAILAND
Merck Ltd
PO Box 128
Pratunam Post Office
Bangkok 10409
Thailand
Tel: +66-2-3080218
Fax: +66-2-3080218

TURKEY
Alfred Paluka & Co.
PO Box 532
Istanbul-Karakoy
Turkey
Tel: +90-1-451246/47
Fax: +90-1-524458

Telex: 24971

UNITED ARAB EMIRATES
Gulf Drug Establishment
PO Box 3264
Dubai
United Arab Emirates
Tel: +971-4-20512
Fax: +971-4-22267
Telex: 46553

UNITED KINGDOM
Merck Ltd
Merck House
Poole
Dorset, BH15 1TD
United Kingdom
Tel: +44-01202-669700
Fax: +44-01202-665599
Telex: 41186

URUGUAY
Quimica Oriental SA
Casilla de correo 1443
Montevideo
Uruguay
Tel: +598-2-250627
Fax: +598-2-253334
Telex: 23045

USA
EM Diagnostics Systems Inc.
480 Democrat Road
Gibbstown, NY 08027
USA
Tel: +1-609-423-6300
Fax: +1-609-423-0671

VENEZUELA
Merck SA
Apartado 2020
Caracas 1010 A
Venezuela
Tel: +58-2-213455
Fax: +58-2-217164
Telex: 25236

Enteric Products

Corporate
Enteric Products
25 East Loop Drive, Rm 102
Stony Brook, NY 11790-3355
USA
Tel: +1-516-444-8872
Fax: +1-516-444-8855

Euro-Diagnostica

Corporate
Euro-Diagnostica
PO Box 2820
7303 GC Apeldoorn
Netherlands
Tel: +31-55-422433
Fax: +31-55-425017

By region
BELGIUM
Organon Teknika Int'l
Veedijk 58
2300 Turnhout
Belgium

CANADA
Immunocorp Sciences Inc
5800 Royalmount
Montreal
Quebec, H4P 1K5
Canada
Tel: +1-514-733-3000
Fax: +1-514-733-1212

FRANCE
Unipath SA
6 route de Paisy
Boite Postale 13
69572 Dardilly Cedex
France
Tel: +33-78351731
Fax: +33-78660376

GERMANY
Laboserv GmbH
Am Zollstock 2
35392 Giessen
Germany
Tel: +49-641-2674
Fax: +49-641-28535

GREECE
Biodiagnostics
8 Thetidos Street
11528 Athens
Greece
Tel: +30-1-7211185
Fax: +30-1-7214227

Biodynamics SA
Alexandras Avenue & 62
Koniari St.
11521 Athens
Greece
Tel: +30-1-6449421
Fax: +30-1-6442266

ITALY
IFCI Clone Systems Spa
Via Del Fornacial 24
40129 Bologna
Italy
Tel: +39-51-326267
Fax: +39-51-323136

Tema Ricerca Srl
Via della Repubblica 20
40068 Dan Lazaro di Savena
Bologna
Italy
Tel: +39-51-454324
Fax: +39-51-464685

SWEDEN
Euro-Diagnostica AB
Box 30561
200 62 Malmo
Sweden

SWITZERLAND
Bio Science AG
Gerliswillstrasse 43
6020 Emmenbruche
Switzerland
Tel: +41-41-555075
Fax: +41-41-558322

TURKEY
Hayat Lab. Maix
Sanavi ve Ticaret AS
Millet cadessl Renk Ap 76/4
D8
34280 Findikzade Istanbul

© *KLUWER ACADEMIC PUBLISHERS 1995, ISSN 1381-5067*

Turkey

UNITED KINGDOM
Euro-Path Ltd
Union Hill
Stratton Bude
Cornwall, EX23 9BL
United Kingdom
Tel: +44-01288-353686
Fax: +44-01288-352866

Gamma SA

Corporate
Gamma SA
Parc de Recherches du Sart
   Tilman
Rues des Chasseurs
Ardennais
Belgium
Tel: +32-41-674784
Fax: +32-41-674784

GenBio Inc

Corporate
GenBio Inc
15222 Avenue of Science,
   Unit A
San Diego, CA 92128
USA
Tel: +1-619-592-9300
Fax: +1-619-592-9400

Gull Laboratories Inc

Corporate
Gull Laboratories Inc
1011 East 4800 South
Salt Lake City, UT 84117
USA
Tel: +1-801-263-3524
Fax: +1-801-265-9268

By region
BELGIUM
Gull Diagnostics
Rue de la Station, 19
1300 Limal (Wavre)
Belgium
Tel: +32-10-419800
Fax: +32-10-417285

USA
Gull Laboratories Inc
1011 East 4800 South
Salt Lake City, Utah 84117
USA
Tel: +1-801-263-3524
Fax: +1-801-265-9268

Biodesign International
105 York Street
Kennebunk, ME 04043
USA
Tel: +1-207-985-1944
Fax: +1-207-985-6322

Hemagen Diagnostics Inc

Corporate
Hemagen Diagnostics Inc
34-40 Bear Hill Road
Waltham, MA 02154
USA
Tel: +1-617-890-3766
Fax: +1-617-890-3748

By region
ITALY
Alfa Biotech SpA
Via Marelli 303
I-20099 Sesto.S. Giovanni
   (MI)
Italy
Tel: +39-2-262761
Fax: +39-2-26276288

USA
Hemagen Diagnostics Inc
34-40 Bear Hill Road
Waltham, MA 02154
USA
Tel: +1-617-890-3766
Fax: +1-617-890-3748

Human GmbH

Corporate
Human GmbH
Silberbachstrasse 9
65232 Taunusstein
Germany
Tel: +49-6128-875-266
Fax: +49-6128-875100

Hycor Biomedical Inc

Corporate
Hycor Biomedical Inc
18800 Von Karmen Avenue
Irvine, CA 92641
USA
Tel: +1-714-440-2000
Fax: +1-714-440-2222
Telex: 403338

IBL Gesellschaft fur Immunchemie und Immunbiologie mbH

Corporate
IBL Gesellschaft fur
   Immunchemie und
   Immunbiologie mbH
Flughafenstrasse 52a
Airport Center, Haus C
22335 Hamburg
Germany
Tel: +49-40-5328910
Fax: +49-40-52389111

IFCI CloneSystems SpA

Corporate
IFCI CloneSystems SpA
Sede e Stabilimento
Via Magnanelli 2
40033 Casalecchio di Reno
Bologna
Italy
Tel: +39-51-575326
Fax: +39-51-592738

By region
ITALY
IFCI CloneSystems SpA
Sede e Stabilimento
Via Magnanelli 2
40033 Casalecchio di Reno
Bologna
Italy
Tel: +39-51-575326
Fax: +39-51-592738

SPAIN
IFCI Clone Systems Espana
Plaza de La Torre 7 - 4 Planta
08006 Barcelona
Spain
Tel: +34-3-2375932
Fax: +34-3-4158300

Immunetics

Corporate
Immunetics
63 Rogers Street
Cambridge, MA 02142
USA
Fax: +1-617-868-7879

Immuno Concepts Inc

Corporate
Immuno Concepts Inc
9779 Business Park Drive
Suite 1
Sacramento, CA 95827
USA
Tel: +1-916-363-2649
Fax: +1-916-363-2843

Immuno Pharmacology Research

Corporate
Immuno Pharmacology
   Research
Via Musumeci 130
95128 Catania
Italy
Tel: +39-95-444239
Fax: +39-95-445359

Incstar Corporation

Corporate
Incstar Corporation

PO Box 285
1990 Industrial Boulevard
Stillwater, MN 55082
USA
Tel: +1-612-439-9710
Fax: +1-612-779-7847
Telex: 02 6879023

By region
ARGENTINA
TecnoGam
Av. Cordoba 966 6 "42"
1054 Buenos Aires
Argentina
Tel: +54-1-3264254
Fax: +54-1-3264264

AUSTRALIA
Baxter Diagnostics Pty Ltd
8 Murdoch Circuit
Acacia Ridge
Queensland 4110
Australia
Tel: +61-7-2737111
Fax: +61-7-2737122

Immuno Diagnostics
12-14 Purkis Street
PO Box 126
Camperdown, NSW 2050
Australia
Tel: +61-2-5199300
Fax: +61-2-5196762

AUSTRIA
Bender MedSystems
Dr. Boehringer-Gasse 5-11
1121 Vienna
Austria
Tel: +43-222-801050
Fax: +43-222-80105488

LD Labor Diagnostika GmbH
Industriestrasse 12
46359 Heiden/W
Germany
Tel: +49-2867-99070
Fax: +49-2867-990719

Dipl. Ing. Zoltan Szabo GmbH
Hernalser Hauptstrasse 86
1170 Vienna
Austria
Tel: +43-1-40939610
Fax: +43-1-40939617

Biomedica Gruppe
Divischgasse 4
1210 Vienna
Austria
Tel: +43-1-2923527
Fax: +43-1-2901361

BAHRAIN
Al-Shaikh Establishment
PO Box 30342
Bld.60, Road 7503
Abusaiba 475
Bahrain
Tel: +973-593346
Fax: +973-590046

BELGIUM
Sorin Biomedica SA
Rue de la Grenouillette 2F
Waterranonkelstraat 2F
1130 Brussels

Belgium
Tel: +32-224-54020
Fax: +32-224-54067

BRAZIL
Fergo Prod. Hosp. Ltda
Rua Tenente Costa 135
20771 Rio de Janeiro
Brazil
Tel: +55-21-5813445
Fax: +55-21-2816293

Labcare do Brasil Produtos
  Hospitalares Ltda
Rua Dom Luiz de Braganca
80 Bairro de Mirandopolis
CEP 04050 Sao Paulo
Brazil
Tel: +55-11-5777955
Fax: +55-11-2766567

Pro-Cirurgica Produtos
  Cirurgicos Ltda
Rua Dr Ramiro D'Avila Nr. 44
Porto Alegre - CEP 90.620
Brazil
Tel: +55-512-238775
Fax: +55-512-230125

BULGARIA/ROMANIA
Biomedica GmbH
Divischgasse 4
1210 Vienna
Austria
Tel: +43-1-2923527
Fax: +43-1-2901361

CANADA
Incstar Corporation
PO Box 285
1990 Industrial Boulevard
Stillwater, MN 55082
USA
Tel: +1-800-328-1428
Fax: +1-612-779-7847

CHILE
Farmanuclear Ltda
Santa Filomena 91
Santiago
Chile
Tel: +56-2-7775042
Fax: +56-2-7378862

COSTA RICA
Biocientifica Internacional
  SRL
PO Box 2733-1000
San Jose
Costa Rica
Tel: +506-2-570618
Fax: +506-2-336952

CROATIA/SLOVENIA
Biomedica GmbH
Joseph Marx Strasse 45
8043 Graz
Austria
Tel: +43-316-381782
Fax: +43-316-3817823

CYPRUS
Advanced Diagnostics Ltd
22 Amohostou Street
Lakatamia, Nicosia
Cyprus
Tel: +357-2-383185
Fax: +357-2-384136

CZECHOSLOVAKIA
Biomedica CS
Vsejanska 5
19700 Prague 9
Czechoslovakia
Tel: +42-2-8501392
Fax: +42-2-8501392

DENMARK
Biotech IgG Aps
Oesterbrogade 95
2100 Copenhagen
Denmark
Tel: +45-31380500
Fax: +45-31387322

DOMINICAN REPUBLIC
Zucean Industrial SA
Apartado de Correos 396-2
Santo Domingo
Dominican Republic
Tel: +809-5678172
Fax: +809-5416352

ECUADOR
Commercial Mersal
Italia 344
PO Box 17-21 - 01093
Quito
Ecuador
Tel: +593-2-524936
Fax: +593-2-561976

EGYPT
Gamma Trade Company
PO Box 12655
61 El Mohandsin
Cairo
Egypt
Tel: +202-3480997
Fax: +202-3492687

FINLAND
Oy Tamro AB/APTA
Rajatorpantie 41B
01640 Vantaa
Finland
Tel: +358-0-85201787
Fax: +358-0-85201770

Scanlab OY
ElectroCity 4.krs
Tykistokatu 4
20520 Turku
Finland
Tel: +358-21-6375772
Fax: +358-21-6375729

FRANCE
Sorin Biomedica France SA
Parc de haute technologie
9 rue Georges Besse
  Batiment 4
Antony 92160
France
Tel: +33-1-46115230
Fax: +33-1-46661463

Bio-Rad SA
94-96 Rue Victor Hugo
94200 Ivry Sur Seine
France
Tel: +33-1-49606834
Fax: +33-1-46712467

GERMANY
Sorin Biomedica Deutschland
Optizstrasse 10

40470 Dusseldorf
Germany
Tel: +49-211-618030
Fax: +49-211-6180319

Baxter Deutschland GmbH
Bereich Diagnostik
Edisonstrasse 3-4
85716 Unterschleissheim
Germany
Tel: +49-89-317010
Fax: +49-89-31701-365

GREECE
Diachel Diagnostics Ltd
S. Merkouri 78 & Alkimachoul
Athens 116 34
Greece
Tel: +30-1-7243911
Fax: +30-1-7219874

GUATEMALA
Comerical Selecta
Boulevard Los Cipresales
#11-40, Zona 5
Guatemala
Tel: +502-2-350303
Fax: +502-2-351523

HONDURAS
Laboratorio "Salgado"
Edificio Lanza, Apartado
  Postal 507
4a Calle, SO No.66
entre 9 y 10 Aves., SO
San Pedro Sula
Honduras
Tel: +504-530497
Fax: +504-528884

HONG KONG
Science International
  Corporation
14/F Gee Tuck Building
16-20 Bonham Strang East
Sheung Wan
Hong Kong
Tel: +852-5437442
Fax: +852-5414089

Wenlin Company
Rooms 302-303 Lap Fei Bldg
6-8 Pottinger Street
Central District
Hong Kong
Tel: +852-5-212439
Fax: +852-2-267053

HUNGARY
Biomedica Hungaria
Hararor ut. 27/1/7
1122 Budapest
Hungary
Tel: +36-1-1559380
Fax: +36-1-1559380

Unilab
Beg ut. 35
1025 Budapest 2
Hungary
Tel: +36-1-1151278
Fax: +36-1-1154072

ICELAND/NORWAY
Biotech - IgG ApS
Oesterbrogade 95
2100 Copenhagen
Denmark

Tel: +45-31380500
Fax: +45-31387322

INDIA
Ezra Brothers
Mustafa Building, 4th Floor
Sir P.M. Road
Fort, Bombay 400 001
India
Tel: +91-22-266-0975
Fax: +91-22-266-5407

INDONESIA
UD Delta
Bona Indah Gardens
Block A4/46-47
Jakarta 12440
Indonesia
Tel: +62-21-7512736
Fax: +62-21-7502905

Multi Medindo Madyatama Pt.
Duta Merlin Block B/22-23
JLN Gajah Mada 3-5 Jakarta
Pusat PO Box 66 JKWA
Indonesia
Tel: +62-21-374152
Fax: +62-21-373025

ISRAEL
Pharmatope Ltd
3 Basel St
Kiriat Arie
Petach Tikvah
Israel
Tel: +972-3-9230048
Fax: +972-3-9232549

ITALY
SORIN-Italy
Via Crescentino
13040 VC Saluggia
Italy
Tel: +39-161-487994
Fax: +39-161-487396

Medical Systems SpA
Via Rio Torbido 40
16165 Genova Struppa
Italy
Tel: +39-1083401
Fax: +39-10804661

JAPAN
Baxter Ltd
4 Rokubancho
Chiyoda-ku
Tokyo 102
Japan
Tel: +81-3-32376616
Fax: +81-3-32376679

International Reagents Corp.
Sannomiya Kokusai Bldg 1-30
Hamabe-Dori 2-Chome
Chuo-ku Kobe 651
Japan
Tel: +81-78-2314053
Fax: +81-78-2320557

JORDAN
Burgan Drugstores
PO Box 773
Amman
Jordan
Tel: +962-6-699170
Fax: +962-6-699171

KOREA
New Korea Industrial Co Ltd
819-3 Yeok Sam-Dong
Kangnam-ku
Seoul
Korea
Tel: +82-2-552-2531
Fax: +82-2-557-0763

Sail Chemical Commercial
590-14 Sinsa Dong
Kangnam-ku
Seoul
Korea
Tel: +82-2-5400473-4
Fax: +82-2-5400475

Hoil Corporation
Dae Dong Building, Suite 402
930-10 Jaegi-Dong
Dong Dae Moon-ku
Seoul
Korea
Tel: +82-2-9682777
Fax: +82-2-9680464

I.K. Song
Sang Chung Industrial Co.
154-18 Garak Dong
Songpa-Gu
Seoul
Korea
Tel: +82-2-4499333
Fax: +82-2-4499335

LATVIA/LITHUANIA
OY Tamro AB/APTA
Rajatorpantie 41B
01640 Vantaa
Finland
Tel: +358-0-85201787
Fax: +358-0-85201770

LEBANON
Medek SARL
Boulevard Jdeideh
Sin-El-Fil Gecco Bldg 6 Floor
PO Box 90-946
Jdeidet El Metn
Lebanon
Tel: +961-1-898147
Fax: +961-1-881042

MALAYSIA
Biomarketing Services (M)
Sdn Bhd
No.31 Jalan Tembaga SD5/
2H
Sri Damansara
52200 Kuala Lumpur
Malaysia
Tel: +60-3-6333068
Fax: +60-3-6320093

MEXICO
Empresa Medica
Internacional SA
Fuentes Brontantes
31 Col Portales
Mexico City, CP 03570
Mexico
Tel: +52-5-5395524
Fax: +52-5-5326274

NETHERLANDS
SORIN Biomedica Nederland
NV
De Pael 37

1351 JG Almers-Haven
Netherlands
Tel: +31-365-314344
Fax: +31-365-349637

NEW ZEALAND
Biotek
Division of Lark International
Unit D, Donnor Place
PO Box 14 - 323 Panmure
Auckland
New Zealand
Tel: +64-9-5276413
Fax: +64-9-5709670

Baxter Diagnostics Pty Ltd
18 Allright Place
Mt. Wellington, Auckland
New Zealand
Tel: +64-9-5703200
Fax: +64-9-5703201

OMAN
IBN SINA Pharmacy LLC
PO Box 169, Muscat
Postal Code 114
Oman
Tel: +968-796367
Fax: +968-703472

PAKISTAN
Moonlight Scientific Traders
Flat No.9, 2nd Floor
Mall Plaza, The Mall
Rawalpindi Cantt.
Pakistan
Tel: +92-51-567953
Fax: +92-51-567351

PANAMA
Servicio y Equipo Medico
Hospitalario SA
PO Box 8556
Panama 5
Panama
Tel: +507-27-0537
Fax: +507-27-1246

PARAGUAY
Teri Scientific SRL
Azara 1529
Paraguay
Tel: +595-21-201455
Fax: +595-21-201455

PERU
AB Chimica Laboratorios SA
Peru Box 466
2898 NW 79th Avenue
Miami, FL 33122
USA

PHILIPPINES
TNC Everlight Philippines Inc
5th Floor, Unit B
Country Space I Building
Sen. Gil J. Puyat Avenue
Makati
Metro Manila 1200
Philippines
Tel: +63-2-8128736
Fax: +63-2-7173655

POLAND
Bellco Biomedica Poland
Plac Zbawiciela 2
00642 Warsaw
Poland

Tel: +48-2-6280543
Fax: +48-2-6211404

PORTUGAL
Isoder, LDA
R. Gregorio Lopes
1513 r/c, CED
1400 Lisbon
Portugal
Tel: +351-1-3020050
Fax: +351-1-3020034

PUERTO RICO
ISLA Lab Products, Corp
PO Box 361810
San Juan, PR 00936-1810
Puerto Rico
Tel: +809-792-2222
Fax: +809-781-4462

QATAR
SciTech Arabia
PO Box 8359
Doha
Qatar
Tel: +974-411605
Fax: +974-411606

RUSSIA
DRG International Moscow
Sovincentr/Firm "Inpred"
Krasnopresnenskaya Nab,
12#508
Moscow 123610
Russia
Tel: +7-095-253-1094
Fax: +7-095-253-1082

DRG Biomedical A/O
6 Graftio, No.30
St Petersburg 197376
Russia
Tel: +7-812-2344481
Fax: +7-812-2342647

SAUDI ARABIA
Abdulla Fouad
PO Box 257
Dammam 31411
Saudi Arabia
Tel: +966-3-8324400
Fax: +966-3-8345722

SINGAPORE
All Eight Marketing Services
Pte Ltd
No.6 Harper Road, #03-02
Leong Huat Bldg
Singapore 1336
Singapore
Tel: +65-2886388
Fax: +65-2849804

Diethelm Singapore Pte Ltd
Healthcare Division
34 Boon Leat Terrace 4th
Floor
Off Pasir Panjang Road
Singapore 0511
Singapore
Tel: +65-4701634
Fax: +65-4795676

SLOVAKIA
Biomedica GmbH
Divischgasse 4
1210 Vienna
Austria

Tel: +43-1-2923527
Fax: +43-1-2901361

PROLAB
Str. Romanova 22
851 02 Bratislava
Slovakia
Tel: +42-7-775628
Fax: +42-7-779186

SOUTH AFRICA
Ridge Diagnostics
PO Box 1216
Randpark Ridge
Johannesburg 2156
South Africa
Tel: +27-11-4651430/1
Fax: +27-11-4651454

Benmore Diagnostics
PO Box 784978
Sandton 2146
South Africa
Tel: +27-11-8042100
Fax: +27-11-8028010

SPAIN
SORIN Espana SA
Dr Esquerdo 70
28007 Madrid
Spain
Tel: +34-1-4096655
Fax: +34-1-4097763

DPC Diagnostic Products
Espana SA
Plaza Nuestra Senora del
Prado, No.2
28034 Madrid
Spain
Tel: +34-1-7304133
Fax: +34-1-7381313

SWEDEN
Boule Nordic AB
Lunastigen 3
141 44 Huddinge
Sweden
Tel: +46-8-7460035
Fax: +46-8-7468496

Kebo Lab AB
Fagerstagatan 18A
16394 Spanga
Sweden
Tel: +46-8-6213400
Fax: +46-8-7604596

SWITZERLAND
Dispolab AG
Altmoosstr 86
8157 Dielsdorf/Zurich
Switzerland
Tel: +41-1-8533626
Fax: +41-1-8532707

Sodiag SA
Via ai Molini 3
6616 Losone
Switzerland
Tel: +41-93-350363
Fax: +41-93-350364

SYRIA
Medical Supply Company
Chamber of Commerce
Building
1st Floor, PO Box 1359

© KLUWER ACADEMIC PUBLISHERS 1995, ISSN 1381-5067

Lattakia
Syria
Tel: +963-41-226497
Fax: +963-41-228488

TAIWAN
New Scientific Equipment Co
Ltd
PO Box 33-022
11F-1 32 Kungyuan Rd
Taipei 100
Taiwan
Tel: +886-2-3315260
Fax: +886-2-3116283

Taiwan Everlight Trading Co.
Ltd
#502/5F Fu-Hsing N. Road
Taipei
Taiwan
Tel: +886-2-5011960
Fax: +886-2-5010973

THAILAND
The Vidhayakom Public Co.
Ltd
Q. House Building, Floor 20-
21
66 Sukhumvit 21 (Asoke)
Klongtoey, Bangkok 10110
Thailand
Tel: +66-2-2642555
Fax: +66-2-2642558

TRINIDAD
Western Scientific Company
Ltd
Freeport Mission Road
Freeport
Trinidad
Tel: +809-6731378
Fax: +809-6730767

TURKEY
Duzen Sanayl Ve Dis Ticaret
Ltd
Necatlbey Caddesi 88/8
Ankara
Turkey
Tel: +90-312-2299323
Fax: +90-312-2300940

Kurtest
Halk Sokak Cilingiroglu
Ishani No. 5/4
Yenisehir, Ankara
Turkey
Tel: +90-312-4354635
Fax: +90-312-4339775

Limeks Tibbi Malzeme Sanayi
Ve Ticaret AS
Buyukdere Cad. Samanyolu
Ap. 33/9
PO Box 382
Sisli
Istanbul 80220
Turkey
Tel: +90-212-2256242
Fax: +90-212-2329354

UNITED ARAB EMIRATES
Bin Naeem Hospital Supply
Co.
PO Box 5264
Mubarak Bin Thamir Building
Tourist Club Area, Abu Dhabi
United Arab Emirates

Tel: +971-2-790671
Fax: +971-2-791970

UNITED KINGDOM
Incstar Ltd
Charles House
Toutley Rd
Wokingham
Berkshire, RG11 5QN
United Kingdom
Tel: +44-01734-772693
Fax: +44-01734-792061

URUGUAY
ENOL SA
Colon 1563, Of. 001
PO Box 6632
Montevideo
Uruguay
Tel: +598-2-962444
Fax: +598-2-963611

USA
Incstar Corporation
PO Box 285
1990 Industrial Boulevard
Stillwater, MN 55082
USA
Tel: +1-612-439-9710
Fax: +1-612-779-7847
Telex: 02 6879023

VENEZUELA
Distribuidora Akron CA
Avda Sucre
Torre Centro Pque, Boyaca
Piso 17, Of. 172
Urb. Los Dos Caminos
Caracas 1071
Venezuela
Tel: +58-2-2831755
Fax: +58-2-2839729

International Immuno-Diagnostics

Corporate
International Immuno-
Diagnostics
1155 Chess Drive #121
Foster City, CA 94404
USA
Tel: +1-415-345-9518
Fax: +1-415-578-1810

Johnson & Johnson Clinical Diagnostics Inc

Corporate
Johnson & Johnson Clinical
Diagnostics Inc
100 Indigo Creek Drive
Rochester, NY 14650
USA

By region
AUSTRALIA
Kodak (Australasia) Pty Ltd
Clinical Diagnostics Division
15 Talavera Road
PO Box 10
North Ryde, NSW 2113

Australia
Tel: +61-2-8704379
Fax: +61-2-8704545

BELGIUM
Johnson & Johnson Clinical
Diagnostics
Antwerpseweg 19-21
2340 Beerse
Belgium
Tel: +32-14-600211
Fax: +32-14-636611

CANADA
Kodak Canada Inc
3500 Eglinton Avenue West
Toronto
Ontario M6M 1V3
Canada
Tel: +1-416-766-8233
Fax: +1-416-760-4487

DENMARK
Johnson & Johnson Clinical
Diagnostics
Dybendal alle 10
2630 Taastrup
Denmark
Tel: +45-820222

FRANCE
Johnson & Johnson Clinical
Diagnostics SA
6 Ave du Canada
BP 232
91943 Les Ulis Cedex
France
Tel: +33-1-69865300
Fax: +33-1-64460464

GERMANY
Johnson & Johnson Clinical
Diagnostics GmbH
Karl-Landsteiner-Strasse 1
69151 Neckargemund
Germany

ITALY
Johnson & Johnson Clinical
Diagnostics SpA
Via le Matteotii 62
20092 Cinisello B
Milan
Italy
Tel: +39-2-660281
Fax: +39-2-66028765

JAPAN
Kodak Japan Diagnostics Ltd
Kinoshige Building 3F
1-10-3 Iwamoto-Cho
Chiyoda-ku
Tokyo 101
Japan
Tel: +81-3-58208200
Fax: +81-3-58208222

NETHERLANDS
Johnson & Johnson Clinical
Diagnostics
Postbus 1276
5004 BG Tilburg
Netherlands
Tel: +31-13-600211
Fax: +31-13-636611

NORWAY
Johnson & Johnson Clinical

Diagnostics
Ravnsborgvn 52
1364 Hvalstad
Norway
Tel: +47-66981030
Fax: +47-66981777

UNITED KINGDOM
Johnson & Johnson Clinical
Diagnostics Ltd
Mandeville House
62 The Broadway
Amersham
Bucks, HP7 0HJ
United Kingdom
Tel: +44-01494-431717
Fax: +44-01494-431165
Telex: 83447

Kreatech Diagnostics

Corporate
Kreatech Diagnostics
PO Box 12756
1100 AT Amsterdam
Netherlands
Tel: +31-20-6919181
Fax: +31-20-6963531

Laboratoire Fumouze

Corporate
Laboratoire Fumouze
26 Rue des Freres Chausson
92600 Asnieres
France
Tel: +33-1-49684142
Fax: +33-1-49684100

Laboratoire Eurobio

Corporate
Laboratoire Eurobio
7 Avenue de Scandanavia
91953 Les Ulis
France
Tel: +33-69079503
Fax: +33-69079534

Labsystems Oy

Corporate
Labsystems Oy
PO Box 8
00880 Helsinki
Finland
Finland
Fax: +358-0-7557610

Labor Diagnostica

Corporate
Labor Diagnostica
Industriestrasse 12
46359 Heiden
Germany
Tel: +49-2867-99070

Fax: +49-2867-990729

LMD Laboratories Inc

Corporate
LMD Laboratories Inc
2792 Loker Avenue West
Suite 103
Carlsbad, CA 92008
USA
Tel: +1-619-929-0110
Fax: +1-619-929-0115

Mast Immunosystems

Corporate
Mast Immunosystems
630 Clyde Court
Mountain View, CA 94043
USA
Tel: +1-415-961-5501
Fax: +1-415-969-2745

Medical Instruments Corporation

Corporate
Medical Instruments
    Corporation
Postfach 706
Solothurn 4502
Switzerland
Tel: +41-65-234355
Fax: +41-65-221792

Medix Biotech Inc

Corporate
Medix Biotech Inc
1531 Industrial Road
San Carlos, CA 94070
USA
Tel: +1-415-594-0513
Fax: +1-415-594-0571
Telex: 1561412

Melotec SA

Corporate
Melotec SA
Parc Tecnologic del Valles
08290 Cerdanyola
Barcelona
Spain
Tel: +34-3-5820166
Fax: +34-3-5801438

Menarini Diagnostics

Corporate
Menarini Diagnostics
Via Sette Santi 3
50131 Firenze
Italy

Tel: +39-55-56801
Fax: +39-55-5680216

Meridian Diagnostics Inc

Corporate
Meridian Diagnostics Inc
3471 River Mills Drive
Cincinnati, OH 45244
USA
Tel: +1-513-271-3700
Fax: +1-513-271-0124

By region
ITALY
Meridian Diagnostics Europe,
    Srl
Via G. Strobino 4
PO Box 33
20025 Legnano
Milan
Italy
Tel: +39-2-544178
Fax: +39-2-544178

USA
Meridian Diagnostics Inc
3471 River Mills Drive
Cincinnati, OH 45244
USA
Tel: +1-513-271-3700
Fax: +1-513-271-0124

Microgen Ltd

Corporate
Microgen Ltd
1 Admiralty Way
Camberley, Surrey
GU15 3DT
United Kingdom
Tel: +44-01276-600081

Murex Diagnostics Ltd

Corporate
Murex Diagnostics Ltd
Central Road
Temple Hill
Dartford, Kent
DA1 5LR
United Kingdom
Tel: +44-01322-277711
Fax: +44-01322-282572

By region
DENMARK
Orion Diagnostica Denmark
    A/S
Ndr. Strandvej 119
3150 Hellebaek
Denmark

FINLAND
Orion Diagnostica
PO Box 83
02101 Espoo
Finland

NORWAY
Orion Diagnostica a.s.
Solbraveien 43
1370 Asker
Norway

SWEDEN
Orion Diagnostica AB
Radhuset
619 00 Trosa
Sweden

UNITED KINGDOM
Murex Diagnostics Ltd
Central Road
Temple Hill
Dartford, Kent
DA1 5LR
United Kingdom
Tel: +44-01322-277711
Fax: +44-01322-282572

MRL Diagnostics

Corporate
MRL Diagnostics
10703 Progress Way
Cypress
USA
Tel: +1-714-445-0185
Fax: +1-714-220-1683

By region
FRANCE
MRL Diagnostics (European
    Division)
Quartier de la Courtisane
01250 Rignat
France
Tel: +33-74518181
Fax: +33-74518185

USA
MRL Diagnostics
10703 Progress Way
Cypress
USA
Tel: +1-714-445-0185
Fax: +1-714-220-1683

Omega Diagnostics Ltd

Corporate
Omega Diagnostics Ltd
Alloa Business Centre
Whins Road
Alloa, FK10 3SA
United Kingdom
Tel: +44-01259-217315
Fax: +44-01259-723251

Organon Teknika nv

Corporate
Organon Teknika nv
AKZO
Veedijk 58
2300 Turnhout
Belgium
Tel: +32-14-404040
Fax: +32-14-421600

By region
ARGENTINA
Organon Teknika SAIC
Gorriti 5143 C P
1414 Buenos Aires
Argentina
Tel: +54-1-726155
Fax: +54-1-8330959

AUSTRALIA
Organon Teknika Australia
Unit 13, 5 Hudson Avenue
Castle Hill, NSW 2154
Australia
Tel: +61-2-8993944
Fax: +61-2-8993984

AUSTRIA
Organon Teknika Austria
Siebenbrunnengasse 21/D/IV
1050 Vienna
Austria
Tel: +43-1-5454030
Fax: +43-1-545403055

BELGIUM
Organon Teknika
AKZO
Veedijk 58
2300 Turnhout
Belgium
Tel: +32-14-404040
Fax: +32-14-421600

BRAZIL
Organon Teknika Brasil
Rua Joao Alfredo, 403
04747 Sao Paulo
Brazil
Tel: +55-11-5229027
Fax: +55-11-5231445

CANADA
Organon Teknika Inc
30, North Wind Place
Scarborough, Ontario
M1S 3R5
Canada
Tel: +1-416-754-4344
Fax: +1-416-754-4488

COLOMBIA
Organon Teknika SA
Calle 79, no 16-32
Bogota
Colombia
Tel: +57-1-6109051
Fax: +57-1-6108560

CROATIA
Organon Teknika Croatia
Representative Office
Ljudevita Posavskog 9/1
41000 Zagreb
Croatia
Tel: +385-41-675108
Fax: +385-41-675108

DENMARK
Organon Teknika
9, Literbuen
PO Box 48
2740 Skovlunde
Denmark
Tel: +45-42846800
Fax: +45-44530181

FINLAND
Oy Organon Teknika AB
PO Box 254
SF-00181 Helsinki
Finland
Tel: +358-0-6949466
Fax: +358-0-6944931

FRANCE
Organon Teknika SA
5, Avenue des Pres
BP 26
94267 Fresnes Cedex
France
Tel: +33-146159015
Fax: +33-146603773

GERMANY
Organon Teknika Med Prod
GmbH
Postfach 1280
69209 Eppelheim
Germany
Tel: +49-6221-7923/0
Fax: +49-6221-763813

GREECE
Organon Teknika Hellas Ltd
PO Box 73893
122, Vouliagmenis Avenue
167 10 Athens
Greece
Tel: +30-1-9648500/1
Fax: +30-1-9648517

INDONESIA
Organon Teknika
Representative Office
Jl Wijaya X/7 Kabayoran Baru
PO Box 6584, JKSDW
12065 Jakarta
Indonesia
Tel: +62-21-7395741
Fax: +62-21-7395741

ITALY
Organon Teknika SpA
Via Ostilla 15
Rome 00184
Italy
Tel: +39-6-701921
Fax: +39-6-7005059

JAPAN
Organon Teknika KK
No. 2 Monarni Building
31-11, Kabukicho, 2-chome
Shinjuku-ku
Tokyo 160
Japan
Tel: +81-3-32324333
Fax: +81-3-32327937

NORTH AFRICA
Organon Teknika SARL
42, Avenue de l'Armee
Royale
PO Box 7520
Casablanca
Morocco
Tel: +212-2-312832/3
Fax: +212-2-312438

NORWAY
Organon Teknika
PO Box 325
Roykenvn 70
1371 Asker

Norway
Tel: +47-66-784365
Fax: +47-66-795172

PHILIPPINES
Organon Teknika Philippines
7th Floor, Philcox Building
172 Salcedo Street
Legaspi Village
Makati, Metro Manila
Philippines
Tel: +63-2-8102131
Fax: +63-2-8120896

POLAND
Organon Teknika
Oddzial w Warszawie
UL Kubickiego 3m 2
02-954 Warsaw
Poland
Tel: +48-2-6420026/7
Fax: +48-2-6424505

PORTUGAL
Organon Teknika
Av Visconde Valmor, 65-A
1000 Lisbon
Portugal
Tel: +351-1-7936565
Fax: +351-1-7966155

RUSSIA
AKZO Moscow
Organon Teknika Russia
69, Ul. Vavilova
117846 Moscow
Russia
Tel: +7-095-9382985
Fax: +7-095-1343365

SPAIN
Organon Teknika Espanola
SA
Carretera de Enlace B-201
Apartado 56
08830 S. Baudilio de
Llobregat
(Barcelona)
Spain
Tel: +34-3-6401462
Fax: +34-3-6541750

SWEDEN
Organon Teknika AB
PO Box 5076
Redegatan 9
426 05 Vastra Frolunda
Sweden
Tel: +46-31-299490
Fax: +46-31-299958

SWITZERLAND
Organon Teknika AG
Postfach 129
Churerstrasse 160
8808 Pfaffikon
Switzerland
Tel: +41-55-486131
Fax: +41-55-486207

TURKEY
Organon Teknika AS
Kayisdagi Caddesi
No. 288, Kat.2 Sahrayicedid
81080 Erenkoy
Istanbul
Turkey
Tel: +90-216-3850633

Fax: +90-216-3850718

UNITED KINGDOM
Organon Teknika Ltd
Cambridge Science Park
Milton Road
Cambridge, CB4 4FL
United Kingdom
Tel: +44-01223-423650
Fax: +44-01223-420264

Orion Corporation, Orion Diagnostica

Corporate
Orion Corporation, Orion
Diagnostica
PO Box 83
02101 Espoo
Finland
Tel: +358-0-4291
Fax: +358-0-4292794

By region
FINLAND
Orion Corporation, Orion
Diagnostica
PO Box 83
02101 Espoo
Finland
Tel: +358-0-4291
Fax: +358-0-4292794

Orion Corporation, Orion
Diagnostica
PO Box 425
20101 Turku
Finland
Tel: +358-21-662011
Fax: +358-21-662546

NORWAY
Orion Diagnostica as
Postboks 321
1371 Asker
Norway
Tel: +47-66904675
Fax: +47-66904788

SWEDEN
Orion Diagnostica AB
Radhuset
S-619 00 Trosa
Sweden
Tel: +46-156-13260
Fax: +46-156-17355

USA
Orion Diagnostica Inc
PO Box 218
Somerset, NJ 08875-0218
USA
Tel: +1-908-246-3366
Fax: +1-908-246-0570

Ortho Diagnostic Systems Ltd

Corporate
Ortho Diagnostic Systems Ltd
PO Box 690
Mandeville House
62 The Broadway
Amersham, Bucks

HP7 0JS
United Kingdom
Tel: +44-01494-545600
Fax: +44-01494-431165

PanBio

Corporate
PanBio
116 Lutwyche Road
Windsor, Queensland 4030
Australia
Tel: +61-7-3571177
Fax: +61-7-3571222

Progen (see Quidel)

Corporate
Progen (see Quidel)

PRO-LAB Diagnostics

Corporate
PRO-LAB Diagnostics
Unit 7, Westwood Court
Clayhill Industrial Estate
Neston, South Wirral
Cheshire, L64 3UH
United Kingdom
Tel: +44-0151-353-1613
Fax: +44-0151-353-1614

Quidel

Corporate
Quidel
10165 McKellar Court
San Diego, CA 92121
USA
Tel: +1-619-552-1100
Fax: +1-619-546-8955

By region
NETHERLANDS
Quidel European Office
Berenkoog 29
1822 BH Alkmaar
Netherlands
Tel: +31-72-643796
Fax: +31-72-643132

USA
Quidel
10165 McKellar Court
San Diego, CA 92121
USA
Tel: +1-619-552-1100
Fax: +1-619-546-8955

Radim

Corporate
Radim
Via del Mare, 125
00040 Pomezia
Rome
Italy
Tel: +39-6-9108364
Fax: +39-6-9106128

By region
BELGIUM
Radim SA
Parc Scientifique du Sart-
Tilman
Avenue Pre Aily 10
4031 Angleur (Liege)
Belgium
Tel: +32-41-674464
Fax: +32-41-670063

ITALY
Radim
Via del Mare, 125
00040 Pomezia
Roma
Italy
Tel: +39-6-9108364
Fax: +39-6-9106128

SEAC srl
Via di Prato, 72/74
50041 Calenzano-Prato (FI)
Italy
Tel: +39-55-8877469
Fax: +39-55-8877771

SPAIN
Radim Iberica SA
C. Lepanto, 339
Bajos, Local 7
08025 Barcelona
Spain
Tel: +34-3-4333921
Fax: +34-3-4333796

R-Biopharm GmbH

Corporate
R-Biopharm GmbH
Rossler Strasse 94
64293 Darmstadt
Germany
Tel: +49-6151-81020
Fax: +49-6151-810220

Remel

Corporate
Remel
12076 Santa Fe Drive
Lenexa, KS 66215
USA
Tel: +1-800-255-6730

F. Hoffmann-La Roche Ltd

Corporate
F. Hoffmann-La Roche Ltd
Roche Diagnostic Systems
4002 Basel
Switzerland
Tel: +41-61-6888726
Fax: +41-61-6814135

By region
AUSTRALIA
Roche Diagnostic Systems
Unit C1, 1-3 Rodborough
Road
French Forest, NSW 2086
Australia
Tel: +61-2-9758150
Fax: +61-2-9755254

BENELUX
Produits Roche SA
75 Rue Dante
1070 Brussels
Belgium
Tel: +32-2-5258211
Fax: +32-2-5258201

BRAZIL
Produtos Roche Quimicos e
Farmaceuticos SA
Caixa Postal 6364
01000 Sao Paulo SP
Brazil
Tel: +55-11-8693322
Fax: +55-11-8698877

CANADA
Hoffmann-La Roche Ltd
2455 Meadowpine Boulevard
Mississauga
Ontario L5N 6L7
Canada
Tel: +1-416-542-5555
Fax: +1-416-542-5649

EASTERN EUROPE
Hoffmann-La Roche Wien
GmbH
Postfach 70
1103 Vienna
Austria
Tel: +43-222-781604
Fax: +43-222-781604253

FRANCE
Produits Roche SA
52 Boulevard du Parc
92521 Neuilly sur Seine
France
Tel: +33-1-46405000
Fax: +33-1-46405292

GERMANY
Hoffmann-La Roche AG
79639 Grenzach-Wyhlen
Germany
Tel: +49-7624-140
Fax: +49-7624-1019

ITALY
Dr G Minola
Prodotti Roche SpA
Piazza Durante 11
20131 Milan
Italy
Tel: +39-2-28841
Fax: +39-2-2884585

JAPAN
Nippon Roche KK
6th Floor, Shin-Onarimon
Bldg
6-17-19, Shinbashi, Minato-ku
Tokyo 105
Japan
Tel: +81-3-54701707
Fax: +81-3-54701720

NEW ZEALAND
Roche Products (New
Zealand) Ltd
PO Box 12-492
Penrose, Auckland
New Zealand

Tel: +64-9-640029
Fax: +64-9-640020

PORTUGAL
Roche Farmaceutica Quimica
Lda
Estrada Nacional 249-1
2700 Amadora
Portugal
Tel: +351-1-4184565
Fax: +351-1-4186677

RUSSIA
DIAplus
Nauchny proezd, 8
Moscow 117246
Russia
Tel: +7-095-332-6440
Fax: +7-095-332-6557

SCANDINAVIA
Roche A/S
Industriholmen 59
2650 Hvidovre-Copenhagen
Denmark
Tel: +45-31787211
Fax: +45-3187215

SOUTH AFRICA
Roche Products (Pty) Ltd
PO Box 4589
Johannesburg 2000
South Africa
Tel: +27-11-9745335
Fax: +27-11-3922338

SPAIN
Productos Roche SA
Apartado de Correos 1.157
28080 Madrid
Spain
Tel: +34-1-2086240
Fax: +34-1-2084442

SWITZERLAND
F. Hoffmann-La Roche Ltd
Roche Diagnostic Systems
4002 Basel
Switzerland
Tel: +41-61-6888726
Fax: +41-61-6814135

UNITED KINGDOM
Roche Products Ltd
Roche Diagnostic Systems
PO Box 8
Welwyn Garden City
Herts, AL7 3AY
United Kingdom
Tel: +44-01707-366000
Fax: +44-01707-373556

USA
Hoffmann-La Roche Inc
Roche Diagnostic Systems
Inc
1080 US Highway 202
Branchburg, NJ 08876
USA
Tel: +1-908-253-7652
Fax: +1-908-253-7200

Sanofi Diagnostics Pasteur, Inc

Corporate
Sanofi Diagnostics Pasteur,
Inc
1000 Lake Hazeltine Drive
Chaska, MN 55318-1084
USA
Tel: +1-612-448-4848
Fax: +1-612-368-1110

By region
AUSTRALIA
Sanofi Diagnostics Pasteur
4th floor Clyde House
140 Arthur Street
North Sydney, NSW 2060
Australia
Tel: +61-2-9575515
Fax: +61-2-9290364

AUSTRIA
Sanofi Diagnostics Pasteur H.
GmbH
Ameisgasse 31, 4/5
1140 Vienna
Austria
Tel: +43-222-8941290
Fax: +43-222-8941299

BELGIUM
Sanofi Diagnostics Pasteur
NV
Woudstraat 25
3600 Genk
Belgium
Tel: +32-89-384292
Fax: +32-89-384392
Telex: 38087

CANADA
Sanofi Diagnostics Pasteur
2403 Guenette Street
Montreal
Quebec, H4R 2E9
Canada
Tel: +1-514-334-4372
Fax: +1-514-334-4415

FRANCE
Diagnostics Pasteur
3 Blvd Raymond Poincare
92430 Marnes la Coquette
France
Tel: +33-1-47956140
Fax: +33-1-47956141

GERMANY
Sanofi Diagnostics Pasteur
GmbH
Sasbacher Strasse 5
79111 Freiburg
Germany
Tel: +49-761-490510
Fax: +49-761-4905199

ITALY
Sanofi Diagnostics Pasteur
Via Carbonera 2
20137 Milan
Italy
Tel: +39-2-739419
Fax: +39-2-57404678

© *KLUWER ACADEMIC PUBLISHERS 1995, ISSN 1381-5067*

MEXICO
Sanofi Diagnostics Pasteur
SA
Av Periferico sur 6677 2nd
Floor
Col Ejidos de Tepepan
Deleg Xochimilco
Mexico 16018
Mexico

NETHERLANDS
Sanofi Diagnostics Pasteur
BV
Govert van Wijnkade 48
3144 EG Maassluis
Netherlands
Tel: +31-18-9917555
Fax: +31-18-9914555

PORTUGAL
Sanofi Diagnostics Pasteur
Lda
Rua Artilharia UM, 63 - r/c
1200 Lisbon
Portugal
Tel: +351-1-3806008
Fax: +351-1-3806099

SPAIN
Sanofi Diagnostics Pasteur
SA
Calle Jarama
28002 Madrid
Spain
Tel: +34-1-5630100
Fax: +34-1-5644383

SWITZERLAND
Sanofi Diagnostics Pasteur
SA
Zuchwillerstrasse 41
PO Box 420
4501 Solothurn
Switzerland
Tel: +41-65-234151
Fax: +41-65234153

THAILAND
Sanofi Pacific Diagnostics
IFCT Building 9th Floor
1770 New Petchbury Road
Bangkok 10310
Thailand
Tel: +66-2-2548070
Fax: +66-2-2548060

UNITED KINGDOM
Sanofi Diagnostics Pasteur
Ltd
PO Box 209
3 Rhodes Way
Watford, Hertfordshire
WD2 4QE
United Kingdom
Tel: +44-01923-212212
Fax: +44-01923-243001

USA
Sanofi Diagnostics Pasteur,
Inc
1000 Lake Hazeltine Drive
Chaska, MN 55318-1084
USA
Tel: +1-612-448-4848
Fax: +1-612-368-1110

Genetic Systems
6565 185th Avenue North

East
Redmond, WA 98052
USA
Tel: +1-206-881-8300
Fax: +1-206-861-5010

Savyon Diagnostics Ltd

Corporate
Savyon Diagnostics Ltd
Kiryat Minrav
3 Habosem St
Ashdod 77101
Israel
Tel: +972-8-562920
Fax: +972-8-563258

By region
ISRAEL
Savyon Diagnostics Ltd
Kiryat Minrav
3 Habosem St
Ashdod 77101
Israel
Tel: +972-8-562920
Fax: +972-8-563258

UNITED KINGDOM
Omni Triage Medical Ltd
131 Tranmere Road
London, SW18 3QP
United Kingdom
Tel: +44-0171-737-7781
Fax: +44-0171-738-4473

SA Scientific

Corporate
SA Scientific
4919 Golden Quail
San Antonio, TX 78240
USA
Tel: +1-210-699-8800
Fax: +1-210-699-6545

Scimedix Corporation

Corporate
Scimedix Corporation
400 Ford Road
Denville, NJ 07834
USA
Tel: +1-201-625-8822
Fax: +1-201-625-8796

Seradyn Inc

Corporate
Seradyn Inc
1200 Madison Avenue
PO Box 1210
Indianapolis, IN 46206
USA
Tel: +1-317-266-2932
Fax: +1-317-266-2991

SFRI Laboratoire

Corporate
SFRI Laboratoire
Berganton
33127 Saint Jean D'Illac
France
Tel: +33-56216195
Fax: +33-56689009

Shield Diagnostics

Corporate
Shield Diagnostics
Technology Park
Dundee, DD2 1SW
United Kingdom
Tel: +44-01382-561000
Fax: +44-01382-561056

Sigma Chemical Company

Corporate
Sigma Chemical Company
PO Box 14508
St Louis, MO 63178
USA
Tel: +1-314-771-5750
Fax: +1-314-771-5757

By region
AUSTRALIA
Sigma-Aldrich Pty Ltd
Unit 2, 10 Anella Avenue
Castle Hill, NSW 2154
Australia
Tel: +61-2-899-9977
Fax: +61-2-899-9742

BELGIUM
Sigma Diagnostics
K. Cardijnplein 8
B-2880 Bornem
Belgium
Tel: +32-3-8991301
Fax: +32-3-8991311

BRAZIL
Sigma-Aldrich Chemical
Representocoes Ltda
Rua Sabara, 566 - Conj. 53
01239-010 Sao Paulo, SP
Brazil
Tel: +55-11-2311866
Fax: +55-11-2579079

CZECHOSLOVAKIA
Sigma-Aldrich sro
Krizikova 27
180 00 Prague 8
Czechoslovakia
Tel: +42-2-2366973
Fax: +42-2-2364141

FINLAND
Scanlab Oy
Elsktracity, 4Krs
Yrityskeskus Dio
Tykistonkatu 2-4
20520 Turku
Finland
Tel: +358-21-63757729

Fax: +358-21-6375729

FRANCE
Sigma Chimi S.a.r.l.
L'Isle d'Abeau Chesnes
BP 701
38070 St Quentin Fallavier
Cedex
France
Tel: +33-74822800
Fax: +33-74956808
Telex: 308215

GERMANY
Sigma Chemie GmbH
Grunwalder Weg 30
82041 Deisenhofen
Germany
Tel: +49-89-61301
Fax: +49-89-6135135
Telex: 528252

HUNGARY
Sigma Aldrich
Terez Krt 39
1.em.11
1067 Budapest
Hungary
Tel: +36-1-2691288
Fax: +36-1-1533391

INDIA
Sigma-Aldrich Corporation
Flat No 4082
Sector B-4/5
Vasant Kuns
New Delhi 110 070
India
Tel: +91-11-6899826
Fax: +91-11-6899827

ISRAEL
Sigma Israel Chemicals Ltd
PO Box 369
Holon 58100
Israel
Tel: +972-3-5596610
Fax: +972-3-5596596

ITALY
Sigma Chimica
Via Gallarate 154
20151 Milan
Italy
Tel: +39-2-33417-30
Fax: +39-2-38010737

POLAND
Sigma Aldrich Sp
Bastionowa 19
61-663 Poznan
Poland
Tel: +48-61-232481
Fax: +48-61-232781

SPAIN
Sigma Quimica
Apt. Correos 161
28100 Alcobendas
Madrid
Spain
Tel: +34-1-6619977
Fax: +34-1-6619642
Telex: 22189

SWITZERLAND
Sigma Chemie
PO Box 260

© KLUWER ACADEMIC PUBLISHERS 1995, ISSN 1381-5067

9470 Buchs
Switzerland
Telex: 855282

UNITED KINGDOM
Sigma Chemical Co. Ltd
Fancy Road
Poole
Dorset, BH17 7NH
United Kingdom
Tel: +44-01202-733114
Fax: +44-01202-715460

USA
Sigma Chemical Company
PO Box 14508
St Louis, MO 63178
USA
Tel: +1-314-771-5750
Fax: +1-314-771-5757

Scanlab Oy
Elsktracity, 4Krs
Yrityskeskus Dio
Tykistonkatu 2-4
20520 Turku
Finland
Tel: +358-21-63757729
Fax: +358-21-6375729

Sorin Biomedica

Corporate
Sorin Biomedica
Via Crescentino
13040 Saluggia
Italy
Tel: +39-161-4871
Fax: +39-161-487545

By region
ARGENTINA
Tecnogam
Avanida Cordoba 950
Piso 13
1054 Buenos Aires
Argentina
Tel: +54-1-3935041
Fax: +54-1-3935068

AUSTRALIA
CSL
45 Poplar Road
Parkville, Victoria
Australia
Tel: +61-3-3891911
Fax: +61-3-3891646

AUSTRIA
Biomedica
   Handelsgesellschaft GmbH
Divischgasse 4
1210 Vienna
Austria
Tel: +43-222-393527
Fax: +43-222-3901361

BRAZIL
Sorin Biomedica Industrial
Rua Robert Bosch 130
Parque Industrial Tomas
   Edison
01141 Sao Paulo
Brazil
Tel: +55-11-8263377
Fax: +55-11-673873

EGYPT
Clinilab
127 Mohammed Farid Street
Cairo
Egypt
Tel: +20-2-3919397
Fax: +20-2-3915247

GREECE
Diachel
Sp. Merkouri 78 & Alkimachou
   1
11634 Athens
Greece
Tel: +30-1-7235523
Fax: +30-1-7219874

INDIA
Ranbaxy Laboratories
10th Floor, Devika Tower
6 Nehru Place
New Delhi 110019
India
Tel: +91-11-6437078
Fax: +91-11-6430633

ITALY
Sorin Biomedica
Via Crescentino
13040 Saluggia
Italy
Tel: +39-161-4871
Fax: +39-161-487545

JORDAN
Burgan Drugstores
PO Box 773
Amman
Jordan
Tel: +962-6-699170
Fax: +962-6-699171

KOREA
Boo Kyung
CPO Box 426
Seoul
Korea
Tel: +82-2-7483331
Fax: +82-2-7845123

LEBANON
Medek SARL
St Joseph Hospital Street
St Joseph Pharmacy Building
Beirut
Lebanon
Tel: +961-1-881042/898147

MALAYSIA
MD Products & Services SDN
   BHD
35A Jalan 1/76, Desa Pandan
Off Jalan Kampung Pandan
55100 Kuala Lumpur
Malaysia
Tel: +60-3-9862939
Fax: +60-3-9862935

NORWAY
Dan Meszanski AS
Holstsgate 6
PO Box 4324
Torshov
Oslo 4
Norway
Tel: +47-2-370788
Fax: +47-2-379585

OMAN
Muscat Pharmacy
PO Box 438
Muscat
Oman
Tel: +968-794501
Fax: +968-795202

PAKISTAN
Scherzo Agencies
1st Floor, Madina Market
32 Abkari Road
La Hore 2
Pakistan
Tel: +92-42-324626
Fax: +92-42-306299

PORTUGAL
Isoder
Rua dos Lusiades, 5-5
Letra H
1300 Lisbon
Portugal
Tel: +351-1-3647208
Fax: +351-1-3638788

SAUDI ARABIA
Al Salehia Medical Est.
PO Box 991
Riyadh 11421
Saudi Arabia
Tel: +966-1-4633205
Fax: +966-1-4634362

SWEDEN
Karo Bio Diagnostics
Lunastigen 3
141 44 Huddinge
Sweden
Tel: +46-8-7460990
Fax: +46-8-7468496

SWITZERLAND
Sodiag
via Locarno 76
6616 Losone
Switzerland
Tel: +41-93-350363
Fax: +41-93-350364

THAILAND
Rapport
4/631 Sahakorn Klongkum
   Soi 26
Sukhaphiban 2 Rd
Bangkapi
Bangkok 10240
Thailand
Tel: +66-2-3746977/6978
Fax: +66-2-3745312

TURKEY
Gokham Laboratuvar AS
1440 Solkak
No. 4 Alsancak
Izmir 35220
Turkey
Tel: +90-51-220055
Fax: +90-51-636904

Duzen Industry & Foreign
   Trade Co. Ltd
Necatlbey Caddesi 88/8
Ankara
Turkey
Tel: +90-4-2298025/9323
Fax: +90-4-2300940

UNITED ARAB EMIRATES
Emirates Medical Co.
PO Box 1286
Ahmed Saif Bei Hasa Bldg
(opposite Dubai Police HQ)
Dubai Sharajah Road
Deira, Dubai
United Arab Emirates
Tel: +971-4-662319
Fax: +971-4-692774

VENEZUELA
Clini-kit srl
Edif. Alpha
Calle Republica Dominicane
Piso 3, Locale 4
Boleita Sur
Caracas 1070
Venezuela
Tel: +58-2-2392858
Fax: +58-2-2398905

Stellar Bio Systems Inc

Corporate
Stellar Bio Systems Inc
9075 Guildford Road
Columbia, MD 21046
USA
Tel: +1-410-381-8550
Fax: +1-410-381-8984

Tecra Diagnostics

Corporate
Tecra Diagnostics
PO Box 20
Roseville
NSW 2069
Australia
Tel: +61-2-4175344
Fax: +61-2-4177858

Unipath Ltd

Corporate
Unipath Ltd
Wade Road
Basingstoke, Hampshire
RG24 8PW
United Kingdom
Tel: +44-01256-841144
Fax: +44-01256-463388

By region
AUSTRALIA
Oxoid Australia Pty Ltd
West Heidelberg
PO Box 220
Melbourne, Victoria 3081
Australia
Tel: +61-3-4581311
Fax: +61-3-4584759

FRANCE
Unipath SA
6 route de Paisy, BP 13
69572 Dardilly Cedex
France
Tel: +33-78351731
Fax: +33-78660376

© KLUWER ACADEMIC PUBLISHERS 1995, ISSN 1381-5067

GERMANY
Unipath GmbH
Am Lippeglacis 6-8
46483 Wesel
Germany
Tel: +49-281-1520
Fax: +49-281-1521

ITALY
Unipath SpA
Via Montenero 180
20024 Garbagnate
Milanese, Milan
Italy
Tel: +39-2-9955651
Fax: +39-2-9958260

NORTH AMERICA
Unipath Inc
217 Colonnade Road,
Nepean
Ontario, K2E 7K3
Canada
Tel: +1-613-226-1318
Fax: +1-613-226-3728

SPAIN
Unipath Espana SA
Via de los Poblados 10
Nave 3-13
Madrid 28033
Spain
Tel: +34-1-7642554
Telex: 45670

UNITED KINGDOM
Unipath Ltd
Wade Road
Basingstoke, Hampshire
RG24 8PW
United Kingdom
Tel: +44-01256-841144
Fax: +44-01256-463388

United Biotech Inc

Corporate
United Biotech Inc
110-C Pioneer Way
Mountain View, CA 94041
USA

Tel: +1-415-961-2910
Fax: +1-415-961-0766

Zeus Scientific Inc

Corporate
Zeus Scientific Inc
PO Box 38
Raritan, NJ 08869
USA
Tel: +1-201-526-3744
Fax: +1-201-526-2058

INDEX BY MANUFACTURER, ASSAY TYPE, ANTIBODY/ANTIGEN DETECTION AND MICROORGANISM

Abbott Laboratories
EIA (non-competitive)
Antibody detection (IgA)
Toxoplasma gondii - 920
Antibody detection (IgG)
Rubella virus - 862
Toxoplasma gondii - 923
Toxoplasma gondii - 923
Antibody detection (IgM)
Rubella virus - 879
Toxoplasma gondii - 942
Toxoplasma gondii - 943
Antigen detection
Rotavirus - 847
Rotavirus - 847

Alexon Inc
EIA (non-competitive)
Antibody detection
Borrelia burgdorferi - 651
Antibody detection (IgG)
Helicobacter pylori - 777
Antigen detection
Clostridium difficile (toxin A) - 682
Cryptosporidium species - 696
Cryptosporidium species - 696
Entamoeba histolytica - 704
Giardia lamblia - 763
Giardia lamblia - 763
Rotavirus - 848

Alfa Biotech
EIA (non-competitive)
Antibody detection (IgG)
Brucella species - 672
Epstein-Barr virus (VCA) - 731
Rubella virus - 863
Toxoplasma gondii - 924
Antibody detection (IgM)
Brucella species - 675
Epstein-Barr virus (VCA) - 741
Rubella virus - 880
Toxoplasma gondii - 943
Antigen detection
Adenovirus - 630

Amico Laboratories Inc
EIA (non-competitive)
Antibody detection (IgG)
Adenovirus - 637
Entamoeba histolytica - 706
Epstein-Barr virus (VCA) - 732
Leishmania species - 810
Mumps virus - 820
Rotavirus - 856
Schistosoma species - 910
Toxoplasma gondii - 924
Toxoplasma gondii - 925
Antibody detection (IgM)
Adenovirus - 640
Entamoeba histolytica - 708
Entamoeba histolytica - 708
Epstein-Barr virus (VCA) - 741
Epstein-Barr virus (VCA) - 742
Leishmania species - 811
Leishmania species - 812
Mumps virus - 828
Mumps virus - 828
Rotavirus - 857
Rotavirus - 858
Rubella virus - 880
Rubella virus - 881
Schistosoma species - 910
Schistosoma species - 911
Toxoplasma gondii - 944
Toxoplasma gondii - 944

Amrad Corporation Ltd
EIA (non-competitive)
Antibody detection (IgG)

Helicobacter pylori - 778

Argene-Biosoft
Immunofluorescence assay (direct)
Antigen detection
Adenovirus - 633
Immunofluorescence assay (indirect)
Antigen detection
Adenovirus - 635

BAG-Biologische Analysensystem GmbH
EIA (non-competitive)
Antibody detection (IgG)
Adenovirus - 638
Borrelia burgdorferi - 655
Epstein-Barr virus (VCA) - 732
Mumps virus - 820
Rubella virus - 863
Antibody detection (IgM)
Adenovirus - 640
Borrelia burgdorferi - 659
Epstein-Barr virus (VCA) - 742
Mumps virus - 829
Rubella virus - 881

Bard Diagnostic Sciences, Inc
EIA (non-competitive)
Antibody detection (IgG)
Helicobacter pylori - 778

Becton Dickinson
EIA (non-competitive)
Antigen detection
Clostridium difficile (toxin A) - 682
Particle agglutination assay
Antibody detection
Rubella virus - 901
Antigen detection
Clostridium difficile - 680
Meningitis pathogens: E. coli, N. meningitis - 983
Meningitis pathogens: E. coli, H. influenzae, N. meningitidis, S. pneumoniae and Strep group B - 983
Neisseria meningitidis - 838

Behringwerke AG
EIA (non-competitive)
Antibody detection (IgG)
Epstein-Barr virus - 714
Helicobacter pylori - 779
Mumps virus - 821
Rubella virus - 864
Toxoplasma gondii - 925
Antibody detection (IgG and/or IgM)
Borrelia burgdorferi - 656
Antibody detection (IgM)
Epstein-Barr virus - 716
Mumps virus - 829
Rubella virus - 882
Toxoplasma gondii - 945

BIOKIT SA
EIA (non-competitive)
Antibody detection (IgG)
Rubella virus - 864
Toxoplasma gondii - 926
Antibody detection (IgM)
Rubella virus - 882
Toxoplasma gondii - 945
Particle agglutination assay
Antibody detection
Heterophile antigen - 755
Rubella virus - 901
Toxoplasma gondii - 968
Antigen detection
Adenovirus - 635
Rotavirus - 852

Biomerica
EIA (non-competitive)
Antibody detection (IgA)
Helicobacter pylori - 773
Antibody detection (IgG)
Helicobacter pylori - 780
Antibody detection (IgM)
Helicobacter pylori - 790

bioMerieux
EIA (non-competitive)
Antibody detection
Borrelia burgdorferi - 651
Antibody detection (IgG)
Mumps virus - 821
Rubella virus - 865
Toxoplasma gondii - 926
Antibody detection (IgM)
Rubella virus - 883
Toxoplasma gondii - 946
Antigen detection
Clostridium difficile (toxin A) - 683
Immunofluorescence assay (indirect)
Antibody detection
Borrelia burgdorferi - 667
Leishmania species - 812
Plasmodium falciparum - 841
Rickettsia conorii - 843
Toxoplasma gondii - 960
Particle agglutination assay
Antibody detection (IgA and/or IgM)
Toxoplasma gondii - 970
Antibody detection (IgM)
Toxoplasma gondii - 970
Antigen detection
Adenovirus - 636
Cryptococcus neoformans - 692
Enterococcus species (Streptococcus beta-haemolytic, group D) - 711
Rotavirus - 853

Bion Enterprises Ltd
Immunofluorescence assay (indirect)
Antibody detection (IgG)
Epstein-Barr virus (VCA) - 749
Mumps virus - 834
Antibody detection (IgM)
Epstein-Barr virus (VCA) - 753

Bio-Rad
EIA (non-competitive)
Antibody detection (IgA)
Helicobacter pylori - 773
Antibody detection (IgG)
Helicobacter pylori - 779
Antibody detection (IgM)
Helicobacter pylori - 790

Biotecx Laboratories Inc
EIA (non-competitive)
Antibody detection (IgG)
Toxoplasma gondii - 927

Biotest Diagnostics
EIA (non-competitive)
Antibody detection (IgG)
Epstein-Barr virus (EA) - 720
Epstein-Barr virus (EBNA) - 724
Antibody detection (IgM)
Epstein-Barr virus (EA) - 721

Biotrin International Ltd
EIA (non-competitive)
Antibody detection (IgG)
Human Herpes virus 6 - 796
Human parvovirus B19 - 799
Antibody detection (IgM)
Human parvovirus B19 - 802
Antigen detection
Adenovirus - 630

Immunoblot assay
  Antibody detection (IgG and/or IgM)
    Human parvovirus B19 - 806
Immunofluorescence assay (indirect)
  Antibody detection (IgG)
    Human Herpes virus 6 - 797
  Antibody detection (IgG and/or IgM)
    Human parvovirus B19 - 808

BioWhittaker Inc
EIA (non-competitive)
  Antibody detection (IgA)
    Helicobacter pylori - 774
  Antibody detection (IgG)
    Epstein-Barr virus (EBNA) - 724
    Epstein Barr virus (EBNA) - 725
    Epstein-Barr virus (VCA) - 733
    Epstein-Barr virus (VCA) - 733
    Helicobacter pylori - 780
    Helicobacter pylori - 781
    Mumps virus - 822
    Mumps virus - 822
    Rubella virus - 865
    Rubella virus - 866
    Toxoplasma gondii - 927
    Toxoplasma gondii - 928
  Antibody detection (IgG and/or IgM)
    Borrelia burgdorferi - 657
    Borrelia burgdorferi - 657
  Antibody detection (IgM)
    Borrelia burgdorferi - 660
    Epstein-Barr virus (VCA) - 743
    Rubella virus - 883
    Toxoplasma gondii - 946

Boule Diagnostics AB
Particle agglutination assay (coagglutination)
  Antigen detection
    Enterococcus species (Streptococcus beta-haemolytic, group D) - 711
    Meningitis pathogens: *H. influenzae, N. meningitidis, S. agalactiae, S. pneumoniae* - 984
    Salmonella species - 907

Bouty SpA
EIA (non-competitive)
  Antibody detection (IgA)
    Toxoplasma gondii - 921
  Antibody detection (IgG)
    Epstein-Barr virus - 714
    Rubella virus - 866
    Toxoplasma gondii - 928
  Antibody detection (IgM)
    Epstein-Barr virus - 717
    Rubella virus - 884
    Toxoplasma gondii - 947

Cambridge Biotech Corporation
EIA (non-competitive)
  Antigen detection
    Adenovirus - 631
    Adenovirus - 631
    Clostridium difficile (toxin A and toxin B) - 685
    Rotavirus - 848

Cambridge Diagnostics Ireland Ltd
Immunoblot assay
  Antibody detection (IgG)
    Borrelia burgdorferi - 663
  Antibody detection (IgG and/or IgM)
    Borrelia burgdorferi - 664

Carter-Wallace Inc
Particle agglutination assay
  Antibody detection
    Rubella virus - 902
  Antigen detection

Cryptococcus neoformans - 693
Gastrointestinal pathogens: Salmonella spp and Shigella spp - 982

Cellabs Pty Ltd
EIA (non-competitive)
  Antibody detection
    Plasmodium falciparum - 840
    Toxocara canis - 917
  Antibody detection (IgG)
    Trypanosoma cruzi - 974
  Antigen detection
    Entamoeba histolytica/dispar - 704
    Entamoeba histolytica - 705
    Giardia lamblia - 764
    Plasmodium falciparum - 840
Immunofluorescence assay (direct)
  Antigen detection
    Cryptosporidium species - 698
    Gastrointestinal pathogens: Cryptosporidium species and Giardia species - 980
    Giardia lamblia - 766
    Toxoplasma gondii - 919

Centocor Inc
EIA (non-competitive)
  Antibody detection (IgG)
    Rubella virus - 860
    Rubella virus - 860
    Toxoplasma gondii - 929
  Antibody detection (IgM)
    Rubella virus - 861
    Toxoplasma gondii - 947

Chimica Diagnostica
EIA (non-competitive)
  Antibody detection (IgA)
    Helicobacter pylori - 774
    Toxoplasma gondii - 921
  Antibody detection (IgG)
    Epstein-Barr virus - 715
    Helicobacter pylori - 781
    Mumps virus - 823
    Rubella virus - 867
    Toxoplasma gondii - 929
  Antibody detection (IgM)
    Epstein-Barr virus - 717
    Helicobacter pylori - 791
    Mumps virus - 830
    Rubella virus - 884
    Toxoplasma gondii - 948

Clark Laboratories
EIA (non-competitive)
  Antibody detection (IgG)
    Brucella species - 672
    Entamoeba histolytica - 707
    Epstein-Barr virus (VCA) - 734
    Mumps virus - 823
    Rubella virus - 867
    Toxoplasma gondii - 930
  Antibody detection (IgM)
    Brucella species - 675
    Entamoeba histolytica - 709
    Epstein-Barr virus (VCA) - 743
    Mumps virus - 830
    Rubella virus - 885
    Toxoplasma gondii - 948

Cortecs Diagnostics
EIA (non-competitive)
  Antibody detection (IgG)
    Helicobacter pylori - 782
    Helicobacter pylori - 782

Dade International Inc (Bartels Division)
EIA (non-competitive)
  Antibody detection (IgG)
    Epstein-Barr virus (VCA) - 734

Toxoplasma gondii - 930
  Antibody detection (IgM)
    Borrelia burgdorferi - 660
    Rubella virus - 885
    Toxoplasma gondii - 949
  Antigen detection
    Clostridium difficile (toxin A) - 683

Dako A/S
EIA (non-competitive)
  Antibody detection (IgA and/or IgG)
    Yersinia enterocolitica - 978
  Antibody detection (IgG)
    Borrelia burgdorferi - 655
    Helicobacter pylori - 783
    Human parvovirus B19 - 799
  Antibody detection (IgG and/or IgM)
    Borrelia burgdorferi - 658
  Antibody detection (IgM)
    Borrelia burgdorferi - 661
    Human parvovirus B19 - 803
  Antigen detection
    Adenovirus - 632
    Rotavirus - 849
Immunofluorescence assay (direct)
  Antigen detection
    Adenovirus - 634

Denka Seiken Co. Ltd
EIA (non-competitive)
  Antibody detection (IgG)
    Mumps virus - 824
    Rubella virus - 868
    Toxoplasma gondii - 931
  Antibody detection (IgM)
    Mumps virus - 831
    Rubella virus - 886
    Toxoplasma gondii - 949
Particle agglutination assay
  Antigen detection
    Bacillus cereus (diarrhoeal enterotoxin) - 649
    Clostridium perfringens (enterotoxin, type A) - 687
    Escherichia coli (verotoxin 1 and verotoxin 2) - 761
    Gastrointestinal pathogens: *E. coli* (enterotoxin, LT) and *V. cholerae* (enterotoxin, CT) - 981
    Vibrio parahaemolyticus (haemolytic toxin) - 976

Diagast Laboratories
EIA (non-competitive)
  Antibody detection (IgG and/or IgM)
    Borrelia burgdorferi - 658
Immunoblot assay
  Antibody detection (IgG and/or IgM)
    Borrelia burgdorferi - 665
Immunofluorescence assay (indirect)
  Antibody detection
    Borrelia burgdorferi - 668

Diagnostic Products Corporation
Luminometric immunoassay (non-competitive)
  Antibody detection (IgG)
    Rubella virus - 898
    Rubella virus - 898
    Toxoplasma gondii - 965
    Toxoplasma gondii - 966
Particle agglutination assay
  Antigen detection
    Enterococcus species (Streptococcus beta-haemolytic, group D) - 712

Diamedix Corporation
EIA (non-competitive)
  Antibody detection

© KLUWER ACADEMIC PUBLISHERS 1995, ISSN 1381-5067

Entamoeba histolytica - 705
Antibody detection (IgG)
Rubella virus - 868
Antibody detection (IgM)
Rubella virus - 886

Diesse
EIA (competitive)
Antibody detection
Toxoplasma gondii - 919
EIA (non-competitive)
Antibody detection
Rubella virus - 861
Antibody detection (IgA)
Helicobacter pylori - 775
Toxoplasma gondii - 922
Antibody detection (IgG)
Epstein-Barr virus (EBNA) - 725
Epstein-Barr virus (VCA) - 735
Helicobacter pylori - 783
Toxoplasma gondii - 931
Antibody detection (IgM)
Epstein-Barr virus (VCA) - 744
Rubella virus - 887
Toxoplasma gondii - 950
Particle agglutination assay
Antibody detection
Heterophile antigen - 756

E. Merck
EIA (non-competitive)
Antibody detection (IgG)
Rubella virus - 869
Toxoplasma gondii - 932
Antibody detection (IgM)
Toxoplasma gondii - 950

Eiken Chemical Co. Ltd
Particle agglutination assay
Antibody detection
Toxoplasma gondii - 969

Enteric Products Inc
EIA (non-competitive)
Antibody detection (IgG)
Helicobacter pylori - 784
Immunochromatographic assay
Antibody detection (IgG)
Helicobacter pylori - 792

Euro-Diagnostica B.V.
EIA (non-competitive)
Antibody detection (IgG)
Human parvovirus B19 - 800
Antibody detection (IgM)
Human parvovirus B19 - 803

Gamma SA
EIA (non-competitive)
Antibody detection (IgG)
Clostridium tetani (toxin) - 689
Mumps virus - 824
Antibody detection (IgM)
Mumps virus - 831

GenBio
EIA (non-competitive)
Antibody detection
Borrelia burgdorferi - 652
Borrelia burgdorferi - 652
Borrelia burgdorferi - 653
Infectious Mononucleosis Syndrome
Test: CMV, EBV, heterophile Ag and
T. gondii - 982
Rubella virus - 862
TECH Screening Panel: *T. gondii*, EBV,
CMV and HSV - 984
TORCH Screening Panel: *T. gondii*,
Rubella virus, CMV and HSV - 985
Immunofluorescence assay (indirect)
Antibody detection (IgG)

Toxoplasma gondii - 961
Antibody detection (IgM)
Toxoplasma gondii - 964

Gull Laboratories Inc
EIA (non-competitive)
Antibody detection (IgG)
Epstein-Barr virus (EA) - 720
Epstein-Barr virus (EBNA) - 726
Epstein-Barr virus (VCA) - 735
Rubella virus - 869
Toxoplasma gondii - 932
Antibody detection (IgM)
Borrelia burgdorferi - 661
Epstein-Barr virus (VCA) - 744
Rubella virus - 887
Toxoplasma gondii - 951
Immunofluorescence assay (direct)
Antigen detection
Adenovirus - 634
Immunofluorescence assay (indirect)
Antibody detection
Epstein-Barr virus (EBNA) - 730
Antibody detection (IgG)
Epstein-Barr virus (EA) - 722
Epstein-Barr virus (VCA) - 750
Toxoplasma gondii - 962
Antibody detection (IgM)
Epstein-Barr virus (VCA) - 753
Toxoplasma gondii - 964

Hemagen Diagnostics Inc
Immunofluorescence assay (indirect)
Antibody detection (IgG)
Epstein-Barr virus (VCA) - 750
Mumps virus - 835
Rubella virus - 896
Toxoplasma gondii - 962

Human GmbH
EIA (non-competitive)
Antibody detection (IgG)
Epstein-Barr virus (VCA) - 736
Mumps virus - 825
Rubella virus - 870
Toxoplasma gondii - 933
Antibody detection (IgM)
Epstein-Barr virus (VCA) - 745
Mumps virus - 832
Rubella virus - 888
Rubella virus - 888
Toxoplasma gondii - 951
Toxoplasma gondii - 952

Hycor Biomedical Inc
EIA (non-competitive)
Antibody detection (IgA)
Helicobacter pylori - 775
Antibody detection (IgG)
Helicobacter pylori - 784

IFCI Clone Systems
EIA (non-competitive)
Antibody detection (IgA)
Helicobacter pylori - 776
Antibody detection (IgG)
Epstein-Barr virus (EBNA) - 726
Epstein-Barr virus (VCA) - 736
Helicobacter pylori - 785
Rubella virus - 870
Toxoplasma gondii - 933
Antibody detection (IgM)
Epstein-Barr virus (VCA) - 745
Rubella virus - 889
Toxoplasma gondii - 952

Immunetics
Immunoblot assay
Antibody detection (IgG)
Borrelia burgdorferi - 663
Borrelia burgdorferi - 664

Echinococcus granulosus - 702
Taenia solium - 915
Antibody detection (IgM)
Borrelia burgdorferi - 666
Borrelia burgdorferi - 666
Human parvovirus B19 - 807

Immunobiological Laboratories
EIA (non-competitive)
Antibody detection (IgA)
Helicobacter pylori - 777
Antibody detection (IgG)
Adenovirus - 638
Borrelia burgdorferi - 656
Brucella species - 673
Clostridium tetani (toxin) - 690
Epstein-Barr virus (VCA) - 737
Helicobacter pylori - 786
Human parvovirus B19 - 800
Mumps virus - 825
Toxoplasma gondii - 934
Antibody detection (IgM)
Adenovirus - 641
Borrelia burgdorferi - 662
Brucella species - 676
Epstein-Barr virus (VCA) - 746
Human parvovirus B19 - 804
Mumps virus - 832

Immuno Concepts Inc
EIA (non-competitive)
Antibody detection
Epstein-Barr virus (EBNA) - 723
Antibody detection (IgG)
Epstein-Barr virus (EA) - 721
Epstein-Barr virus (VCA) - 737
Antibody detection (IgM)
Epstein-Barr virus (VCA) - 746
Immunofluorescence assay (indirect)
Antibody detection
Epstein-Barr virus (EBNA) - 730
Antibody detection (IgG)
Epstein-Barr virus (EA) - 722
Epstein-Barr virus (VCA) - 751
Antibody detection (IgM)
Epstein-Barr virus (VCA) - 754

Immuno Pharmacology Research
EIA (non-competitive)
Antibody detection (IgA)
Helicobacter pylori - 776
Antibody detection (IgG)
Clostridium tetani - 689
Echinococcus species - 701
Epstein-Barr virus - 715
Helicobacter pylori - 785
Leishmania species - 810
Rotavirus - 857
Rubella virus - 871
Toxoplasma gondii - 934
Antibody detection (IgG and/or IgM))
Borrelia burgdorferi - 659
Brucella species - 674
Leptospira species - 815
Rickettsia conorii - 843
Salmonella typhi - 908
Antibody detection (IgM)
Epstein-Barr virus - 718
Helicobacter pylori - 791
Rubella virus - 889
Toxoplasma gondii - 953
Immunofluorescence assay (indirect)
Antibody detection
Leishmania species - 813
Rickettsia conorii - 844

Incstar
EIA (non-competitive)
Antibody detection (IgG)

Epstein-Barr virus (EBNA) - 727
Epstein-Barr virus (VCA) - 738
Rubella virus - 871
Toxoplasma gondii - 935
Antibody detection (IgM)
Epstein-Barr virus (EBNA) - 729
Epstein-Barr virus (VCA) - 747
Rubella virus - 890
Toxoplasma gondii - 953

International Immunodiagnostics
EIA (non-competitive)
Antibody detection (IgG)
Adenovirus - 639
Antibody detection (IgM)
Adenovirus - 641
Antigen detection
Adenovirus - 632
Particle agglutination assay
Antigen detection
Cryptococcus neoformans - 693

Johnson & Johnson Clinical Diagnostics Inc
Luminometric immunoassay (non-competitive)
Antibody detection (IgG)
Rubella virus - 899
Toxoplasma gondii - 966
Antibody detection (IgM)
Rubella virus - 900
Toxoplasma gondii - 967

Kreatech Diagnostics
EIA (non-competitive)
Antibody detection (IgG)
Rubella virus - 872
Toxoplasma gondii - 935
Antibody detection (IgM)
Rubella virus - 890
Toxoplasma gondii - 954

Laboratoire Eurobio
EIA (non-competitive)
Antibody detection (IgG)
Human parvovirus B19 - 801
Mumps virus - 826
Rubella virus - 872
Antibody detection (IgM)
Human parvovirus B19 - 804
Mumps virus - 833
Rubella virus - 891

Laboratoire Fumouze
Particle agglutination assay
Antibody detection
Entamoeba histolytica - 709
Toxoplasma gondii - 969

Labsystems Oy
EIA (non-competitive)
Antibody detection (IgG)
Rubella virus - 873
Rubella virus - 873
Toxoplasma gondii - 936
Toxoplasma gondii - 936
Antibody detection (IgM)
Rubella virus - 891
Toxoplasma gondii - 954

LD, Labor Diagnostika GmbH
Immunofluorescence assay (indirect)
Antibody detection (IgG)
Borrelia burgdorferi - 669
Toxoplasma gondii - 963

LMD Laboratories Inc
EIA (non-competitive)
Antibody detection
Echinococcus species - 701
Entamoeba histolytica - 706

Taenia solium - 914
Toxocara canis - 917
Toxoplasma gondii - 920
Trichinella spiralis - 972
Antigen detection
Cryptosporidium species - 697
Escherichia coli - 758
Giardia lamblia - 764
Rotavirus - 849

Mast Diagnostics Ltd
EIA (non-competitive)
Antigen detection
Salmonella species - 905

Medical Instruments Corporation
Immunochromatographic assay
Antibody detection (IgG)
Helicobacter pylori - 792

Medix Biotech Inc
EIA (non-competitive)
Antibody detection (IgG)
Mumps virus - 826
Rubella virus - 874
Toxoplasma gondii - 937
Antibody detection (IgM)
Mumps virus - 833
Rubella virus - 892
Toxoplasma gondii - 955

Melotec S.A.
EIA (non-competitive)
Antibody detection
Trichinella spiralis - 972
Antibody detection (IgG)
Echinococcus species - 702
Entamoeba histolytica - 707
Epstein-Barr virus (VCA) - 738
Leishmania species - 811
Mumps virus - 827
Rubella virus - 874
Taenia solium - 914
Toxoplasma gondii - 937
Antibody detection (IgM)
Rubella virus - 892
Toxoplasma gondii - 955
Antigen detection
Cryptosporidium species - 697
Giardia lamblia - 765
Rotavirus - 850

Menarini Diagnostics
EIA (non-competitive)
Antibody detection (IgG)
Rubella virus - 875
Toxoplasma gondii - 938
Antibody detection (IgM)
Rubella virus - 893
Toxoplasma gondii - 956

Meridian Diagnostics Inc
EIA (non-competitive)
Antibody detection (IgG)
Helicobacter pylori - 786
Antibody detection (IgG and/or IgM)
Epstein-Barr virus (EBNA) - 728
Antigen detection
Clostridium difficile (toxin A) - 684
Cryptococcus neoformans - 692
Escherichia coli (verotoxins) - 760
Immunochromatographic assay
Antibody detection (IgG)
Helicobacter pylori - 793
Antigen detection
Clostridium difficile - 680
Rotavirus - 852
Immunofluorescence assay (direct)
Antigen detection
Gastrointestinal pathogens: Cryptosporidium species and Giardia species

- 981
Particle agglutination assay
Antigen detection
Campylobacter species - 678
Clostridium difficile - 681
Cryptococcus neoformans - 694
Rotavirus - 853

Microgen Bioproducts
EIA (non-competitive)
Antigen detection
Rotavirus - 850
Particle agglutination assay
Antigen detection
Clostridium difficile - 681
Escherichia coli - 759
Listeria species - 818
Rotavirus - 854
Salmonella species - 907

MRL Diagnostics
EIA (non-competitive)
Antibody detection
Borrelia burgdorferi - 653
Antibody detection (IgG)
Human parvovirus B19 - 801
Antibody detection (IgM)
Human parvovirus B19 - 805
Immunofluorescence assay (indirect)
Antibody detection (IgG)
Borrelia burgdorferi - 669
Encephalitis bio-group: EEEV, WEEV, SLEV and CEV - 646
Epstein-Barr virus (EA) - 723
Epstein-Barr virus (VCA) - 751
Spotted Fever bio-group and Typhus Fever bio-group - 844
Antibody detection (IgM)
Borrelia burgdorferi - 670
Encephalitis bio-group: EEEV, WEEV, SLEV and CEV - 646
Epstein-Barr virus (VCA) - 754
Spotted Fever bio-group and Typhus Fever bio-group - 845

Murex Diagnostics Limited
Particle agglutination assay
Antigen detection
Cryptococcus neoformans - 694
Rotavirus - 854

Omega Diagnostics Ltd
Particle agglutination assay
Antigen detection
Rotavirus - 855

Organon Teknika NV
EIA (non-competitive)
Antibody detection (IgG)
Toxoplasma gondii - 938
Antibody detection (IgM)
Toxoplasma gondii - 956

Orion Diagnostica
EIA (non-competitive)
Antibody detection (IgG)
Helicobacter pylori - 787
Particle agglutination assay
Antibody detection
Helicobacter pylori - 793
Rubella virus - 902
Antigen detection
Adenovirus - 636
Adenovirus - 637
Gastrointestinal pathogens: Adenovirus and Rotavirus - 980
Rotavirus - 855
Rotavirus - 856

© *KLUWER ACADEMIC PUBLISHERS 1995, ISSN 1381-5067*

Ortho Diagnostic Systems
EIA (non-competitive)
Antibody detection (IgG)
Epstein-Barr virus (EBNA) - 727
Epstein-Barr virus (VCA) - 739
Antibody detection (IgM)
Epstein-Barr virus (VCA) - 747

PanBio
EIA (non-competitive)
Antibody detection (IgG)
Barmah Forest virus - 644
Brucella species - 673
Dengue fever virus - 647
Epstein-Barr virus (VCA) - 739
Human Herpes virus 6 - 796
Ross River virus - 645
Rubella virus - 875
Antibody detection (IgM)
Barmah Forest virus - 644
Brucella species - 676
Dengue fever virus - 647
Epstein-Barr virus (VCA) - 748
Leptospira species - 815
Ross River virus - 645
Immunochromatographic assay
Antigen detection
Salmonella species - 906

PRO-LAB Diagnostics
Particle agglutination assay
Antigen detection
Escherichia coli - 759

Progen
EIA (non-competitive)
Antibody detection (IgG)
Hantaan virus - 768
Puumala virus - 769
Antibody detection (IgM)
Hantaan virus - 768
Puumala virus - 770
Immunofluorescence assay (indirect)
Antibody detection
Hantaan virus - 769
Puumala virus - 770
Seoul virus - 771

Quidel Corporation
EIA (non-competitive)
Antibody detection (IgG)
Helicobacter pylori - 787
Immunochromatographic assay
Antibody detection (IgM)
Epstein-Barr virus - 719

r-biopharm GmbH
EIA (non-competitive)
Antibody detection (IgG)
Human parvovirus B19 - 802
Antibody detection (IgM)
Human parvovirus B19 - 805
Antigen detection
Adenovirus - 633
Clostridium difficile (toxin A and toxin B) - 685
Rotavirus - 851
Immunoblot assay
Antibody detection (IgG and/or IgM)
Human parvovirus B19 - 806

Radim
EIA (non-competitive)
Antibody detection (IgG)
Helicobacter pylori - 788
Mumps virus - 827
Rubella virus - 876
Toxoplasma gondii - 939
Antibody detection (IgM)
Mumps virus - 834
Rubella virus - 893

Toxoplasma gondii - 957

remel
Immunochromatographic assay
Antibody detection
Borrelia burgdorferi - 667

Roche Diagnostic Systems
EIA (non-competitive)
Antibody detection (IgG)
Helicobacter pylori - 788
Rubella virus - 876
Toxoplasma gondii - 939
Antibody detection (IgM)
Rubella virus - 894
Toxoplasma gondii - 957

S.A. Scientific
Particle agglutination assay
Antibody detection
Rubella virus - 903

Sanofi Diagnostics Pasteur
EIA (non-competitive)
Antibody detection (IgG)
Rubella virus - 877
Toxoplasma gondii - 940
Antibody detection (IgM)
Toxoplasma gondii - 958
Antigen detection
Rotavirus - 851
Luminometric immunoassay (non-competitive)
Antibody detection (IgG)
Rubella virus - 899
Toxoplasma gondii - 967
Antibody detection (IgM)
Rubella virus - 900
Toxoplasma gondii - 968

Savyon Diagnostics Ltd
EIA (non-competitive)
Antibody detection (IgA)
Epstein-Barr virus (VCA) - 731
Antibody detection (IgG)
Epstein-Barr virus - 716
Epstein-Barr virus (VCA) - 740
Antibody detection (IgM)
Epstein-Barr virus - 718
Epstein-Barr virus (VCA) - 748

Scimedix Corporation
EIA (non-competitive)
Antibody detection (IgG)
Adenovirus - 639
Antibody detection (IgG and/or IgM)
Brucella species - 674
Antibody detection (IgM)
Adenovirus - 642
Immunoblot assay
Antibody detection (IgG and/or IgM)
Borrelia burgdorferi - 665
Epstein-Barr virus - 719

Seradyn
EIA (non-competitive)
Antibody detection
Borrelia burgdorferi - 654
Antigen detection
Cryptosporidium species - 698
Giardia lamblia - 765
Immunoblot assay
Antibody detection
Borrelia burgdorferi - 662
Particle agglutination assay
Antibody detection
Rubella virus - 903

SFRI Laboratoire
EIA (non-competitive)
Antibody detection (IgA)
Toxoplasma gondii - 922

Antibody detection (IgG)
Toxoplasma gondii - 940
Antibody detection (IgM)
Toxoplasma gondii - 958

Shield Diagnostics
EIA (non-competitive)
Antibody detection (IgG)
Helicobacter pylori - 789
Antigen detection
Clostridium difficile (toxin A) - 684
Immunofluorescence assay (direct)
Antigen detection
Cryptosporidium species - 699

Sigma Diagnostics
EIA (non-competitive)
Antibody detection
Borrelia burgdorferi - 654
Antibody detection (IgG)
Epstein-Barr virus (EBNA) - 728
Epstein-Barr virus (VCA) - 740
Rubella virus - 877
Toxoplasma gondii - 941
Antibody detection (IgM)
Epstein-Barr virus (EBNA) - 729
Epstein-Barr virus (VCA) - 749
Rubella virus - 894
Rubella virus - 895
Toxoplasma gondii - 959
Toxoplasma gondii - 959

Sorin Biomedica
EIA (non-competitive)
Antibody detection (IgG)
Rubella virus - 878
Toxoplasma gondii - 941
Antibody detection (IgM)
Rubella virus - 895
Toxoplasma gondii - 960

Stellar Bio Systems Inc
Immunofluorescence assay (indirect)
Antibody detection (IgG)
Epstein-Barr virus (VCA) - 752
Human Herpes virus 6 - 797
Human parvovirus B19 - 807
Mumps virus - 835
Rubella virus - 897
Toxoplasma gondii - 963
Antibody detection (IgM)
Epstein-Barr virus (VCA) - 755
Human parvovirus B19 - 808
Mumps virus - 836
Rubella virus - 897
Toxoplasma gondii - 965

Tecra Diagnostics
EIA (non-competitive)
Antigen detection
Bacillus cereus (diarrhoeal enterotoxin) - 649
Escherichia coli - 758
Listeria species - 817
Listeria species - 817
Salmonella species - 905
Salmonella species - 906

Unipath Limited
Immunochromatographic assay
Antigen detection
Listeria species - 818
Particle agglutination assay
Antibody detection
Helicobacter pylori - 794
Heterophile antigen - 756
Antigen detection
Escherichia coli - 760
Salmonella species - 908

United Biotech Inc

EIA (non-competitive)
  Antibody detection (IgG)
    Helicobacter pylori - 789
    Rubella virus - 878
    Toxoplasma gondii - 942
  Antibody detection (IgM)
    Rubella virus - 896

Zeus Scientific Inc

EIA (non-competitive)
  Antibody detection (IgG)
    Rubella virus - 879

Immunofluorescence assay (indirect)
  Antibody detection
    Borrelia burgdorferi - 668
    Toxoplasma gondii - 961
  Antibody detection (IgG)
    Epstein-Barr virus (VCA) - 752

© KLUWER ACADEMIC PUBLISHERS 1995, ISSN 1381-5067

INDEX BY ASSAY TYPE, MICROORGANISM, ANTIGEN/ANTIBODY DETECTION AND MANUFACTURER

EIA (competitive)

Toxoplasma gondii
Antibody detection
Diesse - 919

EIA (non-competitive)

Adenovirus
Antibody detection (IgG)
Amico Laboratories Inc - 637
BAG-Biologische Analysensystem GmbH - 638
Immunobiological Laboratories - 638
International Immunodiagnostics - 639
Scimedix Corporation - 639
Antibody detection (IgM)
Amico Laboratories Inc - 640
BAG-Biologische Analysensystem GmbH - 640
Immunobiological Laboratories - 641
International Immunodiagnostics - 641
Scimedix Corporation - 642
Antigen detection
Alfa Biotech - 630
Biotrin International Ltd - 630
Cambridge Biotech Corporation - 631
Cambridge Biotech Corporation - 631
Dako A/S - 632
International Immunodiagnostics - 632
r-biopharm GmbH - 633

Bacillus cereus (diarrhoeal enterotoxin)
Antigen detection
Tecra Diagnostics - 649

Barmah Forest virus
Antibody detection (IgG)
PanBio - 644
Antibody detection (IgM)
PanBio - 644

Borrelia burgdorferi
Antibody detection
Alexon Inc - 651
bioMerieux - 651
GenBio - 652
GenBio - 652
GenBio - 653
MRL Diagnostics - 653
Seradyn - 654
Sigma Diagnostics - 654
Antibody detection (IgG)
BAG-Biologische Analysensystem GmbH - 655
Dako A/S - 655
Immunobiological Laboratories - 656
Antibody detection (IgG and/or IgM)
Behringwerke AG - 656
BioWhittaker Inc - 657
BioWhittaker Inc - 657
Dako A/S - 658
Diagast Laboratories - 658
Immuno Pharmacology Research - 659
Antibody detection (IgM)
BAG-Biologische Analysensystem GmbH - 659
BioWhittaker Inc - 660
Dade International Inc (Bartels Division) - 660
Dako A/S - 661
Gull Laboratories Inc - 661
Immunobiological Laboratories - 662

Brucella species
Antibody detection (IgG)
Alfa Biotech - 672
Clark Laboratories - 672
Immunobiological Laboratories - 673
PanBio - 673
Antibody detection (IgG and/or IgM))
Immuno Pharmacology Research - 674
Scimedix Corporation - 674
Antibody detection (IgM)

Alfa Biotech - 675
Clark Laboratories - 675
Immunobiological Laboratories - 676
PanBio - 676

Clostridium difficile (toxin A)
Antigen detection
Alexon Inc - 682
Becton Dickinson - 682
bioMerieux - 683
Dade International Inc (Bartels Division) - 683
Meridian Diagnostics Inc - 684
Shield Diagnostics - 684

Clostridium difficile (toxin A and toxin B)
Antigen detection
Cambridge Biotech Corporation - 685
r-biopharm GmbH - 685

Clostridium tetani (toxin)
Antibody detection (IgG)
Gamma SA - 689
Immuno Pharmacology Research - 689
Immunobiological Laboratories - 690

Cryptococcus neoformans
Antigen detection
Meridian Diagnostics Inc - 692

Cryptosporidium species
Antigen detection
Alexon Inc - 696
Alexon Inc - 696
LMD Laboratories Inc - 697
Melotec S.A. - 697
Seradyn - 698

Dengue fever virus
Antibody detection (IgG)
PanBio - 647
Antibody detection (IgM)
PanBio - 647

Echinococcus species
Antibody detection
LMD Laboratories Inc - 701
Antibody detection (IgG)
Immuno Pharmacology Research - 701
Melotec S.A. - 702

Entamoeba histolytica
Antibody detection
Diamedix Corporation - 705
LMD Laboratories Inc - 706
Antibody detection (IgG)
Amico Laboratories Inc - 706
Clark Laboratories - 707
Melotec S.A. - 707
Antibody detection (IgM)
Amico Laboratories Inc - 708
Amico Laboratories Inc - 708
Clark Laboratories - 709
Antigen detection
Alexon Inc - 704
Cellabs Pty Ltd - 704
Cellabs Pty Ltd - 705

Epstein-Barr virus
Antibody detection (IgG)
Behringwerke AG - 714
Bouty SpA - 714
Chimica Diagnostica - 715
Immuno Pharmacology Research - 715
Savyon Diagnostics Ltd - 716
Antibody detection (IgM)
Behringwerke AG - 716
Bouty SpA - 717
Chimica Diagnostica - 717
Immuno Pharmacology Research - 718
Savyon Diagnostics Ltd - 718

Epstein-Barr virus (EA)
Antibody detection (IgG)
Biotest Diagnostics - 720
Gull Laboratories Inc - 720
Immuno Concepts Inc - 721
Antibody detection (IgM)

Biotest Diagnostics - 721

Epstein-Barr virus (EBNA)
Antibody detection
Immuno Concepts Inc - 723
Antibody detection (IgG)
Biotest Diagnostics - 724
BioWhittaker Inc - 724
BioWhittaker Inc - 725
Diesse - 725
Gull Laboratories Inc - 726
IFCI Clone Systems - 726
Incstar - 727
Ortho Diagnostic Systems - 727
Sigma Diagnostics - 728
Antibody detection (IgG and/or IgM)
Meridian Diagnostics Inc - 728
Antibody detection (IgM)
Incstar - 729
Sigma Diagnostics - 729

Epstein-Barr virus (VCA)
Antibody detection (IgA)
Savyon Diagnostics Ltd - 731
Antibody detection (IgG)
Alfa Biotech - 731
Amico Laboratories Inc - 732
BAG-Biologische Analysensystem GmbH - 732
BioWhittaker Inc - 733
BioWhittaker Inc - 733
Clark Laboratories - 734
Dade International Inc (Bartels Division) - 734
Diesse - 735
Gull Laboratories Inc - 735
Human GmbH - 736
IFCI Clone Systems - 736
Immuno Concepts Inc - 737
Immunobiological Laboratories - 737
Incstar - 738
Melotec S.A. - 738
Ortho Diagnostic Systems - 739
PanBio - 739
Savyon Diagnostics Ltd - 740
Sigma Diagnostics - 740
Antibody detection (IgM)
Alfa Biotech - 741
Amico Laboratories Inc - 741
Amico Laboratories Inc - 742
BAG-Biologische Analysensystem GmbH - 742
BioWhittaker Inc - 743
Clark Laboratories - 743
Diesse - 744
Gull Laboratories Inc - 744
Human GmbH - 745
IFCI Clone Systems - 745
Immuno Concepts Inc - 746
Immunobiological Laboratories - 746
Incstar - 747
Ortho Diagnostic Systems - 747
PanBio - 748
Savyon Diagnostics Ltd - 748
Sigma Diagnostics - 749

Escherichia coli
Antigen detection
LMD Laboratories Inc - 758
Tecra Diagnostics - 758

Escherichia coli (verotoxins)
Antigen detection
Meridian Diagnostics Inc - 760

Giardia lamblia
Antigen detection
Alexon Inc - 763
Alexon Inc - 763
Cellabs Pty Ltd - 764
LMD Laboratories Inc - 764
Melotec S.A. - 765
Seradyn - 765

© KLUWER ACADEMIC PUBLISHERS 1995, ISSN 1381-5067

Hantaan virus
  Antibody detection (IgG)
    Progen - 768
  Antibody detection (IgM)
    Progen - 768
Helicobacter pylori
  Antibody detection (IgA)
    Bio-Rad - 773
    Biomerica - 773
    BioWhittaker Inc - 774
    Chimica Diagnostica - 774
    Diesse - 775
    Hycor Biomedical Inc - 775
    IFCI Clone Systems - 776
    Immuno Pharmacology Research - 776
    Immunobiological Laboratories - 777
  Antibody detection (IgG)
    Alexon Inc - 777
    Amrad Corporation Ltd - 778
    Bard Diagnostic Sciences, Inc - 778
    Behringwerke AG - 779
    Bio-Rad - 779
    Biomerica - 780
    BioWhittaker Inc - 780
    BioWhittaker Inc - 781
    Chimica Diagnostica - 781
    Cortecs Diagnostics - 782
    Cortecs Diagnostics - 782
    Dako A/S - 783
    Diesse - 783
    Enteric Products Inc - 784
    Hycor Biomedical Inc - 784
    IFCI Clone Systems - 785
    Immuno Pharmacology Research - 785
    Immunobiological Laboratories - 786
    Meridian Diagnostics Inc - 786
    Orion Diagnostica - 787
    Quidel Corporation - 787
    Radim - 788
    Roche Diagnostic Systems - 788
    Shield Diagnostics - 789
    United Biotech Inc - 789
  Antibody detection (IgM)
    Bio-Rad - 790
    Biomerica - 790
    Chimica Diagnostica - 791
    Immuno Pharmacology Research - 791
Human Herpes virus 6
  Antibody detection (IgG)
    Biotrin International Ltd - 796
    PanBio - 796
Human parvovirus B19
  Antibody detection (IgG)
    Biotrin International Ltd - 799
    Dako A/S - 799
    Euro-Diagnostica B.V. - 800
    Immunobiological Laboratories - 800
    Laboratoire Eurobio - 801
    MRL Diagnostics - 801
    r-biopharm GmbH - 802
  Antibody detection (IgM)
    Biotrin International Ltd - 802
    Dako A/S - 803
    Euro-Diagnostica B.V. - 803
    Immunobiological Laboratories - 804
    Laboratoire Eurobio - 804
    MRL Diagnostics - 805
    r-biopharm GmbH - 805
Infectious Mononucleosis Syndrome Test: CMV, EBV, heterophile Ag and T. gondii
  Antibody detection
    GenBio - 982
Leishmania species
  Antibody detection (IgG)
    Amico Laboratories Inc - 810
    Immuno Pharmacology Research - 810
    Melotec S.A. - 811
  Antibody detection (IgM)

    Amico Laboratories Inc - 811
    Amico Laboratories Inc - 812
Leptospira species
  Antibody detection (IgG and/or IgM))
    Immuno Pharmacology Research - 815
  Antibody detection (IgM)
    PanBio - 815
Listeria species
  Antigen detection
    Tecra Diagnostics - 817
    Tecra Diagnostics - 817
Mumps virus
  Antibody detection (IgG)
    Amico Laboratories Inc - 820
    BAG-Biologische Analysensystem
      GmbH - 820
    Behringwerke AG - 821
    bioMerieux - 821
    BioWhittaker Inc - 822
    BioWhittaker Inc - 822
    Chimica Diagnostica - 823
    Clark Laboratories - 823
    Denka Seiken Co. Ltd - 824
    Gamma SA - 824
    Human GmbH - 825
    Immunobiological Laboratories - 825
    Laboratoire Eurobio - 826
    Medix Biotech Inc - 826
    Melotec S.A. - 827
    Radim - 827
  Antibody detection (IgM)
    Amico Laboratories Inc - 828
    Amico Laboratories Inc - 828
    BAG-Biologische Analysensystem
      GmbH - 829
    Behringwerke AG - 829
    Chimica Diagnostica - 830
    Clark Laboratories - 830
    Denka Seiken Co. Ltd - 831
    Gamma SA - 831
    Human GmbH - 832
    Immunobiological Laboratories - 832
    Laboratoire Eurobio - 833
    Medix Biotech Inc - 833
    Radim - 834
Plasmodium falciparum
  Antibody detection
    Cellabs Pty Ltd - 840
  Antigen detection
    Cellabs Pty Ltd - 840
Puumala virus
  Antibody detection (IgG)
    Progen - 769
  Antibody detection (IgM)
    Progen - 770
Rickettsia conorii
  Antibody detection (IgG and/or IgM))
    Immuno Pharmacology Research - 843
Ross River virus
  Antibody detection (IgG)
    PanBio - 645
  Antibody detection (IgM)
    PanBio - 645
Rotavirus
  Antibody detection (IgG)
    Amico Laboratories Inc - 856
    Immuno Pharmacology Research - 857
  Antibody detection (IgM)
    Amico Laboratories Inc - 857
    Amico Laboratories Inc - 858
  Antigen detection
    Abbott Laboratories - 847
    Abbott Laboratories - 847
    Alexon Inc - 848
    Cambridge Biotech Corporation - 848
    Dako A/S - 849
    LMD Laboratories Inc - 849
    Melotec S.A. - 850

    Microgen Bioproducts - 850
    r-biopharm GmbH - 851
    Sanofi Diagnostics Pasteur - 851
Rubella virus
  Antibody detection
    Diesse - 861
    GenBio - 862
  Antibody detection (IgG)
    Abbott Laboratories - 862
    Alfa Biotech - 863
    BAG-Biologische Analysensystem
      GmbH - 863
    Behringwerke AG - 864
    BIOKIT SA - 864
    bioMerieux - 865
    BioWhittaker Inc - 865
    BioWhittaker Inc - 866
    Bouty SpA - 866
    Centocor Inc - 860
    Centocor Inc - 860
    Chimica Diagnostica - 867
    Clark Laboratories - 867
    Denka Seiken Co. Ltd - 868
    Diamedix Corporation - 868
    E. Merck - 869
    Gull Laboratories Inc - 869
    Human GmbH - 870
    IFCI Clone Systems - 870
    Immuno Pharmacology Research - 871
    Incstar - 871
    Kreatech Diagnostics - 872
    Laboratoire Eurobio - 872
    Labsystems Oy - 873
    Labsystems Oy - 873
    Medix Biotech Inc - 874
    Melotec S.A. - 874
    Menarini Diagnostics - 875
    PanBio - 875
    Radim - 876
    Roche Diagnostic Systems - 876
    Sanofi Diagnostics Pasteur - 877
    Sigma Diagnostics - 877
    Sorin Biomedica - 878
    United Biotech Inc - 878
    Zeus Scientific Inc - 879
  Antibody detection (IgM)
    Abbott Laboratories - 879
    Alfa Biotech - 880
    Amico Laboratories Inc - 880
    Amico Laboratories Inc - 881
    BAG-Biologische Analysensystem
      GmbH - 881
    Behringwerke AG - 882
    BIOKIT SA - 882
    bioMerieux - 883
    BioWhittaker Inc - 883
    Bouty SpA - 884
    Centocor Inc - 861
    Chimica Diagnostica - 884
    Clark Laboratories - 885
    Dade International Inc (Bartels Division)
      - 885
    Denka Seiken Co. Ltd - 886
    Diamedix Corporation - 886
    Diesse - 887
    Gull Laboratories Inc - 887
    Human GmbH - 888
    Human GmbH - 888
    IFCI Clone Systems - 889
    Immuno Pharmacology Research - 889
    Incstar - 890
    Kreatech Diagnostics - 890
    Laboratoire Eurobio - 891
    Labsystems Oy - 891
    Medix Biotech Inc - 892
    Melotec S.A. - 892
    Menarini Diagnostics - 893
    Radim - 893

© KLUWER ACADEMIC PUBLISHERS 1995, ISSN 1381-5067

Roche Diagnostic Systems - 894
Sigma Diagnostics - 894
Sigma Diagnostics - 895
Sorin Biomedica - 895
United Biotech Inc - 896
Salmonella species
  Antigen detection
    Mast Diagnostics Ltd - 905
    Tecra Diagnostics - 905
    Tecra Diagnostics - 906
Salmonella typhi
  Antibody detection (IgG and/or IgM))
    Immuno Pharmacology Research - 908
Schistosoma species
  Antibody detection (IgG)
    Amico Laboratories Inc - 910
  Antibody detection (IgM)
    Amico Laboratories Inc - 910
    Amico Laboratories Inc - 911
Taenia solium
  Antibody detection
    LMD Laboratories Inc - 914
  Antibody detection (IgG)
    Melotec S.A. - 914
TECH Screening Panel: T. gondii, EBV,
CMV and HSV
  Antibody detection
    GenBio - 984
TORCH Screening Panel: T. gondii, Rubella
virus, CMV and HSV
  Antibody detection
    GenBio - 985
Toxocara canis
  Antibody detection
    Cellabs Pty Ltd - 917
    LMD Laboratories Inc - 917
Toxoplasma gondii
  Antibody detection
    LMD Laboratories Inc - 920
  Antibody detection (IgA)
    Abbott Laboratories - 920
    Bouty SpA - 921
    Chimica Diagnostica - 921
    Diesse - 922
    SFRI Laboratoire - 922
  Antibody detection (IgG)
    Abbott Laboratories - 923
    Abbott Laboratories - 923
    Alfa Biotech - 924
    Amico Laboratories Inc - 924
    Amico Laboratories Inc - 925
    Behringwerke AG - 925
    BIOKIT SA - 926
    bioMerieux - 926
    Biotecx Laboratories Inc - 927
    BioWhittaker Inc - 927
    BioWhittaker Inc - 928
    Bouty SpA - 928
    Centocor Inc - 929
    Chimica Diagnostica - 929
    Clark Laboratories - 930
    Dade International Inc (Bartels Division)
      - 930
    Denka Seiken Co. Ltd - 931
    Diesse - 931
    E. Merck - 932
    Gull Laboratories Inc - 932
    Human GmbH - 933
    IFCI Clone Systems - 933
    Immuno Pharmacology Research - 934
    Immunobiological Laboratories - 934
    Incstar - 935
    Kreatech Diagnostics - 935
    Labsystems Oy - 936
    Labsystems Oy - 936
    Medix Biotech Inc - 937
    Melotec S.A. - 937
    Menarini Diagnostics - 938

Organon Teknika NV - 938
Radim - 939
Roche Diagnostic Systems - 939
Sanofi Diagnostics Pasteur - 940
SFRI Laboratoire - 940
Sigma Diagnostics - 941
Sorin Biomedica - 941
United Biotech Inc - 942
Antibody detection (IgM)
  Abbott Laboratories - 942
  Abbott Laboratories - 943
  Alfa Biotech - 943
  Amico Laboratories Inc - 944
  Amico Laboratories Inc - 944
  Behringwerke AG - 945
  BIOKIT SA - 945
  bioMerieux - 946
  BioWhittaker Inc - 946
  Bouty SpA - 947
  Centocor Inc - 947
  Chimica Diagnostica - 948
  Clark Laboratories - 948
  Dade International Inc (Bartels Division)
    - 949
  Denka Seiken Co. Ltd - 949
  Diesse - 950
  E. Merck - 950
  Gull Laboratories Inc - 951
  Human GmbH - 951
  Human GmbH - 952
  IFCI Clone Systems - 952
  Immuno Pharmacology Research - 953
  Incstar - 953
  Kreatech Diagnostics - 954
  Labsystems Oy - 954
  Medix Biotech Inc - 955
  Melotec S.A. - 955
  Menarini Diagnostics - 956
  Organon Teknika NV - 956
  Radim - 957
  Roche Diagnostic Systems - 957
  Sanofi Diagnostics Pasteur - 958
  SFRI Laboratoire - 958
  Sigma Diagnostics - 959
  Sigma Diagnostics - 959
  Sorin Biomedica - 960
Trichinella spiralis
  Antibody detection
    LMD Laboratories Inc - 972
    Melotec S.A. - 972
Trypanosoma cruzi
  Antibody detection (IgG)
    Cellabs Pty Ltd - 974
Yersinia enterocolitica
  Antibody detection (IgA and/or IgG)
    Dako A/S - 978

Immunoblot assay
Borrelia burgdorferi
  Antibody detection
    Seradyn - 662
  Antibody detection (IgG)
    Cambridge Diagnostics Ireland Ltd - 663
    Immunetics - 663
    Immunetics - 664
  Antibody detection (IgG and/or IgM)
    Cambridge Diagnostics Ireland Ltd - 664
    Diagast Laboratories - 665
    Scimedix Corporation - 665
  Antibody detection (IgM)
    Immunetics - 666
    Immunetics - 666
Echinococcus granulosus
  Antibody detection (IgG)
    Immunetics - 702
Epstein-Barr virus
  Antibody detection (IgG and/or IgM)
    Scimedix Corporation - 719

Human parvovirus B19
  Antibody detection (IgG and/or IgM)
    Biotrin International Ltd - 806
    r-biopharm GmbH - 806
  Antibody detection (IgM)
    Immunetics - 807
Taenia solium
  Antibody detection (IgG)
    Immunetics - 915

Immunochromatographic assay
Borrelia burgdorferi
  Antibody detection
    remel - 667
Clostridium difficile
  Antigen detection
    Meridian Diagnostics Inc - 680
Epstein-Barr virus
  Antibody detection (IgM)
    Quidel Corporation - 719
Helicobacter pylori
  Antibody detection (IgG)
    Enteric Products Inc - 792
    Medical Instruments Corporation - 792
    Meridian Diagnostics Inc - 793
Listeria species
  Antigen detection
    Unipath Limited - 818
Rotavirus
  Antigen detection
    Meridian Diagnostics Inc - 852
Salmonella species
  Antigen detection
    PanBio - 906

Immunofluorescence assay (direct)
Adenovirus
  Antigen detection
    Argene-Biosoft - 633
    Dako A/S - 634
    Gull Laboratories Inc - 634
Cryptosporidium species
  Antigen detection
    Cellabs Pty Ltd - 698
    Shield Diagnostics - 699
Gastrointestinal pathogens: Cryptosporidium
species and Giardia species
  Antigen detection
    Cellabs Pty Ltd - 980
    Meridian Diagnostics Inc - 981
Giardia lamblia
  Antigen detection
    Cellabs Pty Ltd - 766
Toxoplasma gondii
  Antigen detection
    Cellabs Pty Ltd - 919

Immunofluorescence assay (indirect)
Adenovirus
  Antigen detection
    Argene-Biosoft - 635
Borrelia burgdorferi
  Antibody detection
    bioMerieux - 667
    Diagast Laboratories - 668
    Zeus Scientific Inc - 668
  Antibody detection (IgG)
    LD, Labor Diagnostika GmbH - 669
    MRL Diagnostics - 669
  Antibody detection (IgM)
    MRL Diagnostics - 670
Encephalitis bio-group: EEEV, WEEV, SLEV
and CEV
  Antibody detection (IgG)
    MRL Diagnostics - 646
  Antibody detection (IgM)
    MRL Diagnostics - 646

Epstein-Barr virus (EA)
Antibody detection (IgG)
Gull Laboratories Inc - 722
Immuno Concepts Inc - 722
MRL Diagnostics - 723
Epstein-Barr virus (EBNA)
Antibody detection
Gull Laboratories Inc - 730
Immuno Concepts Inc - 730
Epstein-Barr virus (VCA)
Antibody detection (IgG)
Bion Enterprises Ltd - 749
Gull Laboratories Inc - 750
Hemagen Diagnostics Inc - 750
Immuno Concepts Inc - 751
MRL Diagnostics - 751
Stellar Bio Systems Inc - 752
Zeus Scientific Inc - 752
Antibody detection (IgM)
Bion Enterprises Ltd - 753
Gull Laboratories Inc - 753
Immuno Concepts Inc - 754
MRL Diagnostics - 754
Stellar Bio Systems Inc - 755
Hantaan virus
Antibody detection
Progen - 769
Human Herpes virus 6
Antibody detection (IgG)
Biotrin International Ltd - 797
Stellar Bio Systems Inc - 797
Human parvovirus B19
Antibody detection (IgG)
Stellar Bio Systems Inc - 807
Antibody detection (IgG and/or IgM)
Biotrin International Ltd - 808
Antibody detection (IgM)
Stellar Bio Systems Inc - 808
Leishmania species
Antibody detection
bioMerieux - 812
Immuno Pharmacology Research - 813
Mumps virus
Antibody detection (IgG)
Bion Enterprises Ltd - 834
Hemagen Diagnostics Inc - 835
Stellar Bio Systems Inc - 835
Antibody detection (IgM)
Stellar Bio Systems Inc - 836
Plasmodium falciparum
Antibody detection
bioMerieux - 841
Puumala virus
Antibody detection
Progen - 770
Rickettsia conorii
Antibody detection
bioMerieux - 843
Immuno Pharmacology Research - 844
Rubella virus
Antibody detection (IgG)
Hemagen Diagnostics Inc - 896
Stellar Bio Systems Inc - 897
Antibody detection (IgM)
Stellar Bio Systems Inc - 897
Seoul virus
Antibody detection
Progen - 771
Spotted Fever bio-group and Typhus Fever bio-group
Antibody detection (IgG)
MRL Diagnostics - 844
Antibody detection (IgM)
MRL Diagnostics - 845

Toxoplasma gondii
Antibody detection (Total Ab and/or IgG and/or IgM)
bioMerieux - 960
Zeus Scientific Inc - 961
Antibody detection (IgG)
GenBio - 961
Gull Laboratories Inc - 962
Hemagen Diagnostics Inc - 962
LD, Labor Diagnostika GmbH - 963
Stellar Bio Systems Inc - 963
Antibody detection (IgM)
GenBio - 964
Gull Laboratories Inc - 964
Stellar Bio Systems Inc - 965

Luminometric immunoassay (non-competitive)
Rubella virus
Antibody detection (IgG)
Diagnostic Products Corporation - 898
Diagnostic Products Corporation - 898
Johnson & Johnson Clinical Diagnostics Inc - 899
Sanofi Diagnostics Pasteur - 899
Antibody detection (IgM)
Johnson & Johnson Clinical Diagnostics Inc - 900
Sanofi Diagnostics Pasteur - 900
Toxoplasma gondii
Antibody detection (IgG)
Diagnostic Products Corporation - 965
Diagnostic Products Corporation - 966
Johnson & Johnson Clinical Diagnostics Inc - 966
Sanofi Diagnostics Pasteur - 967
Antibody detection (IgM)
Johnson & Johnson Clinical Diagnostics Inc - 967
Sanofi Diagnostics Pasteur - 968

Particle agglutination assay
Adenovirus
Antigen detection
BIOKIT SA - 635
bioMerieux - 636
Orion Diagnostica - 636
Orion Diagnostica - 637
Bacillus cereus (diarrhoeal enterotoxin)
Antigen detection
Denka Seiken Co. Ltd - 649
Campylobacter species
Antigen detection
Meridian Diagnostics Inc - 678
Clostridium difficile
Antigen detection
Becton Dickinson - 680
Meridian Diagnostics Inc - 681
Microgen Bioproducts - 681
Clostridium perfringens (enterotoxin, type A)
Antigen detection
Denka Seiken Co. Ltd - 687
Cryptococcus neoformans
Antigen detection
bioMerieux - 692
Carter-Wallace Inc - 693
International Immunodiagnostics - 693
Meridian Diagnostics Inc - 694
Murex Diagnostics Limited - 694
Entamoeba histolytica
Antibody detection
Laboratoire Fumouze - 709
Enterococcus species (Streptococcus beta-haemolytic, group D)
Antigen detection

bioMerieux - 711
Boule Diagnostics AB - 711
Diagnostic Products Corporation - 712
Escherichia coli
Antigen detection
Microgen Bioproducts - 759
PRO-LAB Diagnostics - 759
Unipath Limited - 760
Escherichia coli (verotoxin 1 and verotoxin 2)
Antigen detection
Denka Seiken Co. Ltd - 761
Gastrointestinal pathogens: Salmonella spp and Shigella spp
Antigen detection
Carter-Wallace Inc - 982
Denka Seiken Co. Ltd - 981
Orion Diagnostica - 980
Helicobacter pylori
Antibody detection
Orion Diagnostica - 793
Unipath Limited - 794
Heterophile antigen
Antibody detection
BIOKIT SA - 755
Diesse - 756
Unipath Limited - 756
Listeria species
Antigen detection
Microgen Bioproducts - 818
Meningitis pathogens: E. coli, N. meningitis
Antigen detection
Becton Dickinson - 983
Becton Dickinson - 983
Boule Diagnostics AB - 984
Neisseria meningitidis
Antigen detection
Becton Dickinson - 838
Rotavirus
Antigen detection
BIOKIT SA - 852
bioMerieux - 853
Meridian Diagnostics Inc - 853
Microgen Bioproducts - 854
Murex Diagnostics Limited - 854
Omega Diagnostics Ltd - 855
Orion Diagnostica - 855
Orion Diagnostica - 856
Rubella virus
Antibody detection
Becton Dickinson - 901
BIOKIT SA - 901
Carter-Wallace Inc - 902
Orion Diagnostica - 902
S.A. Scientific - 903
Seradyn - 903
Salmonella species
Antigen detection
Boule Diagnostics AB - 907
Microgen Bioproducts - 907
Unipath Limited - 908
Toxoplasma gondii
Antibody detection
BIOKIT SA - 968
Eiken Chemical Co. Ltd - 969
Laboratoire Fumouze - 969
Antibody detection (IgA and/or IgM)
bioMerieux - 970
Antibody detection (IgM)
bioMerieux - 970
Vibrio parahaemolyticus (haemolytic toxin)
Antigen detection
Denka Seiken Co. Ltd - 976

GPSR Compliance

The European Union's (EU) General Product Safety Regulation (GPSR) is a set of rules that requires consumer products to be safe and our obligations to ensure this.

If you have any concerns about our products, you can contact us on ProductSafety@springernature.com

In case Publisher is established outside the EU, the EU authorized representative is:

Springer Nature Customer Service Center GmbH
Europaplatz 3
69115 Heidelberg, Germany

Batch number: 09625678

Printed by Printforce, the Netherlands